Vulkan 3D Graphi
Cookbook

Second Edition

Implement expert-level techniques for high-performance graphics with Vulkan

Sergey Kosarevsky
Alexey Medvedev
Viktor Latypov

Vulkan 3D Graphics Rendering Cookbook

Second Edition

Portfolio Director: Rohit Rajkumar
Relationship Lead: Kaustubh Manglurkar
Content Engineer: Anuradha Vishwas Joglekar
Program Manager: Sandip Tadge
Growth Lead: Namita Velgekar
Copy Editor: Safis Editing
Technical Editor: Tejas Mhasvekar
Proofreader: Anuradha Vishwas Joglekar
Indexer: Manju Arasan
Presentation Designer: Ajay Patule
Marketing Owner: Nivedita Pandey

First published: August 2021
Second edition: February 2025

Production reference: 2090725

Published by Packt Publishing Ltd.
Grosvenor House
11 St Paul's Square
Birmingham
B3 1RB, UK.

ISBN 978-1-80324-811-0

www.packt.com

Dedicated to my grandmother, Liudmila Fedorovna Sitorkina, and my mother, Irina Leonidovna Kosarevskaya.

– Sergey Kosarevsky

Dedicated to my family, friends, colleagues and Khronos fellows. I'm thankful, grateful and blessed to have you all in my life!

– Alexey Medvedev

Foreword

In the modern ocean of GPU APIs, Vulkan is gaining popularity for rendering, thanks to its high performance and cross-platform nature. Thanks to the dedication of the authors, this book covers a lot of ground, from the very first steps to modern and advanced topics. A graphics programmer who wants to start with Vulkan can find an easy-to-follow guide on how to set up the system and get to the first frame. A seasoned practitioner can find modern practices of physically based rendering, advanced resource management, and GPU-side optimizations. What I really like is the "There's more..." section at the end of every section, where the authors make the topic truly open-ended, leaving useful references, suggestions, and room for you to improve upon the material in the future.

One thing that makes this book special is its practicality, with the abundance of code available and explained right in the book. This way, you can both immediately see how Vulkan works and read about the reasons behind it. The book also goes way beyond just the API level, introducing higher-level rendering pipeline architecture and related concepts, such as the material system or geometry and asset formats. This enables you to be able to write not just a Vulkan-based renderer but also a small end-to-end experience with user interaction and advanced content.

The authors are well-renowned experts in the field with a lot of experience writing real-time engines and shipping AAA game titles, so you are in good hands. I was delighted to see the second edition of this book coming out with more elaborate and advanced topics. If you were looking for a comprehensive handbook on Vulkan and modern real-time rendering, you are holding it!

Anton Kaplanyan

VP, Graphics Research at Intel Corp.

Contributors

About the authors

Sergey Kosarevsky is a former rendering lead at Ubisoft RedLynx. He currently leads Vulkan development at Meta. He worked in the mobile industry at SPB Software, Yandex, Layar and Blippar, TWNKLS, and DAQRI, where he designed and implemented real-time rendering technology. He has more than 20 years of software development experience and more than 12 years of mobile and embedded 3D graphics experience. In his Ph.D. thesis, Sergey employed computer vision to solve mechanical engineering problems. He is also a co-author of several books on 3D graphics and mobile software development in C++, including *3D Graphics Rendering Cookbook*.

I want to express my heartfelt gratitude to my dedicated reviewer and friend, Alexander Pavlov, for his incredible commitment over the past 13 years, meticulously reviewing all our books. I also extend my sincere thanks to all my friends, colleagues, and reviewers whose support made this book possible!

Alexey Medvedev is the AR Tech Lead at Meta, with over 20 years of experience in software development, primarily in game development. He has worked as an engine, graphics, and rendering engineer at renowned companies like Crytek, Blizzard, and Hangar 13, contributing to the release of several AAA games. At the time of writing this book, Alexey also serves as the Khronos Chair of the 3D Formats Working Group, which develops the glTF standards.

I would like to first and foremost thank my family for their continued support, patience, and encouragement throughout the long process of writing this book.

Viktor Latypov is a software engineer specializing in embedded C/C++, 3D graphics, and computer vision. With more than 15 years of software development experience and a Ph.D. in applied mathematics, he has implemented a number of real-time renderers for medical and automotive applications over the last 10 years. Together with Sergey, he has co-authored two books on mobile software development in C++.

About the reviewers

Marco Castorina is the co-author of Mastering Graphics Programming with Vulkan. He has been working on graphics since completing his Masters in Digital Games at the Dublin Institute of Technology. First as a hobby and then as his day to day job when he joined Samsung as a driver developer. Here he got familiar with Vulkan and mobile rendering architecture. Later he developed a 2D and 3D renderer in Vulkan from scratch for a leading media-server company. He currently works on the games graphics performance team at AMD, focusing on DX12. In his spare time he keeps up to date with the latest techniques in real-time graphics, likes to cook and make music.

David Sena is a Graphics Technnical Director at NaturalMotion where he works on the CSR Racing game series. Before that, he was a Senior Graphics Software Engineer at Samsung R&D UK where he worked on solutions for different mobile devices and platforms. While at Samsung, David did research on solutions for global illumination, point clouds and VR/AR. David has been working as a graphics software engineer for over 12 years, with a focus on real-time rendering as well as developing and researching novel graphics solutions for mobile and handheld console platforms.

Alexander Pavlov is a Senior Software Engineer at Google with over 20 years of experience in industrial software development. He specializes in large-scale system design, many of the systems he has worked on are used daily by millions of people. His areas of interest include compiler theory and 3D graphics. Alexander also contributed as a technical reviewer and proofreader for the *Android NDK Game Development Cookbook* and the *3D Graphic Rendering Cookbook* (*1st ed.*), published by Packt.

Table of Contents

Chapter 10: Image-Based Techniques 543

Chapter 11: Advanced Rendering Techniques and Optimizations 619

Chapter 12: Unlock Your Book's Exclusive Benefits 689

Other Books You May Enjoy 693

Index 697

Preface

The Vulkan 3D Graphics Rendering Cookbook is a practical, all-in-one guide to mastering modern graphics rendering techniques and algorithms using C++ and Vulkan 1.3. You'll begin by setting up your Vulkan development environment, then move on to key aspects of graphics programming, such as working with graphics debugging tools, creating physically-based rendering pipelines, and handling large geometric data.

As you progress, the book walks you through building a 3D rendering engine step by step, presenting a series of small, self-contained recipes. Each recipe allows you to incrementally expand your codebase while integrating various 3D graphics techniques into a cohesive project. Along the way, you'll explore essential rendering methods, including glTF 2.0 shading model, image-based techniques, and GPU-driven rendering. You'll also learn how to manage large datasets for 3D rendering, apply optimization techniques, and develop high-performance, feature-rich graphics applications. By the end of the book, you'll have the skills to create fast and flexible 3D rendering frameworks and a solid understanding of best practices in modern Vulkan development. Rather than focusing on individual Vulkan API features in isolation, this book emphasizes integrating multiple Vulkan capabilities to create fully realized rendering demos. Throughout, we'll use Vulkan 1.3 along with the bindless rendering approach. To do this, we introduce LightweightVK `https://github.com/corporateshark/lightweightvk`, a standalone framework designed for Vulkan development, which we'll explore in depth as we progress.

Who this book is for

We expect our readers to have a solid understanding of real-time 3D graphics based on older rendering APIs. The first few chapters will cover what you need to get started with Vulkan, but we won't dwell on the basics for long. Instead, we'll quickly move on to more advanced topics. If you're familiar with OpenGL 4 or OpenGL ES 3 and want to explore modern rendering techniques and migration paths to current rendering APIs, this book is likely a great fit for you. While graphics programming may seem like an easy and fun entry point into software development, it actually requires mastering many advanced programming concepts. Readers should have a strong grasp of modern C++ and some foundational math skills, such as basic linear algebra and computational geometry.

What this book covers

This book is structured into distinct chapters, each focusing on a specific aspect of 3D rendering. As you progress, you'll gradually build a set of versatile 3D graphics demos, starting with the fundamentals, then exploring more complex techniques, and finally incorporating advanced rendering methods into your code.

Chapter 1, Establishing a Build Environment, guides you through setting up a Vulkan 1.3 development environment. You'll learn which tools and dependencies are needed to work with the book's source code and how to configure them. This chapter also introduces essential Vulkan recipes, including compiling Vulkan shaders from GLSL at runtime.

Chapter 2, Getting Started with Vulkan, introduces the fundamental components of the Vulkan API, including instance and device creation, swapchain management, debugging setup, and command buffer usage. You'll also learn how to create Vulkan rendering pipelines and explore a collection of recipes for quickly building minimal graphical applications from scratch using open-source libraries like GLFW, GLM, STB, and LightweightVK.

Chapter 3, Working with Vulkan Objects, explores handling various buffers and textures in Vulkan, as well as organizing a staging buffer. You'll learn how to wrap low-level Vulkan objects into user-friendly abstractions and get introduced to descriptor indexing.

Chapter 4, Adding User Interaction and Productivity Tools, focuses on debugging, profiling, and user interaction mechanisms. You'll learn various techniques for debugging and profiling graphical applications, starting with on-screen counters and graphs, then exploring open-source instrumenting profiler capabilities, and finally implementing helper classes for interactive application debugging.

Chapter 5, Working with Geometry Data, covers handling geometry in a modern 3D rendering pipeline and introduces concepts like Level-of-Detail (LOD) and tessellation. You'll also explore GLSL techniques for implementing various utility functions for geometry rendering, and introduce you to Vulkan compute shaders.

Chapter 6, Physically Based Rendering Using the glTF 2.0 Shading Model, introduces the glTF 2.0 physically based shading model and its implementation using GLSL in Vulkan. You'll explore various data preprocessing techniques, including the precalculation of Bidirectional Reflectance Distribution Function (BRDF) look-up tables and irradiance maps, with all necessary tooling built from scratch.

Chapter 7, Advanced PBR Extensions, explores advanced glTF PBR extensions from Khronos that extend the base metallic-roughness model. You'll learn how to integrate each of these extensions into GLSL shader code.

Chapter 8, Graphics Rendering Pipeline, goes the representation of complex 3D scene data with multiple dependencies and cross-references. You'll learn how to apply performance-oriented techniques, such as data-oriented design, to build a high-performance 3D rendering system. This chapter marks the beginning of real 3D engine design, demonstrating how to scale a scene graph approach to develop a practical graphics engine.

Chapter 9, glTF Animations, introduces a framework for supporting glTF animations in your rendering code. You'll learn the fundamentals of node-based animations, skeletal animations, morph targets, and animation blending.

Chapter 10, Image-Based Techniques, presents a series of recipes for enhancing rendering realism using image-based techniques, such as screen space ambient occlusion, high dynamic range rendering with light adaptation, and projective shadow mapping.

Chapter 11, Advanced Rendering Techniques and Optimizations, dives deeper into constructing GPU-driven rendering pipelines, multi-threaded resources loading, and other advanced techniques for feature-rich graphics applications. The book concludes by integrating various recipes and techniques into a single application.

To get the most out of this book

You'll need a machine that supports Vulkan 1.3 with the latest GPU drivers. All code examples in this book have been tested with Vulkan SDK 1.4.304.1, using Visual Studio 2022 on Windows 10 and 11, and GCC 12 on Ubuntu. While macOS is not officially supported, its users should be able to run some very first demos from this book.

Download the example code files

The code bundle for the book is hosted on GitHub at `https://github.com/PacktPublishing/3D-Graphics-Rendering-Cookbook-Second-Edition`. We also have other code bundles from our rich catalog of books and videos available at `https://github.com/PacktPublishing/`. Check them out!

Download the color images

We also provide a PDF file that has color images of the screenshots/diagrams used in this book. You can download it here: `https://static.packt-cdn.com/downloads/9781803248110_ColorImages.pdf`.

Conventions used

There are a number of text conventions used throughout this book.

`CodeInText`: Indicates code words in text, database table names, folder names, filenames, file extensions, pathnames, dummy URLs, user input, and Twitter handles. For example: "In OpenGL, presenting an offscreen buffer to the visible area of a window is done using system-dependent functions, such as `wglSwapBuffers()` on Windows, `eglSwapBuffers()` on OpenGL ES embedded systems, `glXSwapBuffers()` on Linux, or automatically on macOS. Vulkan, however, gives us much more fine-grained control."

A block of code is set as follows:

```
while (!glfwWindowShouldClose(window)) {
  glfwPollEvents();
  glfwGetFramebufferSize(window, &width, &height);
  if (!width || !height) continue;
```

```
    lvk::ICommandBuffer& buf = ctx->acquireCommandBuffer();
    ctx->submit(buf, ctx->getCurrentSwapchainTexture());
  }
```

Bold: Indicates a new term, an important word, or words that you see on the screen. For instance, words in menus or dialog boxes appear in the text like this. For example: "Here, we use the **SPIRV-Reflect** library to introspect the SPIR-V code and retrieve the size of the push constants from it."

Warnings or important notes appear like this.

Tips and tricks appear like this.

Get in touch

Feedback from our readers is always welcome.

General feedback: Email `feedback@packtpub.com` and mention the book's title in the subject of your message. If you have questions about any aspect of this book, please email us at `questions@packtpub.com`.

Errata: Although we have taken every care to ensure the accuracy of our content, mistakes do happen. If you have found a mistake in this book, we would be grateful if you reported this to us. Please visit `http://www.packtpub.com/submit-errata`, click **Submit Errata**, and fill in the form.

Piracy: If you come across any illegal copies of our works in any form on the internet, we would be grateful if you would provide us with the location address or website name. Please contact us at `copyright@packtpub.com` with a link to the material.

If you are interested in becoming an author: If there is a topic that you have expertise in and you are interested in either writing or contributing to a book, please visit `http://authors.packtpub.com`.

Share your thoughts

Once you've read *Vulkan 3D Graphics Rendering Cookbook, Second edition*, we'd love to hear your thoughts! Scan the QR code below to go straight to the Amazon review page for this book and share your feedback.

https://packt.link/r/1803248114

Your review is important to us and the tech community and will help us make sure we're delivering excellent quality content.

1

Establishing a Build Environment

In this chapter, you will learn how to set up a 3D graphics development environment on your computer for Windows and Linux operating systems. You will learn which software tools are needed to run the demos from this book's source code bundle: `https://github.com/PacktPublishing/3D-Graphics-Rendering-Cookbook-Second-Edition`.

We will cover the following topics:

- Setting up our development environment on Microsoft Windows
- Setting up our development environment on Linux
- Installing the Vulkan SDK for Microsoft Windows and Linux
- Managing dependencies
- Getting the demo data
- Creating utilities for CMake projects
- Using the `GLFW` library
- Multithreading with `Taskflow`
- Compiling Vulkan shaders at runtime
- Compressing textures into the `BC7` format

Getting the most out of this book – get to know your free benefits

Unlock exclusive **free** benefits that come with your purchase, thoughtfully crafted to supercharge your learning journey and help you learn without limits.

Here's a quick overview of what you get with this book:

Next-gen reader

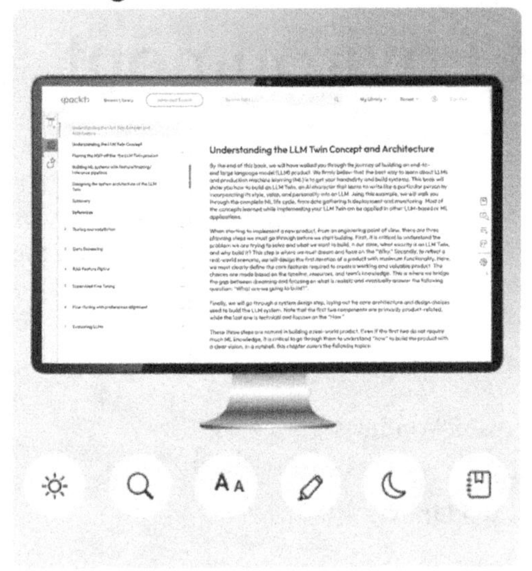

Figure 1.1: Illustration of the next-gen Packt Reader's features

Our web-based reader, designed to help you learn effectively, comes with the following features:

Multi-device progress sync: Learn from any device with seamless progress sync.

Highlighting and notetaking: Turn your reading into lasting knowledge.

Bookmarking: Revisit your most important learnings anytime.

Dark mode: Focus with minimal eye strain by switching to dark or sepia mode.

Interactive AI assistant (beta)

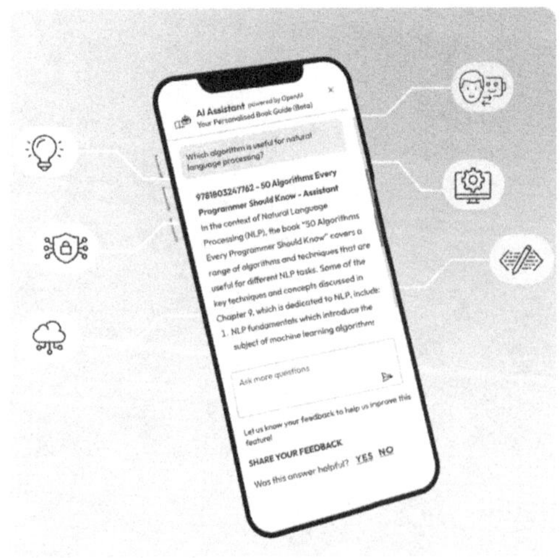

Figure 1.2: Illustration of Packt's AI assistant

Our interactive AI assistant has been trained on the content of this book, so it can help you out if you encounter any issues. It comes with the following features:

Summarize it: Summarize key sections or an entire chapter.

AI code explainers: In the next-gen Packt Reader, click the **Explain** button above each code block for AI-powered code explanations.

Note: The AI assistant is part of next-gen Packt Reader and is still in beta.

DRM-free PDF or ePub version

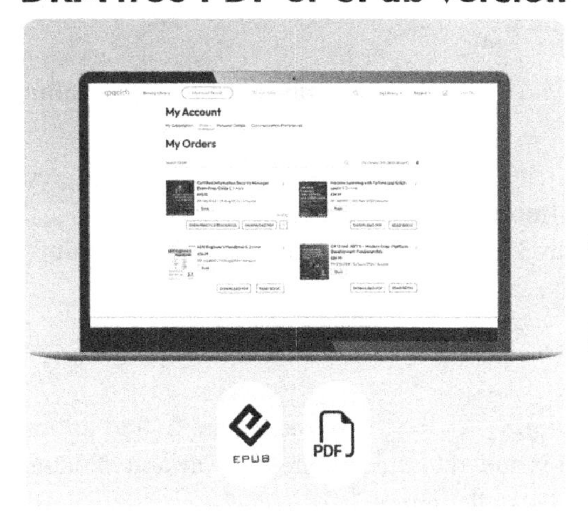

Figure 1.3: Free PDF and ePub

Learn without limits with the following perks included with your purchase:

Learn from anywhere with a DRM-free PDF copy of this book.

Use your favorite e-reader to learn using a DRM-free ePub version of this book.

Setting up our development environment on Microsoft Windows

In this recipe, we will get started by setting up our development environment on Windows. We will go through the installation of each of the required tools individually and in detail.

Getting ready

In order to start working with the examples from this book in a Microsoft Windows environment, you will need some essential tools to be installed in your system.

The most important one is Microsoft Visual Studio 2022. Additional tools include the **Git** version control system, the **CMake** build tool, and the **Python** programming language. Throughout this book, we use these tools on the command line only, so no GUI add-ons will be required.

How to do it...

Let's install each of the required tools individually.

Microsoft Visual Studio 2022

Follow the given steps to install Microsoft Visual Studio 2022:

1. Open `https://visualstudio.microsoft.com` and download the **Visual Studio 2022 Community Edition** installer.
2. Start the installer and follow the on-screen instructions. For the purposes of this book, you need to have a native C++ compiler for the 64-bit Intel platform. Other components of the Visual Studio development environment are not required to run this book's bundled sample code.

Git

Follow the given steps to install Git:

1. Download the latest Git installer from `https://git-scm.com/downloads`, run it, and follow the on-screen instructions. We assume that Git is added to the system PATH variable. Enable the option shown in the following image during installation:

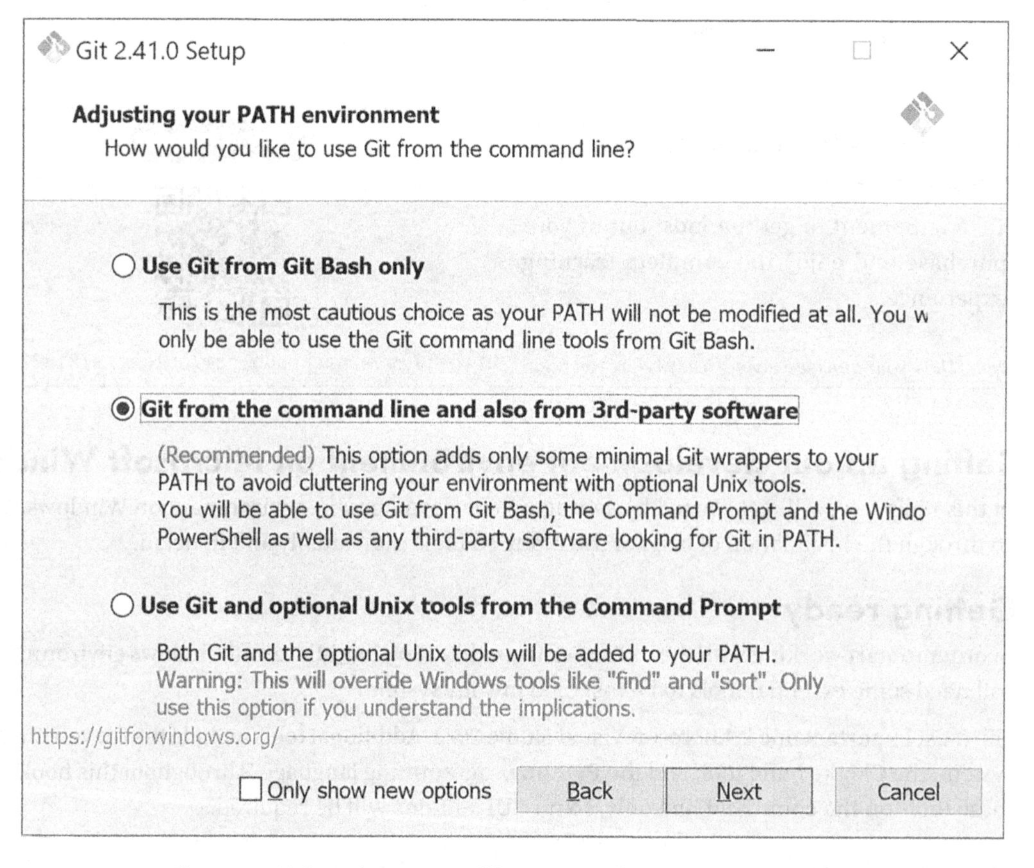

Figure 1.4: Git from the command line and also from third-party software

2. Select **Use Windows' default console window**, as shown in the next screenshot. This option will allow you to build the scripts in this book from any directory on your computer:

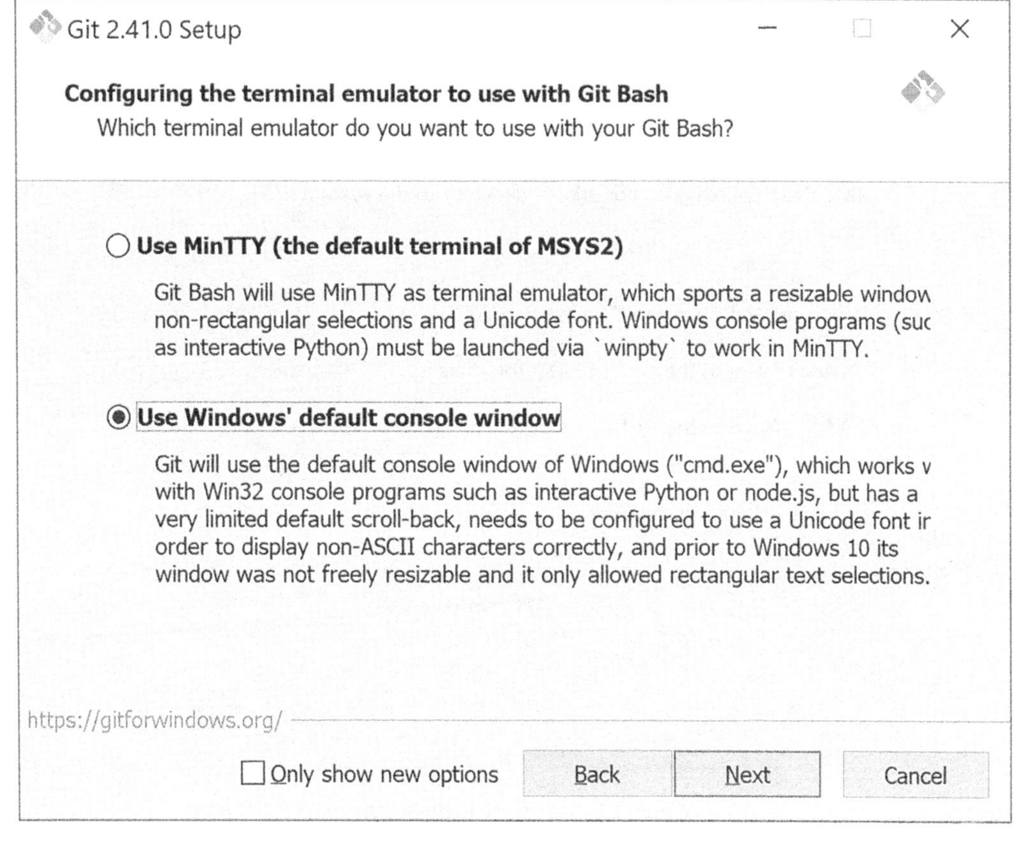

Figure 1.5: Use Windows' default console window

Read more

Git is complex software and a huge topic in itself. We recommend the book *Mastering Git, 2nd edition* written by Jakub Narębski and published by Packt Publishing, `https://www.packtpub.com/en-us/product/mastering-git-9781835080054`, along with *Git Essentials: Developer's Guide to Git* by François Dupire and the downloadable ebook *ProGit, Second Edition*, by Scott Chacon and Ben Straub, `https://git-scm.com/book/en/v2`.

CMake

To install CMake, please follow the given steps:

1. Download the latest 64-bit CMake installer from `https://cmake.org/download/`.
2. Run it and follow the on-screen instructions. If you already have an earlier version of CMake installed, it is recommended to uninstall it first.

3. Select the **Add CMake to the system PATH for all users** option, as shown here:

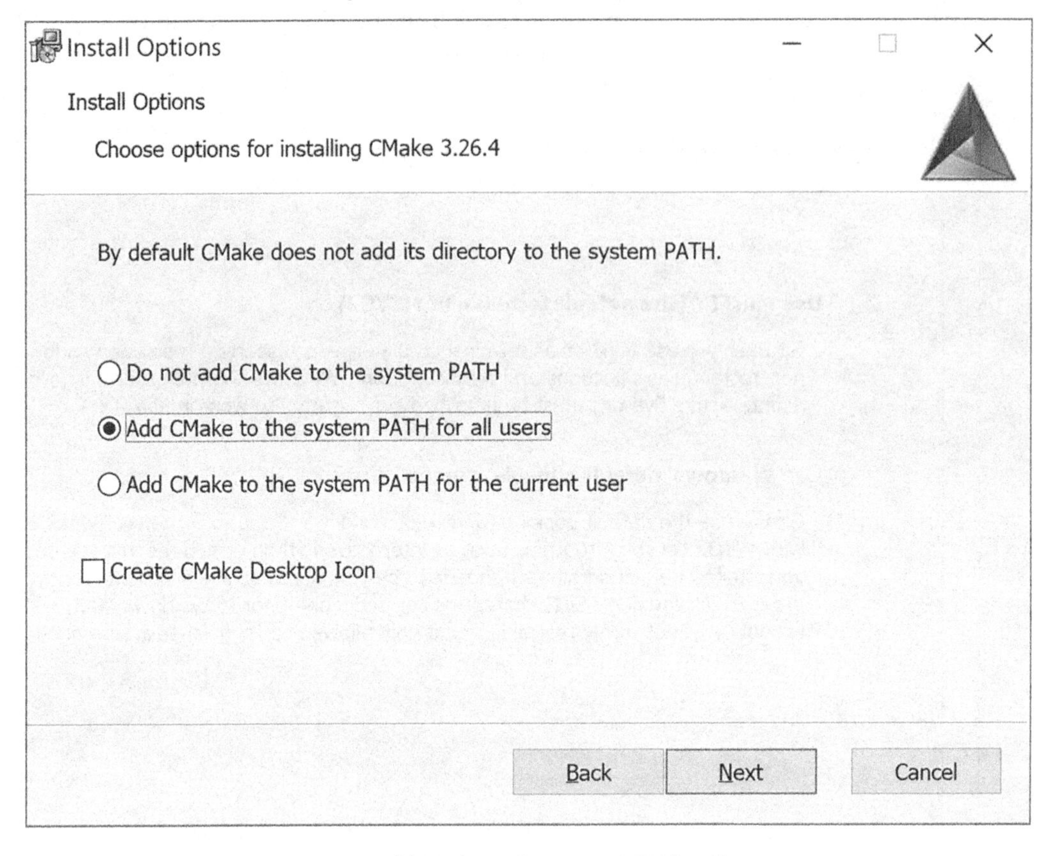

Figure 1.6: Add CMake to the system PATH for all users

Python

To install Python, please follow the given steps:

1. Download the latest Python 3 installer for 64-bit systems from `https://www.python.org/downloads/`.
2. Run it and follow the on-screen instructions.
3. During the installation, you also need to install the `pip` feature. Choose **Custom Installation** and make sure that the **pip** checkbox is checked, as shown:

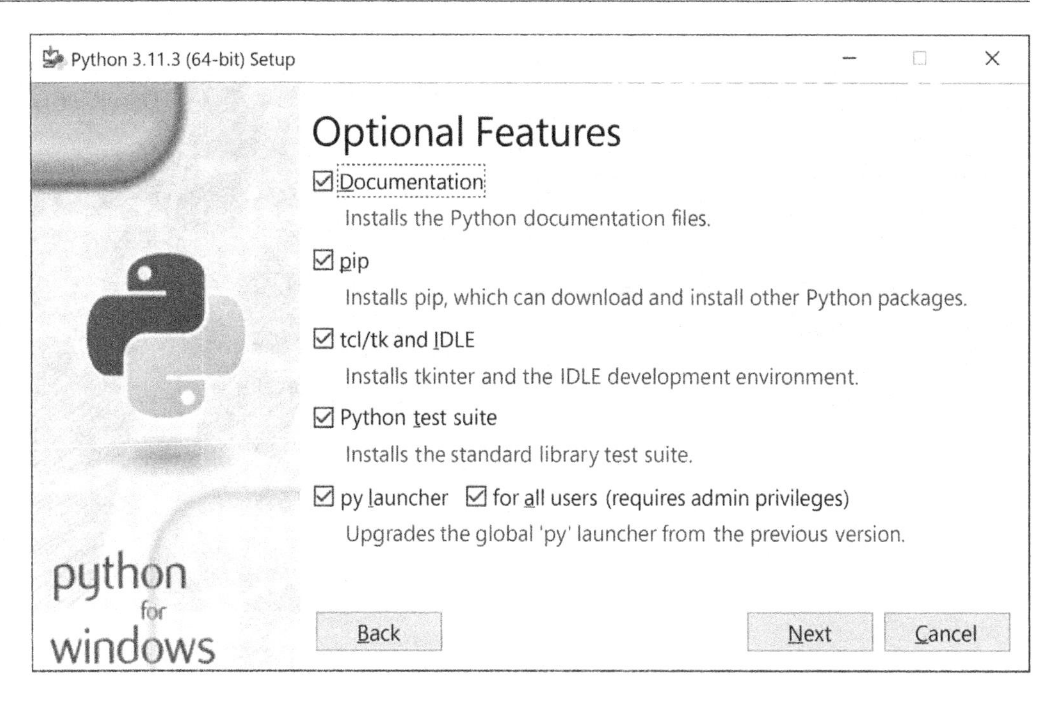

Figure 1.7: Custom installation

4. Once the installation has completed, make sure to add the folder containing `python.exe` to the `PATH` environment variable.

There's more...

Besides Git, there are other popular version control systems, like SVN and Mercurial. While developing large software systems, you will inevitably face the need to download some libraries from a non-Git repository. We recommend getting familiar with Mercurial.

While working in the command-line environment, it is useful to have some tools from the Unix environment, like `wget`, `grep`, `find`, etc. The **GnuWin32** project provides precompiled binaries of these tools, which can be downloaded from `http://gnuwin32.sourceforge.net`.

Furthermore, in the Windows environment, orthodox file managers make file manipulation a lot easier. We definitely recommend giving the open-source Far Manager a try. You can download it from `https://farmanager.com`. It looks like this:

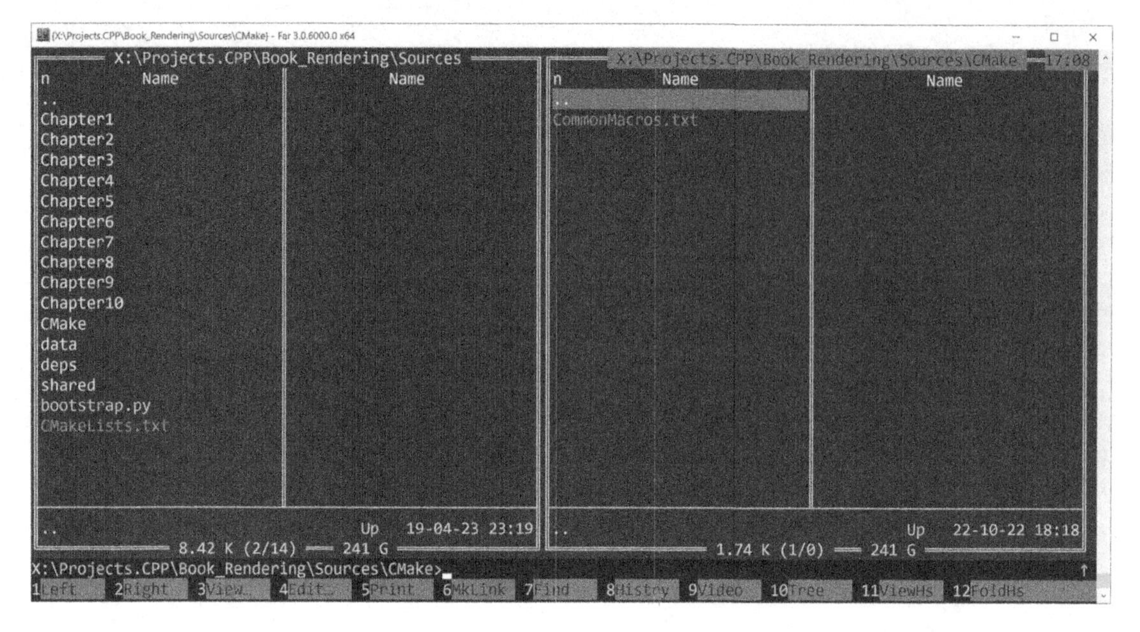

Figure 1.8: The look and feel of Far Manager

Setting up our development environment on Linux

Linux is becoming more and more attractive for 3D graphics development, including gaming technology. Let's go through a list of tools necessary to start working with this book on Linux.

Getting ready

We assume you have a desktop computer with a Debian-based GNU/Linux operating system installed. We also assume you are familiar with the apt package manager.

To start developing modern graphics programs on Linux, you need to have up-to-date video card drivers installed that support Vulkan 1.3. To build examples from this book, a C++ compiler with C++20 support is required. We tested our code with Clang and the GNU Compiler Collection.

How to do it...

On a Debian-based system, the installation process is straightforward; however, before installing any of the required packages, we recommend running the following command to ensure your system is up to date:

```
sudo apt-get update
```

Let us go through the list of essential software and install whatever is missing.

1. **GCC Compiler**: Assuming you have a properly configured `apt` package manager, run the following command to install the GCC compiler and related tools. We tested GCC 12:

```
sudo apt-get install build-essential
```

Quick tip: Enhance your coding experience with the **AI Code Explainer** and **Quick Copy** features. Open this book in the next-gen Packt Reader. Click the **Copy** button (**1**) to quickly copy code into your coding environment, or click the **Explain** button (**2**) to get the AI assistant to explain a block of code to you.

```
function calculate(a, b) {
  return {sum: a + b};
};
```

Copy (1) Explain (2)

The next-gen Packt Reader is included for free with the purchase of this book. Unlock it by scanning the QR code below or visiting `https://www.packtpub.com/unlock/9781803248110`.

2. **CMake**: The CMake build tool is also available in the standard repositories. To install CMake, type the following command:

```
sudo apt-get install cmake
```

Note

CMake 3.19 or above is sufficient for the code samples in this book.

3. **Git**: To install the Git version control system, run the following command:

```
sudo apt-get install git
```

4. **Python 3**: To install the Python 3 package, run the following command:

```
sudo apt-get install python3.7
```

The exact version of Python may vary between Linux distributions. Any version of Python 3 will suffice for the scripts in this book.

Now we are done with the basic packages and can install graphics-related software. Let us move on to the next recipe to learn how to set up the Vulkan SDK.

Installing the Vulkan SDK for Windows and Linux

In this recipe, you will learn how to get started with the Vulkan SDK. We will describe the requirements and procedure for installing the LunarG Vulkan SDK for Windows and Linux.

In principle, it is possible to write Vulkan applications without the Vulkan SDK, using only C/C++ header files provided by Khronos. You can get these header files by cloning the Git repository: `https://github.com/KhronosGroup/Vulkan-Headers`. However, it is advised to install the complete Vulkan SDK to be able to use Vulkan validation layers and a standalone GLSL compiler.

Getting ready

Make sure you have the latest up-to-date video card drivers for your operating system. On Windows, you can download video drivers from your GPU vendor's website. For Ubuntu, refer to the documentation: `https://ubuntu.com/server/docs/nvidia-drivers-installation`.

How to do it...

To install Vulkan 1.3 on Linux, follow these steps:

1. Open the `https://www.lunarg.com/vulkan-sdk/` page in a browser and download the latest Vulkan SDK for Windows or Linux.
2. After the download has finished, run the Windows installer file and follow the on-screen instructions. If you have Ubuntu 22.04 installed, use the following commands provided on LunarG's website:

```
wget -qO- https://packages.lunarg.com/lunarg-signing-key-pub.asc | sudo
tee /etc/apt/trusted.gpg.d/lunarg.asc
sudo wget -qO /etc/apt/sources.list.d/lunarg-vulkan-1.3.296-jammy.list
https://packages.lunarg.com/vulkan/1.3.296/lunarg-vulkan-1.3.296-jammy.
list
sudo apt update
sudo apt install vulkan-sdk
```

3. For other Linux distributions, you may need to download the `.tar.gz` SDK archive from `https://vulkan.lunarg.com/sdk/home#linux` and unpack it manually. You need to set environment variables to locate the Vulkan SDK components. Use the `source` command to run a config script that will do it for you:

```
source ~/vulkan/1.3.296.1/setup-env.sh
```

There's more...

When developing cross-platform applications, it is good to use similar tools for each platform. Since Linux supports GCC and Clang compilers, using GCC or Clang on Windows ensures that you avoid the most common portability issues. A complete package of C and C++ compilers can be downloaded from `http://www.equation.com/servlet/equation.cmd?fa=fortran`.

An alternative way to use GCC on Windows is to install the MSYS2 environment from `https://www.msys2.org`. It features the package management system used in Arch Linux, **Pacman**.

Managing dependencies

This book's examples use multiple open-source libraries. To manage these dependencies, we use a free and open-source tool called **Bootstrap**. The tool is similar to Google's repo tool and works on both Windows and Linux, as well as on macOS for that matter.

In this recipe, we will learn how to use Bootstrap to download libraries using the Vulkan Headers repository as an example.

Getting ready

Make sure you have Git and Python installed as described in the previous recipes. After that, clone the Bootstrap repository from GitHub:

```
git clone https://github.com/corporateshark/bootstrapping
```

How to do it...

Let's look into the source code bundle and run the `bootstrap.py` script:

```
bootstrap.py
```

The script will start downloading all the third-party libraries required to compile and run the source code bundle for this book. On Windows, the tail of the output should look as follows.

```
Cloning into 'M:\Projects.CPP\Book_Rendering2\Sources\deps\src\assimp'...
remote: Enumerating objects: 25, done.
remote: Counting objects: 100% (25/25), done.
remote: Compressing objects: 100% (24/24), done.
remote: Total 51414 (delta 2), reused 10 (delta 1), pack-reused 51389
Receiving objects: 100% (51414/51414), 148.46 MiB | 3.95 MiB/s, done.
Resolving deltas: 100% (36665/36665), done.
Checking out files: 100% (2163/2163), done.
```

Once the download process is complete, we are ready to build the project.

How it works...

Bootstrap takes a JSON file as input, opening `bootstrap.json` from the current directory by default. It contains metadata of libraries we want to download; for example, their names, where to retrieve them from, a specific version to download, and so on. Besides that, each used library can have some additional instructions on how to build it. Those can be patches applied to the original library, unpacking instructions, SHA hashes to check archive integrity, and many others.

The source code for each library can be represented by either a URL of a version control system repository or by an archive file with the library source files.

A typical JSON file entry corresponding to one library looks like this snippet:

```
[{
  "name": "vulkan",
  "source": {
    "type": "git",
    "url" : "https://github.com/KhronosGroup/Vulkan-Headers.git",
    "revision": "v1.3.296"
 }
}]
```

The field `type` can have one of these values: `archive`, `git`, `hg`, or `svn`. The first value corresponds to an archive file, such as `.zip`, `.tar.gz`, or `.tar.bz2`, while the last three types describe different version control system repositories. The `url` field contains a URL of the archive file to be downloaded or a URL of the repository. The `revision` field can specify a particular revision, tag, or branch to check out.

The complete JSON file is a comma-separated list of such entries. For this recipe, we have only one library to download. We will add more libraries in the next chapters. The accompanying source code bundle contains a JSON file with all the libraries used in this book.

There's more...

There is comprehensive documentation for this tool that describes other command-line options and JSON fields in great detail. It can be downloaded from `https://github.com/corporateshark/bootstrapping`.

The Bootstrap tool does not differentiate between source code and binary assets. All the textures, 3D models, and other resources for your application can also be downloaded and kept up to date and organized in an automated way.

Getting the demo data

This book makes use of free 3D graphics datasets as much as possible. The comprehensive list of large 3D datasets is maintained by Morgan McGuire – Computer Graphics Archive, July 2017 (`https://casual-effects.com/data`). We will use some large 3D models from his archive for demonstration purposes in this book. Let us download one of them.

How to do it...

The bundled source code contains a Python script, `deploy_deps.py`, which will download all the required 3D models automatically. To download the entire Bistro dataset manually, which is not recommended, follow these simple steps:

1. Open the `https://casual-effects.com/data/` page in a browser and find the **Amazon Lumberyard Bistro** dataset.
2. Click on the **Download** link and allow the browser to download all the data files. Below is a screenshot of Morgan McGuire's site with the download link.

Download 2.4 GB

Triangles: 3858088
Vertices: 3672567
Updated: 2019-05-07
License: CC-BY 4.0
© 2017 Amazon Lumberyard

Amazon Lumberyard Bistro

Created by Amazon Lumberyard for a 2017 GDC demo. Released publicly in the NVIDIA ORCA collection. The exterior contains 2,837,181 triangles and 2,910,304 vertices. The interior contains 1,020,907 triangles and 762,263 vertices.

This version has some manually remastered materials by Morgan McGuire to correct for limitations of the original OBJ export from Lumberyard, and it was split across multiple zipfiles to make downloading easier.Unzip each file into a directory of the same name or load the compressed files directly using the G3D Innovation Engine.

Cite this model as:

```
@misc{ORCAAmazonBistro,
   title = {Amazon Lumberyard Bistro, Open Research Content Archive (ORCA)},
   author = {Amazon Lumberyard},
   year = {2017},
   month = {July},
   note = {\small \texttt{http://developer.nvidia.com/orca/amazon-lumberyard-bistro}},
   url = {http://developer.nvidia.com/orca/amazon-lumberyard-bistro}
}
```

Figure 1.9: Amazon Lumberyard Bistro as pictured on casualeffects.com as a 2.4 GB download

Creating utilities for CMake projects

In this recipe, we will see how CMake is used to configure all the code examples in this book and learn some small tricks along the way.

Read more

For those who are just starting with CMake, we recommend reading the books *CMake Cookbook* (Radovan Bast and Roberto Di Remigio) by Packt Publishing and *Mastering CMake* (Ken Martin and Bill Hoffman) by Kitware.

Getting ready

For a start, let's create a minimalistic C++ application with a trivial `main()` function and build it using CMake:

```
int main() {
  printf("Hello World!\n");
  return 0;
}
```

How to do it...

Let's introduce two helper macros for CMake. You can find them in the `CMake/CommonMacros.txt` file of our source code bundle at `https://github.com/PacktPublishing/3D-Graphics-Rendering-Cookbook-Second-Edition`.

1. The `SETUP_GROUPS` macro iterates over a space-delimited list of C and C++ files, whether it is a header or a source file, and assigns each of them to a separate group. The group name is constructed based on the path of each individual file. This way, we end up with a nice structure similar to a filesystem within a directory in the Visual Studio Solution Explorer window, as we can see on the right in the following figure:

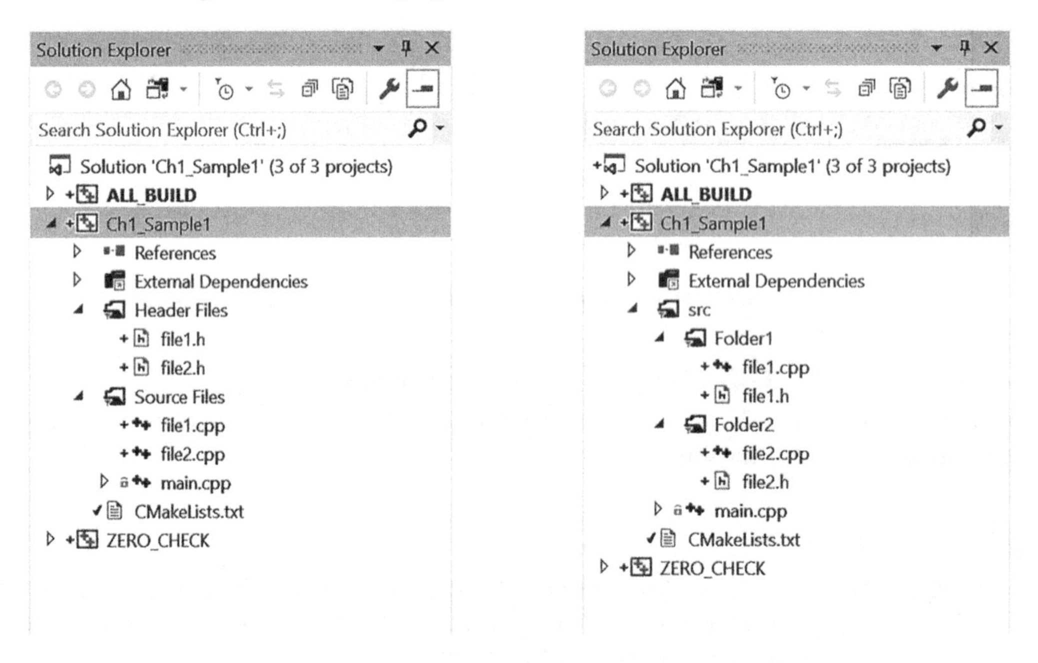

Figure 1.10: Without groups (left) and with groups (right)

2. The macro starts by iterating over a list of files passed in the `src_files` parameter:

```
macro(SETUP_GROUPS src_files)
  foreach(FILE ${src_files})
    get_filename_component(PARENT_DIR "${FILE}" PATH)
```

3. We store the parent directory name as a default group name. Regardless of the operating system, replace all forward slashes with backslashes:

```
    set(GROUP "${PARENT_DIR}")
    string(REPLACE "/" "\\" GROUP "${GROUP}")
```

4. Then, we can tell CMake to assign the current file to a source group with this name.

```
    source_group("${GROUP}" FILES "${FILE}")
  endforeach()
endmacro()
```

5. The second macro, `SETUP_APP`, is used as a shortcut to create a new CMake project with all the standard properties we want it to have. It is very convenient when having to deal with a number of very similar subprojects, for example, like in this book.

```
macro(SETUP_APP projname chapter)
  set(FOLDER_NAME ${chapter})
  set(PROJECT_NAME ${projname})
  project(${PROJECT_NAME} CXX)
```

6. After setting the project name, this macro uses the `GLOB_RECURSE` function to collect all source and header files into the `SRC_FILES` and `HEADER_FILES` variables.

```
  file(GLOB_RECURSE SRC_FILES LIST_DIRECTORIES false
       RELATIVE ${CMAKE_CURRENT_SOURCE_DIR} src/*.c??)
  file(GLOB_RECURSE HEADER_FILES LIST_DIRECTORIES false
       RELATIVE ${CMAKE_CURRENT_SOURCE_DIR} src/*.h)
```

7. In all our code samples, we use the directory `src` containing the source files as an `include` directory, too.

```
  include_directories(src)
```

8. All enumerated source and header files are added to an executable inside the current project.

```
  add_executable(${PROJ_NAME} ${SRC_FILES} ${HEADER_FILES})
```

9. We use the `SETUP_GROUP` macro from *Step 1* to place each source and header file into an appropriate group inside the project.

```
  SETUP_GROUPS("${SRC_FILES}")
  SETUP_GROUPS("${HEADER_FILES}")
```

10. The next three properties set different executable file names for each supported build configuration. These lines are optional, yet they are really useful when using CMake with the Visual Studio IDE. The reason is that Visual Studio can change build configurations (or "build types", as they are called in CMake) dynamically directly from the IDE, and each build configuration can have its own output file name. We add suffixes to these file names so that they can co-exist in a single output folder.

```
  set_target_properties(${PROJ_NAME}
    PROPERTIES OUTPUT_NAME_DEBUG ${PROJ_NAME}_Debug)
  set_target_properties(${PROJ_NAME}
    PROPERTIES OUTPUT_NAME_RELEASE ${PROJ_NAME}_Release)
  set_target_properties(${PROJ_NAME}
    PROPERTIES OUTPUT_NAME_RELWITHDEBINFO ${PROJ_NAME}_ReleaseDebInfo)
```

11. Since we use C++20 throughout this book, we require CMake to enable it.

```
  set_property(
    TARGET ${PROJ_NAME} PROPERTY CXX_STANDARD 20)
  set_property(
    TARGET ${PROJ_NAME} PROPERTY CXX_STANDARD_REQUIRED ON)
```

12. To ease the debugging with Visual Studio, we enable console output by changing the application type to `Console`. We also set the local debugger working directory to `CMAKE_SOURCE_DIR`, which will make finding assets a lot more straightforward and consistent. There are some Apple-specific properties to allow building the source code on Mac machines.

```
  if(MSVC)
    add_definitions(-D_CONSOLE)
    set_property(TARGET ${PROJ_NAME} PROPERTY
      VS_DEBUGGER_WORKING_DIRECTORY "${CMAKE_SOURCE_DIR}")
  endif()
  if(APPLE)
    set_target_properties(${PROJECT_NAME} PROPERTIES
      XCODE_GENERATE_SCHEME TRUE
      XCODE_SCHEME_WORKING_DIRECTORY "${CMAKE_SOURCE_DIR}")
  endif()
endmacro()
```

13. Finally, the top-level `CMakeLists.txt` file of our first project will look like this:

```
cmake_minimum_required(VERSION 3.19)
project(Chapter01)
include(../../CMake/CommonMacros.txt)
SETUP_APP(Ch01_Sample01_CMake "Chapter 01")
```

Note

You may notice that the line `project(Chapter01)` above is overridden by a call to `project()` inside the `SETUP_APP` macro. This is due to the following CMake warning, which will be emitted if we do not declare a new project right from the get-go.

```
CMake Warning (dev) in CMakeLists.txt:
No project() command is present. The top-level CMakeLists.
txt file must contain a literal, direct call to the
project() command. Add a line of project(ProjectName) near
the top of the file, but after cmake_minimum_required().
```

14. To build and test the executable, create the `build` subfolder, change the working directory to `build`, and run CMake as follows:
    - For Windows and Visual Studio 2022, run the following command to configure our project for the 64-bit target platform architecture.

      ```
      cmake .. -G "Visual Studio 17 2022" -A x64
      ```

    - For Linux, we can use the `Unix Makefiles` CMake generator as follows:

      ```
      cmake .. -G "Unix Makefiles"
      ```

15. To build an executable for the `release` build type, you can use the following command on any platform. To build a debug version, use `--config Debug` or skip that parameter entirely.

```
cmake --build . --config Release
```

All the demo applications from the source code bundle should be run from the folder where the `data/` subfolder is located.

There's more...

Alternatively, you can use the cross-platform build system Ninja along with CMake. It is possible to do so simply by changing the CMake project generator name.

```
cmake .. -G "Ninja"
```

Invoke Ninja from the command line to compile the project.

```
ninja
[2/2] Linking CXX executable Ch01_Sample01_CMake.exe
```

Notice how fast everything gets built now, compared to the classic `cmake --build` command. See `https://ninja-build.org` for more details.

Now let's take a look at how to work with some basic open source libraries.

Using the GLFW library

The GLFW library hides all the complexity of creating windows, graphics contexts, and surfaces and getting input events from the operating system. In this recipe, we build a minimalistic application with GLFW to get an empty window onto the screen.

Getting ready

We build our examples with GLFW 3.4. Here is a JSON snippet for the Bootstrap script so that you can download the proper library version:

```
{
        "name": "glfw",
        "source": {
                "type": "git",
                "url": "https://github.com/glfw/glfw.git",
                "revision": "3.4"
        }
}
```

The complete source code for this recipe can be found in the source code bundle under the name of `Chapter01/02_GLFW`.

How to do it...

Let's write a minimalistic application that creates a window and waits for an exit command from the user – pressing the *Esc* key. This functionality will be used in all of our subsequent demos, so we have wrapped it into a helper function `initWindow()` declared in `shared/HelpersGLFW.h`. Let's take a look at how to use it to create an empty GLFW window:

1. Include all necessary headers and decide on the initial window dimensions:

    ```
    #include <shared/HelpersGLFW.h>
    int main() {
      uint32_t width = 1280;
      uint32_t height = 800;
    ```

2. Invoke the `initWindow()` function to create a window. The `width` and `height` parameters are passed by reference and, after the call, will contain the actual working area of the created window. If we pass the initial values of `0`, the window will be created to span the entire desktop working area without overlapping the system taskbar.

    ```
      GLFWwindow* window =
        initWindow("GLFW example", width, height);
    ```

3. For this application, the main loop and cleanup are trivial:

    ```
        while (!glfwWindowShouldClose(window)) {
    ```

```
    glfwPollEvents();
  }
  glfwDestroyWindow(window);
  glfwTerminate();
  return 0;
}
```

Now we will take a look at the internals of `initWindow()` for some interesting details.

How it works...

Let's use this library to create an application that opens an empty window:

1. First, we set the GLFW error callback via a lambda to catch potential errors and then initialize GLFW:

    ```
    GLFWwindow* initWindow(const char* windowTitle,
      uint32_t& outWidth, uint32_t& outHeight) {
      glfwSetErrorCallback([](int error,
                              const char* description) {
        printf("GLFW Error (%i): %s\n", error, description);
      });
      if (!glfwInit()) return nullptr;
    ```

2. Let's decide if we want to make a desktop full-screen window. Set the resizable flag for windows that aren't full-screen and retrieve the desired window dimensions. We are going to initialize Vulkan manually, so no graphics API initialization is required to be done by GLFW. The flag `wantsWholeArea` determines if we want a true full-screen window or a window that does not overlap the system taskbar.

    ```
      const bool wantsWholeArea = !outWidth || !outHeight;
      glfwWindowHint(GLFW_CLIENT_API, GLFW_NO_API);
      glfwWindowHint(GLFW_RESIZABLE,
        wantsWholeArea ? GLFW_FALSE : GLFW_TRUE);
      GLFWmonitor* monitor     = glfwGetPrimaryMonitor();
      const GLFWvidmode* mode = glfwGetVideoMode(monitor);
      int x = 0;
      int y = 0;
      int w = mode->width;
      int h = mode->height;
      if (wantsWholeArea) {
        glfwGetMonitorWorkarea(monitor, &x, &y, &w, &h);
      } else {
        w = outWidth;
        h = outHeight;
      }
    ```

3. Create a window and retrieve the actual window dimensions:

```
  GLFWwindow* window = glfwCreateWindow(
    w, h, windowTitle, nullptr, nullptr);
  if (!window) {
    glfwTerminate();
    return nullptr;
  }
  if (wantsWholeArea) glfwSetWindowPos(window, x, y);
  glfwGetWindowSize(window, &w, &h);
  outWidth  = (uint32_t)w;
  outHeight = (uint32_t)h;
```

4. Set a default keyboard callback to handle the *Esc* key. A simple lambda will do this job for us.

```
  glfwSetKeyCallback(window, [](GLFWwindow* window,
    int key, int, int action, int) {
    if (key == GLFW_KEY_ESCAPE && action == GLFW_PRESS) {
      glfwSetWindowShouldClose(window, GLFW_TRUE);
    }
  });
  return window;
}
```

If you run this tiny application, it will create an empty window, as in the following screenshot:

Figure 1.11: Our first application

There's more...

Further details about how to use GLFW can be found at `https://www.glfw.org/documentation.html`.

Multithreading with Taskflow

Modern graphical applications require us to harness the power of multiple CPUs to be performant. **Taskflow** is a fast C++ header-only library that can help you write parallel programs with complex task dependencies quickly. This library is extremely useful as it allows you to jump into the development of multithreaded graphical applications that make use of advanced rendering concepts, such as frame graphs and multithreaded command buffer generation.

Getting ready

Here, we use Taskflow version 3.8.0. You can download it using the following Bootstrap snippet:

```
{
  "name": "taskflow",
  "source": {
    "type": "git",
    "url" : "https://github.com/taskflow/taskflow.git",
    "revision": "v3.8.0"
  }
}
```

To debug dependency graphs produced by Taskflow, it is recommended that you install the **GraphViz** tool from `https://www.graphviz.org`.

The complete source code for this recipe can be found in `Chapter01/03_Taskflow`.

How to do it...

Let's create and run a set of concurrent dependent tasks via the `for_each_index()` algorithm. Each task will print a single value from an array in a concurrent fashion. The processing order can vary between different runs of the program:

1. Include the `taskflow.hpp` header file. The `tf::Taskflow` class is the main place to create a task dependency graph. Declare an instance and a data vector to process.

```
#include <taskflow/taskflow.hpp>
int main() {
  tf::Taskflow taskflow;
  std::vector<int> items{ 1, 2, 3, 4, 5, 6, 7, 8 };
```

2. The `for_each_index()` member function returns a task that implements a parallel for loop algorithm. We specify the range `0..items.size()` and the step `1`. The returned `task` can be used for synchronization purposes:

```
auto task = taskflow.for_each_index(
  0u, uint32_t(items.size()), 1u, [&](int i) {
    printf("%i", items[i]);
  }).name("for_each_index");
```

3. Let's attach some work before and after the parallel for task so that we can view `Start` and `End` messages in the output. Let's call the new `S` and `T` tasks accordingly:

```
taskflow.emplace([]() {
  printf("\nS - Start\n"); }).name("S").precede(task);
taskflow.emplace([]() {
  printf("\nT - End\n"); }).name("T").succeed(task);
```

4. Save the generated tasks dependency graph in `.dot` format so that we can process it later with the GraphViz dot tool:

```
std::ofstream os(".cache/taskflow.dot");
taskflow.dump(os);
```

5. Now we can create a `tf::executor` object and run the constructed Taskflow graph:

```
  tf::Executor executor;
  executor.run(taskflow).wait();
  return 0;
}
```

An important point to note is that the Taskflow dependency graph is built only once. After that, it can be reused in every frame to efficiently execute concurrent tasks.

The output from the preceding program should look similar to the following listing:

```
S - Start
18345672
T - End
```

Here, we can see our `S` and `T` tasks. Between them, there are multiple threads with different IDs processing different elements of the `items[]` vector in parallel. Your output may vary due to concurrency.

There's more...

The application saved the dependency graph inside the `.cache/taskflow.dot` file. It can be converted into a visual representation by GraphViz, `https://graphviz.org`, using the following command:

```
dot -Tpng taskflow.dot > output.png
```

The resulting .png image should look similar to the following screenshot:

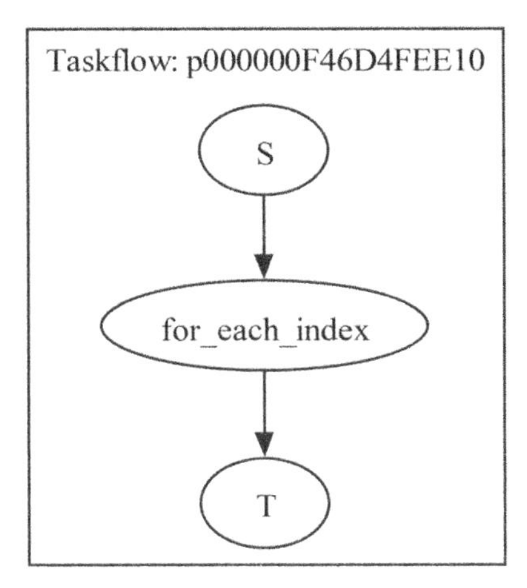

Figure 1.12: The Taskflow dependency graph for for_each_index()

This functionality is extremely useful when you are debugging complex dependency graphs (and producing complex-looking images for your books and papers).

The `Taskflow` library functionality is vast and provides implementations for numerous parallel algorithms and profiling capabilities. Please refer to the official documentation for in-depth coverage at `https://taskflow.github.io/taskflow/index.html`.

Compiling Vulkan shaders at runtime

Before we can start working with Vulkan, it's important to speed up the iterative process of writing shaders. Vulkan uses shaders in the binary form called Standard Portable Intermediate Representation, known as **SPIR-V**, and typically relies on a standalone shader compiler to precompile shaders offline. While this is ideal for a released product, it can slow down early stages of development and rapid prototyping, where shaders are frequently changed and need to be recompiled with every run. In this recipe, we'll show you how to compile Vulkan shaders at runtime using Khronos' reference shader compiler **glslang**.

Getting ready

Our application is using **LightweightVK** which is statically linked to the **glslang** shader compiler. The compiler version used in this recipe was downloaded using the following Bootstrap snippet.

```
{
  "name": "glslang",
  "source": {
    "type": "git",
    "url" : "https://github.com/KhronosGroup/glslang.git",
    "revision": "15.1.0"
  }
}
```

The complete source code for this recipe can be found in the source code bundle under the name `Chapter01/04_GLSLang`. The implementation of shader compilation in **LightweightVK** is located in `deps/src/lightweightvk/lvk/vulkan/VulkanUtils.cpp` in the function `lvk::compileShader()`.

How to do it...

Let us go through the steps necessary to compile GLSL shaders to SPIR-V using **glslang**. In this book text, we've omitted most error checking for clarity, though it is included in the actual source code.

1. We should use this helper function to compile a GLSL shader from its source code for a specific Vulkan pipeline stage and return the resulting SPIR-V binary as a vector bytes. Initializing the compiler input structure can be a bit verbose. We specify the shader's source language as `GLSLANG_SOURCE_GLSL` and set the correct targets to generate SPIR-V 1.6 for Vulkan 1.3. With C++20's designated initializers, this task becomes much simpler. This feature is also helpful in general Vulkan development, where a common task for developers is to populate structure members with specific values.

```
lvk::Result compileShader(
  glslang_stage_t stage,
  const char* code,
  std::vector<uint8_t>* outSPIRV,
  const glslang_resource_t* glslLangResource)
{
  const glslang_input_t input = {
    .language                = GLSLANG_SOURCE_GLSL,
    .stage                   = stage,
    .client                  = GLSLANG_CLIENT_VULKAN,
    .client_version          = GLSLANG_TARGET_VULKAN_1_3,
    .target_language         = GLSLANG_TARGET_SPV,
    .target_language_version = GLSLANG_TARGET_SPV_1_6,
    .code                    = code,
    .default_version         = 100,
    .default_profile         = GLSLANG_NO_PROFILE,
    .force_default_version_and_profile = false,
    .forward_compatible      = false,
    .messages                = GLSLANG_MSG_DEFAULT_BIT,
    .resource                = glslLangResource,
  };
```

2. Let`s create a **glslang** shader object using the abovementioned input. The `SCOPE_EXIT` macro ensures the shader object is properly deallocated whenever we exit the current scope. Its implementation is beyond the scope of this book. However, we recommend watching the *CppCon 2015* presentation *Declarative Control Flow* by Andrei Alexandrescu which dives deep into the implementation details. The implementation we use is located in `deps/src/lightweightvk/third-party/deps/src/ldrutils/lutils/ScopeExit.h`.

```
glslang_shader_t* shader = glslang_shader_create(&input);
SCOPE_EXIT {
  glslang_shader_delete(shader);
};
```

3. The shader needs to be preprocessed by the compiler. The following function returns `true` if all extensions, pragmas and version strings mentioned in the shader source code are valid. The `logShaderSource()` function prints to the console the source code of the shader together with line numbers and is very handy for debugging.

```
if (!glslang_shader_preprocess(shader, &input)) {
  LLOGW("Shader preprocessing failed:\n");
  LLOGW("  %s\n", glslang_shader_get_info_log(shader));
  LLOGW("  %s\n",
    glslang_shader_get_info_debug_log(shader));
  lvk::logShaderSource(code);
  return Result(Result::Code::RuntimeError);
}
```

4. Then the shader gets parsed into an internal representation inside the compiler.

```
if (!glslang_shader_parse(shader, &input)) {
  LLOGW("Shader parsing failed:\n");
  LLOGW("  %s\n", glslang_shader_get_info_log(shader));
  LLOGW("  %s\n",
    glslang_shader_get_info_debug_log(shader));
  lvk::logShaderSource(
    glslang_shader_get_preprocessed_code(shader));
  return Result(Result::Code::RuntimeError);
}
```

5. If everything went fine during the previous stages, now we can link the shader into a program and proceed with the binary code generation stage. The same `SCOPE_EXIT` is applied to `program` to prevent memory leaks.

```
glslang_program_t* program = glslang_program_create();
glslang_program_add_shader(program, shader);
SCOPE_EXIT {
  glslang_program_delete(program);
};
if (!glslang_program_link(program,
      GLSLANG_MSG_SPV_RULES_BIT |
      GLSLANG_MSG_VULKAN_RULES_BIT)) {
  LLOGW("Shader linking failed:\n");
  LLOGW("  %s\n", glslang_program_get_info_log(program));
```

```
    LLOGW("  %s\n",
      glslang_program_get_info_debug_log(program));
    return Result(Result::Code::RuntimeError);
  }
```

6. Now we are ready to generate some binary SPIR-V code. The **glslang** compiler supports multiple code generation flags to control debug info generation. This is useful if you want to inspect your shaders at run-time with a tool such as **RenderDoc**.

```
  glslang_spv_options_t options = {
    .generate_debug_info                 = true,
    .strip_debug_info                    = false,
    .disable_optimizer                   = false,
    .optimize_size                       = true,
    .disassemble                         = false,
    .validate                            = true,
    .emit_nonsemantic_shader_debug_info   = false,
    .emit_nonsemantic_shader_debug_source = false,
  };
  glslang_program_SPIRV_generate_with_options(
    program, input.stage, &options);
```

7. There might be some messages produced by the code generator. Check and print them if there are any. The resulting SPIR-V code can be retrieved in the following way. We return it as a binary blob of `uint8_t` values.

```
  if (glslang_program_SPIRV_get_messages(program)) {
    LLOGW("%s\n",
      glslang_program_SPIRV_get_messages(program));
  }
  const uint8_t* spirv = reinterpret_cast<const uint8_t*>(
    glslang_program_SPIRV_get_ptr(program));
  const size_t numBytes =
    glslang_program_SPIRV_get_size(program) *
    sizeof(uint32_t);
  *outSPIRV = std::vector(spirv, spirv + numBytes);

  return Result();
}
```

Let's inspect the demo application which loads GLSL shader code from files, compiles it using the abovementioned function, and then saves SPIR-V binaries into files.

How it works...

The application is straightforward. It loads the shader source code from a text file and uses the `compileShader()` function we have just written to compile it into SPIR-V.

```
void testShaderCompilation(
  const char* sourceFilename, const char* destFilename)
{
  std::string shaderSource = readShaderFile(sourceFilename);
  std::vector<uint8_t> spirv;
  lvk::Result res = lvk::compileShader(
    vkShaderStageFromFileName(sourceFilename),
    shaderSource.c_str(),
    &spirv,
    glslang_default_resource());
  saveSPIRVBinaryFile(destFilename, spirv.data(), spirv.size());
}
```

Each generated SPIR-V binary blob is saved to a file for further inspection.

```
void saveSPIRVBinaryFile(
  const char* filename, const uint8_t* code, size_t size)
{
  FILE* f = fopen(filename, "wb");
  fwrite(code, sizeof(uint8_t), size, f);
  fclose(f);
}
```

The `main()` function, which drives the demo application, initializes the **glslang** compiler and runs the tests.

```
int main()
{
  glslang_initialize_process();
  testShaderCompilation("Chapter01/04_GLSLang/src/main.vert",
                        ".cache/04_GLSLang.vert.bin");
  testShaderCompilation("Chapter01/04_GLSLang/src/main.frag",
                        ".cache/04_GLSLang.frag.bin");
  glslang_finalize_process();
  return 0;
}
```

The abovementioned program produces the same SPIR-V output as the following console commands:

```
glslangValidator
  -g -Os --target-env vulkan1.3 main.vert -o main.vert.bin
glslangValidator
  -g -Os --target-env vulkan1.3 main.frag -o main.frag.bin
```

Binary SPIR-V modules can be used to create Vulkan shader modules as follows. One key detail to note in this example is how to determine the size of the push constants used by the shader. This information will be important in later chapters when we build Vulkan pipelines:

```
ShaderModuleState VulkanContext::createShaderModuleFromSPIRV(
  const void* spirv,
  size_t numBytes,
  const char* debugName,
  Result* outResult) const
{
  VkShaderModule vkShaderModule = VK_NULL_HANDLE;

  const VkShaderModuleCreateInfo ci = {
    .sType = VK_STRUCTURE_TYPE_SHADER_MODULE_CREATE_INFO,
    .codeSize = numBytes,
    .pCode = (const uint32_t*)spirv,
  };

  const VkResult result =
    vkCreateShaderModule(vkDevice_, &ci, nullptr, &vkShaderModule);
  lvk::setResultFrom(outResult, result);
  if (result != VK_SUCCESS) return {.sm = VK_NULL_HANDLE};

  VK_ASSERT(lvk::setDebugObjectName(
    vkDevice_,
    VK_OBJECT_TYPE_SHADER_MODULE,
    (uint64_t)vkShaderModule,
    debugName));
```

After creating the Vulkan shader module, we can determine the size of the push constants used by the shader. To do this, we use the `SPIRV-Reflect` library, which makes it easy to extract this information:

```
  SpvReflectShaderModule mdl;
  SpvReflectResult result = spvReflectCreateShaderModule(
    numBytes, spirv, &mdl);
  SCOPE_EXIT {
    spvReflectDestroyShaderModule(&mdl);
```

```
  };

  uint32_t pushConstantsSize = 0;

  for (uint32_t i = 0; i < mdl.push_constant_block_count; ++i) {
    const SpvReflectBlockVariable& block =
      mdl.push_constant_blocks[i];
    pushConstantsSize =
      std::max(pushConstantsSize, block.offset + block.size);
  }
  return { .sm = vkShaderModule,
           .pushConstantsSize = pushConstantsSize };
}
```

This function is used throughout the `LightweightVK` library to create all Vulkan shader modules. It is located in `deps/src/lightweightvk/lvk/vulkan/VulkanClasses.cpp`.

There's more...

While it may be convenient to include a full compiler during the development phase, shipping a large compiler with the release version of your application is generally not recommended. Unless your application specifically needs to compile shaders at runtime, it's better to include precompiled SPIR-V shader binaries in the release version. One way to handle this is by implementing a shader caching mechanism. When a shader is needed, the application can first check if a compiled version already exists. If it doesn't, the app can load the **glslang** compiler from a `.dll` or `.so` file at runtime to compile the shader. Once all shaders are cached, the cached shaders can be shipped with the app without any GLSL source code. This approach ensures that compiled shaders are always available in the release version without bundling the compiler's shared libraries.

Compressing textures into the BC7 format

A major downside of high-resolution texture data is that it uses a lot of GPU memory for storage. To address this, all modern real-time rendering APIs offer texture compression, which lets us store textures in compressed formats on the GPU. One common format is `BC7`, a standard texture compression format for Vulkan that is supported on many devices.

`bc7enc` is an open-source library that can compress RGBA bitmaps into the `BC7` format. In this recipe, you will learn how to integrate this library into your own applications to create tools for your custom graphics pre-processing pipelines.

The `BC7` format is described in great detail in `https://learn.microsoft.com/en-us/windows/win32/direct3d11/bc7-format` and we will use it in our book to store textures for the large *Lumberyard Bistro* dataset.

Let's learn how to compress 2D `.jpg` or `.png` images into `BC7`.

Getting ready

The source code for this recipe is located at `Chapter01/05_BC7Compression`. We use an the `STB_Image` library by Sean Barrett `https://github.com/nothings/stb` to load `.jpg` and `.png` files, and an open-source library `KTX-Software` from Khronos `https://github.com/KhronosGroup/KTX-Software` to compress images and save the compressed `BC7` images in the `.ktx` file format, making them suitable for runtime use.

How to do it...

Let's check the code in `Chapter01/05_BC7Compression/src/main.cpp` to load a `.jpg` file `data/wood.jpg`, compress it and save into `.cache/image.ktx`:

1. First, we load the file using the `STB_Image` library.

```
int main()
{
  const int numChannels = 4;
  int origW, origH;
  const uint8_t* pixels = stbi_load(
    "data/Chapter02/03_STB.jpg",
    &origW, &origH, nullptr, numChannels);
```

2. Next, we create a KTX texture object to store the compressed image. We calculate the number of mip-levels to generate the full `BC7`-compressed mip-pyramid for this image.

```
  const uint32_t numMipLevels =
    lvk::calcNumMipLevels(origW, origH);
  ktxTextureCreateInfo createInfo = {
    .glInternalformat = GL_COMPRESSED_RGBA_BPTC_UNORM,
    .vkFormat         = VK_FORMAT_BC7_UNORM_BLOCK,
    .baseWidth        = (uint32_t)origW,
    .baseHeight       = (uint32_t)origH,
    .baseDepth        = 1u,
    .numDimensions    = 2u,
    .numLevels        = numMipLevels,
    .numLayers        = 1u,
    .numFaces         = 1u,
    .generateMipmaps  = KTX_FALSE,
  };
  ktxTexture1* texture = nullptr;
  ktxTexture1_Create(&createInfo,
    KTX_TEXTURE_CREATE_ALLOC_STORAGE, &texture);
```

3. Let's go through each mip-level, generate a down-sampled version of the image for each one, and store it in the KTX texture.

```
int w = origW;
int h = origH;
for (uint32_t i = 0; i != numMipLevels; ++i) {
  size_t offset = 0;
  ktxTexture_GetImageOffset(
    ktxTexture(textureKTX2), i, 0, 0, &offset);
  stbir_resize_uint8_linear(
    (const unsigned char*)pixels, origW, origH, 0,
    ktxTexture_GetData(ktxTexture(textureKTX2)) + offset,
    w, h, 0, STBIR_RGBA);
  h = h > 1 ? h >> 1 : 1;
  w = w > 1 ? w >> 1 : 1;
}
```

4. Use KTX-Software to compress the image, including its entire mip-pyramid, into the `Basis` format. Then, transcode the `Basis` format to `BC7`.

```
ktxTexture2_CompressBasis(textureKTX2, 255);
ktxTexture2_TranscodeBasis(
  textureKTX2, KTX_TTF_BC7_RGBA, 0);
```

5. Use **KTX-Software** to save the image in the `KTX1` file format. We choose the older `KTX1` format because viewing tools for it are more widely available than those for `KTX2`. For instance, you can use **PicoPixel** to view the resulting files.

```
ktxTextureCreateInfo createInfoKTX1 = {
  .glInternalformat = GL_COMPRESSED_RGBA_BPTC_UNORM,
  .vkFormat         = VK_FORMAT_BC7_UNORM_BLOCK,
  .baseWidth        = (uint32_t)origW,
  .baseHeight       = (uint32_t)origH,
  .baseDepth        = 1u,
  .numDimensions    = 2u,
  .numLevels        = numMipLevels,
  .numLayers        = 1u,
  .numFaces         = 1u,
  .generateMipmaps  = KTX_FALSE,
};
ktxTexture1* textureKTX1 = nullptr;
ktxTexture1_Create(&createInfoKTX1,
  KTX_TEXTURE_CREATE_ALLOC_STORAGE, &textureKTX1);

for (uint32_t i = 0; i != numMipLevels; ++i) {
  size_t offset1 = 0;
  ktxTexture_GetImageOffset(
```

```
      ktxTexture(textureKTX1), i, 0, 0, &offset1);
    size_t offset2 = 0;
    ktxTexture_GetImageOffset(ktxTexture(
      textureKTX2), i, 0, 0, &offset2);
    memcpy(
      ktxTexture_GetData(ktxTexture(textureKTX1)) + offset1,
      ktxTexture_GetData(ktxTexture(textureKTX2)) + offset2,
      ktxTexture_GetImageSize(ktxTexture(textureKTX1), i));
  }

  ktxTexture_WriteToNamedFile(
    ktxTexture(textureKTX1), ".cache/image.ktx");
  ktxTexture_Destroy(ktxTexture(textureKTX1));
  ktxTexture_Destroy(ktxTexture(textureKTX2));
  stbi_image_free(pixels);
  return 0;
}
```

The image is now saved in the `cache/image.ktx` file. In the following chapters, we'll load it into a Vulkan texture.

There's more...

The KTX File Format Specification is maintained by Khronos and is located at `https://registry.khronos.org/KTX/specs/1.0/ktxspec.v1.html`. We use KTX Version 1.0 throughout this book for compatibility with PicoPixel.

PicoPixel is a great tool that you can use to view `.ktx` files and other texture formats `https://pixelandpolygon.com`. It is freeware but not open source. There is a publicly available issues tracker on GitHub. You can find it at `https://github.com/inalogic/pico-pixel-public/issues`.

For those who want to jump into the latest state-of-the-art texture compression techniques, check out please refer to the Basis project from Binomial at `https://github.com/BinomialLLC/basis_universal`.

Let's move on to the next chapter and learn how to start working with Vulkan.

2

Getting Started with Vulkan

In this chapter, we'll take our first steps with Vulkan, focusing on swapchains, shaders, and pipelines. The recipes in this chapter will guide you through getting your first triangle on the screen using Vulkan. The Vulkan implementation we'll use is based on the open-source library *LightweightVK* `https://github.com/corporateshark/lightweightvk`, which we'll explore throughout the book.

In the chapter, we will cover the following recipes:

- Initializing Vulkan instance and graphical device
- Initializing Vulkan swapchain
- Setting up Vulkan debugging capabilities
- Using Vulkan command buffers
- Initializing Vulkan shader modules
- Initializing Vulkan pipelines

Technical requirements

To run the recipes from this chapter, you have to use a Windows or Linux computer with a video card and drivers supporting Vulkan 1.3. Read the *Chapter 1* if you want to learn how to configure it properly.

Initializing Vulkan instance and graphical device

As some readers may recall from the first edition of our book, the Vulkan API is significantly more verbose than OpenGL. To make things more manageable, we've broken down the process of creating our first graphical demo apps into a series of smaller, focused recipes. In this recipe, we'll cover how to create a Vulkan instance, enumerate all physical devices in the system capable of 3D graphics rendering, and initialize one of these devices to create a window with an attached surface.

Getting ready

We recommend starting with beginner-friendly Vulkan books, such as *The Modern Vulkan Cookbook* by Preetish Kakkar and Mauricio Maurer (published by Packt) or *Vulkan Programming Guide: The Official Guide to Learning Vulkan* by Graham Sellers and John Kessenich (Addison-Wesley Professional).

The most challenging aspect of transitioning from OpenGL to Vulkan—or to any similar modern graphics API—is the extensive amount of explicit code required to set up the rendering process, which, fortunately, only needs to be done once. It's also helpful to familiarize yourself with Vulkan's object model. A great starting point is Adam Sawicki's article, Understanding Vulkan Objects `https://gpuopen.com/understanding-vulkan-objects`. In the recipes that follow, our goal is to start rendering 3D scenes with the minimal setup needed, demonstrating how modern bindless Vulkan can be wrapped into a more user-friendly API.

All our Vulkan recipes rely on the *LightweightVK* library, which can be downloaded from `https://github.com/corporateshark/lightweightvk` using the provided Bootstrap snippet. This library implements all the low-level Vulkan wrapper classes, which we will discuss in detail throughout this book.

```
{
  "name": "lightweightvk",
  "source": {
    "type": "git",
    "url" : "https://github.com/corporateshark/lightweightvk.git",
    "revision": "v1.3"
  }
}
```

The complete Vulkan example for this recipe can be found in `Chapter02/01_Swapchain`.

How to do it...

Before diving into the actual implementation, let's take a look at some scaffolding code that makes debugging Vulkan backends a bit easier. We will begin with error-checking facilities.

1. Any function call from a complex API can fail. To handle failures, or at least provide the developer with the exact location of the failure, *LightweightVK* wraps most Vulkan calls in the `VK_ASSERT()` and `VK_ASSERT_RETURN()` macros, which check the results of Vulkan operations. When starting a new Vulkan implementation from scratch, having something like this in place from the beginning can be very helpful.

```
#define VK_ASSERT(func) {                                          \
  const VkResult vk_assert_result = func;                          \
  if (vk_assert_result != VK_SUCCESS) {                            \
    LLOGW("Vulkan API call failed: %s:%i\n  %s\n  %s\n",           \
      __FILE__, __LINE__, #func,                                   \
    ivkGetVulkanResultString(vk_assert_result));                   \
    assert(false);                                                 \
  }                                                                \
}
```

2. The `VK_ASSERT_RETURN()` macro is very similar and returns the control to the calling code.

```
#define VK_ASSERT_RETURN(func) {                                  \
  const VkResult vk_assert_result = func;                        \
  if (vk_assert_result != VK_SUCCESS) {                          \
    LLOGW("Vulkan API call failed: %s:%i\n  %s\n  %s\n",        \
      __FILE__, __LINE__, #func,                                 \
    ivkGetVulkanResultString(vk_assert_result));                 \
    assert(false);                                               \
    return getResultFromVkResult(vk_assert_result);              \
  }                                                              \
}
```

Now we can start creating our first Vulkan application. Let's explore what is going on in the sample application `Chapter02/01_Swapchain` which creates a window, a Vulkan instance and device together with a Vulkan swapchain, which will be explained in a few moments. The application code is very simple:

1. We start by initializing the *Minilog* logging library and creating a GLFW window as we discussed in the recipe *Using the GLFW library* from the *Chapter 1*. All the Vulkan setup magic, including creating a context and swapchain, is handled by the `lvk::createVulkanContextWithSwapchain()` helper function, which we will examine shortly.

```
int main(void) {
  minilog::initialize(nullptr, { .threadNames = false });
  int width  = 960;
  int height = 540;
  GLFWwindow* window = lvk::initWindow(
    "Simple example", width, height);
  std::unique_ptr<lvk::IContext> ctx =
    lvk::createVulkanContextWithSwapchain(
      window, width, height, {});
```

2. The application's main loop handles updates to the framebuffer size if the window is sized, acquires a command buffer, submits it, and presents the current swapchain image, or texture as it is called in *LightweightVK*.

```
  while (!glfwWindowShouldClose(window)) {
    glfwPollEvents();
    glfwGetFramebufferSize(window, &width, &height);
    if (!width || !height) continue;
    lvk::ICommandBuffer& buf = ctx->acquireCommandBuffer();
    ctx->submit(buf, ctx->getCurrentSwapchainTexture());
  }
```

3. The shutdown process is straightforward. The `IDevice` object should be destroyed before the GLFW window.

```
  ctx.reset();
  glfwDestroyWindow(window);
  glfwTerminate();
  return 0;
}
```

The application should render an empty black window as in the following screenshot:

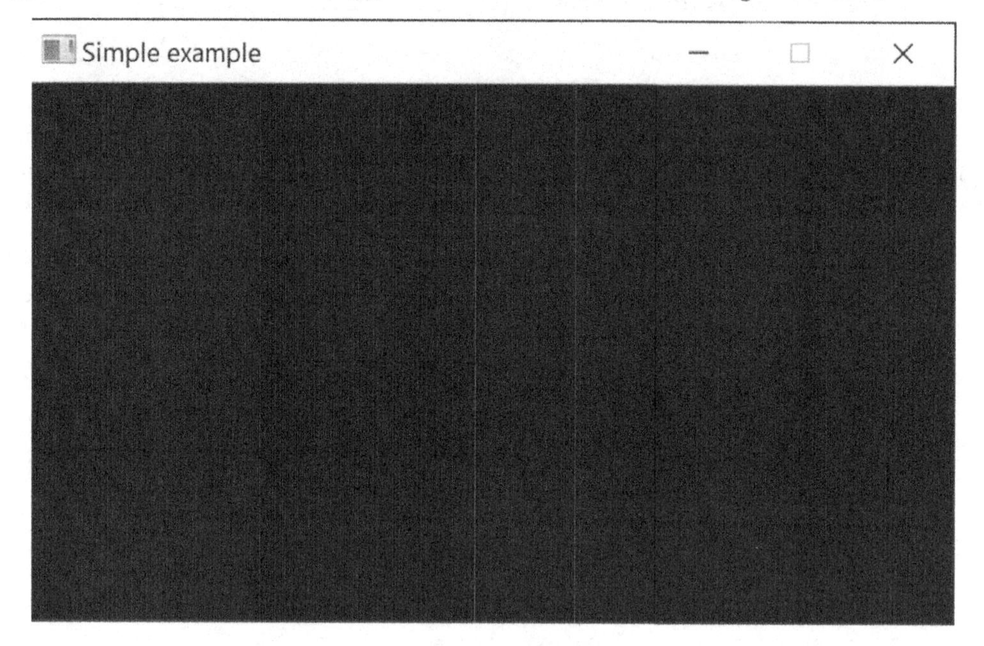

Figure 2.1: The main loop and swapchain

Let's explore `lvk::createVulkanContextWithSwapchain()` and take a sneak peek at its implementation. As before, we will skip most of the error checking in the book text where it doesn't contribute to the overall understanding:

1. This helper function calls *LightweightVK* to create a *VulkanContext* object, taking the provided GLFW window and display properties for our operating system into account. *LightweightVK* includes additional code paths for macOS/MoltenVK and Android initialization. We'll skip them here for the sake of brevity and because not all the demos in this book are compatible with MoltenVK or Android.

```
std::unique_ptr<lvk::IContext> createVulkanContextWithSwapchain(
  GLFWwindow* window, uint32_t width, uint32_t height,
  const lvk::vulkan::VulkanContextConfig& cfg,
  lvk::HWDeviceType preferredDeviceType)
{
```

```
  std::unique_ptr<vulkan::VulkanContext> ctx;
#if defined(_WIN32)
  ctx = std::make_unique<VulkanContext>(cfg,
    (void*)glfwGetWin32Window(window));
#elif defined(__linux__)
  #if defined(LVK_WITH_WAYLAND)
  wl_surface* waylandWindow = glfwGetWaylandWindow(window);
  if (!waylandWindow) {
    LVK_ASSERT_MSG(false, "Wayland window not found");
    return nullptr;
  }
  ctx = std::make_unique<VulkanContext>(cfg,
    (void*)waylandWindow, (void*)glfwGetWaylandDisplay());
  #else
  ctx = std::make_unique<VulkanContext>(cfg,
    (void*)glfwGetX11Window(window), (void*)glfwGetX11Display());
  #endif // LVK_WITH_WAYLAND
#else
#  error Unsupported OS
#endif
```

2. Next, we enumerate Vulkan physical devices and attempt to select the most preferred one. We prioritize choosing a discrete GPU first, and if none is available, we opt for an integrated GPU.

```
  HWDeviceDesc device;
  uint32_t numDevices =
    ctx->queryDevices(preferredDeviceType, &device, 1);
  if (!numDevices) {
    if (preferredDeviceType == HWDeviceType_Discrete) {
      numDevices =
        ctx->queryDevices(HWDeviceType_Integrated, &device);
    } else if (preferredDeviceType == HWDeviceType_Integrated) {
      numDevices =
        ctx->queryDevices(HWDeviceType_Discrete, &device);
    }
  }
```

3. Once a physical device is selected, we call `VulkanContext::initContext()`, which creates all Vulkan and *LightweightVK* internal data structures.

```
  if (!numDevices) return nullptr;
  Result res = ctx->initContext(device);
  if (!res.isOk()) return nullptr;
```

4. If we have a non-empty viewport, initialize a Vulkan swapchain. The swapchain creation process will be explained in detail in the next recipe *Initializing Vulkan swapchain*.

```
    if (width > 0 && height > 0) {
      res = ctx->initSwapchain(width, height);
      if (!res.isOk()) return nullptr;
    }
    return std::move(ctx);
  }
```

That covers the high-level code. Now, let's dive deeper and explore the internals of *LightweightVK* to see how the actual Vulkan interactions work.

How it works...

There are several helper functions involved in getting Vulkan up and running. It all begins with the creation of a Vulkan instance in `VulkanContext::createInstance()`. Once the Vulkan instance is created, we can use it to acquire a list of physical devices with the required properties.

1. First, we need to check if the required Vulkan Validation Layers are available on our system. This ensures we have the flexibility to manually disable validation if no validation layers are present.

```
const char* kDefaultValidationLayers[] =
  {"VK_LAYER_KHRONOS_validation"};
void VulkanContext::createInstance() {
  vkInstance_ = VK_NULL_HANDLE;
  uint32_t numLayerProperties = 0;
  vkEnumerateInstanceLayerProperties(
    &numLayerProperties, nullptr);
  std::vector<VkLayerProperties>
    layerProperties(numLayerProperties);
  vkEnumerateInstanceLayerProperties(
    &numLayerProperties, layerProperties.data());
```

2. We use a local C++ lambda to iterate through the available validation layers and update `VulkanContextConfig::enableValidation` accordingly if none are found.

```
  [this, &layerProperties]() -> void {
    for (const VkLayerProperties& props : layerProperties) {
      for (const char* layer : kDefaultValidationLayers) {
        if (!strcmp(props.layerName, layer)) return;
      }
    }
    config_.enableValidation = false;
  }();
```

3. Then, we need to specify the names of all Vulkan instance extensions required to run our Vulkan graphics backend. We need `VK_KHR_surface` and another platform-specific extension which takes an OS window handle and attaches a rendering surface to it. On Linux, we support both libXCB-based window creation and the Wayland protocol. Here is how Wayland support was added to LightweightVK by Roman Kuznetsov: `https://github.com/corporateshark/lightweightvk/pull/13`.

```
std::vector<const char*> instanceExtensionNames = {
  VK_KHR_SURFACE_EXTENSION_NAME,
  VK_EXT_DEBUG_UTILS_EXTENSION_NAME,
#if defined(_WIN32)
  VK_KHR_WIN32_SURFACE_EXTENSION_NAME,
#elif defined(VK_USE_PLATFORM_ANDROID_KHR)
  VK_KHR_ANDROID_SURFACE_EXTENSION_NAME,
#elif defined(__linux__)
  #if defined(VK_USE_PLATFORM_WAYLAND_KHR)
  VK_KHR_WAYLAND_SURFACE_EXTENSION_NAME,
  #else
  VK_KHR_XLIB_SURFACE_EXTENSION_NAME,
  #endif // VK_USE_PLATFORM_WAYLAND_KHR
#endif
};
```

4. We add `VK_EXT_validation_features` when validation features are requested and available. Additionally, a headless rendering extension, `VK_EXT_headless_surface`, can also be added here together with all custom instance extensions from `VulkanContextConfig::extensionsInstance[]`.

```
if (config_.enableValidation)
  instanceExtensionNames.push_back(
    VK_EXT_VALIDATION_FEATURES_EXTENSION_NAME);
if (config_.enableHeadlessSurface)
  instanceExtensionNames.push_back(
    VK_EXT_HEADLESS_SURFACE_EXTENSION_NAME);
for (const char* ext : config_.extensionsInstance) {
  if (ext) instanceExtensionNames.push_back(ext);
}
```

5. Next, we specify the enabled Vulkan validation features when validation is enabled.

```
VkValidationFeatureEnableEXT validationFeaturesEnabled[] = {
  VK_VALIDATION_FEATURE_ENABLE_GPU_ASSISTED_EXT,
  VK_VALIDATION_FEATURE_ENABLE_GPU_ASSISTED_
    RESERVE_BINDING_SLOT_EXT,
};
```

```
  const VkValidationFeaturesEXT features = {
    .sType = VK_STRUCTURE_TYPE_VALIDATION_FEATURES_EXT,
    .enabledValidationFeatureCount = config_.enableValidation ?
      (uint32_t)LVK_ARRAY_NUM_ELEMENTS(validationFeaturesEnabled) : 0u,
    .pEnabledValidationFeatures = config_.enableValidation ?
      validationFeaturesEnabled : nullptr,
  };
```

6. The next code snippet might be particularly interesting. Sometimes, we need to disable specific Vulkan validation checks, either for performance reasons or due to bugs in the Vulkan validation layers. Here's how LightweightVK handles this to work around some known issues with the validation layers (these were the issues at the time of writing this book, of course).

```
  VkBool32 gpuav_descriptor_checks = VK_FALSE;
  VkBool32 gpuav_indirect_draws_buffers = VK_FALSE;
  VkBool32 gpuav_post_process_descriptor_indexing = VK_FALSE;
#define LAYER_SETTINGS_BOOL32(name, var)                \
  VkLayerSettingEXT {                                    \
    .pLayerName = kDefaultValidationLayers[0],           \
    .pSettingName = name,                                \
    .type = VK_LAYER_SETTING_TYPE_BOOL32_EXT,            \
    .valueCount = 1,                                     \
    .pValues = var }
  const VkLayerSettingEXT settings[] = {
    LAYER_SETTINGS_BOOL32("gpuav_descriptor_checks",
      &gpuav_descriptor_checks),
    LAYER_SETTINGS_BOOL32("gpuav_indirect_draws_buffers",
      &gpuav_indirect_draws_buffers),
    LAYER_SETTINGS_BOOL32(
      "gpuav_post_process_descriptor_indexing",
      &gpuav_post_process_descriptor_indexing),
  };
#undef LAYER_SETTINGS_BOOL32
  const VkLayerSettingsCreateInfoEXT layerSettingsCreateInfo = {
    .sType = VK_STRUCTURE_TYPE_LAYER_SETTINGS_CREATE_INFO_EXT,
    .pNext = config_.enableValidation ? &features : nullptr,
    .settingCount = (uint32_t)LVK_ARRAY_NUM_ELEMENTS(settings),
    .pSettings = settings
  };
```

7. After constructing the list of instance-related extensions, we need to fill in some mandatory information about our application. Here, we request the required Vulkan version, `VK_API_VERSION_1_3`.

```
const VkApplicationInfo appInfo = {
  .sType = VK_STRUCTURE_TYPE_APPLICATION_INFO,
  .pApplicationName = "LVK/Vulkan",
  .applicationVersion = VK_MAKE_VERSION(1, 0, 0),
  .pEngineName = "LVK/Vulkan",
  .engineVersion = VK_MAKE_VERSION(1, 0, 0),
  .apiVersion = VK_API_VERSION_1_3,
};
```

8. To create a `VkInstance` object, we need to populate the `VkInstanceCreateInfo` structure. We use pointers to the previously mentioned `appInfo` constant and `layerSettingsCreateInfo` we created earlier. We also use a list of requested Vulkan layers stored in the global variable `kDefaultValidationLayers[]`, which will allow us to enable debugging output for every Vulkan call. The only layer we use in this book is the Khronos validation layer, `VK_LAYER_KHRONOS_validation`. Then, we use the *Volk* library to load all instance-related Vulkan functions for the created `VkInstance`.

Note

Volk is a meta-loader for Vulkan. It allows you to dynamically load entry points required to use Vulkan without linking to `vulkan-1.dll` or statically linking the Vulkan loader. Volk simplifies the use of Vulkan extensions by automatically loading all associated entry points. Besides that, Volk can load Vulkan entry points directly from the driver which can increase performance by skipping loader dispatch overhead. https://github.com/zeux/volk

```
const VkInstanceCreateInfo ci = {
  .sType = VK_STRUCTURE_TYPE_INSTANCE_CREATE_INFO,
  .pNext = &layerSettingsCreateInfo,
  .pApplicationInfo = &appInfo,
  .enabledLayerCount = config_.enableValidation ?
    (uint32_t)LVK_ARRAY_NUM_ELEMENTS(kDefaultValidationLayers) : 0u,
  .ppEnabledLayerNames = config_.enableValidation ?
    kDefaultValidationLayers : nullptr,
  .enabledExtensionCount =
    (uint32_t)instanceExtensionNames.size(),
  .ppEnabledExtensionNames = instanceExtensionNames.data(),
};
VK_ASSERT(vkCreateInstance(&ci, nullptr, &vkInstance_));
volkLoadInstance(vkInstance_);
```

9. Last but not least, let's print a neatly formatted list of all available Vulkan instance extensions. The function `vkEnumerateInstanceExtensionProperties()` is called twice: first to get the number of available extensions, and second to retrieve information about them.

```
    uint32_t count = 0;
    vkEnumerateInstanceExtensionProperties(
      nullptr, &count, nullptr);
    std::vector<VkExtensionProperties>
      allInstanceExtensions(count);
    vkEnumerateInstanceExtensionProperties(
      nullptr, &count, allInstanceExtensions.data()));
    LLOGL("\nVulkan instance extensions:\n");
    for (const VkExtensionProperties& extension : allInstanceExtensions)
      LLOGL("  %s\n", extension.extensionName);
  }
```

Note

If you've looked at the actual source code in `VulkanClasses.cpp`, you'll have noticed that we skipped the Debug Messenger initialization code here. It will be covered later in the recipe *Setting up Vulkan debugging capabilities.*

Once we've created a Vulkan instance, we can access the list of Vulkan physical devices, which are necessary to continue setting up our Vulkan context. Here's how we can enumerate Vulkan physical devices and choose a suitable one:

1. The function `vkEnumeratePhysicalDevices()` is called twice: first to get the number of available physical devices and allocate `std::vector` storage for it, and second to retrieve the actual physical device data.

```
uint32_t lvk::VulkanContext::queryDevices(
  HWDeviceType deviceType,
  HWDeviceDesc* outDevices,
  uint32_t maxOutDevices)
{
  uint32_t deviceCount = 0;
  vkEnumeratePhysicalDevices(vkInstance_, &deviceCount, nullptr);
  std::vector<VkPhysicalDevice> vkDevices(deviceCount);
  vkEnumeratePhysicalDevices(
    vkInstance_, &deviceCount, vkDevices.data());
```

2. We iterate through the vector of devices to retrieve their properties and filter out non-suitable ones. The local lambda function `convertVulkanDeviceTypeToLVK()` converts a Vulkan enum, `VkPhysicalDeviceType`, into a LightweightVK enum, `HWDeviceType`.

More information

```
enum HWDeviceType {
  HWDeviceType_Discrete = 1,
  HWDeviceType_External = 2,
  HWDeviceType_Integrated = 3,
  HWDeviceType_Software = 4,
};
```

```
  const HWDeviceType desiredDeviceType = deviceType;
  uint32_t numCompatibleDevices = 0;
  for (uint32_t i = 0; i < deviceCount; ++i) {
    VkPhysicalDevice physicalDevice = vkDevices[i];
    VkPhysicalDeviceProperties deviceProperties;
    vkGetPhysicalDeviceProperties(
      physicalDevice, &deviceProperties);
    const HWDeviceType deviceType =
      convertVulkanDeviceTypeToLVK(deviceProperties.deviceType);
    if (desiredDeviceType != HWDeviceType_Software &&
        desiredDeviceType != deviceType) continue;
    if (outDevices && numCompatibleDevices < maxOutDevices) {
      outDevices[numCompatibleDevices] =
        {.guid = (uintptr_t)vkDevices[i], .type = deviceType};
      strncpy(outDevices[numCompatibleDevices].name,
              deviceProperties.deviceName,
              strlen(deviceProperties.deviceName));
      numCompatibleDevices++;
    }
  }
  return numCompatibleDevices;
}
```

Once we've selected a suitable Vulkan physical device, we can create a logical representation of a single GPU, or more precisely, a device `VkDevice`. We can think of Vulkan devices as collections of queues and memory heaps. To use a device for rendering, we need to specify a queue capable of executing graphics-related commands, along with a physical device that has such a queue. Let's explore *LightweightVK* and some parts of the function `VulkanContext::initContext()`, which, among many other things we'll cover later, detects suitable queue families and creates a Vulkan device. As before, most of the error checking will be omitted here in the text.

1. The first thing we do in `VulkanContext::initContext()` is retrieve all supported extensions of the physical device we selected earlier and the Vulkan driver. We store them in `allDeviceExtensions` to later decide which features we can enable. Note how we iterate over the validation layers to check which extensions they bring in.

```
lvk::Result VulkanContext::initContext(const HWDeviceDesc& desc)
{
  vkPhysicalDevice_ = (VkPhysicalDevice)desc.guid;
  std::vector<VkExtensionProperties> allDeviceExtensions;
  getDeviceExtensionProps(
    vkPhysicalDevice_, allDeviceExtensions);
  if (config_.enableValidation) {
    for (const char* layer : kDefaultValidationLayers)
      getDeviceExtensionProps(
        vkPhysicalDevice_, allDeviceExtensions, layer);
  }
```

2. Then, we can retrieve all Vulkan features and properties for this physical device.

```
  vkGetPhysicalDeviceFeatures2(
    vkPhysicalDevice_, &vkFeatures10_);
  vkGetPhysicalDeviceProperties2(
    vkPhysicalDevice_, &vkPhysicalDeviceProperties2_);
```

3. The class member variables `vkFeatures10_`, `vkFeatures11_`, `vkFeatures12_`, and `vkFeatures13_` are declared in `VulkanClasses.h` and correspond to the Vulkan features for Vulkan versions 1.0 to 1.3. These structures are chained together using their `pNext` pointers as follows:

```
  // lightweightvk/lvk/vulkan/VulkanClasses.h
  VkPhysicalDeviceVulkan13Features vkFeatures13_ = {
    .sType = VK_STRUCTURE_TYPE_PHYSICAL_DEVICE_VULKAN_1_3_FEATURES};
  VkPhysicalDeviceVulkan12Features vkFeatures12_ = {
    .sType = VK_STRUCTURE_TYPE_PHYSICAL_DEVICE_VULKAN_1_2_FEATURES,
    .pNext = &vkFeatures13_};
  VkPhysicalDeviceVulkan11Features vkFeatures11_ = {
    .sType = VK_STRUCTURE_TYPE_PHYSICAL_DEVICE_VULKAN_1_1_FEATURES,
    .pNext = &vkFeatures12_};
  VkPhysicalDeviceFeatures2 vkFeatures10_ = {
    .sType = VK_STRUCTURE_TYPE_PHYSICAL_DEVICE_FEATURES_2,
    .pNext = &vkFeatures11_};
  // ...
```

4. Let's get back to `initContext()` and print some information related to the Vulkan physical device and a list of all supported extensions. This is very useful for debugging.

```
const uint32_t apiVersion =
  vkPhysicalDeviceProperties2_.properties.apiVersion;
LLOGL("Vulkan physical device: %s\n",
      vkPhysicalDeviceProperties2_.properties.deviceName);
LLOGL("           API version: %i.%i.%i.%i\n",
      VK_API_VERSION_MAJOR(apiVersion),
      VK_API_VERSION_MINOR(apiVersion),
      VK_API_VERSION_PATCH(apiVersion),
      VK_API_VERSION_VARIANT(apiVersion));
LLOGL("           Driver info: %s %s\n",
      vkPhysicalDeviceDriverProperties_.driverName,
      vkPhysicalDeviceDriverProperties_.driverInfo);
LLOGL("Vulkan physical device extensions:\n");
for (const VkExtensionProperties& ext : allDeviceExtensions) {
  LLOGL("  %s\n", ext.extensionName);
}
```

5. Before creating a `VkDevice` object, we need to find the queue family indices and create queues. This code block creates one or two device queues—graphical and compute—based on the actual queue availability on the provided physical device. The helper function `lvk::findQueueFamilyIndex()`, implemented in `lvk/vulkan/VulkanUtils.cpp`, returns the first dedicated queue family index that matches the requested queue flag. It's recommended to take a look at it to see how it ensures the selection of dedicated queues first.

Note

In Vulkan, `queueFamilyIndex` is the index of the queue family to which the queue belongs. A queue family is a collection of Vulkan queues with similar properties and functionality. Here `deviceQueues_` is member field of `VulkanContext` holding a structure with queues information:

```
struct DeviceQueues {
  const static uint32_t INVALID = 0xFFFFFFFF;
  uint32_t graphicsQueueFamilyIndex = INVALID;
  uint32_t computeQueueFamilyIndex = INVALID;
  VkQueue graphicsQueue = VK_NULL_HANDLE;
  VkQueue computeQueue = VK_NULL_HANDLE;
};
```

```
deviceQueues_.graphicsQueueFamilyIndex =
  lvk::findQueueFamilyIndex(vkPhysicalDevice_,
    VK_QUEUE_GRAPHICS_BIT);
deviceQueues_.computeQueueFamilyIndex =
  lvk::findQueueFamilyIndex(vkPhysicalDevice_,
    VK_QUEUE_COMPUTE_BIT);
const float queuePriority = 1.0f;
const VkDeviceQueueCreateInfo ciQueue[2] = {
  { .sType = VK_STRUCTURE_TYPE_DEVICE_QUEUE_CREATE_INFO,
    .queueFamilyIndex = deviceQueues_.graphicsQueueFamilyIndex,
    .queueCount = 1,
    .pQueuePriorities = &queuePriority, },
  { .sType = VK_STRUCTURE_TYPE_DEVICE_QUEUE_CREATE_INFO,
    .queueFamilyIndex = deviceQueues_.computeQueueFamilyIndex,
    .queueCount = 1,
    .pQueuePriorities = &queuePriority, },
};
```

6. Sometimes, especially on mobile GPUs, graphics and compute queues might be the same. Here we take care of such corner cases.

```
const uint32_t numQueues =
  ciQueue[0].queueFamilyIndex == ciQueue[1].queueFamilyIndex ?
    1 : 2;
```

7. Let's construct a list of extensions that our logical device is required to support. A device must support a swapchain object, which allows us to present rendered frames onto the screen. We use Vulkan 1.3, which includes all the necessary functionality, so no extra extensions are required. However, users can provide additional custom extensions via `VulkanContextConfig::extensionsDevice[]`.

```
std::vector<const char*> deviceExtensionNames = {
  VK_KHR_SWAPCHAIN_EXTENSION_NAME,
};
for (const char* ext : config_.extensionsDevice) {
  if (ext) deviceExtensionNames.push_back(ext);
}
```

8. Let's request all the necessary Vulkan `1.0–1.3` features we'll be using in our Vulkan implementation. The most important features are descriptor indexing from Vulkan `1.2` and dynamic rendering from Vulkan `1.3`, which we'll discuss in subsequent chapters. Take a look at how to request these and other features we'll be using.

Note

Descriptor indexing is a set of Vulkan 1.2 features that enable applications to access all of their resources and select among them using integer indices in shaders.

Dynamic rendering is a Vulkan 1.3 feature that allows applications to render directly into images without the need to create render pass objects or framebuffers.

```
VkPhysicalDeviceFeatures deviceFeatures10 = {
    .geometryShader = vkFeatures10_.features.geometryShader,
    .sampleRateShading = VK_TRUE,
    .multiDrawIndirect = VK_TRUE,
  // ...
};
```

9. The structures are chained together using their `pNext` pointers. Note how we access the `vkFeatures10_` through `vkFeatures13_` structures here to enable optional features only if they are actually supported by the physical device. The complete list is quite long, so we skip some parts of it here.

```
VkPhysicalDeviceVulkan11Features deviceFeatures11 = {
  .sType = VK_STRUCTURE_TYPE_PHYSICAL_DEVICE_VULKAN_1_1_FEATURES,
  .pNext = config_.extensionsDeviceFeatures,
  .storageBuffer16BitAccess = VK_TRUE,
  .samplerYcbcrConversion = vkFeatures11_.samplerYcbcrConversion,
  .shaderDrawParameters = VK_TRUE,
};
VkPhysicalDeviceVulkan12Features deviceFeatures12 = {
  .sType = VK_STRUCTURE_TYPE_PHYSICAL_DEVICE_VULKAN_1_2_FEATURES,
  .pNext = &deviceFeatures11,
  .drawIndirectCount = vkFeatures12_.drawIndirectCount,
  // ...
  .descriptorIndexing = VK_TRUE,
  .shaderSampledImageArrayNonUniformIndexing = VK_TRUE,
  .descriptorBindingSampledImageUpdateAfterBind = VK_TRUE,
  .descriptorBindingStorageImageUpdateAfterBind = VK_TRUE,
  .descriptorBindingUpdateUnusedWhilePending = VK_TRUE,
  .descriptorBindingPartiallyBound = VK_TRUE,
  .descriptorBindingVariableDescriptorCount = VK_TRUE,
  .runtimeDescriptorArray = VK_TRUE,
  // ...
};
```

```
VkPhysicalDeviceVulkan13Features deviceFeatures13 = {
  .sType = VK_STRUCTURE_TYPE_PHYSICAL_DEVICE_VULKAN_1_3_FEATURES,
  .pNext = &deviceFeatures12,
  .subgroupSizeControl = VK_TRUE,
  .synchronization2 = VK_TRUE,
  .dynamicRendering = VK_TRUE,
  .maintenance4 = VK_TRUE,
};
```

10. A few more steps before we can create the actual `VkDevice` object. We check our list of requested device extensions against the list of available extensions. Any missing extensions are printed into the log, and the initialization function returns. This is very convenient for debugging.

```
std::string missingExtensions;
for (const char* ext : deviceExtensionNames) {
  if (!hasExtension(ext, allDeviceExtensions))
    missingExtensions += "\n    " + std::string(ext);
}
if (!missingExtensions.empty()) {
  MINILOG_LOG_PROC(minilog::FatalError,
    "Missing Vulkan device extensions: %s\n",
    missingExtensions.c_str());
  return Result(Result::Code::RuntimeError);
}
```

11. One last important thing worth mentioning before we proceed with creating a device: because some Vulkan features are mandatory for our code, we enable them unconditionally. We should check all the requested Vulkan features against the actual available features. With the help of C-macros, we can do this in a very clean way. When we're missing some Vulkan features, this code will print a neatly formatted list of missing features, each marked with the corresponding Vulkan version. This is invaluable for debugging and makes your Vulkan backend adjustable to fit different devices.

```
  std::string missingFeatures;
#define CHECK_VULKAN_FEATURE(                             \
  reqFeatures, availFeatures, feature, version)         \
  if ((reqFeatures.feature == VK_TRUE) &&               \
      (availFeatures.feature == VK_FALSE))              \
        missingFeatures.append("\n    " version " ." #feature);
#define CHECK_FEATURE_1_0(feature)                                  \
  CHECK_VULKAN_FEATURE(deviceFeatures10, vkFeatures10_.features, \
  feature, "1.0 ");
    CHECK_FEATURE_1_0(robustBufferAccess);
```

```
    CHECK_FEATURE_1_0(fullDrawIndexUint32);
    CHECK_FEATURE_1_0(imageCubeArray);
    … // omitted a lot of other Vulkan 1.0 features here
#undef CHECK_FEATURE_1_0
#define CHECK_FEATURE_1_1(feature)                      \
  CHECK_VULKAN_FEATURE(deviceFeatures11, vkFeatures11_, \
    feature, "1.1 ");
    CHECK_FEATURE_1_1(storageBuffer16BitAccess);
    CHECK_FEATURE_1_1(uniformAndStorageBuffer16BitAccess);
    CHECK_FEATURE_1_1(storagePushConstant16);
    … // omitted a lot of other Vulkan 1.1 features here
#undef CHECK_FEATURE_1_1
#define CHECK_FEATURE_1_2(feature)                      \
  CHECK_VULKAN_FEATURE(deviceFeatures12, vkFeatures12_, \
  feature, "1.2 ");
    CHECK_FEATURE_1_2(samplerMirrorClampToEdge);
    CHECK_FEATURE_1_2(drawIndirectCount);
    CHECK_FEATURE_1_2(storageBuffer8BitAccess);
    … // omitted a lot of other Vulkan 1.2 features here
#undef CHECK_FEATURE_1_2
#define CHECK_FEATURE_1_3(feature)                      \
  CHECK_VULKAN_FEATURE(deviceFeatures13, vkFeatures13_, \
  feature, "1.3 ");
    CHECK_FEATURE_1_3(robustImageAccess);
    CHECK_FEATURE_1_3(inlineUniformBlock);
    … // omitted a lot of other Vulkan 1.3 features here
#undef CHECK_FEATURE_1_3
  if (!missingFeatures.empty()) {
    MINILOG_LOG_PROC(minilog::FatalError,
      "Missing Vulkan features: %s\n", missingFeatures.c_str());
    return Result(Result::Code::RuntimeError);
  }
```

12. Finally, we are ready to create the Vulkan device, load all related Vulkan functions with *Volk*, and retrieve the actual device queues based on the queue family indices we selected earlier in this recipe.

```
  const VkDeviceCreateInfo ci = {
    .sType = VK_STRUCTURE_TYPE_DEVICE_CREATE_INFO,
    .pNext = createInfoNext,
    .queueCreateInfoCount = numQueues,
    .pQueueCreateInfos = ciQueue,
```

```
    .enabledExtensionCount = deviceExtensionNames.size(),
    .ppEnabledExtensionNames = deviceExtensionNames.data(),
    .pEnabledFeatures = &deviceFeatures10,
  };
  vkCreateDevice(vkPhysicalDevice_, &ci, nullptr, &vkDevice_);
  volkLoadDevice(vkDevice_);
  vkGetDeviceQueue(vkDevice_,
    deviceQueues_.graphicsQueueFamilyIndex, 0,
    &deviceQueues_.graphicsQueue);
  vkGetDeviceQueue(vkDevice_,
    deviceQueues_.computeQueueFamilyIndex, 0,
    &deviceQueues_.computeQueue);
  // ... other code in initContext() is unrelated to this recipe
}
```

The `VkDevice` object is now ready to be used, but the initialization of the Vulkan rendering pipeline is far from complete. The next step is to create a swapchain object. Let's proceed to the next recipe to learn how to do this.

Initializing Vulkan swapchain

Normally, each frame is rendered into an offscreen image. After the rendering process is finished, the offscreen image should be made visible or "presented." A swapchain is an object that holds a collection of available offscreen images, or more specifically, a queue of rendered images waiting to be presented to the screen. In OpenGL, presenting an offscreen buffer to the visible area of a window is done using system-dependent functions, such as `wglSwapBuffers()` on Windows, `eglSwapBuffers()` on OpenGL ES embedded systems, `glXSwapBuffers()` on Linux, or automatically on macOS. Vulkan, however, gives us much more fine-grained control. We need to select a presentation mode for swapchain images and specify various flags.

In this recipe, we will show how to create a Vulkan swapchain object using the Vulkan instance and device initialized in the previous recipe.

Getting ready

Revisit the previous recipe *Initializing Vulkan instance and graphical device,* which covers the initial steps necessary to initialize Vulkan. The source code discussed in this recipe is implemented in the class `lvk::VulkanSwapchain`.

How to do it...

In the previous recipe, we began learning how Vulkan instances and devices are created by exploring the helper function `lvk::createVulkanContextWithSwapchain()`. This led us to the function `VulkanContext::initContext()`, which we discussed in detail. Let's continue our journey by exploring `VulkanContext::initSwapchain()` and the related class `VulkanSwapchain` from *LightweightVK.*

1. First, let us take a look at a function which retrieves various surface format support capabilities and stores them in the member fields of `VulkanContext`. The function also checks depth format support, but only for those depth formats that might be used by *LightweightVK*.

```
void lvk::VulkanContext::querySurfaceCapabilities() {
  const VkFormat depthFormats[] = {
    VK_FORMAT_D32_SFLOAT_S8_UINT,
    VK_FORMAT_D24_UNORM_S8_UINT,
    VK_FORMAT_D16_UNORM_S8_UINT, VK_FORMAT_D32_SFLOAT,
    VK_FORMAT_D16_UNORM };
  for (const auto& depthFormat : depthFormats) {
    VkFormatProperties formatProps;
    vkGetPhysicalDeviceFormatProperties(
      vkPhysicalDevice_, depthFormat, &formatProps);
    if (formatProps.optimalTilingFeatures)
      deviceDepthFormats_.push_back(depthFormat);
  }
  if (vkSurface_ == VK_NULL_HANDLE) return;
```

2. All the surface capabilities and surface formats are retrieved and stored. First, we get the number of supported formats, then allocate the storage to hold them and read the actual properties.

```
  vkGetPhysicalDeviceSurfaceCapabilitiesKHR(
    vkPhysicalDevice_, vkSurface_, &deviceSurfaceCaps_);
  uint32_t formatCount;
  vkGetPhysicalDeviceSurfaceFormatsKHR(
    vkPhysicalDevice_, vkSurface_, &formatCount, nullptr);
  if (formatCount) {
    deviceSurfaceFormats_.resize(formatCount);
    vkGetPhysicalDeviceSurfaceFormatsKHR(
      vkPhysicalDevice_, vkSurface_,
      &formatCount, deviceSurfaceFormats_.data());
  }
```

3. In a similar way, store surface present modes as well.

```
  uint32_t presentModeCount;
  vkGetPhysicalDeviceSurfacePresentModesKHR(
    vkPhysicalDevice_, vkSurface_, &presentModeCount, nullptr);
  if (presentModeCount) {
    devicePresentModes_.resize(presentModeCount);
    vkGetPhysicalDeviceSurfacePresentModesKHR(
      vkPhysicalDevice_, vkSurface_,
```

```
      &presentModeCount, devicePresentModes_.data());
  }
}
```

Knowing all supported color surface formats, we can choose a suitable one for our swapchain. Let's take a look at the `chooseSwapSurfaceFormat()` helper function to see how it's done. The function takes a list of available formats and a desired color space as input.

1. First, it selects a preferred surface format based on the desired color space and the RGB/BGR native swapchain image format. RGB or BGR is determined by going through all available color formats returned by Vulkan and picking the one—RGB or BGR—that appears first in the list. If BGR is encountered earlier, it will be chosen. Once the preferred image format and color space are selected, the function goes through the list of supported formats to try to find an exact match. Here, `colorSpaceToVkSurfaceFormat()` and `isNativeSwapChainBGR()` are local C++ lambdas. Check the full source code to see their implementations.

```
VkSurfaceFormatKHR chooseSwapSurfaceFormat(
  const std::vector<VkSurfaceFormatKHR>& formats,
  lvk::ColorSpace colorSpace)
{
  const VkSurfaceFormatKHR preferred =
    colorSpaceToVkSurfaceFormat(
      colorSpace, isNativeSwapChainBGR(formats));
  for (const VkSurfaceFormatKHR& fmt : formats) {
    if (fmt.format == preferred.format &&
        fmt.colorSpace == preferred.colorSpace) return fmt;
  }
```

2. If we cannot find both a matching format and color space, try matching only the format. If we cannot match the format, default to the first available format. On many systems, it will be `VK_FORMAT_R8G8B8A8_UNORM` or a similar format.

```
  for (const VkSurfaceFormatKHR& fmt : formats) {
    if (fmt.format == preferred.format) return fmt;
  }
  return formats[0];
}
```

This function is called from the constructor of `VulkanSwapchain`. Once the format has been selected, we need to do a few more checks before we can create an actual Vulkan swapchain.

1. The first check is to ensure that the selected format supports presentation operation on the graphics queue family used to create the swapchain.

```
lvk::VulkanSwapchain::VulkanSwapchain(
  VulkanContext& ctx, uint32_t width, uint32_t height) :
  ctx_(ctx),
  device_(ctx.vkDevice_),
  graphicsQueue_(ctx.deviceQueues_.graphicsQueue),
  width_(width), height_(height)
{
  surfaceFormat_ = chooseSwapSurfaceFormat(
    ctx.deviceSurfaceFormats_, ctx.config_.swapChainColorSpace);
  VkBool32 queueFamilySupportsPresentation = VK_FALSE;
  vkGetPhysicalDeviceSurfaceSupportKHR(ctx.getVkPhysicalDevice(),
    ctx.deviceQueues_.graphicsQueueFamilyIndex, ctx.vkSurface_,
    &queueFamilySupportsPresentation));
```

2. The second check is necessary to choose usage flags for swapchain images. Usage flags define if swapchain images can be used as color attachments, in transfer operations, or as storage images to allow compute shaders to operate directly on them. Different devices have different capabilities and storage images are not always supported, especially on mobile GPUs. Here's a C++ local lambda to do it:

```
auto chooseUsageFlags = [](VkPhysicalDevice pd,
  VkSurfaceKHR surface, VkFormat format) -> VkImageUsageFlags
{
  VkImageUsageFlags usageFlags =
    VK_IMAGE_USAGE_COLOR_ATTACHMENT_BIT |
    VK_IMAGE_USAGE_TRANSFER_DST_BIT |
    VK_IMAGE_USAGE_TRANSFER_SRC_BIT;
  VkSurfaceCapabilitiesKHR caps;
  vkGetPhysicalDeviceSurfaceCapabilitiesKHR(pd, surface, &caps);
  const bool isStorageSupported =
    (caps.supportedUsageFlags & VK_IMAGE_USAGE_STORAGE_BIT) > 0;
  VkFormatProperties props;
  vkGetPhysicalDeviceFormatProperties(pd, format, &props);
  const bool isTilingOptimalSupported =
    (props.optimalTilingFeatures & VK_IMAGE_USAGE_STORAGE_BIT) > 0;
  if (isStorageSupported && isTilingOptimalSupported) {
    usageFlags |= VK_IMAGE_USAGE_STORAGE_BIT;
  }
  return usageFlags;
}
```

3. Now we should select the presentation mode. The preferred presentation mode is VK_PRESENT_MODE_MAILBOX_KHR which specifies that the Vulkan presentation system should wait for the next vertical blanking period to update the current image. Visual tearing will not be observed in this case. However, this presentation mode is not guaranteed to be supported. In this situation, we can try picking VK_PRESENT_MODE_IMMEDIATE_KHR for the fastest frames-per-second without V-sync, or we can always fall back to VK_PRESENT_MODE_FIFO_KHR. The differences between all possible presentation mode are described in the Vulkan specification https://www.khronos.org/registry/vulkan/specs/1.3-extensions/man/html/VkPresentModeKHR.html

```
auto chooseSwapPresentMode = [](
  const std::vector<VkPresentModeKHR>& modes) -> VkPresentModeKHR
{
#if defined(__linux__) || defined(_M_ARM64)
    if (std::find(modes.cbegin(), modes.cend(),
        VK_PRESENT_MODE_IMMEDIATE_KHR) != modes.cend()) {
      return VK_PRESENT_MODE_IMMEDIATE_KHR;
    }
#endif
    if (std::find(modes.cbegin(), modes.cend(),
        VK_PRESENT_MODE_MAILBOX_KHR) != modes.cend()) {
      return VK_PRESENT_MODE_MAILBOX_KHR;
    }
    return VK_PRESENT_MODE_FIFO_KHR;
  };
```

4. The last helper lambda we need will choose the number of images in the swapchain object. It is based on the surface capabilities we retrieved earlier. Instead of using minImageCount directly, we request one additional image to make sure we are not waiting on the GPU to complete any operations.

```
auto chooseSwapImageCount = [](
  const VkSurfaceCapabilitiesKHR& caps) -> uint32_t
{
  const uint32_t desired = caps.minImageCount + 1;
  const bool exceeded = caps.maxImageCount > 0 &&
                        desired > caps.maxImageCount;
  return exceeded ? caps.maxImageCount : desired;
};
```

5. Let's go back to the constructor VulkanSwapchain::VulkanSwapchain() and explore how it uses all abovementioned helper functions to create a Vulkan swapchain object. The code here becomes rather short and consists only of filling in the VkSwapchainCreateInfoKHR structure.

```
const VkImageUsageFlags usageFlags = chooseUsageFlags(
  ctx.getVkPhysicalDevice(), ctx.vkSurface_,
  surfaceFormat_.format);
const bool isCompositeAlphaOpaqueSupported =
  (ctx.deviceSurfaceCaps_.supportedCompositeAlpha &
   VK_COMPOSITE_ALPHA_OPAQUE_BIT_KHR) != 0;
const VkSwapchainCreateInfoKHR ci = {
  .sType = VK_STRUCTURE_TYPE_SWAPCHAIN_CREATE_INFO_KHR,
  .surface = ctx.vkSurface_,
  .minImageCount = chooseSwapImageCount(ctx.deviceSurfaceCaps_),
  .imageFormat = surfaceFormat_.format,
  .imageColorSpace = surfaceFormat_.colorSpace,
  .imageExtent = {.width = width, .height = height},
  .imageArrayLayers = 1,
  .imageUsage = usageFlags,
  .imageSharingMode = VK_SHARING_MODE_EXCLUSIVE,
  .queueFamilyIndexCount = 1,
  .pQueueFamilyIndices = &ctx.deviceQueues_.graphicsQueueFamilyIndex,
  .preTransform = ctx.deviceSurfaceCaps_.currentTransform,
  .compositeAlpha = isCompositeAlphaOpaqueSupported ?
    VK_COMPOSITE_ALPHA_OPAQUE_BIT_KHR :
    VK_COMPOSITE_ALPHA_INHERIT_BIT_KHR,
  .presentMode = chooseSwapPresentMode(ctx.devicePresentModes_),
  .clipped = VK_TRUE,
  .oldSwapchain = VK_NULL_HANDLE,
};
vkCreateSwapchainKHR(device_, &ci, nullptr, &swapchain_);
```

6. After the swapchain object has been created, we can retrieve swapchain images.

```
VkImage swapchainImages[LVK_MAX_SWAPCHAIN_IMAGES];
vkGetSwapchainImagesKHR(
  device_, swapchain_, &numSwapchainImages_, nullptr);
if (numSwapchainImages_ > LVK_MAX_SWAPCHAIN_IMAGES) {
  numSwapchainImages_ = LVK_MAX_SWAPCHAIN_IMAGES;
}
vkGetSwapchainImagesKHR(
  device_, swapchain_, &numSwapchainImages_, swapchainImages);
```

The retrieved `VkImage` objects can be used to create `VkImageView` objects for textures and attachments. This topic will be discussed in the recipe *Using texture data in Vulkan* in the next chapter.

With Vulkan now initialized, we can run our first application, `Chapter02/01_Swapchain`, which displays an empty black window. In the next recipe, we'll explore Vulkan's built-in debugging capabilities to move closer to actual rendering.

Setting up Vulkan debugging capabilities

After creating a Vulkan instance, we can start monitoring all potential errors and warnings generated by the validation layers. This is done by using the `VK_EXT_debug_utils` extension to create a callback function and register it with the Vulkan instance. In this recipe, we'll learn how to set up and use this feature.

Getting ready

Please revising the first recipe *Initializing Vulkan instance and graphical device* for details how to initialize Vulkan in your applications and enable the instance extension `VK_EXT_debug_utils`.

How to do it...

We have to provide a callback function to Vulkan to catch the debug output. In *LightweightVK* it is called `vulkanDebugCallback()`. Here's how it can be passed into Vulkan to intercept logs.

1. Let's create a debug messenger to forward debug messages to an application-provided callback function, `vulkanDebugCallback()`. This can be done right after the `VkInstance` object has been created.

```
...
vkCreateInstance(&ci, nullptr, &vkInstance_);
volkLoadInstance(vkInstance_);
const VkDebugUtilsMessengerCreateInfoEXT ci = {
  .sType =
    VK_STRUCTURE_TYPE_DEBUG_UTILS_MESSENGER_CREATE_INFO_EXT,
  .messageSeverity =
    VK_DEBUG_UTILS_MESSAGE_SEVERITY_VERBOSE_BIT_EXT |
    VK_DEBUG_UTILS_MESSAGE_SEVERITY_INFO_BIT_EXT |
    VK_DEBUG_UTILS_MESSAGE_SEVERITY_WARNING_BIT_EXT |
    VK_DEBUG_UTILS_MESSAGE_SEVERITY_ERROR_BIT_EXT,
  .messageType = VK_DEBUG_UTILS_MESSAGE_TYPE_GENERAL_BIT_EXT |
                 VK_DEBUG_UTILS_MESSAGE_TYPE_VALIDATION_BIT_EXT |
                 VK_DEBUG_UTILS_MESSAGE_TYPE_PERFORMANCE_BIT_EXT,
  .pfnUserCallback = &vulkanDebugCallback,
  .pUserData = this,
};
vkCreateDebugUtilsMessengerEXT(
  vkInstance_, &ci, nullptr, &vkDebugUtilsMessenger_);
```

2. The callback code is more elaborate and can provide information about the Vulkan object causing an error or warning. However, we won't cover tagged object allocation or associating custom data. Some performance warnings are suppressed to keep the debug output easier to read.

```
VKAPI_ATTR VkBool32 VKAPI_CALL vulkanDebugCallback(
  VkDebugUtilsMessageSeverityFlagBitsEXT msgSeverity,
  VkDebugUtilsMessageTypeFlagsEXT msgType,
  const VkDebugUtilsMessengerCallbackDataEXT* cbData,
  void* userData)
{
  if (msgSeverity < VK_DEBUG_UTILS_MESSAGE_SEVERITY_INFO_BIT_EXT)
    return VK_FALSE;
  const bool isError = (msgSeverity &
    VK_DEBUG_UTILS_MESSAGE_SEVERITY_ERROR_BIT_EXT) != 0;
  const bool isWarning = (msgSeverity &
    VK_DEBUG_UTILS_MESSAGE_SEVERITY_WARNING_BIT_EXT) != 0;
  lvk::VulkanContext* ctx = static_cast<lvk::VulkanContext*>(userData);
  minilog::eLogLevel level = minilog::Log;
  if (isError) {
    level = ctx->config_.terminateOnValidationError ?
      minilog::FatalError : minilog::Warning;
  }
  MINILOG_LOG_PROC(
    level, "%sValidation layer:\n%s\n", isError ?
    "\nERROR:\n" : "", cbData->pMessage);
  if (isError) {
    lvk::VulkanContext* ctx =
      static_cast<lvk::VulkanContext*>(userData);
    if (ctx->config_.terminateOnValidationError) {
      std::terminate();
    }
  }
  return VK_FALSE;
}
```

This code is enough to get you started with reading validation layer messages and debugging your Vulkan applications. Remember to destroy the validation layer callbacks just before destroying the Vulkan instance. Refer to the full source code for all the details `https://github.com/corporateshark/lightweightvk/blob/master/lvk/vulkan/VulkanClasses.cpp`.

There's more...

The extension `VK_EXT_debug_utils` provides the ability to identify specific Vulkan objects using a textual name or tag to improve Vulkan objects tracking and debugging experience.

For example, in *LightweightVK*, we can assign a name to our `VkDevice` object.

```
lvk::setDebugObjectName(vkDevice_, VK_OBJECT_TYPE_DEVICE,
  (uint64_t)vkDevice_, "Device: VulkanContext::vkDevice_");
```

This helper function is implemented in `lvk/vulkan/VulkanUtils.cpp` and looks as follows:

```
VkResult lvk::setDebugObjectName(VkDevice device, VkObjectType type,
  uint64_t handle, const char* name)
{
  if (!name || !*name) return VK_SUCCESS;
  const VkDebugUtilsObjectNameInfoEXT ni = {
    .sType = VK_STRUCTURE_TYPE_DEBUG_UTILS_OBJECT_NAME_INFO_EXT,
    .objectType = type,
    .objectHandle = handle,
    .pObjectName = name,
  };
  return vkSetDebugUtilsObjectNameEXT(device, &ni);
}
```

Using Vulkan command buffers

In the previous recipes, we learned how to create a Vulkan instance, a device for rendering, and a swapchain. In this recipe, we will learn how to manage command buffers and submit them using command queues which will bring us a bit closer to rendering our first image with Vulkan.

Vulkan command buffers are used to record Vulkan commands which can be then submitted to a device queue for execution. Command buffers are allocated from pools which allow the Vulkan implementation to amortize the cost of resource creation across multiple command buffers. Command pools are be externally synchronized which means one command pool should not be used between multiple threads. Let's learn how to make a convenient user-friendly wrapper on top of Vulkan command buffers and pools.

Getting ready...

We are going to explore the command buffers management code from the LightweightVK library. Take a look at the class `VulkanImmediateCommands` from `lvk/vulkan/VulkanClasses.h`. In the previous edition of our book, we used very rudimentary command buffers management code which did not suppose any synchronization because every frame was "synchronized" with `vkDeviceWaitIdle()`. Here we are going to explore a more pragmatic solution with some facilities for synchronization.

Let's go back to our demo application from the recipe *Initializing Vulkan swapchain* which renders a black empty window `Chapter02/01_Swapchain`. The main loop of the application looks as follows:

```
while (!glfwWindowShouldClose(window)) {
  glfwPollEvents();
  glfwGetFramebufferSize(window, &width, &height);
  if (!width || !height) continue;
  lvk::ICommandBuffer& buf = ctx->acquireCommandBuffer();
  ctx->submit(buf, ctx->getCurrentSwapchainTexture());
}
```

Here we acquire a next command buffer and then submit it without writhing any commands into it so that `LightweightVK` can run its swapchain presentation code and render a black window. Let's dive deep into the implementation and learn how `lvk::VulkanImmediateCommands` does all the heavy lifting behind the scenes.

How to do it...

1. First, we need a helper struct, `SubmitHandle`, to identify previously submitted command buffers. It will be essential for implementing synchronization when scheduling work that depends on the results of a previously submitted command buffer. The struct includes an internal index for the submitted buffer and an integer ID for the submission. For convenience, handles can be converted to and from 64-bit integers.

```
struct SubmitHandle {
  uint32_t bufferIndex_ = 0;
  uint32_t submitId_ = 0;
  SubmitHandle() = default;
  explicit SubmitHandle(uint64_t handle) :
    bufferIndex_(uint32_t(handle & 0xffffffff)),
    submitId_(uint32_t(handle >> 32)) {}
  bool empty() const { return submitId_ == 0; }
  uint64_t handle() const
  { return (uint64_t(submitId_) << 32) + bufferIndex_; }
};
```

2. Another helper struct, `CommandBufferWrapper`, is needed to encapsulate all Vulkan objects associated with a single Vulkan command buffer. This struct stores the originally allocated and currently active command buffers, the most recent `SubmitHandle` linked to the command buffer, a Vulkan fence, and a Vulkan semaphore. The fence is used for GPU-CPU synchronization, while the semaphore ensures that command buffers are processed by the GPU in the order they were submitted. This sequential processing, enforced by *LightweightVK*, simplifies many aspects of rendering.

```
struct CommandBufferWrapper {
  VkCommandBuffer cmdBuf_ = VK_NULL_HANDLE;
  VkCommandBuffer cmdBufAllocated_ = VK_NULL_HANDLE;
  SubmitHandle handle_ = {};
  VkFence fence_ = VK_NULL_HANDLE;
  VkSemaphore semaphore_ = VK_NULL_HANDLE;
  bool isEncoding_ = false;
};
```

Now let's take a look at the interface of `lvk::VulkanImmediateCommands`.

1. Vulkan command buffers are preallocated and used in a round-robin manner. The maximum number of preallocated command buffers is defined by `kMaxCommandBuffers`. If all buffers are in use, `VulkanImmediateCommands` waits for an existing command buffer to become available by waiting on a fence. Typically, `64` command buffers are sufficient to ensure non-blocking operation in most cases. The constructor takes a `queueFamilyIdx` parameter to retrieve the appropriate Vulkan queue.

```
class VulkanImmediateCommands final {
 public:
   static constexpr uint32_t kMaxCommandBuffers = 64;
  VulkanImmediateCommands(VkDevice device,
    uint32_t queueFamilyIdx, const char* debugName);
  ~VulkanImmediateCommands();
```

2. The `acquire()` method returns a reference to the next available command buffer. If all command buffers are in use, it waits on a fence until one becomes available. The `submit()` method submits a command buffer to the assigned Vulkan queue.

```
  const CommandBufferWrapper& acquire();
  SubmitHandle submit(const CommandBufferWrapper& wrapper);
```

3. The next three methods provide GPU-GPU and GPU-CPU synchronization mechanisms. The `waitSemaphore()` method ensures the current command buffer waits on a given semaphore before execution. A common use case is using an "acquire semaphore" from our `VulkanSwapchain` object, which signals a semaphore when acquiring a swapchain image, ensuring the command buffer waits for it before starting to render into the swapchain image. The `signalSemaphore()` method signals a corresponding Vulkan timeline semaphore when the current command buffer finishes execution. The `acquireLastSubmitSemaphore()` method retrieves the semaphore signaled when the last submitted command buffer completes. This semaphore can be used by the swapchain before presentation to ensure that rendering into the image is complete. We'll take a closer look at how this works in a moment.

```
  void waitSemaphore(VkSemaphore semaphore);
  void signalSemaphore(VkSemaphore semaphore, uint64_t signalValue);
  VkSemaphore acquireLastSubmitSemaphore();
```

4. The next set of methods manages GPU-CPU synchronization. As we'll see later in this recipe, submit handles are implemented using Vulkan fences and can be used to wait for specific GPU operations to complete.

```
SubmitHandle getLastSubmitHandle() const;
bool isReady(SubmitHandle handle) const;
void wait(SubmitHandle handle);
void waitAll();
```

5. The private section of the class contains all the local state, including an array of preallocated `CommandBufferWrapper` objects called `buffers_[]`.

```
private:
 void purge();
 VkDevice device_ = VK_NULL_HANDLE;
 VkQueue queue_ = VK_NULL_HANDLE;
 VkCommandPool commandPool_ = VK_NULL_HANDLE;
 uint32_t queueFamilyIndex_ = 0;
 const char* debugName_ = "";
 CommandBufferWrapper buffers_[kMaxCommandBuffers];
```

6. Note how the `VkSemaphoreSubmitInfo` structures are preinitialized with generic `stageMask` values. For submitting Vulkan command buffers, we use the function `vkQueueSubmit2()` introduced in Vulkan 1.3, which requires pointers to these structures.

```
 VkSemaphoreSubmitInfo lastSubmitSemaphore_ = {
   .sType = VK_STRUCTURE_TYPE_SEMAPHORE_SUBMIT_INFO,
   .stageMask = VK_PIPELINE_STAGE_ALL_COMMANDS_BIT};
 VkSemaphoreSubmitInfo waitSemaphore_ = {
   .sType = VK_STRUCTURE_TYPE_SEMAPHORE_SUBMIT_INFO,
   .stageMask = VK_PIPELINE_STAGE_ALL_COMMANDS_BIT};
 VkSemaphoreSubmitInfo signalSemaphore_ = {
   .sType = VK_STRUCTURE_TYPE_SEMAPHORE_SUBMIT_INFO,
   .stageMask = VK_PIPELINE_STAGE_ALL_COMMANDS_BIT};
 uint32_t numAvailableCommandBuffers_ = kMaxCommandBuffers;
 uint32_t submitCounter_ = 1;
};
```

The `VulkanImmediateCommands` class is central to the entire operation of our Vulkan backend. Let's dive into its implementation, examining each method in detail.

Let's begin with the class constructor and destructor. The constructor preallocates all command buffers. For simplicity, error checking and debugging code will be omitted here; please refer to the *LightweightVK* library source code for full error-checking details.

1. First, we should retrieve a Vulkan device queue and allocate a command pool. The `VK_COMMAND_POOL_CREATE_RESET_COMMAND_BUFFER_BIT` flag is used to specify that any command buffers allocated from this pool can be individually reset to their initial state using the Vulkan function `vkResetCommandBuffer()`. To indicate that command buffers allocated from this pool will have a short lifespan, we use the `VK_COMMAND_POOL_CREATE_TRANSIENT_BIT` flag, meaning they will be reset or freed within a relatively short timeframe.

```
lvk::VulkanImmediateCommands::VulkanImmediateCommands(
  VkDevice device,
  uint32_t queueFamilyIndex, const char* debugName) :
  device_(device), queueFamilyIndex_(queueFamilyIndex),
  debugName_(debugName)
{
  vkGetDeviceQueue(device, queueFamilyIndex, 0, &queue_);
  const VkCommandPoolCreateInfo ci = {
      .sType = VK_STRUCTURE_TYPE_COMMAND_POOL_CREATE_INFO,
      .flags = VK_COMMAND_POOL_CREATE_RESET_COMMAND_BUFFER_BIT |
               VK_COMMAND_POOL_CREATE_TRANSIENT_BIT,
      .queueFamilyIndex = queueFamilyIndex,
  };
  vkCreateCommandPool(device, &ci, nullptr, &commandPool_);
```

2. Now, we can preallocate all the command buffers from the command pool. In addition, we create one semaphore and one fence for each command buffer to enable our synchronization system.

```
  const VkCommandBufferAllocateInfo ai = {
      .sType = VK_STRUCTURE_TYPE_COMMAND_BUFFER_ALLOCATE_INFO,
      .commandPool = commandPool_,
      .level = VK_COMMAND_BUFFER_LEVEL_PRIMARY,
      .commandBufferCount = 1,
  };
  for (uint32_t i = 0; i != kMaxCommandBuffers; i++) {
    CommandBufferWrapper& buf = buffers_[i];
    char fenceName[256] = {0};
    char semaphoreName[256] = {0};
    if (debugName) {
      // ... assign debug names to fenceName and semaphoreName
    }
    buf.semaphore_ = lvk::createSemaphore(device, semaphoreName);
```

```
    buf.fence_ = lvk::createFence(device, fenceName);
    vkAllocateCommandBuffers(
      device, &ai, &buf.cmdBufAllocated_);
    buffers_[i].handle_.bufferIndex_ = i;
  }
}
```

3. The destructor is almost trivial. We simply wait for all command buffers to be processed before destroying the command pool, fences, and semaphores.

```
lvk::VulkanImmediateCommands::~VulkanImmediateCommands() {
  waitAll();
  for (CommandBufferWrapper& buf : buffers_) {
    vkDestroyFence(device_, buf.fence_, nullptr);
    vkDestroySemaphore(device_, buf.semaphore_, nullptr);
  }
  vkDestroyCommandPool(device_, commandPool_, nullptr);
}
```

Now, let's examine the implementation of our most important function `acquire()`. All error checking code is omitted again to keep the explanation clear and focused.

1. Before we can find an available command buffer, we need to ensure there is one. This busy-wait loop checks the number of currently available command buffers and calls the `purge()` function, which recycles processed command buffers and resets them to their initial state, until at least one buffer becomes available. In practice, this loop almost never runs.

```
const lvk::VulkanImmediateCommands::CommandBufferWrapper&
  lvk::VulkanImmediateCommands::acquire()
{
  while (!numAvailableCommandBuffers_) purge();
```

2. Once we know there's at least one command buffer available, we can find it by going through the array of all buffers and selecting the first available one. At this point, we decrement `numAvailableCommandBuffers` to ensure proper busy-waiting on the next call to `acquire()`. The `isEncoding` member field is used to prevent the reuse of a command buffer that has already been acquired but has not yet been submitted.

```
  VulkanImmediateCommands::CommandBufferWrapper*
    current = nullptr;
  for (CommandBufferWrapper& buf : buffers_) {
    if (buf.cmdBuf_ == VK_NULL_HANDLE) {
      current = &buf;
      break;
    }
  }
```

```
    current->handle_.submitId_ = submitCounter_;
    numAvailableCommandBuffers_--;
    current->cmdBuf_ = current->cmdBufAllocated_;
    current->isEncoding_ = true;
```

3. After completing all the bookkeeping on our side, we can call the Vulkan API to begin recording the current command buffer.

```
    const VkCommandBufferBeginInfo bi = {
        .sType = VK_STRUCTURE_TYPE_COMMAND_BUFFER_BEGIN_INFO,
        .flags = VK_COMMAND_BUFFER_USAGE_ONE_TIME_SUBMIT_BIT,
    };
    VK_ASSERT(vkBeginCommandBuffer(current->cmdBuf_, &bi));
    nextSubmitHandle_ = current->handle_;
    return *current;
  }
```

4. Before we dive into the next series of functions, let's take a look at a short helper function, `purge()`, which was mentioned earlier in `acquire()`. This function calls `vkWaitForFences()` with a Vulkan fence and a timeout value of `0`, which causes it to return the current status of the fence without waiting. If the fence is signaled, we can reset the command buffer and increment `numAvailableCommandBuffers`. We always begin checking with the oldest submitted buffer and then wrap around.

```
  void lvk::VulkanImmediateCommands::purge() {
    const uint32_t numBuffers = LVK_ARRAY_NUM_ELEMENTS(buffers_);
    for (uint32_t i = 0; i != numBuffers; i++) {
      const uint32_t index = i + lastSubmitHandle_.bufferIndex_+1;
      CommandBufferWrapper& buf = buffers_[index % numBuffers];
      if (buf.cmdBuf_ == VK_NULL_HANDLE || buf.isEncoding_)
        continue;
      const VkResult result =
        vkWaitForFences(device_, 1, &buf.fence_, VK_TRUE, 0);
      if (result == VK_SUCCESS) {
        vkResetCommandBuffer(
          buf.cmdBuf_, VkCommandBufferResetFlags{0});
        vkResetFences(device_, 1, &buf.fence_);
        buf.cmdBuf_ = VK_NULL_HANDLE;
        numAvailableCommandBuffers_++;
      } else {
        if (result != VK_TIMEOUT) VK_ASSERT(result);
      }
    }
  }
```

Another crucial function is submit(), which submits a command buffer to a queue. Let's take a look.

1. First, we should call vkEndCommandBuffer() to finish recording a command buffer.

```
SubmitHandle lvk::VulkanImmediateCommands::submit(
  const CommandBufferWrapper& wrapper) {
  vkEndCommandBuffer(wrapper.cmdBuf_);
```

2. Then we should prepare semaphores. We can set two optional semaphores to be waited on before GPU processes our command buffer. The first one is the semaphore we injected with the waitSemaphore() function. It can be an "acquire semaphore" from a swapchain or any other user-provided semaphore if we want to organize a frame graph of some sort. The second semaphore lastSubmitSemaphore_ is the semaphore signaled by a previously submitted command buffer. This ensures all command buffers are processed sequentially one by one.

```
  VkSemaphoreSubmitInfo waitSemaphores[] = {{}, {}};
  uint32_t numWaitSemaphores = 0;
  if (waitSemaphore_.semaphore)
    waitSemaphores[numWaitSemaphores++] = waitSemaphore_;
  if (lastSubmitSemaphore_.semaphore)
    waitSemaphores[numWaitSemaphores++] = lastSubmitSemaphore_;
```

3. The signalSemaphores[] are signaled when the command buffer finishes execution. There are two of them: The first is the one we allocated along with our command buffer and is used for chaining command buffers together. The second is an optional timeline semaphore, injected by the signalSemaphore() function. It is injected at the end of the frame, before presenting the final image to the screen, and is used to orchestrate the swapchain presentation.

```
  VkSemaphoreSubmitInfo signalSemaphores[] = {
    VkSemaphoreSubmitInfo{
      .sType = VK_STRUCTURE_TYPE_SEMAPHORE_SUBMIT_INFO,
      .semaphore = wrapper.semaphore_,
      .stageMask = VK_PIPELINE_STAGE_ALL_COMMANDS_BIT},
    {},
  };
  uint32_t numSignalSemaphores = 1;
  if (signalSemaphore_.semaphore) {
    signalSemaphores[numSignalSemaphores++] = signalSemaphore_;
  }
```

4. Once we have all the data in place, calling vkQueueSubmit2() is straightforward. We populate the VkCommandBufferSubmitInfo structure using VkCommandBuffer from the current CommandBufferWrapper object and add all the semaphores to VkSubmitInfo2, allowing us to synchronize on them during the next submit() call.

```
  const VkCommandBufferSubmitInfo bufferSI = {
    .sType = VK_STRUCTURE_TYPE_COMMAND_BUFFER_SUBMIT_INFO,
    .commandBuffer = wrapper.cmdBuf_,
  };
  const VkSubmitInfo2 si = {
    .sType = VK_STRUCTURE_TYPE_SUBMIT_INFO_2,
    .waitSemaphoreInfoCount = numWaitSemaphores,
    .pWaitSemaphoreInfos = waitSemaphores,
    .commandBufferInfoCount = 1u,
    .pCommandBufferInfos = &bufferSI,
    .signalSemaphoreInfoCount = numSignalSemaphores,
    .pSignalSemaphoreInfos = signalSemaphores,
  };
  vkQueueSubmit2(queue_, 1u, &si, wrapper.fence_);
  lastSubmitSemaphore_.semaphore = wrapper.semaphore_;
  lastSubmitHandle_ = wrapper.handle_;
```

5. Once the `waitSemaphore_` and `signalSemaphore_` objects have been used, they should be discarded. They are meant to be used with exactly one command buffer. The `submitCounter_` variable is used to set the `submitId` value in the next `SubmitHandle`. Here's a trick we use: a `SubmitHandle` is considered empty when its command buffer and `submitId` are both zero. A simple way to achieve this is to always skip the zero value of `submitCounter`, hence double incrementing when we encounter zero.

```
  waitSemaphore_.semaphore   = VK_NULL_HANDLE;
  signalSemaphore_.semaphore = VK_NULL_HANDLE;
  const_cast<CommandBufferWrapper&>(wrapper).isEncoding_ = false;
  submitCounter_++;
  if (!submitCounter_) submitCounter_++;
  return lastSubmitHandle_;
}
```

This code is already sufficient to manage command buffers in an application. However, let's take a look at other methods of `VulkanImmediateCommands` that simplify working with Vulkan fences by hiding them behind `SubmitHandle`. The next most useful method is `isReady()`, which serves as our high-level equivalent of `vkWaitForFences()` with the timeout set to `0`.

1. First, we perform a trivial check for an empty submit handle.

```
bool VulkanImmediateCommands::isReady(
  const SubmitHandle handle) const
{
  if (handle.empty()) return true;
```

2. Next, we inspect the actual command buffer wrapper and check if its command buffer has already been recycled by the `purge()` method we explored earlier.

```
const CommandBufferWrapper& buf =
  buffers_[handle.bufferIndex_];
if (buf.cmdBuf_ == VK_NULL_HANDLE) return true;
```

3. Another scenario occurs when a command buffer has been recycled and then reused. Reuse can only happen after the command buffer has finished execution. In this case, the `submitId` values would be different. Only after this comparison can we invoke the Vulkan API to check the status of our `VkFence` object.

```
  if (buf.handle_.submitId_ != handle.submitId_) return true;
  return vkWaitForFences(device_, 1, &buf.fence_, VK_TRUE, 0) ==
    VK_SUCCESS;
}
```

The `isReady()` method provides a simple interface to Vulkan fences, which can be exposed to applications using the *LightweightVK* library without revealing the actual `VkFence` objects or the entire mechanism of how `VkCommandBuffer` objects are submitted and reset.

There is a pair of similar methods that allow us to wait for a specific `VkFence` hidden behind `SubmitHandle`.

1. The first method is `wait()`, and it waits for a single fence to be signaled. Two important points to mention here: We can detect a wait operation on a non-submitted command buffer using the `isEncoding_` flag. Also, we call `purge()` at the end of the function because we are sure there is now at least one command buffer available to be reclaimed. There's a special shortcut here: if we call `wait()` with an empty `SubmitHandle`, it will invoke `vkDeviceWaitIdle()`, which is often useful for debugging.

```
void lvk::VulkanImmediateCommands::wait(
  const SubmitHandle handle) {
  if (handle.empty()) {
    vkDeviceWaitIdle(device_);
    return;
  }
  if (isReady(handle)) return;
  if (!LVK_VERIFY(!buffers_[handle.bufferIndex_].isEncoding_))
    return;
  VK_ASSERT(vkWaitForFences(device_, 1,
    &buffers_[handle.bufferIndex_].fence_, VK_TRUE, UINT64_MAX));
  purge();
}
```

2. The second function waits for all submitted command buffers to be completed, and it is useful when we want to delete all resources, such as in the destructor. The implementation is straightforward, and we call purge() again to reclaim all completed command buffers.

```
void lvk::VulkanImmediateCommands::waitAll() {
  VkFence fences[kMaxCommandBuffers];
  uint32_t numFences = 0;
  for (const CommandBufferWrapper& buf : buffers_) {
    if (buf.cmdBuf_ != VK_NULL_HANDLE && !buf.isEncoding_)
      fences[numFences++] = buf.fence_;
  }
  if (numFences) VK_ASSERT(vkWaitForFences(
    device_, numFences, fences, VK_TRUE, UINT64_MAX));
  purge();
}
```

Those are all the details about the low-level command buffers implementation. Now, let's take a look at how this code works together with our demo application.

How it works...

Let's go all the way back to our demo application `Chapter02/01_Swapchain` and its main loop. We call the function `VulkanContext::acquireCommandBuffer()`, which returns a reference to a high-level interface `lvk::ICommandBuffer`. Then, we call `VulkanContext::submit()` to submit that command buffer.

```
while (!glfwWindowShouldClose(window)) {
  glfwPollEvents();
  glfwGetFramebufferSize(window, &width, &height);
  if (!width || !height) continue;
  lvk::ICommandBuffer& buf = ctx->acquireCommandBuffer();
  ctx->submit(buf, ctx->getCurrentSwapchainTexture());
}
```

Here's what is going on inside those functions.

1. The first function `VulkanContext::acquireCommandBuffer()` is very simple. It stores a new `lvk::CommandBuffer` object inside `VulkanContext` and returns a referent to it. This lightweight object implements the `lvk::ICommandBuffer` interface and, in the constructor, just calls `VulkanImmediateCommands::acquire()` we explored earlier.

```
ICommandBuffer& VulkanContext::acquireCommandBuffer() {
  LVK_ASSERT_MSG(!pimpl_->currentCommandBuffer_.ctx_,
    "Cannot acquire more than 1 command buffer simultaneously");
  pimpl_->currentCommandBuffer_ = CommandBuffer(this);
  return pimpl_->currentCommandBuffer_;
}
```

2. The function `VulkanContext::submit()` is more elaborate. Besides submitting a command buffer, it takes an optional argument of a swapchain texture to be presented. For now, we will skip this part and focus only on the command buffer submission.

```
void VulkanContext::submit(
  const lvk::ICommandBuffer& commandBuffer, TextureHandle present) {
  vulkan::CommandBuffer* vkCmdBuffer =
    const_cast<vulkan::CommandBuffer*>(
      static_cast<const vulkan::CommandBuffer*>(&commandBuffer));
  if (present) {
    // … do proper layout transitioning for the Vulkan image
  }
```

3. If we are presenting a swapchain image to the screen, we need to signal our timeline semaphore. Our timeline semaphore orchestrates the swapchain and works as follows: There is a `uint64_t` frame counter `VulkanSwapchain::currentFrameIndex_`, which increments monotonically with each presented frame. We have a specific number of frames in the swapchain—let's say `3` for example. Then, we can calculate different timeline signal values for each swapchain image so that we wait on these values every `3` frames. We wait for these corresponding timeline values when we want to acquire the same swapchain image the next time, before calling `vkAcquireNextImageKHR()`. For example, we render frame `0`, and the next time we want to acquire it, we wait until the signal semaphore value reaches at least `3`. Here, we call the function `signalSemaphore()` mentioned earlier to inject this timeline signal into our command buffer submission.

```
  const bool shouldPresent = hasSwapchain() && present;
  if (shouldPresent) {
   const uint64_t signalValue = swapchain_->currentFrameIndex_ +
                                swapchain_->getNumSwapchainImages();
    swapchain_->timelineWaitValues_[
      swapchain_->currentImageIndex_] = signalValue;
    immediate_->signalSemaphore(timelineSemaphore_, signalValue);
  }
  vkCmdBuffer->lastSubmitHandle_ =
    immediate_->submit(*vkCmdBuffer->wrapper_);
```

4. After submission, we retrieve the last submit semaphore and pass it into the swapchain so it can wait on it before the image to be presented is fully rendered by the GPU.

```
  if (shouldPresent) {
    swapchain_->present(
      immediate_->acquireLastSubmitSemaphore());
  }
```

5. Then we call abovementioned `VulkanImmediateCommands::submit()` and use its last submit semaphore to tell the swapchain to wait until the rendering is completed.

```
    vkCmdBuffer->lastSubmitHandle_ =
      immediate_->submit(*vkCmdBuffer->wrapper_);
    if (shouldPresent) {
      swapchain_->present(immediate_->acquireLastSubmitSemaphore());
    }
```

6. On every submit operation, we process so-called deferred tasks. Our deferred task is an `std::packaged_task` that should only be run when an associated `SubmitHandle`, also known as `VkFence`, is ready. This mechanism is very helpful for managing or deallocating Vulkan resources that might still be in use by the GPU, and will be discussed in subsequent chapters.

```
    processDeferredTasks();
    SubmitHandle handle = vkCmdBuffer->lastSubmitHandle_;
    pimpl_->currentCommandBuffer_ = {};
    return handle;
  }
```

7. Last but not least, let's take a quick look at `VulkanSwapchain::getCurrentTexture()` to see how `vkAcquireNextImageKHR()` interacts with all the aforementioned semaphores. Here, we wait on the timeline semaphore using the specific signal value for the current swapchain image, which we calculated in the code above. If you're confused, the pattern here is that for rendering frame `N`, we wait for the signal value `N`. After submitting GPU work, we signal the value `N+numSwapchainImages`.

```
  lvk::TextureHandle lvk::VulkanSwapchain::getCurrentTexture() {
    if (getNextImage_) {
      const VkSemaphoreWaitInfo waitInfo = {
        .sType = VK_STRUCTURE_TYPE_SEMAPHORE_WAIT_INFO,
        .semaphoreCount = 1,
        .pSemaphores = &ctx_.timelineSemaphore_,
        .pValues = &timelineWaitValues_[currentImageIndex_],
      };
      vkWaitSemaphores(device_, &waitInfo, UINT64_MAX);
```

8. Then, we can pass the corresponding acquire semaphore to `vkAcquireNextImageKHR()`. After this call, we pass this `acquireSemaphore` to `VulkanImmediateCommands::waitSemaphore()` so that we wait on it before submitting the next command buffer that renders into this swapchain image.

```
      VkSemaphore acquireSemaphore =
        acquireSemaphore_[currentImageIndex_];
      vkAcquireNextImageKHR(device_, swapchain_, UINT64_MAX,
```

```
    acquireSemaphore, VK_NULL_HANDLE, &currentImageIndex_);
  getNextImage_ = false;
  ctx_.immediate_->waitSemaphore(acquireSemaphore);
}
if (LVK_VERIFY(currentImageIndex_ < numSwapchainImages_))
  return swapchainTextures_[currentImageIndex_];
return {};
}
```

Now we have a working subsystem to wrangle Vulkan command buffers and expose `VkFence` objects to user applications in a clean and straightforward way. We didn't cover the `ICommandBuffer` interface in this recipe, but we will address it shortly in this chapter while working on our first Vulkan rendering demo. Before we start rendering, let's learn how to use compiled SPIR-V shaders from the recipe *Compiling Vulkan shaders at runtime* in *Chapter 1*.

There's more...

We recommend referring to *Vulkan Cookbook* by Packt for in-depth coverage of swapchain creation and command queues management.

Initializing Vulkan shader modules

The Vulkan API consumes shaders in the form of compiled SPIR-V binaries. We already learned how to compile shaders from GLSL source code to SPIR-V using the open-source **glslang** compiler from Khronos. In this recipe, we will learn how to use GLSL shaders and precompiled SPIR-V binaries in Vulkan.

Getting ready

We recommend reading the recipe *Compiling Vulkan shaders at runtime* in *Chapter 1* before you proceed.

How to do it...

Let's take a look at our next demo application, `Chapter02/02_HelloTriangle`, to learn the high-level *LightweightVK API* for shader modules. There's a method `createShaderModule()` in `IContext` that does the work and a helper function `loadShaderModule()` which makes it easier to use.

1. The helper function `loadShaderModule()` is defined in `shared/Utils.cpp`. It detects the shader stage type from the file name extension and calls `createShaderModule()` with the appropriate parameters.

```
Holder<ShaderModuleHandle> loadShaderModule(
  const std::unique_ptr<lvk::IContext>& ctx,
  const char* fileName)
{
  const std::string code = readShaderFile(fileName);
  const lvk::ShaderStage stage =
```

```
    lvkShaderStageFromFileName(fileName);
  Holder<ShaderModuleHandle> handle =
    ctx->createShaderModule({ code.c_str(), stage,
     (std::string("Shader Module: ") + fileName).c_str() });
  return handle;
}
```

2. In this way, given a pointer to `IContext`, Vulkan shader modules can be created from GLSL shaders as follows, where `codeVS` and `codeFS` are null-terminated strings holding the vertex and fragment shader source code, respectively.

```
Holder<ShaderModuleHandle> vert = loadShaderModule(
  ctx, "Chapter02/02_HelloTriangle/src/main.vert");
Holder<ShaderModuleHandle> frag = loadShaderModule(
  ctx, "Chapter02/02_HelloTriangle/src/main.frag");
```

3. The parameter of `createShaderModule()` is a structure `ShaderModuleDesc` containing all properties required to create a Vulkan shader module. If the `dataSize` member field is non-zero, the `data` field is treated as a binary SPIR-V blob. If `dataSize` is zero, `data` is treated as a null-terminated string containing GLSL source code.

```
struct ShaderModuleDesc {
  ShaderStage stage = Stage_Frag;
  const char* data = nullptr;
  size_t dataSize = 0;
  const char* debugName = "";
  ShaderModuleDesc(const char* source, lvk::ShaderStage stage,
    const char* debugName) : stage(stage), data(source),
    debugName(debugName) {}
  ShaderModuleDesc(const void* data, size_t dataLength,
    lvk::ShaderStage stage, const char* debugName) :
    stage(stage), data(static_cast<const char*>(data)),
    dataSize(dataLength), debugName(debugName) {}
};
```

4. Inside `VulkanContext::createShaderModule()`, we handle the branching for textual GLSL and binary SPIR-V shaders. An actual `VkShaderModule` object is stored in a pool, which we will discuss in subsequent chapters.

```
struct ShaderModuleState final {
  VkShaderModule sm = VK_NULL_HANDLE;
  uint32_t pushConstantsSize = 0;
};
Holder<ShaderModuleHandle>
  VulkanContext::createShaderModule(const ShaderModuleDesc& desc)
```

```
{
  Result result;
  ShaderModuleState sm = desc.dataSize ?
    createShaderModuleFromSPIRV(
      desc.data, desc.dataSize, desc.debugName, &result) :
    createShaderModuleFromGLSL(
      desc.stage, desc.data, desc.debugName, &result); // text
  return { this, shaderModulesPool_.create(std::move(sm)) };
}
```

5. The creation of a Vulkan shader module from a binary SPIR-V blob looks as follows. Error checking is omitted for simplicity.

```
ShaderModuleState VulkanContext::createShaderModuleFromSPIRV(
  const void* spirv,
  size_t numBytes,
  const char* debugName,
  Result* outResult) const
{
  VkShaderModule vkShaderModule = VK_NULL_HANDLE;
  const VkShaderModuleCreateInfo ci = {
    .sType = VK_STRUCTURE_TYPE_SHADER_MODULE_CREATE_INFO,
    .codeSize = numBytes,
    .pCode = (const uint32_t*)spirv,
  };
  vkCreateShaderModule(vkDevice_, &ci, nullptr, &vkShaderModule);
```

6. There's one important trick here. We will need the size of push constants in the shader to initialize our Vulkan pipelines later. Here, we use the **SPIRV-Reflect** library to introspect the SPIR-V code and retrieve the size of the push constants from it.

```
  SpvReflectShaderModule mdl;
  SpvReflectResult result =
    spvReflectCreateShaderModule(numBytes, spirv, &mdl);
  LVK_ASSERT(result == SPV_REFLECT_RESULT_SUCCESS);
  SCOPE_EXIT {
    spvReflectDestroyShaderModule(&mdl);
  };
  uint32_t pushConstantsSize = 0;
  for (uint32_t i = 0; i < mdl.push_constant_block_count; ++i) {
    const SpvReflectBlockVariable& block =
      mdl.push_constant_blocks[i];
    pushConstantsSize = std::max(pushConstantsSize, block.offset + block.size);
```

```
  }
  return {
    .sm = vkShaderModule,
    .pushConstantsSize = pushConstantsSize,
  };
}
```

7. The `VulkanContext::createShaderModuleFromGLSL()` function invokes `compileShader()`, which we learned about in the recipe *Compiling Vulkan shaders at runtime* in *Chapter 1* to create a SPIR-V binary blob. It then calls the aforementioned `createShaderModuleFromSPIRV()` to create an actual `VkShaderModule`. Before doing so, it injects a bunch of textual source code into the provided GLSL code. This is done to reduce code duplication in the shader. Things like declaring GLSL extensions and helper functions for bindless rendering are injected here. The injected code is quite large, and we will explore it step by step in subsequent chapters. For now, you can find it in `lightweightvk/lvk/vulkan/VulkanClasses.cpp`.

```
ShaderModuleState VulkanContext::createShaderModuleFromGLSL(
  ShaderStage stage,
  const char* source,
  const char* debugName,
  Result* outResult) const
{
  const VkShaderStageFlagBits vkStage = shaderStageToVkShaderStage(stage);
  std::string sourcePatched;
```

8. The automatic GLSL code injection happens only when the provided GLSL shader does not contain the `#version` directive. This allows you to override the code injection and provide complete GLSL shaders manually.

```
  if (strstr(source, "#version ") == nullptr) {
    if (vkStage == VK_SHADER_STAGE_TASK_BIT_EXT ||
        vkStage == VK_SHADER_STAGE_MESH_BIT_EXT) {
      sourcePatched += R"(
      #version 460
      #extension GL_EXT_buffer_reference : require
      // ... skipped a lot of injected code
    }
    sourcePatched += source;
    source = sourcePatched.c_str();
  }
  const glslang_resource_t glslangResource =
    lvk::getGlslangResource(
      getVkPhysicalDeviceProperties().limits);
```

```
  std::vector<uint8_t> spirv;
  const Result result = lvk::compileShader(
    vkStage, source, &spirv, &glslangResource);
  return createShaderModuleFromSPIRV(
    spirv.data(), spirv.size(), debugName, outResult);
}
```

Now that our Vulkan shader modules are ready to be used with Vulkan pipelines, let's learn how to do that in the next recipe.

Initializing Vulkan pipelines

A Vulkan pipeline is an implementation of an abstract graphics pipeline, which is a sequence of operations that transform vertices and rasterize the resulting image. Essentially, it's like a single snapshot of a "frozen" OpenGL state. Vulkan pipelines are mostly immutable, meaning multiple Vulkan pipelines should be created to allow different data paths through the graphics pipeline. In this recipe, we will learn how to create a Vulkan pipeline suitable for rendering a colorful triangle and explore how low-level and verbose Vulkan can be wrapped into a simple, high-level interface.

Getting ready...

To get all the basic information about Vulkan pipelines, we recommend reading Vulkan Cookbook by Pawel Lapinski which was published by Packt, or the Vulkan Tutorial series by Alexander Overvoorde `https://vulkan-tutorial.com/Drawing_a_triangle/Graphics_pipeline_basics/Introduction`.

For additional information on descriptor set layouts, check out the chapter `https://vulkan-tutorial.com/Uniform_buffers/Descriptor_layout_and_buffer`.

Vulkan pipelines require Vulkan shader modules. Check the previous recipe *Initializing Vulkan shader modules* before going through this recipe.

How to do it...

Let's dive into how to set up a Vulkan pipeline suitable for our triangle rendering application. Due to the verbosity of the Vulkan API, this recipe will be one of the longest. We will begin with the high-level code in our demo application, `Chapter02/02_HelloTriangle`, and work our way through to the internals of *LightweightVK* down to the Vulkan API. In the following chapters, we will explore more details, such as dynamic states, multisampling, vertex input, and more.

Let's take a look at the initialization and the main loop of `Chapter02/02_HelloTriangle`.

1. First, we create a window and a Vulkan context as described in the previous recipes.

```
GLFWwindow* window =
  lvk::initWindow("Simple example", width, height);
std::unique_ptr<lvk::IContext> ctx =
  lvk::createVulkanContextWithSwapchain(window, width, height, {});
```

2. Next, we need to create a rendering pipeline. *LightweightVK* uses opaque handles to work with resources, so `lvk::RenderPipelineHandle` is an opaque handle that manages a collection of `VkPipeline` objects, and `lvk::Holder` is a RAII wrapper that automatically disposes of handles when they go out of scope. The method `createRenderPipeline()` takes a `RenderPipelineDesc` structure, which contains the data necessary to configure a rendering pipeline. For our first triangle demo, we aim to keep things as minimalistic as possible, so we simply set the vertex and fragment shaders, and define the format of a color attachment. This is the absolute minimum required to render something into a swapchain image.

```
Holder<lvk::ShaderModuleHandle> vert = loadShaderModule(
  ctx, "Chapter02/02_HelloTriangle/src/main.vert");
Holder<lvk::ShaderModuleHandle> frag = loadShaderModule(
  ctx, "Chapter02/02_HelloTriangle/src/main.frag");
Holder<lvk::RenderPipelineHandle> rpTriangle =
  ctx->createRenderPipeline({
    .smVert = vert,
    .smFrag = frag,
    .color  = { { .format = ctx->getSwapchainFormat() } } });
```

3. Inside the main loop, we acquire a command buffer as described in the recipe *Using Vulkan command buffers* and issue some drawing commands.

```
while (!glfwWindowShouldClose(window)) {
  glfwPollEvents();
  glfwGetFramebufferSize(window, &width, &height);
  if (!width || !height) continue;
  lvk::ICommandBuffer& buf = ctx->acquireCommandBuffer();
```

4. The member function `cmdBeginRendering()` wraps Vulkan 1.3 dynamic rendering functionality, which enables rendering directly into Vulkan images without explicitly creating any render passes or framebuffer objects. It takes a description of a render pass `lvk::RenderPass` and a description of a framebuffer `lvk::Framebuffer`. We will explore it in more detail in subsequent chapters. Here, we use the current swapchain texture as the first color attachment and clear it to a white color before rendering using the attachment load operation `LoadOp_Clear`, which corresponds to `VK_ATTACHMENT_LOAD_OP_CLEAR` in Vulkan. The store operation is set to `StoreOp_Store` by default.

```
  buf.cmdBeginRendering(
    {.color = {{ .loadOp = LoadOp_Clear, .clearColor = {1,1,1,1} }}},
    {.color = {{ .texture = ctx->getCurrentSwapchainTexture() }}});
```

5. The render pipeline can be bound to the command buffer in one line. Then we can issue a drawing command `cmdDraw()`, which is a wrapper on top of `vkCmdDraw()`. You may have noticed that we did not use any index or vertex buffers at all. We will see why in a moment as we look into GLSL shaders. The command `cmdEndRendering()` corresponds to `vkCmdEndRendering()` from Vulkan 1.3.

```
    buf.cmdBindRenderPipeline(rpTriangle);
    buf.cmdPushDebugGroupLabel("Render Triangle", 0xff0000ff);
    buf.cmdDraw(3);
    buf.cmdPopDebugGroupLabel();
    buf.cmdEndRendering();
    ctx->submit(buf, ctx->getCurrentSwapchainTexture());
  }
```

Let's take a look at the GLSL shaders.

1. As we do not provide any vertex input, the vertex shader has to generate vertex data for a triangle. We use the built-in GLSL variable `gl_VertexIndex`, which gets incremented automatically for every subsequent vertex, to return hardcoded values for positions and vertex colors.

```
#version 460
layout (location=0) out vec3 color;
const vec2 pos[3] = vec2[3](
  vec2(-0.6, -0.4), vec2(0.6, -0.4), vec2(0.0, 0.6) );
const vec3 col[3] = vec3[3](
  vec3(1.0, 0.0, 0.0), vec3(0.0, 1.0, 0.0), vec3(0.0, 0.0, 1.0) );
void main() {
  gl_Position = vec4(pos[gl_VertexIndex], 0.0, 1.0);
  color = col[gl_VertexIndex];
}
```

2. The GLSL fragment shader is trivial and just outputs the interpolated fragment color.

```
#version 460
layout (location=0) in vec3 color;
layout (location=0) out vec4 out_FragColor;
void main() {
  out_FragColor = vec4(color, 1.0);
}
```

The application should render a colorful triangle as in the following picture.

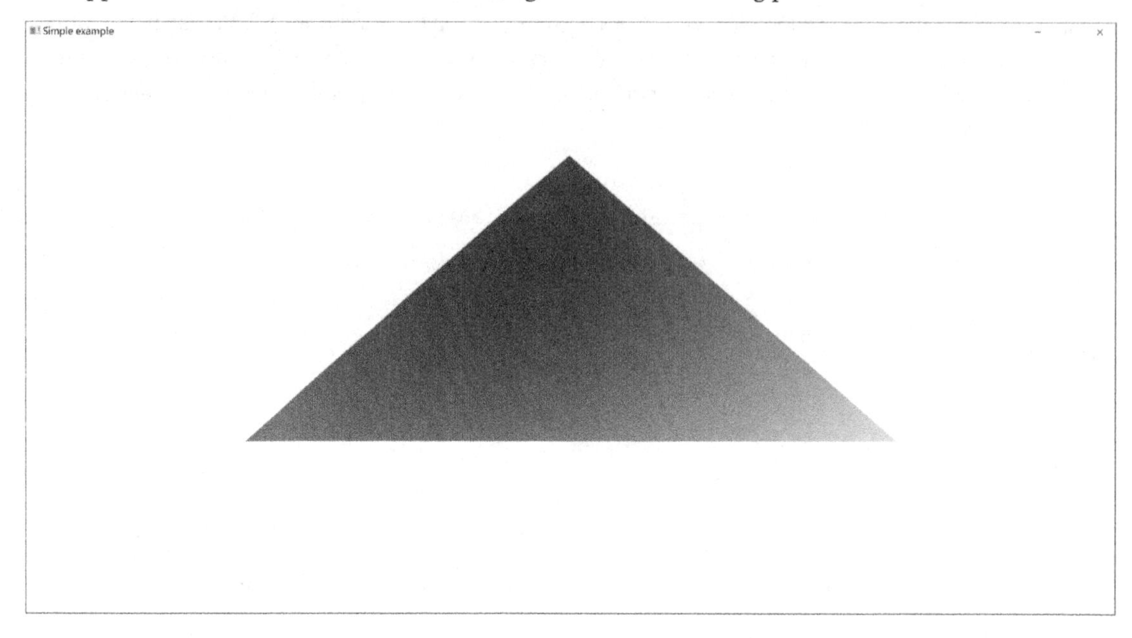

Figure 2.2: Hello Triangle

We learned how to draw a triangle with Vulkan using *LightweightVK*. It is time to look under the hood and find out how this high-level render pipeline management interface is implemented via Vulkan.

How it works...

To get to the underlying Vulkan implementation, we have to peel a few layers one by one. When we want to create a graphics pipeline in our application, we call the member function `IContext::createRenderPipeline()` which is implemented in `VulkanContext`. This function takes in a structure `lvk::RenderPipelineDesc` which describes our rendering pipeline. Let's take a closer look at it.

1. The structure contains a subset of information necessary to create a valid graphics `VkPipeline` object. It starts with the topology and vertex input descriptions, followed by shader modules for all supported shader stages. While *LightweightVK* supports mesh shaders, in this book, we will use only vertex, fragment, geometry, and tessellation shaders.

```
struct RenderPipelineDesc final {
  Topology topology = Topology_Triangle;
  VertexInput vertexInput;
  ShaderModuleHandle smVert;
  ShaderModuleHandle smTesc;
  ShaderModuleHandle smTese;
  ShaderModuleHandle smGeom;
  ShaderModuleHandle smTask;
```

```
  ShaderModuleHandle smMesh;
  ShaderModuleHandle smFrag;
```

2. Specialization constants allow Vulkan shader modules to be specialized after their compilation, at the pipeline creation time. We will demonstrate how to use them in the next chapter.

```
  SpecializationConstantDesc specInfo = {};
  const char* entryPointVert = "main";
  const char* entryPointTesc = "main";
  const char* entryPointTese = "main";
  const char* entryPointGeom = "main";
  const char* entryPointTask = "main";
  const char* entryPointMesh = "main";
  const char* entryPointFrag = "main";
```

3. The maximum number of color attachments is set to 8. We do not store the number of used attachments here. Instead, we use a short helper function to calculate how many attachments we actually have.

```
  ColorAttachment color[LVK_MAX_COLOR_ATTACHMENTS] = {};
  uint32_t getNumColorAttachments() const {
    uint32_t n = 0;
    while (n < LVK_MAX_COLOR_ATTACHMENTS &&
      color[n].format != Format_Invalid) n++;
    return n;
  }
```

4. Other member fields represent a typical rendering state with a cull mode, face winding, polygon mode, and so on.

```
  Format depthFormat = Format_Invalid;
  Format stencilFormat = Format_Invalid;
  CullMode cullMode = lvk::CullMode_None;
  WindingMode frontFaceWinding = lvk::WindingMode_CCW;
  PolygonMode polygonMode = lvk::PolygonMode_Fill;
  StencilState backFaceStencil = {};
  StencilState frontFaceStencil = {};
  uint32_t samplesCount = 1u;
  uint32_t patchControlPoints = 0;
  float minSampleShading = 0.0f;
  const char* debugName = "";
};
```

When we call `VulkanContext::createRenderPipeline()`, it performs sanity checks on `RenderPipelineDesc` and stores all the values in the `RenderPipelineState` struct. *LightweightVK* pipelines cannot be directly mapped one-to-one to `VkPipeline` objects because the actual `VkPipeline` objects have to be created lazily. For example, *LightweightVK* manages Vulkan descriptor set layouts automatically. Vulkan requires a descriptor set layout to be specified for a pipeline object. To address this, the data stored in `RenderPipelineState` is used to lazily create actual `VkPipeline` objects through the function `VulkanContext::getVkPipeline()`. Let's take a look at how this mechanism works, with error checking and some minor details omitted for simplicity.

1. The `RenderPipelineState` structure contains some precached values to avoid reinitializing them every time a new Vulkan pipeline object is created. All shader modules must remain alive as long as any pipeline that uses them is still in use.

```
class RenderPipelineState final {
  RenderPipelineDesc desc_;
  uint32_t numBindings_ = 0;
  uint32_t numAttributes_ = 0;
  VkVertexInputBindingDescription
    vkBindings_[VertexInput::LVK_VERTEX_BUFFER_MAX] = {};
  VkVertexInputAttributeDescription
    vkAttributes_[VertexInput::LVK_VERTEX_ATTRIBUTES_MAX] = {};
  VkDescriptorSetLayout lastVkDescriptorSetLayout_ =
    VK_NULL_HANDLE;
  VkShaderStageFlags shaderStageFlags_ = 0;
```

2. Each `RenderPipelineState` owns a pipeline layout, a pipeline, and memory storage for specialization constants.

```
  VkPipelineLayout pipelineLayout_ = VK_NULL_HANDLE;
  VkPipeline pipeline_ = VK_NULL_HANDLE;
  void* specConstantDataStorage_ = nullptr;
};
```

With all data structures in place, we are now ready to go through the implementation of `VulkanContext::createRenderPipeline()`. Most of the error checking code is skipped for the sake of brevity.

1. The constructor iterates over vertex input attributes, and precaches all necessary data into Vulkan structures for further use.

```
Holder<RenderPipelineHandle> VulkanContext::createRenderPipeline(
  const RenderPipelineDesc& desc, Result* outResult)
{
  const bool hasColorAttachments =
    desc.getNumColorAttachments() > 0;
  const bool hasDepthAttachment =
    desc.depthFormat != Format_Invalid;
```

```
const bool hasAnyAttachments =
  hasColorAttachments || hasDepthAttachment;
if (!LVK_VERIFY(hasAnyAttachments)) return {};
if (!LVK_VERIFY(desc.smVert.valid())) return {};
if (!LVK_VERIFY(desc.smFrag.valid())) return {};
RenderPipelineState rps = {.desc_ = desc};
```

2. Iterate and cache vertex input bindings and attributes. Vertex buffer bindings are tracked in `bufferAlreadyBound`. Everything else is a very trivial conversion code from our high-level data structures to Vulkan.

```
const lvk::VertexInput& vstate = rps.desc_.vertexInput;
bool bufferAlreadyBound[LVK_VERTEX_BUFFER_MAX] = {};
rps.numAttributes_ = vstate.getNumAttributes();
for (uint32_t i = 0; i != rps.numAttributes_; i++) {
  const VertexInput::VertexAttribute& attr = vstate.attributes[i];
  rps.vkAttributes_[i] = { .location = attr.location,
                           .binding = attr.binding,
                           .format =
                             vertexFormatToVkFormat(attr.format),
                           .offset = (uint32_t)attr.offset };
  if (!bufferAlreadyBound[attr.binding]) {
    bufferAlreadyBound[attr.binding] = true;
    rps.vkBindings_[rps.numBindings_++] = {
      .binding = attr.binding,
      .stride = vstate.inputBindings[attr.binding].stride,
      .inputRate = VK_VERTEX_INPUT_RATE_VERTEX };
  }
}
```

3. If specialization constants data is provided, copy it out of `RenderPipelineDesc` into local memory storage owned by the pipeline. This simplifies `RenderPipelineDesc` management on the application side, allowing it to be destroyed after the pipeline is created.

```
if (desc.specInfo.data && desc.specInfo.dataSize) {
  rps.specConstantDataStorage_ =
    malloc(desc.specInfo.dataSize);
  memcpy(rps.specConstantDataStorage_,
    desc.specInfo.data, desc.specInfo.dataSize);
  rps.desc_.specInfo.data = rps.specConstantDataStorage_;
}
return {this, renderPipelinesPool_.create(std::move(rps))};
}
```

Now we can create actual Vulkan pipelines. Well, almost. A couple of very long code snippets await us. These are the longest functions in the entire book, but we have to go through them at least once. Though, error checking is skipped to simplify things a bit.

1. The `getVkPipeline()` functions retrieves a `RenderPipelineState` struct from a pool using a provided pipeline handle.

```
VkPipeline VulkanContext::getVkPipeline(
  RenderPipelineHandle handle)
{
  lvk::RenderPipelineState* rps =
    renderPipelinesPool_.get(handle);
  if (!rps) return VK_NULL_HANDLE;
```

2. Then we check if the descriptor set layout used to create a pipeline layout for this `VkPipeline` object has changed. Our implementation uses Vulkan descriptor indexing to manage all textures in a large descriptor set and creates a descriptor set layout to store them. When new textures are created, there might not be enough space to store them, requiring the creation of a new descriptor set layout. Every time this happens, we have to delete the old `VkPipeline` and `VkPipelineLayout` objects and create new ones.

```
  if (rps->lastVkDescriptorSetLayout_ != vkDSL_) {
    deferredTask(std::packaged_task<void()>(
      [device = getVkDevice(), pipeline = rps->pipeline_]() {
        vkDestroyPipeline(device, pipeline, nullptr); }));
    deferredTask(std::packaged_task<void()>(
      [device = getVkDevice(), layout = rps->pipelineLayout_]() {
        vkDestroyPipelineLayout(device, layout, nullptr); }));
    rps->pipeline_ = VK_NULL_HANDLE;
    rps->lastVkDescriptorSetLayout_ = vkDSL_;
  }
```

3. If there is already a valid Vulkan graphics pipeline compatible with the current descriptor set layout, we can simply return it.

```
  if (rps->pipeline_ != VK_NULL_HANDLE) {
    return rps->pipeline_;
  }
```

4. Let's prepare to build a new Vulkan pipeline object. We need to create color blend attachments only for active color attachments. Helper functions, such as `formatToVkFormat()`, convert *LightweightVK* enumerations to Vulkan.

```
  VkPipelineLayout layout = VK_NULL_HANDLE;
  VkPipeline pipeline = VK_NULL_HANDLE;
  const RenderPipelineDesc& desc = rps->desc_;
  const uint32_t numColorAttachments =
```

```
    desc_.getNumColorAttachments();
  VkPipelineColorBlendAttachmentState
    colorBlendAttachmentStates[LVK_MAX_COLOR_ATTACHMENTS] = {};
  VkFormat colorAttachmentFormats[LVK_MAX_COLOR_ATTACHMENTS] ={};
  for (uint32_t i = 0; i != numColorAttachments; i++) {
    const lvk::ColorAttachment& attachment = desc_.color[i];
    colorAttachmentFormats[i] =
      formatToVkFormat(attachment.format);
```

5. Setting up blending states for color attachments is tedious but very simple.

```
    if (!attachment.blendEnabled) {
      colorBlendAttachmentStates[i] =
        VkPipelineColorBlendAttachmentState{
          .blendEnable = VK_FALSE,
          .srcColorBlendFactor = VK_BLEND_FACTOR_ONE,
          .dstColorBlendFactor = VK_BLEND_FACTOR_ZERO,
          .colorBlendOp = VK_BLEND_OP_ADD,
          .srcAlphaBlendFactor = VK_BLEND_FACTOR_ONE,
          .dstAlphaBlendFactor = VK_BLEND_FACTOR_ZERO,
          .alphaBlendOp = VK_BLEND_OP_ADD,
          .colorWriteMask = VK_COLOR_COMPONENT_R_BIT |
                            VK_COLOR_COMPONENT_G_BIT |
                            VK_COLOR_COMPONENT_B_BIT |
                            VK_COLOR_COMPONENT_A_BIT,
      };
    } else {
      colorBlendAttachmentStates[i] =
        VkPipelineColorBlendAttachmentState{
          .blendEnable = VK_TRUE,
          .srcColorBlendFactor = blendFactorToVkBlendFactor(
            attachment.srcRGBBlendFactor),
          .dstColorBlendFactor = blendFactorToVkBlendFactor(
            attachment.dstRGBBlendFactor),
          .colorBlendOp = blendOpToVkBlendOp(attachment.rgbBlendOp),
          .srcAlphaBlendFactor = blendFactorToVkBlendFactor(
            attachment.srcAlphaBlendFactor),
          .dstAlphaBlendFactor = blendFactorToVkBlendFactor(
            attachment.dstAlphaBlendFactor),
          .alphaBlendOp =
            blendOpToVkBlendOp(attachment.alphaBlendOp),
          .colorWriteMask = VK_COLOR_COMPONENT_R_BIT |
                            VK_COLOR_COMPONENT_G_BIT |
```

```
                            VK_COLOR_COMPONENT_B_BIT |
                            VK_COLOR_COMPONENT_A_BIT,
    };
  }
}
```

6. Retrieve `VkShaderModule` objects from the pool using opaque handles. We will discuss how pools work in the next chapters. Here all we have to know is that they allow fast conversion of an integer handle into the actual data associated with it. The geometry shader is optional. We skip all other shaders here for the sake of brevity.

```
const VkShaderModule* vert =
  ctx_->shaderModulesPool_.get(desc_.smVert);
const VkShaderModule* geom =
  ctx_->shaderModulesPool_.get(desc_.smGeom);
const VkShaderModule* frag =
  ctx_->shaderModulesPool_.get(desc_.smFrag);
```

7. Prepare the vertex input state.

```
const VkPipelineVertexInputStateCreateInfo ciVertexInputState =
{
  .sType = VK_STRUCTURE_TYPE_PIPELINE_VERTEX_INPUT_STATE_CREATE_INFO,
  .vertexBindingDescriptionCount = rps->numBindings_,
  .pVertexBindingDescriptions = rps->numBindings_ ?
    rps->vkBindings_ : nullptr,
  .vertexAttributeDescriptionCount = rps->numAttributes_,
  .pVertexAttributeDescriptions =
    rps->numAttributes_ ? rps->vkAttributes_ : nullptr,
};
```

8. Populate the `VkSpecializationInfo` structure to describe specialization constants for this graphics pipeline.

```
VkSpecializationMapEntry
  entries[LVK_SPECIALIZATION_CONSTANTS_MAX] = {};
const VkSpecializationInfo si =
  lvk::getPipelineShaderStageSpecializationInfo(
    desc.specInfo, entries);
```

9. Create a suitable `VkPipelineLayout` object for this pipeline. Use the current descriptor set layout stored in `VulkanContext`. Here one descriptor set layout `vkDSL_` is duplicated multiple times to create a pipeline layout. This is necessary to ensure compatibility with MoltenVK which does not allow aliasing of different descriptor types. Push constant sizes are retrieved from precompiled shader modules as was described in the previous recipe *Initializing Vulkan shader modules*.

```
  const VkDescriptorSetLayout dsls[] =
    { vkDSL_, vkDSL_, vkDSL_, vkDSL_ };
  const VkPushConstantRange range = {
    .stageFlags = rps->shaderStageFlags_,
    .offset = 0,
    .size = pushConstantsSize,
  };
  const VkPipelineLayoutCreateInfo ci = {
    .sType = VK_STRUCTURE_TYPE_PIPELINE_LAYOUT_CREATE_INFO,
    .setLayoutCount = (uint32_t)LVK_ARRAY_NUM_ELEMENTS(dsls),
    .pSetLayouts = dsls,
    .pushConstantRangeCount = pushConstantsSize ? 1u : 0u,
    .pPushConstantRanges = pushConstantsSize ? &range:nullptr,
};
vkCreatePipelineLayout(vkDevice_, &ci, nullptr, &layout);
```

More information

Here's a snippet to retrieve precalculated push constant sizes from shader modules:

```
#define UPDATE_PUSH_CONSTANT_SIZE(sm, bit) if (sm) { \
  pushConstantsSize = std::max(pushConstantsSize,    \
  sm->pushConstantsSize);                            \
  rps->shaderStageFlags_ |= bit; }
rps->shaderStageFlags_ = 0;
uint32_t pushConstantsSize = 0;
UPDATE_PUSH_CONSTANT_SIZE(vertModule,
  VK_SHADER_STAGE_VERTEX_BIT);
UPDATE_PUSH_CONSTANT_SIZE(tescModule,
  VK_SHADER_STAGE_TESSELLATION_CONTROL_BIT);
UPDATE_PUSH_CONSTANT_SIZE(teseModule,
  VK_SHADER_STAGE_TESSELLATION_EVALUATION_BIT);
UPDATE_PUSH_CONSTANT_SIZE(geomModule,
  VK_SHADER_STAGE_GEOMETRY_BIT);
UPDATE_PUSH_CONSTANT_SIZE(fragModule,
  VK_SHADER_STAGE_FRAGMENT_BIT);
#undef UPDATE_PUSH_CONSTANT_SIZE
```

10. As we peel more and more implementation layers, here is yet another level to peel. However, it is the last one. For convenience, the creation of actual `VkPipeline` objects is encapsulated in `VulkanPipelineBuilder`, which provides reasonable default values for all the numerous Vulkan data members that we do not want to set. Those familiar with Java will recognize a typical *Builder* design pattern here.

```
lvk::vulkan::VulkanPipelineBuilder()
    // from Vulkan 1.0
    .dynamicState(VK_DYNAMIC_STATE_VIEWPORT)
    .dynamicState(VK_DYNAMIC_STATE_SCISSOR)
    .dynamicState(VK_DYNAMIC_STATE_DEPTH_BIAS)
    .dynamicState(VK_DYNAMIC_STATE_BLEND_CONSTANTS)
    // from Vulkan 1.3
    .dynamicState(VK_DYNAMIC_STATE_DEPTH_TEST_ENABLE)
    .dynamicState(VK_DYNAMIC_STATE_DEPTH_WRITE_ENABLE)
    .dynamicState(VK_DYNAMIC_STATE_DEPTH_COMPARE_OP)
    .dynamicState(VK_DYNAMIC_STATE_DEPTH_BIAS_ENABLE)
    .primitiveTopology(
      topologyToVkPrimitiveTopology(desc.topology))
    .rasterizationSamples(
      getVulkanSampleCountFlags(desc.samplesCount,
        getFramebufferMSAABitMask()), desc.minSampleShading)
    .polygonMode(polygonModeToVkPolygonMode(desc_.polygonMode))
    .stencilStateOps(VK_STENCIL_FACE_FRONT_BIT,
      stencilOpToVkStencilOp(desc_.frontFaceStencil.stencilFailureOp),
      stencilOpToVkStencilOp(desc_.frontFaceStencil.depthStencilPassOp),
      stencilOpToVkStencilOp(desc_.frontFaceStencil.depthFailureOp),
      compareOpToVkCompareOp(desc_.frontFaceStencil.stencilCompareOp))
    .stencilStateOps(VK_STENCIL_FACE_BACK_BIT,
      stencilOpToVkStencilOp(desc_.backFaceStencil.stencilFailureOp),
      stencilOpToVkStencilOp(desc_.backFaceStencil.depthStencilPassOp),
      stencilOpToVkStencilOp(desc_.backFaceStencil.depthFailureOp),
      compareOpToVkCompareOp(desc_.backFaceStencil.stencilCompareOp))
    .stencilMasks(VK_STENCIL_FACE_FRONT_BIT, 0xFF,
      desc_.frontFaceStencil.writeMask,
      desc_.frontFaceStencil.readMask)
    .stencilMasks(VK_STENCIL_FACE_BACK_BIT, 0xFF,
      desc_.backFaceStencil.writeMask,
      desc_.backFaceStencil.readMask)
```

11. Shader modules are provided one by one. Only the vertex and fragment shaders are mandatory. We skip the other shaders for the sake of brevity.

```
        .shaderStage(lvk::getPipelineShaderStageCreateInfo(
          VK_SHADER_STAGE_VERTEX_BIT,
          vertModule->sm, desc.entryPointVert, &si))
        .shaderStage(lvk::getPipelineShaderStageCreateInfo(
          VK_SHADER_STAGE_FRAGMENT_BIT,
          fragModule->sm, desc.entryPointFrag, &si))
        .shaderStage(geomModule ?
          lvk::getPipelineShaderStageCreateInfo(
            VK_SHADER_STAGE_GEOMETRY_BIT,
            geomModule->sm, desc.entryPointGeom, &si) :
          VkPipelineShaderStageCreateInfo{ .module = VK_NULL_HANDLE } )
        .cullMode(cullModeToVkCullMode(desc_.cullMode))
        .frontFace(windingModeToVkFrontFace(desc_.frontFaceWinding))
        .vertexInputState(vertexInputStateCreateInfo_)
        .colorAttachments(colorBlendAttachmentStates,
          colorAttachmentFormats, numColorAttachments)
        .depthAttachmentFormat(formatToVkFormat(desc.depthFormat))
        .stencilAttachmentFormat(formatToVkFormat(desc.stencilFormat))
        .patchControlPoints(desc.patchControlPoints)
```

12. Finally, we call the `VulkanPipelineBuilder::build()` method, which creates a `VkPipeline` object and we can store it in our `RenderPipelineState` structure together with the pipeline layout.

```
        .build(vkDevice_, pipelineCache_, layout, &pipeline, desc.debugName);
    rps->pipeline_ = pipeline;
    rps->pipelineLayout_ = layout;
    return pipeline;
  }
```

The last method we want to explore here is `VulkanPipelineBuilder::build()` which is pure Vulkan. Let's take a look at it to conclude the pipeline creation process.

1. First, we put provided dynamic states into `VkPipelineDynamicStateCreateInfo`.

```
VkResult VulkanPipelineBuilder::build(
  VkDevice device,
  VkPipelineCache pipelineCache,
  VkPipelineLayout pipelineLayout,
  VkPipeline* outPipeline,
  const char* debugName)
{
  const VkPipelineDynamicStateCreateInfo dynamicState = {
    .sType = VK_STRUCTURE_TYPE_PIPELINE_DYNAMIC_STATE_CREATE_INFO,
```

```
      .dynamicStateCount = numDynamicStates_,
      .pDynamicStates = dynamicStates_,
    };
```

2. The Vulkan specification says viewport and scissor can be `nullptr` if the viewport and scissor states are dynamic. We are definitely happy to make the most of this opportunity.

```
    const VkPipelineViewportStateCreateInfo viewportState = {
      .sType = VK_STRUCTURE_TYPE_PIPELINE_VIEWPORT_STATE_CREATE_INFO,
      .viewportCount = 1,
      .pViewports = nullptr,
      .scissorCount = 1,
      .pScissors = nullptr,
    };
```

3. Use the color blend states and attachments we prepared earlier in this recipe. The `VkPipelineRenderingCreateInfo` is necessary for the Vulkan 1.3 dynamic rendering feature.

```
    const VkPipelineColorBlendStateCreateInfo colorBlendState = {
      .sType = VK_STRUCTURE_TYPE_PIPELINE_COLOR_BLEND_STATE_CREATE_INFO,
      .logicOpEnable = VK_FALSE,
      .logicOp = VK_LOGIC_OP_COPY,
      .attachmentCount = numColorAttachments_,
      .pAttachments = colorBlendAttachmentStates_,
    };
    const VkPipelineRenderingCreateInfo renderingInfo = {
      .sType = VK_STRUCTURE_TYPE_PIPELINE_RENDERING_CREATE_INFO_KHR,
      .colorAttachmentCount = numColorAttachments_,
      .pColorAttachmentFormats = colorAttachmentFormats_,
      .depthAttachmentFormat   = depthAttachmentFormat_,
      .stencilAttachmentFormat = stencilAttachmentFormat_,
    };
```

4. Put everything together into `VkGraphicsPipelineCreateInfo` and call `vkCreateGraphicsPipelines()`.

```
    const VkGraphicsPipelineCreateInfo ci = {
      .sType = VK_STRUCTURE_TYPE_GRAPHICS_PIPELINE_CREATE_INFO,
      .pNext = &renderingInfo,
      .flags = 0,
      .stageCount = numShaderStages_,
      .pStages = shaderStages_,
      .pVertexInputState = &vertexInputState_,
      .pInputAssemblyState = &inputAssembly_,
      .pTessellationState = &tessellationState_,
```

```
    .pViewportState = &viewportState,
    .pRasterizationState = &rasterizationState_,
    .pMultisampleState = &multisampleState_,
    .pDepthStencilState = &depthStencilState_,
    .pColorBlendState = &colorBlendState,
    .pDynamicState = &dynamicState,
    .layout = pipelineLayout,
    .renderPass = VK_NULL_HANDLE,
    .subpass = 0,
    .basePipelineHandle = VK_NULL_HANDLE,
    .basePipelineIndex = -1,
  };
  vkCreateGraphicsPipelines(
    device, pipelineCache, 1, &ci, nullptr, outPipeline);
  numPipelinesCreated_++;
}
```

This code concludes the pipeline creation process. In addition to the very simple example in `Chapter02/02_HelloTriangle`, we created a slightly more elaborate app in `Chapter02/03_GLM` to demonstrate how to use multiple render pipelines to render a rotating cube with a wireframe overlay. This app uses the GLM library for matrix math. You can check it out in `Chapter02/03_GLM/src/main.cpp`, where it uses `cmdPushConstants()` to animate the cube and specialization constants to use the same set of shaders for both solid and wireframe rendering. It should look as shown in the following screenshot.

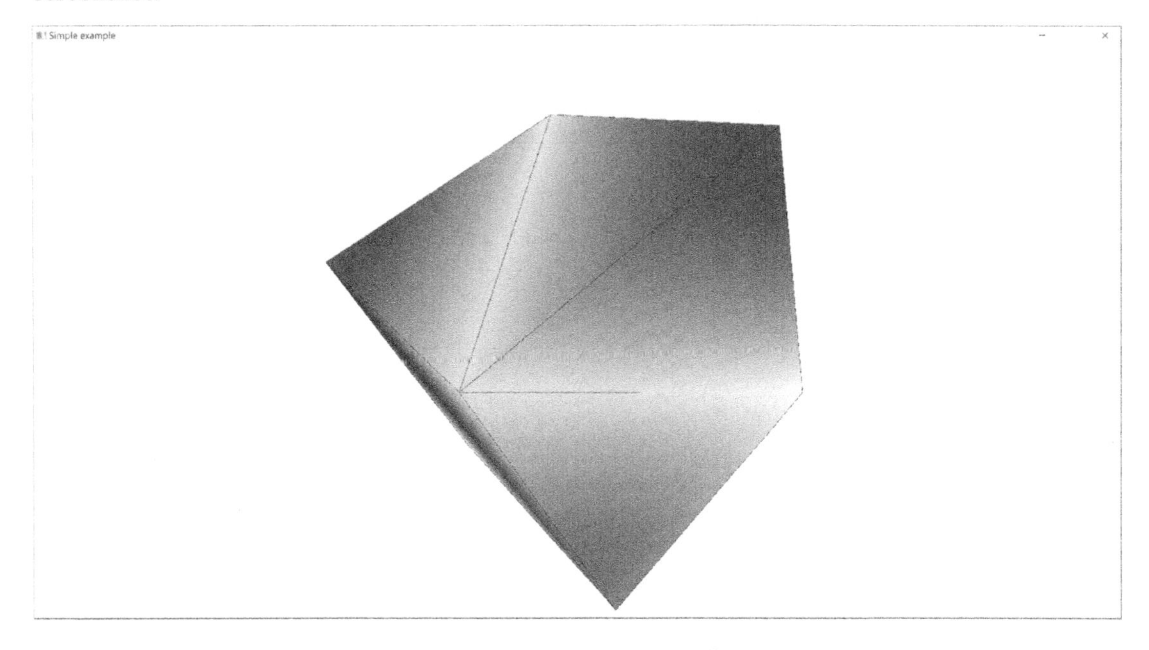

Figure 2.3: GLM usage example

There's more...

If you are familiar with older versions of Vulkan, you might have noticed that in this recipe we completely left out any references to render passes. They are also not mentioned in any of the data structures. The reason for that is that we use Vulkan 1.3 dynamic rendering functionality, which allows `VkPipeline` objects to operate without needing a render pass.

In case you want to implement a similar wrapper for older versions of Vulkan and without using the *VK_KHR_dynamic_rendering* extension, you can maintain a "global" collection of render passes in an array inside `VulkanContext` and add an integer index of a corresponding render pass as a data member to `RenderPipelineDynamicState`. Since we can use only a very restricted number of distinct rendering passes — let's say a maximum of 256 — the index can be saved as `uint8_t`. This would enable us to store them in array inside `VulkanContext`.

If you want to explore an actual working implementation of this approach, take a look at Meta's IGL library `https://github.com/facebook/igl/blob/main/src/igl/vulkan/RenderPipelineState.h` and check out how `renderPassIndex` is handled there.

Now let's jump to the next *Chapter 3* to learn how to use Vulkan in a user-friendly way to build more interesting examples.

3

Working with Vulkan Objects

In the previous chapter, we learned how to get our first triangle on the screen using Vulkan. Let's go forward and learn how to deal with textures and buffers to build a modern Vulkan wrapper. The recipes of this chapter will not be focusing solely on the graphics APIs, but on various tips and tricks necessary to improve graphical applications development and various 3D graphics algorithms. On the Vulkan side, we will cover the basic stuff to get it up and running. The underlying Vulkan implementation is based on the *LightweightVK* library `https://github.com/corporateshark/lightweightvk`.

In the chapter we will cover the following recipes:

- Dealing with buffers in Vulkan
- Implementing staging buffers
- Using texture data in Vulkan
- Storing Vulkan objects
- Using Vulkan descriptor indexing

Technical requirements

To run the recipes from this chapter, you have to use a Windows or Linux computer with a video card and drivers supporting Vulkan 1.3. Read the previous *Chapter 2* to learn the basics necessary to get you started with Vulkan.

Dealing with buffers in Vulkan

Buffers in Vulkan are essentially memory regions that store data for GPU access. More specifically, a Vulkan buffer refers to a `VkBuffer` object that is associated with memory regions allocated through `VkDeviceMemory`. To render a 3D scene with Vulkan, we must convert the scene data into a format the GPU can process. In this recipe, we'll explain how to create GPU buffers and upload vertex data into them. We'll use the open-source asset-loading library **Assimp** `https://github.com/assimp/assimp` to load a 3D model from an `.obj` file and render it using *LightweightVK* and Vulkan. Additionally, this recipe includes a basic introduction to using the **Vulkan Memory Allocator** (**VMA**) library.

Getting ready

In Vulkan, uploading data to GPU buffers is done using command buffers, similar to many other Vulkan operations. This requires a command queue that supports transfer operations. The process of creating and using command buffers was covered in the previous chapter in the recipe *Using Vulkan command buffers*.

How to do it...

Let's begin with the high-level code in our sample application, `Chapter03/01_Assimp`, and work our way down to the Vulkan API from there.

1. First, we need to load the model from an `.obj` file using the *Assimp* library. Here's some basic code to achieve this. Keep in mind that calling `reserve()` on vectors can improve performance. For simplicity, we'll load only the first mesh and extract vertex positions and indices.

```
const aiScene* scene = aiImportFile(
  "data/rubber_duck/scene.gltf", aiProcess_Triangulate);
const aiMesh* mesh = scene->mMeshes[0];
std::vector<vec3> positions;
std::vector<uint32_t> indices;
for (unsigned int i = 0; i != mesh->mNumVertices; i++) {
  const aiVector3D v = mesh->mVertices[i];
  positions.push_back(vec3(v.x, v.y, v.z));
}
for (unsigned int i = 0; i != mesh->mNumFaces; i++) {
  for (int j = 0; j != 3; j++) {
    indices.push_back(mesh->mFaces[i].mIndices[j]);
  }
}
aiReleaseImport(scene);
```

2. Next, we need to create buffers for the loaded vertex positions and indices. The vertex buffer will use the `BufferUsageBits_Vertex` flag, and we'll instruct *LightweightVK* to upload the initial data from `positions.data()` right away. The C++20 designated initializer syntax is particularly convenient for working with high-level APIs like this. Similarly, the index buffer will use the `BufferUsageBits_Index` flag. Both buffers will be stored in GPU memory to optimize performance. This is achieved by specifying the storage type as `StorageType_Device`, which *LightweightVK* interprets to select the appropriate Vulkan memory type for these buffers.

```
Holder<BufferHandle> vertexBuffer = ctx->createBuffer({
  .usage     = lvk::BufferUsageBits_Vertex,
  .storage   = lvk::StorageType_Device,
  .size      = sizeof(vec3) * positions.size(),
  .data      = positions.data(),
```

```
    .debugName = "Buffer: vertex" });
  Holder<BufferHandle> indexBuffer = ctx->createBuffer({
    .usage     = lvk::BufferUsageBits_Index,
    .storage   = lvk::StorageType_Device,
    .size      = sizeof(uint32_t) * indices.size(),
    .data      = indices.data(),
    .debugName = "Buffer: index" });
```

3. To render a concave mesh in Vulkan, we need to use a depth buffer, which we'll create as follows. Here, we specify `Format_Z_F32`, but the *LightweightVK* Vulkan backend will automatically substitute it with the closest available format supported by the current Vulkan implementation. The `width` and `height` values match the dimensions of the framebuffer. Since the depth texture will only be used as a depth buffer and not for sampling, specifying the usage flags as `TextureUsageBits_Attachment` is sufficient.

```
Holder<TextureHandle> depthTexture = ctx->createTexture({
  .type       = lvk::TextureType_2D,
  .format     = lvk::Format_Z_F32,
  .dimensions = {(uint32_t)width, (uint32_t)height},
  .usage      = lvk::TextureUsageBits_Attachment,
  .debugName  = "Depth buffer",
});
```

4. Before proceeding to create rendering pipelines, as described in the previous chapter in the recipe *Initializing Vulkan pipelines*, we need to define the vertex input state using the previously mentioned vertex buffer. Here's how to do it. The `.location = 0` corresponds to the input location in the GLSL vertex shader that will be used to render this mesh.

```
const lvk::VertexInput vdesc = {
  .attributes    = { { .location = 0,
                       .format = lvk::VertexFormat::Float3 } },
  .inputBindings = { { .stride = sizeof(vec3) } },
};
```

5. Now, we can create two rendering pipelines: one for rendering a solid mesh and another for rendering a wireframe mesh on top of it. Note that the pipeline `depthFormat` is set to the actual format of the depth texture we created earlier.

```
Holder<ShaderModuleHandle> vert =
  loadShaderModule(ctx, "Chapter03/01_Assimp/src/main.vert");
Holder<ShaderModuleHandle> frag =
  loadShaderModule(ctx, "Chapter03/01_Assimp/src/main.frag");
Holder<RenderPipelineHandle> pipelineSolid =
  ctx->createRenderPipeline({
    .vertexInput = vdesc,
```

```
      .smVert      = vert,
      .smFrag      = frag,
      .color       = { { .format = ctx->getSwapchainFormat() } },
      .depthFormat = ctx->getFormat(depthTexture),
      .cullMode    = lvk::CullMode_Back,
  });
```

6. The second rendering pipeline performs wireframe rendering by setting the `polygonMode` field to `PolygonMode_Line`. Both pipelines share the same set of shaders, with a specialization constant used to adjust the shader behavior without modifying the underlying GLSL code.

```
const uint32_t isWireframe = 1;
Holder<RenderPipelineHandle> pipelineWireframe =
  ctx->createRenderPipeline({
    .vertexInput = vdesc,
    .smVert      = vert,
    .smFrag      = frag,
    .specInfo = {
      .entries = { { .constantId = 0,
                     .size = sizeof(uint32_t) } },
      .data = &isWireframe,
      .dataSize = sizeof(isWireframe) },
    .color       = { { .format = ctx->getSwapchainFormat() } },
    .depthFormat = ctx->getFormat(depthTexture),
    .cullMode    = lvk::CullMode_Back,
    .polygonMode = lvk::PolygonMode_Line,
});
```

Now let's take a look at the application's main loop. We'll omit the GLFW event polling and framebuffer size update code here, but you can find it in `Chapter03/01_Assimp/src/main.cpp`.

1. In the main loop, the projection matrix `p` is updated based on the current framebuffer's aspect ratio.

```
while (!glfwWindowShouldClose(window)) {
  // … skipped GLFW code
  const float ratio = width / height;
  const mat4 p = glm::perspective(45.0f, ratio, 0.1f, 1000.0f);
```

2. Set the model-view matrix to enable a gradual rotation of the model around the vertical axis. The model matrix `m` aligns the model's "up" direction with Vulkan's vertical axis. The view matrix `v` defines the 3D camera's orientation and viewing direction, which slowly rotates around the vertical axis `Y`.

```
const mat4 m = glm::rotate(
  mat4(1.0f), glm::radians(-90.0f), vec3(1, 0, 0));
const mat4 v = glm::rotate(glm::translate(mat4(1.0f),
  vec3(0.0f, -0.5f, -1.5f)), (float)glfwGetTime(),
  vec3(0.0f, 1.0f, 0.0f));
```

3. The render pass description now includes a load operation and a clear value for the depth buffer. The framebuffer contains only one color attachment—the current swapchain image.

```
const lvk::RenderPass renderPass = {
  .color = {
    { .loadOp = LoadOp_Clear, .clearColor = { 1, 1, 1, 1 }}},
  .depth = { .loadOp = LoadOp_Clear, .clearDepth = 1. }
};
const lvk::Framebuffer framebuffer = {
  .color = {{ .texture = ctx->getCurrentSwapchainTexture()}},
  .depthStencil = { .texture = depthTexture },
};
```

4. With all preparations complete, we can acquire a command buffer as described in the recipe *Using Vulkan command buffers* from *Chapter 2* and begin rendering.

```
lvk::ICommandBuffer& buf = ctx->acquireCommandBuffer();
buf.cmdBeginRendering(renderPass, framebuffer);
```

5. Both the vertex and index buffers should be bound. The vertex buffer is bound to binding point `0`. The index buffer uses unsigned 32-bit integer values as indices.

```
buf.cmdBindVertexBuffer(0, vertexBuffer);
buf.cmdBindIndexBuffer(indexBuffer, lvk::IndexFormat_UI32);
```

6. Let's render a solid mesh using the first rendering pipeline. The model-view-projection matrix is passed to the shader using Vulkan push constants. Push constants are a performant mechanism for passing small amounts of data to shaders. Vulkan 1.3 guarantees at least `128` bytes for push constants, which is enough to store two 4x4 matrices or `16` arbitrary 64-bit GPU buffer addresses. The depth-stencil state is configured to enable the use of our depth buffer.

```
buf.cmdBindRenderPipeline(pipelineSolid);
buf.cmdBindDepthState({ .compareOp = lvk::CompareOp_Less,
                        .isDepthWriteEnabled = true });
buf.cmdPushConstants(p * v * m);
buf.cmdDrawIndexed(indices.size());
```

7. Next, we render a wireframe copy of the mesh on top of the solid one. We set a depth bias to ensure the wireframe edges are rendered correctly and without flickering. This is necessary to prevent Z-fighting between the rendered solid mesh and the wireframe overlay.

```
    buf.cmdBindRenderPipeline(pipelineWireframe);
    buf.cmdSetDepthBiasEnable(true);
    buf.cmdSetDepthBias(0.0f, -1.0f, 0.0f);
    buf.cmdDrawIndexed(indices.size());
```

8. Now the command buffer can be submitted for execution.

```
    buf.cmdEndRendering();
    ctx->submit(buf, ctx->getCurrentSwapchainTexture());
  }
```

The demo application should render a colored rotating ducky with a wireframe overlay, as shown in the following screenshot.

Figure 3.1: Rendering a mesh loaded with Assimp

The high-level part was easy. Now, let's dive into the underlying implementation and explore how to build this streamlined buffer management interface using the Vulkan API.

How it works...

Let's examine the low-level Vulkan code to understand how buffers work. Our deep dive begins with exploring `IContext::createBuffer()`, which takes a buffer description structure `BufferDesc` as input.

1. The declaration of `BufferDesc` looks as follows. The storage type can be one of three enum values: `StorageType_Device`, `StorageType_HostVisible`, and `StorageType_Memoryless`. These correspond to GPU local memory (which is usually not visible from the CPU side), host-visible memory, and memory that is not used for storage. The underlying *LightweightVK* code and the **Vulkan Memory Allocator** (**VMA**) library precisely select the appropriate Vulkan memory type.

```
struct BufferDesc final {
  uint8_t usage = 0;
  StorageType storage = StorageType_HostVisible;
  size_t size = 0;
  const void* data = nullptr;
  const char* debugName = "";
};
```

2. The buffer usage mode is a combination of the following flags. These flags are highly flexible, allowing us to request any necessary combination, with the exception that uniform and storage buffers are mutually exclusive. *LightweightVK* also supports additional buffer usage modes for ray tracing, which are not explored in our book.

```
enum BufferUsageBits : uint8_t {
  BufferUsageBits_Index    = 1 << 0,
  BufferUsageBits_Vertex   = 1 << 1,
  BufferUsageBits_Uniform  = 1 << 2,
  BufferUsageBits_Storage  = 1 << 3,
  BufferUsageBits_Indirect = 1 << 4,
};
```

Let's now examine the implementation of `VulkanContext::createBuffer()`, which converts the requested LightweightVK buffer properties into the corresponding supported Vulkan flags.

1. First, we need to check if a staging buffer should be used to upload data into the new buffer. If the staging buffer is disabled—such as when our GPU has only one shared memory heap that is both host-visible and device-local—we override the requested device-local storage mode with `StorageType_HostVisible`. This is important to avoid the extra copy on GPUs with such memory configurations.

Note

In Vulkan, device-local memory is typically not accessible by the CPU on dedicated graphics cards. To populate such buffers with data, we need to create a host-visible buffer, known as a **staging buffer**, and first populate it from the CPU side. Then, we issue Vulkan commands to copy its contents into the actual device-local buffer.

```
Holder<BufferHandle> VulkanContext::createBuffer(
  const BufferDesc& requestedDesc, Result* outResult) {
  BufferDesc desc = requestedDesc;
  if (!useStaging_ && (desc.storage == StorageType_Device))
    desc.storage = StorageType_HostVisible;
```

2. If the application requires a device-local buffer, we should use a staging buffer to transfer data into the device-local buffer. We need to set the appropriate Vulkan flags to ensure that data can be transferred to and from this buffer.

```
  VkBufferUsageFlags usageFlags =
    desc.storage == StorageType_Device ?
      VK_BUFFER_USAGE_TRANSFER_DST_BIT |
      VK_BUFFER_USAGE_TRANSFER_SRC_BIT : 0;
```

3. For each requested usage flag, enable the corresponding set of necessary Vulkan usage flags. To use the Vulkan 1.2 **buffer device address** feature and access buffers by pointers from shaders, we need to add the flag `VK_BUFFER_USAGE_SHADER_DEVICE_ADDRESS_BIT`.

```
  if (desc.usage & BufferUsageBits_Index)
    usageFlags |= VK_BUFFER_USAGE_INDEX_BUFFER_BIT;
  if (desc.usage & BufferUsageBits_Vertex)
    usageFlags |= VK_BUFFER_USAGE_VERTEX_BUFFER_BIT;
  if (desc.usage & BufferUsageBits_Uniform)
    usageFlags |= VK_BUFFER_USAGE_UNIFORM_BUFFER_BIT |
                  VK_BUFFER_USAGE_SHADER_DEVICE_ADDRESS_BIT;
  if (desc.usage & BufferUsageBits_Storage)
    usageFlags |= VK_BUFFER_USAGE_STORAGE_BUFFER_BIT |
                  VK_BUFFER_USAGE_TRANSFER_DST_BIT |
                  VK_BUFFER_USAGE_SHADER_DEVICE_ADDRESS_BIT;
  if (desc.usage & BufferUsageBits_Indirect)
    usageFlags |= VK_BUFFER_USAGE_INDIRECT_BUFFER_BIT |
                  VK_BUFFER_USAGE_SHADER_DEVICE_ADDRESS_BIT;
```

4. Retrieve the required Vulkan memory properties using a helper function, and then call another overload of `VulkanContext::createBuffer()` that accepts only Vulkan flags. Delegating this step is useful because this function is also used inside the *LightweightVK* Vulkan backend to create internal auxiliary buffers. It checks buffer size limits and creates a `VulkanBuffer` object in the appropriate pool.

```
  const VkMemoryPropertyFlags memFlags =
    storageTypeToVkMemoryPropertyFlags(desc.storage);
  Result result;
  BufferHandle handle = createBuffer(
    desc.size, usageFlags, memFlags, &result, desc.debugName);
```

5. In case some initial buffer data was provided, upload it immediately.

```
    if (desc.data) {
      upload(handle, desc.data, desc.size, 0);
    }
    Result::setResult(outResult, Result());
    return {this, handle};
  }
```

Let's take a look at the declaration of `VulkanBuffer`, which encapsulates the Vulkan buffer management functionality.

1. All previously obtained Vulkan flags are stored in `VulkanBuffer` to be stored in *LightweightVK* object pools. We will discuss these pools in more detail in subsequent chapters. VMA-related fields are used only when the *VMA* library is enabled.

```
  struct VulkanBuffer final {
    VkBuffer vkBuffer_ = VK_NULL_HANDLE;
    VkDeviceMemory vkMemory_ = VK_NULL_HANDLE;
    VmaAllocation vmaAllocation_ = VK_NULL_HANDLE;
    VkDeviceAddress vkDeviceAddress_ = 0;
    VkDeviceSize bufferSize_ = 0;
    VkBufferUsageFlags vkUsageFlags_ = 0;
    VkMemoryPropertyFlags vkMemFlags_ = 0;
    void* mappedPtr_ = nullptr;
    bool isCoherentMemory_ = false;
```

2. A set of methods for transferring data to and from buffers. All host-visible buffers are automatically mapped, allowing us to access their data using normal C++ pointers.

```
    uint8_t* getMappedPtr() const { return (uint8_t*)mappedPtr_; }
    bool isMapped() const { return mappedPtr_ != nullptr;  }
    void bufferSubData(const VulkanContext& ctx,
      size_t offset, size_t size, const void* data);
    void getBufferSubData(const VulkanContext& ctx,
      size_t offset, size_t size, void* data);
```

3. The function `flushMappedMemory()` is required if the buffers on our system do not support coherent memory. This function is called to ensure that the data written into the mapped memory by the host becomes available to the GPU. It acts as a wrapper around `vkFlushMappedMemoryRanges()` and `vmaFlushAllocation()` when the VMA library is enabled. The function `invalidateMappedMemory()` is used for the opposite direction, from GPU to CPU, to ensure that data written by the GPU becomes available to the CPU. It serves as a wrapper around `vkInvalidateMappedMemoryRanges()` and `vmaInvalidateAllocation()`. We will use it in *Chapter 11*.

Note

Using the flushing and invalidation functions is mandatory for non-coherent memory, and their absence is a common mistake made by novice Vulkan developers. This error can often go unnoticed because the program may run correctly on discrete desktop GPUs (which often have coherent memory) but produce incorrect results on mobile GPUs.

```
  void flushMappedMemory(const VulkanContext& ctx,
    VkDeviceSize offset, VkDeviceSize size) const;
  void invalidateMappedMemory(const VulkanContext& ctx,
    VkDeviceSize offset, VkDeviceSize size) const;
};
```

Now we are ready to create the actual `VkBuffer` object. Let's take a look at the code. Error handling is omitted for clarity and better understanding.

1. Here's the corresponding overload of `VulkanContext::createBuffer()` that does this. It begins by populating the `VkBufferCreateInfo` structure.

```
BufferHandle lvk::VulkanContext::createBuffer(
  VkDeviceSize bufferSize,
  VkBufferUsageFlags usageFlags,
  VkMemoryPropertyFlags memFlags,
  lvk::Result* outResult,
  const char* debugName)
{
  VulkanBuffer buf = {
    .bufferSize_ = bufferSize,
    .vkUsageFlags_ = usageFlags,
    .vkMemFlags_ = memFlags,
  };
  const VkBufferCreateInfo ci = {
    .sType = VK_STRUCTURE_TYPE_BUFFER_CREATE_INFO,
    .flags = 0,
    .size = bufferSize,
    .usage = usageFlags,
    .sharingMode = VK_SHARING_MODE_EXCLUSIVE,
    .queueFamilyIndexCount = 0,
    .pQueueFamilyIndices = nullptr,
  };
```

2. Now, we decide whether to use Vulkan directly or let the **Vulkan Memory Allocator** (**VMA**) handle all memory allocation for us. VMA is the primary code path, while direct Vulkan calls are useful for debugging when necessary. If using VMA, we convert the flags once more.

```
  if (LVK_VULKAN_USE_VMA) {
    VmaAllocationCreateInfo vmaAllocInfo = {};
    if (memFlags & VK_MEMORY_PROPERTY_HOST_VISIBLE_BIT) {
      vmaAllocInfo = {
        .flags = VMA_ALLOCATION_CREATE_MAPPED_BIT |
                 VMA_ALLOCATION_CREATE_HOST_ACCESS_RANDOM_BIT,
        .requiredFlags = VK_MEMORY_PROPERTY_HOST_VISIBLE_BIT,
        .preferredFlags = VK_MEMORY_PROPERTY_HOST_COHERENT_BIT |
                          VK_MEMORY_PROPERTY_HOST_CACHED_BIT };
    }
    if (memFlags & VK_MEMORY_PROPERTY_HOST_VISIBLE_BIT) {
      vkCreateBuffer(vkDevice_, &ci, nullptr, &buf.vkBuffer_);
      VkMemoryRequirements requirements = {};
      vkGetBufferMemoryRequirements(
        vkDevice_, buf.vkBuffer_, &requirements);
      vkDestroyBuffer(vkDevice_, buf.vkBuffer_, nullptr);
      buf.vkBuffer_ = VK_NULL_HANDLE;
      if (requirements.memoryTypeBits &
          VK_MEMORY_PROPERTY_HOST_COHERENT_BIT) {
        vmaAllocInfo.requiredFlags |=
          VK_MEMORY_PROPERTY_HOST_COHERENT_BIT;
        buf.isCoherentMemory_ = true;
      }
    }
    vmaAllocInfo.usage = VMA_MEMORY_USAGE_AUTO;
    vmaCreateBuffer((VmaAllocator)getVmaAllocator(), &ci,
      &vmaAllocInfo, &buf.vkBuffer_,
      &buf.vmaAllocation_, nullptr);
```

3. Handle host-visible memory-mapped buffers by using persistent mapping for the entire lifetime of the buffer.

```
    if (memFlags & VK_MEMORY_PROPERTY_HOST_VISIBLE_BIT) {
      vmaMapMemory((VmaAllocator)getVmaAllocator(),
        buf.vmaAllocation_, &buf.mappedPtr_);
    }
  }
```

4. The direct Vulkan code path is straightforward but requires manual memory allocation. For full detailed error checking, refer to `lvk/vulkan/VulkanClasses.cpp`, as it is omitted here for clarity.

```
  else {
    vkCreateBuffer(vkDevice_, &ci, nullptr, &buf.vkBuffer_);
    VkMemoryRequirements requirements = {};
    vkGetBufferMemoryRequirements(
      vkDevice_, buf.vkBuffer_, &requirements);
      if (requirements.memoryTypeBits &
          VK_MEMORY_PROPERTY_HOST_COHERENT_BIT) {
        buf.isCoherentMemory_ = true;
      }
    lvk::allocateMemory(vkPhysicalDevice_, vkDevice_,
      &requirements, memFlags, &buf.vkMemory_);
    vkBindBufferMemory(
      vkDevice_, buf.vkBuffer_, buf.vkMemory_, 0);
```

5. Host-visible buffers are handled in a similar way.

```
    if (memFlags & VK_MEMORY_PROPERTY_HOST_VISIBLE_BIT) {
      vkMapMemory(vkDevice_, buf.vkMemory_, 0,
        buf.bufferSize_, 0, &buf.mappedPtr_);
    }
  }
```

6. After the `VkBuffer` object is created, we set a user-provided debug name for the buffer and obtain a buffer device address, which can be used in shaders to access the buffer.

```
  lvk::setDebugObjectName(vkDevice_, VK_OBJECT_TYPE_BUFFER,
    (uint64_t)buf.vkBuffer_, debugName);
  if (usageFlags & VK_BUFFER_USAGE_SHADER_DEVICE_ADDRESS_BIT) {
    const VkBufferDeviceAddressInfo ai = {
      .sType = VK_STRUCTURE_TYPE_BUFFER_DEVICE_ADDRESS_INFO,
      .buffer = buf.vkBuffer_,
    };
    buf.vkDeviceAddress_ =
      vkGetBufferDeviceAddress(vkDevice_, &ai);
  }
  return buffersPool_.create(std::move(buf));
}
```

The buffer destruction process is important to note because Vulkan buffers should not be deleted while they are still in use by the GPU. In addition to VMA and Vulkan calls for memory unmapping and deallocation, the destructor defers the actual deallocation until the buffer is no longer in use by the GPU.

1. The destruction process occurs in `VulkanContext::destroy(BufferHandle)`, which is invoked from the RAII wrapper `Holder<BufferHandle>`. The `deferredTask()` member function postpones the execution of its lambda argument until a later time, when all previously submitted command buffers have finished processing. We will explore this mechanism in subsequent chapters.

```
void lvk::VulkanContext::destroy(BufferHandle handle) {
  SCOPE_EXIT { buffersPool_.destroy(handle); };
  lvk::VulkanBuffer* buf = buffersPool_.get(handle);
  if (!buf) return;
  if (LVK_VULKAN_USE_VMA) {
    if (buf->mappedPtr_) {
      vmaUnmapMemory(
        (VmaAllocator)getVmaAllocator(), buf->vmaAllocation_);
    }
    deferredTask(std::packaged_task<void()>(
      [vma = getVmaAllocator(),
       buffer = buf->vkBuffer_,
       allocation = buf->vmaAllocation_]() {
      vmaDestroyBuffer((VmaAllocator)vma, buffer, allocation);
    }));
  }
```

2. A similar approach is used when we do not rely on VMA and instead interact with Vulkan directly.

```
  else {
    if (buf->mappedPtr_)
      vkUnmapMemory(vkDevice_, buf->vkMemory_);
    deferredTask(std::packaged_task<void()>(
      [device = vkDevice_,
       buffer = buf->vkBuffer_,
       memory = buf->vkMemory_]() {
      vkDestroyBuffer(device, buffer, nullptr);
      vkFreeMemory(device, memory, nullptr);
    }));
  }
}
```

Three other member functions to mention here before we conclude how we work with Vulkan buffers.

1. The function `flushMappedMemory()` is used to ensure that host writes to the buffer mapped memory become available to the GPU when the coherent memory is not supported.

```
void lvk::VulkanBuffer::flushMappedMemory(
  const VulkanContext& ctx,
  VkDeviceSize offset,
  VkDeviceSize size) const
{
  if (!LVK_VERIFY(isMapped())) return;
  if (LVK_VULKAN_USE_VMA) {
    vmaFlushAllocation((VmaAllocator)ctx.getVmaAllocator(),
      vmaAllocation_, offset, size);
  } else {
    const VkMappedMemoryRange range = {
      .sType = VK_STRUCTURE_TYPE_MAPPED_MEMORY_RANGE,
      .memory = vkMemory_,
      .offset = offset,
      .size = size,
    };
    vkFlushMappedMemoryRanges(ctx.getVkDevice(), 1, &range);
  }
}
```

2. The function `getBufferSubData()` wraps a `memcpy()` operation in a convenient way. It only works with memory-mapped host-visible buffers, while device-local buffers are handled separately by `VulkanContext` using a staging buffer. We will discuss this mechanism in subsequent chapters.

```
void lvk::VulkanBuffer::getBufferSubData(
  const VulkanContext& ctx,
  size_t offset,
  size_t size,
  void* data)
{
  if (!mappedPtr_) return;
  if (!isCoherentMemory_) {
    invalidateMappedMemory(ctx, offset, size);
  }
  memcpy(data, (const uint8_t*)mappedPtr_ + offset, size);
}
```

3. The function `bufferSubData()` is a similar wrapper. It is straightforward for host-visible buffers. Note how `memset()` is used here to set the content of the buffer to `0`.

```
void lvk::VulkanBuffer::bufferSubData(
  const VulkanContext& ctx,
  size_t offset,
  size_t size,
  const void* data)
{
  if (!mappedPtr_) return;
  if (data) {
    memcpy((uint8_t*)mappedPtr_ + offset, data, size);
  } else {
    memset((uint8_t*)mappedPtr_ + offset, 0, size);
  }
  if (!isCoherentMemory_) {
    flushMappedMemory(ctx, offset, size);
  }
}
```

Now that we've covered all the Vulkan code necessary to run the app in `Chapter03/01_Assimp`, which renders a `.obj` 3D model loaded via Assimp, there are two small functions to mention. These functions bind the vertex and index buffers, respectively, and are part of the `ICommandBuffer` interface.

1. The first function binds a vertex buffer via `vkCmdBindVertexBuffers()` to be used for vertex input. A few checks are necessary to ensure the correct usage of buffers. In subsequent chapters, we will learn how to omit vertex buffers altogether and explore the **Programmable-Vertex-Pulling** (**PVP**) approach.

```
void lvk::CommandBuffer::cmdBindVertexBuffer(
  uint32_t index, BufferHandle buffer, uint64_t bufferOffset)
{
  lvk::VulkanBuffer* buf = ctx_->buffersPool_.get(buffer);
  LVK_ASSERT(buf->vkUsageFlags_ & VK_BUFFER_USAGE_VERTEX_BUFFER_BIT);
  vkCmdBindVertexBuffers(
    wrapper_->cmdBuf_, index, 1, &buf->vkBuffer_, &bufferOffset);
}
```

2. The second function binds an index buffer using `vkCmdBindIndexBuffer()`. In addition to assertions, it includes some `enum` type conversions from *LightweightVK* to the Vulkan API.

```
void lvk::CommandBuffer::cmdBindIndexBuffer(
  BufferHandle indexBuffer,
  IndexFormat indexFormat, uint64_t indexBufferOffset)
{
```

```
  lvk::VulkanBuffer* buf = ctx_->buffersPool_.get(indexBuffer);
  LVK_ASSERT(buf->vkUsageFlags_ & VK_BUFFER_USAGE_INDEX_BUFFER_BIT);
  const VkIndexType type = indexFormatToVkIndexType(indexFormat);
  vkCmdBindIndexBuffer(
    wrapper_->cmdBuf_, buf->vkBuffer_, indexBufferOffset, type);
}
```

The application `Chapter03/01_Assimp` should render the following image.

Figure 3.2: Rendering a mesh loaded with Assimp

Now that we've covered some very basic Vulkan usage, we are ready to add textures to our examples.

Implementing staging buffers

Vulkan device-local buffers are often not visible from the host, meaning we can upload data into them only through various CPU-GPU-CPU transfer operations. In Vulkan, this is achieved by creating an auxiliary buffer, known as a staging buffer, which is CPU-visible (host-visible). We upload data into this staging buffer from the host and then issue GPU commands to copy the data from the staging buffer into a device-local buffer. Let's learn how to implement this technique in Vulkan.

Getting ready...

Before reading any further, check out the previous recipe *Dealing with buffers in Vulkan* to learn how to create different types of Vulkan buffers.

How to do it...

As usual, let's start with the high-level interface of *LightweightVK* and then dive deep into the implementation. The interface class `IContext`, declared in `lightweightvk/lvk/LVK.h`, exposes the following virtual methods to operate on buffers.

```
class IContext {
public:
  // ...
  virtual Result upload(BufferHandle handle,
    const void* data, size_t size, size_t offset = 0) = 0;
  virtual Result download(BufferHandle handle,
    void* data, size_t size, size_t offset) = 0;
  virtual uint8_t* getMappedPtr(BufferHandle handle) const = 0;
  virtual uint64_t gpuAddress(BufferHandle handle,
    size_t offset = 0) const = 0;
  virtual void flushMappedMemory(BufferHandle handle,
    size_t offset, size_t size) const = 0;
```

These methods are implemented in a subclass `VulkanContext`, and roughly correspond to the implementation of `VulkanBuffer`, which was discussed in detail in the previous recipe. We also saw how `VulkanContext::createBuffer()` calls `VulkanContext::upload()` if there's any initial data to be uploaded into a buffer. Let's take a look at what's inside that method.

1. First, we need to convert a buffer handle into a pointer to a `VulkanBuffer` object. This is done by a pool that stores all `VulkanBuffer` objects in the system. The pool implementation will be discussed later in the recipe *Storing Vulkan objects*. For now, let's treat it as an opaque mechanism that maps an integer handle to a `VulkanBuffer` object pointer.

```
lvk::Result lvk::VulkanContext::upload(
  lvk::BufferHandle handle,
  const void* data, size_t size, size_t offset)
{
  lvk::VulkanBuffer* buf = buffersPool_.get(handle);
```

2. After performing some initial range checking, we delegate the work to the member function `VulkanStagingDevice::bufferSubData()`.

```
  if (!LVK_VERIFY(offset + size <= buf->bufferSize_)) {
    return Result(Code::ArgumentOutOfRange, "Out of range");
  }
  stagingDevice_->bufferSubData(*buf, offset, size, data);
  return Result();
}
```

The class VulkanStagingDevice encapsulates all the functionality necessary to manage Vulkan staging buffers.

1. The staging device provides functionality to access device-local buffers and images. In this recipe, we will focus only on the buffers part and 2D images. While 3D image uploading is supported by *LightweightVK*, it is not used in our book, so we will skip that. If you're interested in those details, you're encouraged to check the actual source code of *LightweightVK* here: lvk/vulkan/VulkanClasses.cpp.

```
class VulkanStagingDevice final {
 public:
  explicit VulkanStagingDevice(VulkanContext& ctx);
  void bufferSubData(VulkanBuffer& buffer,
    size_t dstOffset, size_t size, const void* data);
  void imageData2D(VulkanImage& image,
                   const VkRect2D& imageRegion,
                   uint32_t baseMipLevel,
                   uint32_t numMipLevels,
                   uint32_t layer,
                   uint32_t numLayers,
                   VkFormat format,
                   const void* data);
  // … imageData3D() is skipped
```

2. Each call to bufferSubData() or imageData2D() occupies some space in the staging buffer. The structure MemoryRegionDesc describes such a memory region, including the SubmitHandle used to upload data through it.

```
 private:
  struct MemoryRegionDesc {
    uint32_t offset_ = 0;
    uint32_t size_ = 0;
    SubmitHandle handle_ = {};
  };
```

3. The function getNextFreeOffset() returns the next available memory region suitable for accommodating a specified number of bytes. The function ensureStagingBufferSize() grows the staging buffer size up to an internal limit, trying to accommodate as much of the uploading size as possible. The function waitAndReset() is used to wait until all memory regions become available. The default minimal buffer size minBufferSize_ is chosen to accommodate an RGBA 2K texture.

```
  MemoryRegionDesc getNextFreeOffset(uint32_t size);
  void ensureStagingBufferSize(uint32_t sizeNeeded);
  void waitAndReset();
```

```
private:
  VulkanContext& ctx_;
  lvk::Holder<BufferHandle> stagingBuffer_;
  uint32_t stagingBufferSize_ = 0;
  uint32_t stagingBufferCounter_ = 0;
  uint32_t maxBufferSize_ = 0;
  const uint32_t minBufferSize_ = 4u * 2048u * 2048u;
  std::deque<MemoryRegionDesc> regions_;
};
```

The uploading process is straightforward once we understand how the `getNextFreeOffset()` helper function works. Let's take a look.

1. Ensure the requested buffer size is aligned. Some compressed image formats require the size to be padded to `16` bytes, so we use that value greedily here. A simple binary arithmetic trick is used in `getAlignedSize()` to ensure the size value is aligned as required.

```
MemoryRegionDesc VulkanStagingDevice::getNextFreeOffset(
  uint32_t size)
{
  const uint32_t requestedAlignedSize = getAlignedSize(
    size, kStagingBufferAlignment);
  ensureStagingBufferSize(requestedAlignedSize);
```

2. We track the most suitable memory region and check if we can reuse any previously used memory regions. This may cause some memory fragmentation in the staging buffer, but it's not a concern since these suballocations have a very short lifespan.

```
  auto bestNextIt = regions_.begin();
  for (auto it = regions_.begin(); it != regions_.end(); ++it) {
    if (ctx_.immediate_->isReady(it->handle_)) {
      if (it->size_ >= requestedAlignedSize) {
```

3. Here, we know that this region is available, as `isReady()` returned true, and large enough to accommodate our entire upload size.

```
        const uint32_t unusedSize   = it->size_ - requestedAlignedSize;
        const uint32_t unusedOffset = it->offset_ + requestedAlignedSize;
```

4. Return this region and add the remaining unused size back to the `regions_` deque.

```
        SCOPE_EXIT {
          regions_.erase(it);
          if (unusedSize) {
            regions_.push_front(
              {unusedOffset, unusedSize, SubmitHandle()});
```

```
        }
      };
      return {
        it->offset_, requestedAlignedSize, SubmitHandle()};
    }
```

5. Cache the largest available region that isn't as big as the one we're looking for.

```
    if (it->size_ > bestNextIt->size_) {
      bestNextIt = it;
    }
  }
}
```

6. We found a region that is available but smaller than the requested size. For now, it's the best we can do. Return it, and the upload will proceed in chunks.

```
if (bestNextIt != regions_.end() &&
    ctx_.immediate_->isReady(bestNextIt->handle_)) {
  SCOPE_EXIT {
    regions_.erase(bestNextIt);
  };
  return {
    bestNextIt->offset_, bestNextIt->size_, SubmitHandle()};
};
```

7. At this point, we know that no part of the staging buffer is available, and we cannot grow its size any further due to platform limits enforced in `ensureStagingBufferSize()`. Let's give up and wait for the entire staging buffer to become free. The function `waitAndReset()` adds a region that spans the entire staging buffer. Since we'll be using part of it, we need to replace it with a used block and a new unused portion.

More information

Here's the code of `waitAndReset()` for your convenience:

```
void lvk::VulkanStagingDevice::waitAndReset() {
  for (const MemoryRegionDesc& r : regions_) {
    ctx_.immediate_->wait(r.handle_);
  };
  regions_.clear();
  regions_.push_front(
    {0, stagingBufferSize_, SubmitHandle()});
}
```

```
    waitAndReset();
    regions_.clear();
    const uint32_t unusedSize =
      stagingBufferSize_ > requestedAlignedSize ?
        stagingBufferSize_ - requestedAlignedSize : 0;
    if (unusedSize) {
      const uint32_t unusedOffset =
        stagingBufferSize_ - unusedSize;
      regions_.push_front({
        unusedOffset, unusedSize, SubmitHandle()});
    }
    return {
      .offset_ = 0,
      .size_ = stagingBufferSize_ - unusedSize,
      .handle_ = SubmitHandle() };
  }
```

Now we can implement the `VulkanStagingDevice::bufferSubData()` function. The main complexity here arises when the size of the data to upload is greater than the size of the staging buffer.

1. If the destination buffer is host-visible, we simply perform a memory copy operation into it, as discussed in the previous recipe *Dealing with buffers in Vulkan*.

```
void lvk::VulkanStagingDevice::bufferSubData(
  VulkanBuffer& buffer,
  size_t dstOffset,
  size_t size,
  const void* data)
{
  if (buffer.isMapped()) {
    buffer.bufferSubData(ctx_, dstOffset, size, data);
    return;
  }
  lvk::VulkanBuffer* stagingBuffer =
    ctx_.buffersPool_.get(stagingBuffer_);
```

2. We iterate while there's still data left to upload. In each iteration, we attempt to acquire an available staging buffer memory region large enough to fit the remaining size. Based on the available space, we adjust the next chunk size accordingly. The staging buffer itself is always host-visible, allowing us to directly mem-copy the data into it using `VulkanBuffer::bufferSubData()` as explained earlier.

```
  while (size) {
    MemoryRegionDesc desc = getNextFreeOffset((uint32_t)size);
```

```
const uint32_t chunkSize =
  std::min((uint32_t)size, desc.size_);
stagingBuffer->bufferSubData(
  desc.srcOffset_, chunkSize, data);
```

3. Acquire a command buffer and issue Vulkan commands to copy buffer data between the staging buffer and the destination buffer.

```
const VulkanImmediateCommands::CommandBufferWrapper& wrapper =
  ctx_.immediate_->acquire();
const VkBufferCopy copy = {
  .srcOffset = desc.offset_,
  .dstOffset = dstOffset,
  .size = chunkSize };
vkCmdCopyBuffer(wrapper.cmdBuf_, stagingBuffer->vkBuffer_,
  buffer.vkBuffer_, 1, &copy);
```

4. Before Vulkan can access data transferred by the command `vkCmdCopyBuffer()`, it requires a buffer memory barrier to ensure the transfer results become available. Here's how we deduce different pipeline destination stage flags and memory access masks for the barrier based on the buffer usage flags. This approach is greedy and overspecifies the barrier but simplifies the high-level API and removes much of the burden from us.

```
VkBufferMemoryBarrier barrier = {
  .sType = VK_STRUCTURE_TYPE_BUFFER_MEMORY_BARRIER,
  .srcAccessMask = VK_ACCESS_TRANSFER_WRITE_BIT,
  .dstAccessMask = 0,
  .srcQueueFamilyIndex = VK_QUEUE_FAMILY_IGNORED,
  .dstQueueFamilyIndex = VK_QUEUE_FAMILY_IGNORED,
  .buffer = buffer.vkBuffer_,
  .offset = dstOffset,
  .size = chunkSize,
};
VkPipelineStageFlags dstMask =
  VK_PIPELINE_STAGE_ALL_COMMANDS_BIT;
if (buffer.vkUsageFlags_ &
    VK_BUFFER_USAGE_INDIRECT_BUFFER_BIT) {
  dstMask |= VK_PIPELINE_STAGE_DRAW_INDIRECT_BIT;
  barrier.dstAccessMask |=
    VK_ACCESS_INDIRECT_COMMAND_READ_BIT;
}
if (buffer.vkUsageFlags_ & VK_BUFFER_USAGE_INDEX_BUFFER_BIT) {
  dstMask |= VK_PIPELINE_STAGE_VERTEX_INPUT_BIT;
  barrier.dstAccessMask |= VK_ACCESS_INDEX_READ_BIT;
```

```
    }
    if (buffer.vkUsageFlags_ &VK_BUFFER_USAGE_VERTEX_BUFFER_BIT) {
      dstMask |= VK_PIPELINE_STAGE_VERTEX_INPUT_BIT;
      barrier.dstAccessMask |= VK_ACCESS_VERTEX_ATTRIBUTE_READ_BIT;
    }
    vkCmdPipelineBarrier(wrapper.cmdBuf_,
      VK_PIPELINE_STAGE_TRANSFER_BIT, dstMask,
      VkDependencyFlags{}, 0, nullptr, 1, &barrier, 0, nullptr);
    desc.handle_ = ctx_.immediate_->submit(wrapper);
```

5. As the GPU is doing the copying, we add this memory region – together with its `SubmitHandle` – to the container of occupied memory regions.

```
    regions_.push_back(desc);
    size -= chunkSize;
    data = (uint8_t*)data + chunkSize;
    dstOffset += chunkSize;
  }
}
```

Another crucial role of the staging buffer is to copy pixel data into Vulkan images. Let's take a look at how it can be implemented. This function is significantly more complicated, so we omit all the error checking again here in the text for the sake of better understanding of the code.

1. The `imageData2D()` function can upload multiple layers of an image in one go, starting from a specified layer, along with multiple mip-levels beginning from the `baseMipLevel`. *Lightweight-VK* assumes there's a maximum possible number of mip-levels. We calculate the size in bytes for each mip-level. As we now the base mip-level that we want to update, we can calculate its dimensions from the Vulkan image extents by bit-shifting.

```
void lvk::VulkanStagingDevice::imageData2D(
  VulkanImage& image,
  const VkRect2D& imageRegion,
  uint32_t baseMipLevel,
  uint32_t numMipLevels,
  uint32_t layer,
  uint32_t numLayers,
  VkFormat format,
  const void* data)
{
  uint32_t width  = image.vkExtent_.width  >> baseMipLevel;
  uint32_t height = image.vkExtent_.height >> baseMipLevel;
  const Format texFormat(vkFormatToFormat(format));
```

2. Now let us calculate per-layer storage sizes, which are necessary to accommodate all corresponding mip-levels of the image. The function `getTextureBytesPerLayer()` returns the size in bytes of a layer with the requested image format.

```
uint32_t layerStorageSize = 0;
for (uint32_t i = 0; i < numMipLevels; ++i) {
  const uint32_t mipSize = lvk::getTextureBytesPerLayer(
    image.vkExtent_.width, image.vkExtent_.height,
    texFormat, i);
  layerStorageSize += mipSize;
  width  = width  <= 1 ? 1 : width  >> 1;
  height = height <= 1 ? 1 : height >> 1;
}
const uint32_t storageSize = layerStorageSize * numLayers;
```

3. Now we know the size necessary to store the entire image data. Try to acquire the next memory region from the staging buffer. *LightweightVK* provides no support for copying image data in multiple smaller chunks. If we get a memory region smaller than `storageSize`, we should wait until a bigger memory region becomes available. One consequence of this is that *LightweightVK* is unable to upload images whose memory footprint is larger than the staging buffer size.

```
ensureStagingBufferSize(storageSize);
MemoryRegionDesc desc = getNextFreeOffset(storageSize);
if (desc.size_ < storageSize) {
  waitAndReset();
  desc = getNextFreeOffset(storageSize);
}
LVK_ASSERT(desc.size_ >= storageSize);
```

4. Now we can acquire a command buffer and copy the image data into the selected memory region in the staging buffer.

```
const VulkanImmediateCommands::CommandBufferWrapper& wrapper =
  ctx_.immediate_->acquire();
lvk::VulkanBuffer* stagingBuffer =
  ctx_.buffersPool_.get(stagingBuffer_);
stagingBuffer->bufferSubData(
  ctx_, desc.offset_, storageSize, data);;
```

5. *LightweightVK* supports uploading of multiplanar images, such as YUV images used for video decoding. However, we do not use this functionality in our book and will skip that part here due to the added complexity. We assume the number of image planes to be `1`.

```
uint32_t offset = 0;
const uint32_t numPlanes = 1;
VkImageAspectFlags imageAspect = VK_IMAGE_ASPECT_COLOR_BIT;
```

6. Now we can iterate over the image's mip-levels and layers and transition the image layout to `VK_IMAGE_LAYOUT_TRANSFER_DST_OPTIMAL`, allowing us to use it as a destination in Vulkan transfer operations.

```
    for (uint32_t mipLevel = 0; mipLevel <numMipLevels; mipLevel++) {
      for (uint32_t layer = 0; layer != numLayers; layer++) {
        const uint32_t currentMipLevel = baseMipLevel + mipLevel;
        lvk::imageMemoryBarrier(
          wrapper.cmdBuf_,
          image.vkImage_,
          0,
          VK_ACCESS_TRANSFER_WRITE_BIT,
          VK_IMAGE_LAYOUT_UNDEFINED,
          VK_IMAGE_LAYOUT_TRANSFER_DST_OPTIMAL,
          VK_PIPELINE_STAGE_TOP_OF_PIPE_BIT,
          VK_PIPELINE_STAGE_TRANSFER_BIT,
          VkImageSubresourceRange{
            imageAspect, currentMipLevel, 1, layer, 1});
```

7. Copy the pixel data for this mip-level from the staging buffer into the corresponding Vulkan image subresource. The buffer offset for this mip-level is calculated by adding the size of all previous mip-levels being uploaded to the start of all mip-levels.

```
        const VkExtent2D extent = lvk::getImagePlaneExtent({
          .width  = std::max(1u, imageRegion.extent.width  >> mipLevel),
          .height = std::max(1u, imageRegion.extent.height >> mipLevel),
          },
          vkFormatToFormat(format), 0);
        const VkRect2D region = {
          .offset = {.x = imageRegion.offset.x >> mipLevel,
                     .y = imageRegion.offset.y >> mipLevel},
          .extent = extent,
        };
```

8. The `bufferOffset` value for this level is calculated by adding the size of all previous mip-levels being uploaded to the start of the mip-levels data.

```
        const VkBufferImageCopy copy = {
          .bufferOffset = desc.offset_ + offset,
          .bufferRowLength = 0,
          .bufferImageHeight = 0,
          .imageSubresource = VkImageSubresourceLayers{
            imageAspect, currentMipLevel, layer, 1},
          .imageOffset = { .x = region.offset.x,
```

```
                        .y = region.offset.y,
                        .z = 0},
        .imageExtent = { .width = region.extent.width,
                        .height = region.extent.height,
                        .depth = 1u },
    };
    vkCmdCopyBufferToImage(wrapper.cmdBuf_,
      stagingBuffer->vkBuffer_, image.vkImage_,
      VK_IMAGE_LAYOUT_TRANSFER_DST_OPTIMAL, 1, &copy);
```

9. Once the mip-level and layer are uploaded, we transition the subresource image layout from `VK_IMAGE_LAYOUT_TRANSFER_DST_OPTIMAL` to `VK_IMAGE_LAYOUT_SHADER_READ_ONLY_OPTIMAL`. This ensures the image is ready for use in the normal rendering pipeline, as any subsequent operations in *LightweightVK* expect this layout.

Note

The Vulkan image layout is a property of each image subresource that defines how the data in memory is organized in a manner that is opaque to the user and specific to the Vulkan drivers. Correctly specifying the layout for various use cases is crucial; failing to do so can lead to undefined behavior, such as distorted images, because Vulkan can use this information to optimize memory access when the image is used.

```
    lvk::imageMemoryBarrier(
      wrapper.cmdBuf_,
      image.vkImage_,
      VK_ACCESS_TRANSFER_WRITE_BIT,
      VK_ACCESS_SHADER_READ_BIT,
      VK_IMAGE_LAYOUT_TRANSFER_DST_OPTIMAL,
      VK_IMAGE_LAYOUT_SHADER_READ_ONLY_OPTIMAL,
      VK_PIPELINE_STAGE_TRANSFER_BIT,
      VK_PIPELINE_STAGE_ALL_COMMANDS_BIT,
      VkImageSubresourceRange{
        imageAspect, currentMipLevel, 1, layer, 1});
```

10. Advance the buffer offset to the next mip-level and layer.

```
    offset += lvk::getTextureBytesPerLayer(
      imageRegion.extent.width, imageRegion.extent.height,
      texFormat, currentMipLevel);
  }
}
```

11. Once the Vulkan commands are recorded in the command buffer, we submit it to copy the image data. Before exiting, we set the last image layout to `VK_IMAGE_LAYOUT_SHADER_READ_ONLY_OPTIMAL` to ensure that future layout transitions are based on this information. It's important to note that *LightweightVK* tracks the layout for the entire `VkImage`, not for its individual subresources. It is a tradeoff that simplifies handling and avoids the complexity of managing layouts for every individual subresource, but it can be a performance issue in some situations.

Note

This approach might lead to unnecessary image layout transitions or suboptimal memory usage if the image is accessed in ways that require different layouts for individual subresources. However, for many use cases, this simplification provides a good balance between ease of use and performance.

```
  image.vkImageLayout_ = VK_IMAGE_LAYOUT_SHADER_READ_ONLY_OPTIMAL;
  desc.handle_ = ctx_.immediate_->submit(wrapper);
  regions_.push_back(desc);
}
```

This concludes the discussion on the staging buffers implementation and the process of uploading device-local buffer data and images via them. This technique is a critical part of working with Vulkan, especially when the GPU's device-local memory is not directly accessible by the CPU.

There's more...

LightweightVK includes a function `VulkanStagingDevice::imageData3D()` to upload 3D texture data using a staging buffer. This function, found in `lvk/vulkan/VulkanClasses.cpp`, follows a similar process as the 2D texture upload. Make sure to explore it.

A Vulkan memory heap can have the flags `VK_MEMORY_PROPERTY_DEVICE_LOCAL_BIT | VK_MEMORY_PROPERTY_HOST_VISIBLE_BIT`. Some GPUs have a separate, relatively small memory heap with these properties, while others have the entire device memory marked as host-visible. This feature is known as **Resizable BAR (ReBAR)**, which allows the CPU to access GPU device memory. If your system has such a memory heap, you can use it to directly write data to GPU local memory. Or, for example, you could allocate a staging buffer in that memory. If you want to learn more about Vulkan memory types and how to use them, here's an excellent article by Adam Sawicki: https://asawicki.info/news_1740_vulkan_memory_types_on_pc_and_how_to_use_them.

Now that we have everything we need, let's move on to the next recipe and learn how to use Vulkan images to create textures.

Using texture data in Vulkan

Before we can create meaningful 3D rendering applications with Vulkan, we need to understand how to work with textures. This recipe shows how to implement several functions for creating, destroying, and modifying texture objects, as well as managing the corresponding `VkImage` and `VkImageView` objects using the Vulkan API.

Getting ready

Uploading texture data to the GPU requires a staging buffer. Be sure to read the previous recipe, *Implementing staging buffers*, before continuing.

The source code for this recipe can be found in `Chapter03/02_STB`.

How to do it...

A Vulkan image is a type of object backed by memory, designed to store 1D, 2D, or 3D images, or arrays of these images. Readers familiar with OpenGL might wonder about cube maps. Cube maps are represented as an array of six 2D images and can be created by setting the `VK_IMAGE_CREATE_CUBE_COMPATIBLE_BIT` flag in the `VkImageCreateInfo` structure. We'll revisit that later. For now, let's focus on the basic use case with a 2D image. Again, we'll begin with the high-level application code and work our way down to Vulkan image allocation.

1. The application in `Chapter03/02_STB` loads pixel data from a `.jpg` file using the STB library `https://github.com/nothings/stb`. We enforce a conversion to 4 channels to simplify texture handling, as many Vulkan implementations do not support 3-channel images.

```
int w, h, comp;
const uint8_t* img = stbi_load(
  "data/wood.jpg", &w, &h, &comp, 4);
```

2. A handle to a texture object is created. Internally, our texture consists of a combination of `VkImage` and `VkImageView`, which we'll examine shortly. The texture format is normalized unsigned 8-bit RGBA, corresponding to the Vulkan format `VK_FORMAT_R8G8B8A8_UNORM`. Since we intend to use this texture for sampling in shaders, we specify the texture usage flag `TextureUsageBits_Sampled`. Once the texture is created, do not forget to free the image memory.

```
Holder<TextureHandle> texture = ctx->createTexture({
  .type       = lvk::TextureType_2D,
  .format     = lvk::Format_RGBA_UN8,
  .dimensions = { (uint32_t)w, (uint32_t)h },
  .usage      = lvk::TextureUsageBits_Sampled,
  .data       = img,
  .debugName  = "03_STB.jpg",
});
stbi_image_free((void*)img);
```

3. Let's take a look at the main loop. *LightweightVK* is built around a bindless renderer design. Bindless rendering is a technique that enables more efficient GPU resource management by eliminating the need to explicitly bind resources like textures, buffers, or samplers. We'll explore how this works in the recipe *Using Vulkan descriptor sets* later in this chapter. For now, here's how we can pass texture data into shaders using push constants. After that, we render a quad formed by 4 triangle strip vertices.

```
while (!glfwWindowShouldClose(window)) {
  // ... skipped GLFW code and matrices setup
  const struct PerFrameData {
    mat4 mvp;
    uint32_t textureId;
  } pc = {
    .mvp       = p * m,
    .textureId = texture.index(),
  };
  lvk::ICommandBuffer& buf = ctx->acquireCommandBuffer();
  // ...
  buf.cmdBindRenderPipeline(pipeline);
  buf.cmdPushConstants(pc);
  buf.cmdDraw(4);
  // ...
}
```

4. The vertices are generated directly in the vertex shader, without any vertex input, in `Chapter03/02_STB/src/main.vert`, which looks as follows.

```
#version 460 core
layout(push_constant) uniform PerFrameData {
  uniform mat4 MVP;
  uint textureId;
};
layout (location=0) out vec2 uv;
const vec2 pos[4] = vec2[4](
  vec2( 1.0, -1.0), vec2( 1.0, 1.0),
  vec2(-1.0, -1.0), vec2(-1.0, 1.0)
);
void main() {
  gl_Position = MVP * vec4(0.5 * pos[gl_VertexIndex], 0.0, 1.0);
  uv = 0.5 * (pos[gl_VertexIndex]+vec2(0.5));
}
```

5. The fragment shader is much more interesting. We need to declare an array of 2D textures, `kTextures2D[]`, and an array of samplers, `kSamplers[]`, both provided by *LightweightVK*. These arrays contain all the textures and samplers loaded at the current moment. The element at index `0` in both arrays corresponds to a dummy object, which is useful for safely handling null values as texture identifiers without branching. Our push constant, `textureId`, is simply an index into the `kTextures2D[]` array.

```
#version 460 core
#extension GL_EXT_nonuniform_qualifier : require
layout (set = 0, binding = 0) uniform texture2D kTextures2D[];
layout (set = 0, binding = 1) uniform sampler kSamplers[];
layout (location=0) in vec2 uv;
layout (location=0) out vec4 out_FragColor;
layout(push_constant) uniform PerFrameData {
  uniform mat4 MVP;
  uint textureId;
};
```

6. Here's a handy helper function, `textureBindless2D()`, that allows us to sample from a bindless 2D texture using a bindless sampler. We'll use this function instead of the standard GLSL `texture()` to efficiently sample a texture.

Note

Here, we've provided the entire fragment shader GLSL source code ourselves. If we omit the `#version` directive at the beginning of the shader, *LightweightVK* will automatically inject this and many other helper functions into our GLSL source, along with `kTextures2D[]` and other declarations. We'll use this feature in later chapters to simplify our GLSL code and avoid code duplication in the shaders. We're listing this function here purely for educational purposes.

```
vec4 textureBindless2D(uint textureid, uint samplerid, vec2 uv) {
  return texture(nonuniformEXT(
   sampler2D(kTextures2D[textureid], kSamplers[samplerid])), uv);
}
void main() {
  out_FragColor = textureBindless2D(textureId, 0, uv);
}
```

Note

When our texture indices are not dynamically uniform, the `nonuniformEXT` type qualifier must be used when indexing descriptor bindings, as required by the Vulkan API. It's important to note that, to be fully correct, we must use `nonuniformEXT(sampler2D())`. The final argument in a function call like `texture()` determines whether the access is considered non-uniform.

The resulting application, `Chapter03/02_STB`, should render a textured, rotating quad, as shown in the following screenshot.

Figure 3.3: Rendering a textured quad

The high-level part was brief and straightforward, hiding all the Vulkan complexity from us. Now, let's take a deep breath and prepare for a deep dive into the underlying Vulkan API implementation to understand how it works.

How it works...

Vulkan textures, namely images and image views, are complex. Along with descriptor sets, they are essential for accessing texture data in shaders. The *LightweightVK* implementation of textures involves multiple layers. Let's peel them back and examine each one to understand how they work.

The tip of the iceberg is the function `VulkanContext::createTexture()`, which returns a handle to a texture. The function is quite lengthy, so we're omitting the error-checking code here to make it easier to understand.

1. This function converts a *LightweightVK* texture description, `TextureDesc`, into various Vulkan flags for images and image views. The additional argument, `debugName`, provides a convenient way to override the `TextureDesc::debugName` field. This is particularly useful when creating multiple textures using the same `TextureDesc` object.

```
struct TextureDesc {
  TextureType type = TextureType_2D;
  Format format = Format_Invalid;
  Dimensions dimensions = {1, 1, 1};
  uint32_t numLayers = 1;
  uint32_t numSamples = 1;
  uint8_t usage = TextureUsageBits_Sampled;
  uint32_t numMipLevels = 1;
  StorageType storage = StorageType_Device;
  ComponentMapping swizzle = {};
  const void* data = nullptr;
```

2. This field defines how many mip-levels we want to upload. It's useful when we only want to upload the first mip-level from the `data` member field and generate all the remaining mip-levels automatically. The `generateMipmaps` field forces the generation of all remaining mip-levels and works only when the data pointer is non-null.

```
  uint32_t dataNumMipLevels = 1;
  bool generateMipmaps = false;
  const char* debugName = "";
};
Holder<TextureHandle> VulkanContext::createTexture(
  const TextureDesc& requestedDesc,
  const char* debugName,
  Result* outResult)
{
  TextureDesc desc(requestedDesc);
  if (debugName && *debugName) desc.debugName = debugName;
```

3. Convert a `LightweightVK` image format into the Vulkan format. Vulkan provides stronger guarantees regarding the support of color formats, so depth formats are converted based on actual availability, while color formats are converted as-is.

```
const VkFormat vkFormat =
  lvk::isDepthOrStencilFormat(desc.format) ?
    getClosestDepthStencilFormat(desc.format) :
    formatToVkFormat(desc.format);
const lvk::TextureType type = desc.type;
```

4. If the image is going to be allocated in the GPU device memory, we should set the `VK_IMAGE_USAGE_TRANSFER_DST_BIT` to allow Vulkan to transfer data into it. Other Vulkan image usage flags are set according to the *LightweightVK* usage flags.

```
VkImageUsageFlags usageFlags =
  (desc.storage == StorageType_Device) ?
    VK_IMAGE_USAGE_TRANSFER_DST_BIT : 0;
if (desc.usage & lvk::TextureUsageBits_Sampled) {
  usageFlags |= VK_IMAGE_USAGE_SAMPLED_BIT;
}
if (desc.usage & lvk::TextureUsageBits_Storage) {
  usageFlags |= VK_IMAGE_USAGE_STORAGE_BIT;
}
if (desc.usage & lvk::TextureUsageBits_Attachment) {
  usageFlags |= lvk::isDepthOrStencilFormat(desc.format) ?
    VK_IMAGE_USAGE_DEPTH_STENCIL_ATTACHMENT_BIT :
    VK_IMAGE_USAGE_COLOR_ATTACHMENT_BIT;
```

5. Memoryless images correspond to Vulkan transient attachments and lazily allocated memory, meaning transient attachments used during a single render pass do not require physical storage. This is particularly useful for MSAA multisampling, and we will explore it further in *Chapter 10*. For convenience, we always allow an image to be read back from the GPU to the CPU by setting `VK_IMAGE_USAGE_TRANSFER_SRC_BIT` for everything except transient attachments. However, it's worth verifying if this image usage flag is supported for the specific image format. The memory property flags are chosen in the same way as described earlier in the recipe *Dealing with buffers in Vulkan*.

```
  if (desc.storage == lvk::StorageType_Memoryless) {
    usageFlags |= VK_IMAGE_USAGE_TRANSIENT_ATTACHMENT_BIT;
  }
}
if (desc.storage != lvk::StorageType_Memoryless) {
  usageFlags |= VK_IMAGE_USAGE_TRANSFER_SRC_BIT;
};
const VkMemoryPropertyFlags memFlags =
  storageTypeToVkMemoryPropertyFlags(desc.storage);
```

6. Generate debug names for Vulkan image and image view objects by prefixing the provided debugName string. This is crucial when using debugging tools like *RenderDoc* or *Vulkan Validation Layers*.

```
const bool hasDebugName = desc.debugName && *desc.debugName;
char debugNameImage[256] = {0};
char debugNameImageView[256] = {0};
if (hasDebugName) {
  snprintf(debugNameImage, sizeof(debugNameImage) - 1,
    "Image: %s", desc.debugName);
  snprintf(debugNameImageView, sizeof(debugNameImageView) - 1,
    "Image View: %s", desc.debugName);
}
```

7. Now we can proceed to determine the VkImageCreateFlags and the types of the Vulkan image and image view. Note that 2D images can be multisampled https://en.wikipedia.org/wiki/Multisample_anti-aliasing. We will use multisampling in *Chapter 10*.

```
VkImageCreateFlags vkCreateFlags = 0;
VkImageViewType vkImageViewType;
VkImageType vkImageType;
VkSampleCountFlagBits vkSamples = VK_SAMPLE_COUNT_1_BIT;
uint32_t numLayers = desc.numLayers;
switch (desc.type) {
  case TextureType_2D:
    vkImageViewType = numLayers > 1 ?
      VK_IMAGE_VIEW_TYPE_2D_ARRAY :
      VK_IMAGE_VIEW_TYPE_2D;
    vkImageType = VK_IMAGE_TYPE_2D;
    vkSamples = lvk::getVulkanSampleCountFlags(
      desc.numSamples, getFramebufferMSAABitMask());
  break;
  case TextureType_3D:
    vkImageViewType = VK_IMAGE_VIEW_TYPE_3D;
    vkImageType = VK_IMAGE_TYPE_3D;
    break;
```

8. In Vulkan, cube textures are represented by an image view of type VK_IMAGE_VIEW_TYPE_CUBE and a 2D image with the VK_IMAGE_CREATE_CUBE_COMPATIBLE_BIT flag. The number of layers is multiplied by 6 to account for all cube map faces, as required by the Vulkan API specification.

```
  case TextureType_Cube:
    vkImageViewType = numLayers > 1 ?
      VK_IMAGE_VIEW_TYPE_CUBE_ARRAY : VK_IMAGE_VIEW_TYPE_CUBE;
```

```
        vkImageType = VK_IMAGE_TYPE_2D;
        vkCreateFlags = VK_IMAGE_CREATE_CUBE_COMPATIBLE_BIT;
        numLayers *= 6;
        break;
    }
```

9. Now we can create a wrapper object, `VulkanImage`, which encapsulates all the necessary `VkImage`-related properties. We'll examine its remaining members in a moment.

```
const VkExtent3D vkExtent{
  desc.dimensions.width,
  desc.dimensions.height,
  desc.dimensions.depth };
const uint32_t numLevels = desc.numMipLevels;
lvk::VulkanImage image = {
  .vkUsageFlags_ = usageFlags,
  .vkExtent_ = vkExtent,
  .vkType_ = vkImageType,
  .vkImageFormat_ = vkFormat,
  .vkSamples_ = vkSamples,
  .numLevels_ = numLevels,
  .numLayers_ = numLayers,
  .isDepthFormat_ = VulkanImage::isDepthFormat(vkFormat),
  .isStencilFormat_ = VulkanImage::isStencilFormat(vkFormat),
};
```

More information

Here's the declaration of the `VulkanImage` structure for your convenience, with some minor parts omitted. Setting `numLevels` to a non-zero value will override the number of mip-levels from the original Vulkan image and can be used to create image views with a different number of levels.

```
struct VulkanImage final {
  VkImageView createImageView(
    VkDevice device,
    VkImageViewType type,
    VkFormat format,
    VkImageAspectFlags aspectMask,
    uint32_t baseLevel,
    uint32_t numLevels = VK_REMAINING_MIP_LEVELS,
    uint32_t baseLayer = 0,
```

```
    uint32_t numLayers = 1,
    const VkComponentMapping mapping = {
      .r = VK_COMPONENT_SWIZZLE_IDENTITY,
      .g = VK_COMPONENT_SWIZZLE_IDENTITY,
      .b = VK_COMPONENT_SWIZZLE_IDENTITY,
      .a = VK_COMPONENT_SWIZZLE_IDENTITY },
    const VkSamplerYcbcrConversionInfo* ycbcr = nullptr,
    const char* debugName = nullptr) const;
  VkImageView getOrCreateVkImageViewForFramebuffer(
    VulkanContext& ctx, uint8_t level, uint16_t layer);
  // ...
  VkImage vkImage_ = VK_NULL_HANDLE;
  VkImageUsageFlags vkUsageFlags_ = 0;
  VkDeviceMemory vkMemory_[1] = {VK_NULL_HANDLE };
  VmaAllocation vmaAllocation_ = VK_NULL_HANDLE;
  VkFormatProperties vkFormatProperties_ = {};
  VkExtent3D vkExtent_ = {0, 0, 0};
  VkImageType vkType_ = VK_IMAGE_TYPE_MAX_ENUM;
  VkFormat vkImageFormat_ = VK_FORMAT_UNDEFINED;
  VkSampleCountFlagBits vkSamples_ = VK_SAMPLE_COUNT_1_BIT;
  void* mappedPtr_ = nullptr;
  bool isSwapchainImage_ = false;
  bool isOwningVkImage_ = true;
  uint32_t numLevels_ = 1u;
  uint32_t numLayers_ = 1u;
  bool isDepthFormat_ = false;
  bool isStencilFormat_ = false;
  // current image layout
  mutable VkImageLayout vkImageLayout_ =
    VK_IMAGE_LAYOUT_UNDEFINED;
  // precached image views - owned by this VulkanImage
  VkImageView imageView_ = VK_NULL_HANDLE; // all levels
  VkImageView imageViewStorage_ =
    VK_NULL_HANDLE; // identity swizzle
  VkImageView
    imageViewForFramebuffer_[LVK_MAX_MIP_LEVELS][6] = {};
};
```

As you can see, the `VulkanImage` structure is simply a data container. One interesting part is the `getOrCreateVkImageViewForFramebuffer()` member function. Image views used as framebuffer attachments should have only `1` mip-level and `1` layer. This function precaches such image views inside the array `imageViewForFramebuffer_[][]`. It supports a maximum of `6` layers, which is just enough for rendering to the faces of a cube map.

10. To create a `VkImage` object, we need to fill out the `VkImageCreateInfo` structure and call `vmaCreateImage()` or use the raw Vulkan API directly. The image creation code in *LightweightVK* includes a code path for handling multiplanar images, but we will omit that part in this book. *LightweightVK* does not work with multiple Vulkan queues so it sets the sharing mode to `VK_SHARING_MODE_EXCLUSIVE`.

```
const VkImageCreateInfo ci = {
  .sType = VK_STRUCTURE_TYPE_IMAGE_CREATE_INFO,
  .flags = vkCreateFlags,
  .imageType = vkImageType,
  .format = vkFormat,
  .extent = vkExtent,
  .mipLevels = numLevels,
  .arrayLayers = numLayers,
  .samples = vkSamples,
  .tiling = VK_IMAGE_TILING_OPTIMAL,
  .usage = usageFlags,
  .sharingMode = VK_SHARING_MODE_EXCLUSIVE,
  .queueFamilyIndexCount = 0,
  .pQueueFamilyIndices = nullptr,
  .initialLayout = VK_IMAGE_LAYOUT_UNDEFINED,
};
```

11. Similar to how we handled buffers in the recipe *Dealing with buffers in Vulkan*, we have two code paths for Vulkan images. One uses the Vulkan Memory Allocator library, while the other calls Vulkan directly to allocate memory. This provides a useful option for debugging purposes. We can have memory-mapped images in the same way we have memory-mapped buffers. However, this is only useful for non-tiled image layouts.

```
if (LVK_VULKAN_USE_VMA) {
  VmaAllocationCreateInfo vmaAllocInfo = {
    .usage = memFlags & VK_MEMORY_PROPERTY_HOST_VISIBLE_BIT ?
       VMA_MEMORY_USAGE_CPU_TO_GPU : VMA_MEMORY_USAGE_AUTO,
  };
  vmaCreateImage((VmaAllocator)getVmaAllocator(), &ci,
    &vmaAllocInfo, &image.vkImage_,
```

```
      &image.vmaAllocation_, nullptr);
    if (memFlags & VK_MEMORY_PROPERTY_HOST_VISIBLE_BIT) {
      vmaMapMemory((VmaAllocator)getVmaAllocator(),
        image.vmaAllocation_, &image.mappedPtr_);
    }
  }
```

12. When creating a VkImage object using the raw Vulkan API, we need to manually allocate memory for it. Here, we only use the code path for a single image plane. To learn how to extend this for multiplanar images, refer to the implementation in lvk/vulkan/VulkanClasses.cpp.

```
  else {
    const uint32_t numPlanes = 1; // simplified code path
    vkCreateImage(vkDevice_, &ci, nullptr, &image.vkImage_);
    constexpr uint32_t kNumMaxImagePlanes = 1;
    VkMemoryRequirements2 memRequirements[kNumMaxImagePlanes] = {
      { .sType = VK_STRUCTURE_TYPE_MEMORY_REQUIREMENTS_2 },
    };
    const VkImagePlaneMemoryRequirementsInfo
      planes[kNumMaxImagePlanes] = {
      { .sType = VK_STRUCTURE_TYPE_IMAGE_PLANE_MEMORY_REQUIREMENTS_INFO,
        .planeAspect = VK_IMAGE_ASPECT_PLANE_0_BIT },
    };
    const VkImage img = image.vkImage_;
    const VkImageMemoryRequirementsInfo2
      imgRequirements[kNumMaxImagePlanes] = {
      { .sType = VK_STRUCTURE_TYPE_IMAGE_MEMORY_REQUIREMENTS_INFO_2,
        .pNext = numPlanes > 0 ? &planes[0] : nullptr,
        .image = img },
    };
    for (uint32_t p = 0; p != numPlanes; p++) {
      vkGetImageMemoryRequirements2(vkDevice_,
        &imgRequirements[p], &memRequirements[p]);
      lvk::allocateMemory2(vkPhysicalDevice_, vkDevice_,
        &memRequirements[p], memFlags, &image.vkMemory_[p]));
    }
    const VkBindImagePlaneMemoryInfo
      bindImagePlaneMemoryInfo[kNumMaxImagePlanes] = {
      { .sType = VK_STRUCTURE_TYPE_BIND_IMAGE_PLANE_MEMORY_INFO,
        .planeAspect = VK_IMAGE_ASPECT_PLANE_0_BIT },
    };
    const VkBindImageMemoryInfo bindInfo[kNumMaxImagePlanes] = {
```

```
    lvk::getBindImageMemoryInfo(
      nullptr, img, image.vkMemory_[0]),
  };
```

13. Note that memory can only be bound to a `VkImage` once. Therefore, the code above handles binding disjoint memory locations in a single call to `vkBindMemory2()` by chaining everything together using `pNext` pointers. For details on handling multiplanar images, refer to the implementation in `lvk/vulkan/VulkanClasses.cpp`.

```
  vkBindImageMemory2(vkDevice_, numPlanes, bindInfo);
  if (memFlags & VK_MEMORY_PROPERTY_HOST_VISIBLE_BIT &&
      numPlanes == 1) {
    vkMapMemory(vkDevice_, image.vkMemory_[0], 0,
      VK_WHOLE_SIZE, 0, &image.mappedPtr_);
  }
}
```

14. Once the `VkImage` object is created, we can set its debug name and store its format properties in our `VulkanImage` structure.

```
lvk::setDebugObjectName(vkDevice_, VK_OBJECT_TYPE_IMAGE,
  (uint64_t)image.vkImage_, debugNameImage);
vkGetPhysicalDeviceFormatProperties(vkPhysicalDevice_,
  image.vkImageFormat_, &image.vkFormatProperties_);
```

15. To access a Vulkan image from shaders, we need to create a `VkImageView` object. In doing so, we must decide which image "aspect" should be included in the view. Vulkan images can have multiple aspects simultaneously, such as combined depth-stencil images, where depth and stencil bits are handled separately. Additionally, Vulkan supports image view component swizzling, which we'll explore in *Chapter 10*. For now, here's the code to initialize it.

```
VkImageAspectFlags aspect = 0;
if (image.isDepthFormat_ || image.isStencilFormat_) {
  if (image.isDepthFormat_) {
    aspect |= VK_IMAGE_ASPECT_DEPTH_BIT;
  } else if (image.isStencilFormat_) {
    aspect |= VK_IMAGE_ASPECT_STENCIL_BIT;
  }
} else {
  aspect = VK_IMAGE_ASPECT_COLOR_BIT;
}
const VkComponentMapping mapping = {
  .r = VkComponentSwizzle(desc.swizzle.r),
  .g = VkComponentSwizzle(desc.swizzle.g),
  .b = VkComponentSwizzle(desc.swizzle.b),
```

```
    .a = VkComponentSwizzle(desc.swizzle.a),
  };
```

16. An image view can control which mip-levels and layers are included. Here, we create an image view that includes all levels and layers of the image. Later, we'll need separate image views for framebuffer attachments, which will contain only one layer and one mip-level. Storage images do not support swizzling, so we need to create a separate image view for them.

```
  image.imageView_ = image.createImageView(
    vkDevice_, vkImageViewType, vkFormat, aspect, 0,
    VK_REMAINING_MIP_LEVELS, 0, numLayers,
    mapping, nullptr, debugNameImageView);
  if (image.vkUsageFlags_ & VK_IMAGE_USAGE_STORAGE_BIT) {
    if (!desc.swizzle.identity()) {
      image.imageViewStorage_ = image.createImageView(
        vkDevice_, vkImageViewType, vkFormat, aspect, 0,
        VK_REMAINING_MIP_LEVELS, 0, numLayers,
        {}, nullptr, debugNameImageView);
    }
  }
```

17. *LightweightVK* refers to a pair of objects—`VkImage`, wrapped in the `VulkanImage` struct, and `VkImageView`—as a texture. The Boolean flag `awaitingCreation_` signals to `VulkanContext` that a texture has been created and that the bindless descriptor set needs to be updated. We'll explore this further in the recipe *Using Vulkan descriptor indexing*. Before returning the texture handle of the newly created texture, let's upload the initial texture data.

```
  TextureHandle handle = texturesPool_.create(std::move(image));
  awaitingCreation_ = true;
  if (desc.data) {
    const uint32_t numLayers = desc.type==TextureType_Cube ? 6:1;
    upload(handle, {.dimensions   = desc.dimensions,
                    .numLayers    = numLayers,
                    .numMipLevels = desc.dataNumMipLevels},
      desc.data);
    if (desc.generateMipmaps) this->generateMipmap(handle);
  }
  return {this, handle};
}
```

18. Now that we've created a texture, one major step we omitted is the actual process of creating `VkImageView` objects. Here's the function `VulkanImage::createImageView()` that handles this. This function does not use any wrappers and creates a `VkImageView` object directly. It supports working with Vulkan YCbCr samplers, but this functionality is not used in the book samples, so we'll omit it here (together with the error checking code).

```
VkImageView lvk::VulkanImage::createImageView(
  VkDevice device,
  VkImageViewType type,
  VkFormat format,
  VkImageAspectFlags aspectMask,
  uint32_t baseLevel,
  uint32_t numLevels,
  uint32_t baseLayer,
  uint32_t numLayers,
  const VkComponentMapping mapping,
  const VkSamplerYcbcrConversionInfo* ycbcr,
  const char* debugName) const
 {
  const VkImageViewCreateInfo ci = {
    .sType = VK_STRUCTURE_TYPE_IMAGE_VIEW_CREATE_INFO,
    .pNext = ycbcr,
    .image = vkImage_,
    .viewType = type,
    .format = format,
    .components = mapping,
    .subresourceRange = { aspectMask, baseLevel,
      numLevels ? numLevels : numLevels_, baseLayer, numLayers},
  };
  VkImageView vkView = VK_NULL_HANDLE;
  vkCreateImageView(device, &ci, nullptr, &vkView);
  lvk::setDebugObjectName(device, VK_OBJECT_TYPE_IMAGE_VIEW,
    (uint64_t)vkView, debugName);
  return vkView;
}
```

19. Another important aspect is the destruction of `VulkanImage` objects we created. The `VulkanContext::destroy()` function handles this and is called from the RAII wrapper `Holder<TextureHandle>`. Here it is for completeness, with multiplanar image support and error checking removed.

```
void lvk::VulkanContext::destroy(lvk::TextureHandle handle) {
  SCOPE_EXIT {
```

```
    texturesPool_.destroy(handle);
    awaitingCreation_ = true;
  };
  lvk::VulkanImage* tex = texturesPool_.get(handle);
  if (!tex) return;
  deferredTask(std::packaged_task<void()>(
    [device = getVkDevice(),
     imageView = tex->imageView_]() {
      vkDestroyImageView(device, imageView, nullptr); }));
  if (tex->imageViewStorage_) {
    deferredTask(std::packaged_task<void()>(
      [device = getVkDevice(),
       imageView = tex->imageViewStorage_]() {
        vkDestroyImageView(device, imageView, nullptr); }));
  }
  for (size_t i = 0; i != LVK_MAX_MIP_LEVELS; i++) {
    for (size_t j = 0; j != LVK_ARRAY_NUM_ELEMENTS(
         tex->imageViewForFramebuffer_[0]); j++)
    {
      VkImageView v = tex->imageViewForFramebuffer_[i][j];
      if (v) {
        deferredTask(std::packaged_task<void()>(
          [device = getVkDevice(), imageView = v]() {
            vkDestroyImageView(device, imageView, nullptr); }));
      }
    }
  }
  if (!tex->isOwningVkImage_) return;
  if (LVK_VULKAN_USE_VMA) {
    if (tex->mappedPtr_) vmaUnmapMemory(
      (VmaAllocator)getVmaAllocator(), tex->vmaAllocation_);
    deferredTask(std::packaged_task<void()>(
      [vma = getVmaAllocator(),
       image = tex->vkImage_,
       allocation = tex->vmaAllocation_]() {
      vmaDestroyImage((VmaAllocator)vma, image, allocation);
    }));
  } else {
    if (tex->mappedPtr_)
      vkUnmapMemory(vkDevice_, tex->vkMemory_[0]);
    deferredTask(std::packaged_task<void()>(
```

```
      [device = vkDevice_,
       image = tex->vkImage_,
       memory0 = tex->vkMemory_[0]]() {
        vkDestroyImage(device, image, nullptr);
        if (memory0) vkFreeMemory(device, memory0, nullptr);
      }
    }));
  }
}
```

While technically all of the above code is sufficient to create `VkImage` and `VkImageView` objects, we still cannot access them from shaders. To do that, we need to learn how to store these objects and how to put them into Vulkan descriptor sets. Let's move on to the next recipes to explore this.

There's more...

You may have noticed that the Boolean flag `VulkanImage::isOwningVkImage_` prevents the `VulkanContext::destroy(TextureHandle)` function from destroying the `VkImage` and `VkDeviceMemory` objects. This allows us to create as many non-owning copies of a `VulkanImage` object as we need, which is very useful when we want to create multiple custom image views of the same `VkImage`. Check the method `VulkanContext::createTextureView()`, which is also a part of the `IContext` interface, to learn how to do this. We will use this functionality in *Chapters 10* and *11*.

Storing Vulkan objects

In the previous recipes, we mentioned a plethora of `lvk::…Handle` classes wrapped into a templated unique-pointer-like class `Holder<>`. These are central to how *LightweightVK* manages Vulkan objects and other resources. Handles are lightweight value types that are inexpensive to pass around as integers, and we don't incur the overhead of shared ownership with atomic counters, as we would with `std::shared_ptr` or other reference-counted smart pointers. When ownership of an object is desirable, we wrap handles into the `Holder<>` class, which is a RAII wrapper and conceptually resembles `std::unique_ptr`.

Getting ready

The *LightweightVK* implementation of handles is inspired by Sebastian Aaltonen`s SIGGRAPH 2023 presentation *HypeHype Mobile Rendering Architecture*. If you want to learn more low-level interesting details about an API design using handles, make sure to read it: `https://advances.realtimerendering.com/s2023/AaltonenHypeHypeAdvances2023.pdf`

How to do it...

Let's learn how to implement lightweight handles, which are represented by a templated class `Handle<>`.

1. Handles are designed to serve akin pointers to objects which are stored in arrays. An index into an array is sufficient to identify an object. To handle situations where objects are deallocated and then replaced by new ones, we introduce a value gen_ which represents object`s "generation" and is incremented each time a new object is assign to the same element in the storage array.

```
template<typename ObjectType> class Handle final {
  uint32_t index_ = 0;
  uint32_t gen_ = 0;
```

2. These values are private so that handles can be constructed only by a friendly `Pool<>` class. The `Pool<>` class is templated and is parametrized by two types, one corresponds to the handle`s object type and the other is the type stored inside the implementation array. It is not visible from the `Handle<>` interface and cannot be used without a `Pool<>` object.

```
  Handle(uint32_t index, uint32_t gen) :
    index_(index), gen_(gen){};
  template<typename ObjectType,
           typename ImplObjectType> friend class Pool;
 public:
  Handle() = default;
```

3. The interface contract is that handles with a generation equal to zero are considered empty null-handles.

```
  bool empty() const { return gen_ == 0; }
  bool valid() const { return gen_ != 0; }
  uint32_t index() const { return index_; }
  uint32_t gen() const { return gen_; }
```

4. The `indexAsVoid()` function is helpful when we need to pass a handle through some third-party C-style interface which accepts `void*` parameters. One example used in this book would be *ImGui* integration which is discussed in the next chapter *Adding User Interaction and Productivity Tools*.

```
  void* indexAsVoid() const {
    return reinterpret_cast<void*>(
      static_cast<ptrdiff_t>(index_));
  }
  bool operator==(const Handle<ObjectType>& other) const
  { return index_ == other.index_ && gen_ == other.gen_;  }
  bool operator!=(const Handle<ObjectType>& other) const
  { return index_ != other.index_ || gen_ != other.gen_; }
```

5. The explicit conversion to bool is necessary to allow usage of handles in conditional statements such as `if (handle) {…}`. We also ensure the size of `Handle<>` is 8 bytes.

```
    explicit operator bool() const { return gen_ != 0; }
  };
  static_assert(sizeof(Handle<class Foo>) == sizeof(uint64_t));
```

6. The `Handle<>` template can be parameterized with a forward-declared type that does not have an actual complete definition. Such types are known as "tags." This ensures type safety, preventing heterogeneous handles from being mixed with each other. Here's how *LightweightVK* declares all its `Handle<>` types. The structs `ComputePipeline`, `RenderPipeline`, and others do not exist.

```
using ComputePipelineHandle = Handle<struct ComputePipeline>;
using RenderPipelineHandle = Handle<struct RenderPipeline>;
using ShaderModuleHandle = Handle<struct ShaderModule>;
using SamplerHandle = Handle<struct Sampler>;
using BufferHandle = Handle<struct Buffer>;
using TextureHandle = Handle<struct Texture>;
using QueryPoolHandle = Handle<struct QueryPool>;
```

Handles do not own objects they point to. Only the `Holder<>` class does. Let's take a look at its implementation.

1. The `Holder<>` class is templated with a `Handle` type it can hold. The constructor accepts a handle and a pointer to `lvk::IContext` to ensure the handle can be destroyed properly. The class has move-only semantics similar to `std::unique_ptr`. We skip definitions of the move-constructor and move-assignment for the sake of brevity.

```
template<typename HandleType> class Holder final {
 public:
  Holder() = default;
  Holder(lvk::IContext* ctx, HandleType handle)
    : ctx_(ctx), handle_(handle) {}
```

2. We do not have a declaration of the `IContext` class here. That is why we use an overloaded forward-declared function `lvk::destroy()` to deallocate the handle. There should be one `lvk::destroy()` overload for each type we want to instantiate the `Holder<>` class with.

```
  ~Holder() { lvk::destroy(ctx_, handle_); }
  Holder(const Holder&) = delete;
  Holder(Holder&& other)
  : ctx_(other.ctx_)
  , handle_(other.handle_) { ... }
  Holder& operator=(const Holder&) = delete;
  Holder& operator=(Holder&& other) { … }
```

3. Assigning `nullptr` to this `Holder<>`.

```
Holder& operator=(std::nullptr_t) { reset(); return *this; }
inline operator HandleType() const { return handle_; }
bool valid() const { return handle_.valid(); }
bool empty() const { return handle_.empty(); }
```

4. Manually reset this `Holder<>` and make it deallocate the stored handle or just return the handle and release ownership when necessary.

```
  void reset() {
    lvk::destroy(ctx_, handle_);
    ctx_ = nullptr;
    handle_ = HandleType{};
  }
  HandleType release() {
    ctx_ = nullptr;
    return std::exchange(handle_, HandleType{});
  }
  uint32_t index() const { return handle_.index(); }
  void* indexAsVoid() const { return handle_.indexAsVoid(); }
 private:
  lvk::IContext* ctx_ = nullptr;
  HandleType handle_;
};
```

5. The `Holder` class calls a family of overloaded `destroy()` functions. Here is how *LightweightVK* defines them. One function per each handle type.

```
void destroy(lvk::IContext* ctx, ComputePipelineHandle handle);
void destroy(lvk::IContext* ctx, RenderPipelineHandle handle);
void destroy(lvk::IContext* ctx, ShaderModuleHandle handle);
void destroy(lvk::IContext* ctx, SamplerHandle handle);
void destroy(lvk::IContext* ctx, BufferHandle handle);
void destroy(lvk::IContext* ctx, TextureHandle handle);
void destroy(lvk::IContext* ctx, QueryPoolHandle handle);
```

6. Implementations of these functions are located in `lightweightvk/lvk/LVK.cpp` and they all look very similar. Each function calls a corresponding overloaded method in `IContext`. While it may seem unnecessary, this actually helps to avoid a circular dependency between the `Holder<>` class and `IContext`, making the interface much cleaner.

```
void destroy(lvk::IContext* ctx, ComputePipelineHandle handle) {
  if (ctx) ctx->destroy(handle);
}
// ... same for other handle types
```

That is all there is to discuss about the Holder<> class and the portion of the Handle-Holder mechanism exposed in the interface. Now, let's dive into the implementation and understand how object Pools can be implemented.

How it works...

The implementation starts with a class Pool<> which is located in lightweightvk/lvk/Pool.h. It stores a collection of objects of type ImplObjectType inside std::vector and can manage handles to these objects. Let's look at the implementation details.

1. Every array element is a struct PoolEntry which stores an ImplObjectType object by value together with its generation used to check handles which point to this element. The field nextFree_ is used to maintain a linked list of free elements inside the array. Once a handle is deallocated, a corresponding array element is added to the free list. The field freeListHead_ stores an index of the first free element or kListEndSentinel in case there are no free elements.

```
template<typename ObjectType, typename ImplObjectType>
class Pool {
  static constexpr uint32_t kListEndSentinel = 0xffffffff;
  struct PoolEntry {
    explicit PoolEntry(ImplObjectType& obj)
    : obj_(std::move(obj)) {}
    ImplObjectType obj_ = {};
    uint32_t gen_ = 1;
    uint32_t nextFree_ = kListEndSentinel;
  };
  uint32_t freeListHead_ = kListEndSentinel;
public:
  std::vector<PoolEntry> objects_;
```

Note

Proponents of Data-Oriented Design may argue that this structure minimizes cache utilization by interleaving the payload of ImplObjectType with utility values gen_ and nextFree_. This is indeed true. One approach mitigate to this is to store (memory alias) the nextFree_ element in-place of the unused object in the pool. Another approach to is to maintain two separate arrays. The first array can tightly pack ImplObjectType values, while the second one can store the necessary metadata for bookkeeping. In fact, it can go a step further, as mentioned in the original presentation by Sebastian Aaltonen, by separating a high-frequency accessed "hot" object type from a low-frequency accessed "cold" type, which can be stored in different arrays. However, for the sake of simplicity, we will leave this as an exercise for our readers.

2. The method `create()` takes an R-value reference. It checks the head of the free list. If there is a free element within the array, we can immediately place our object into it and remove the front element from the free list.

```
Handle<ObjectType> create(ImplObjectType&& obj) {
  uint32_t idx = 0;
  if (freeListHead_ != kListEndSentinel) {
    idx = freeListHead_;
    freeListHead_ = objects_[idx].nextFree_;
    objects_[idx].obj_ = std::move(obj);
  } else {
```

3. If there's no space inside, append a new element to the `std::vector` container.

```
    idx = (uint32_t)objects_.size();
    objects_.emplace_back(obj);
  }
  numObjects_++;
  return Handle<ObjectType>(idx, objects_[idx].gen_);
}
```

4. Destruction is straightforward but involves additional error checking. Empty handles should not be destroyed. Attempting to remove a non-empty handle from an empty pool indicates a logic error and should trigger an assertion. If the generation of the handle does not match the generation of the corresponding array element, it means we are attempting double-deletion.

```
void destroy(Handle<ObjectType> handle) {
  if (handle.empty()) return;
  assert(numObjects_ > 0);
  const uint32_t index = handle.index();
  assert(index < objects_.size());
  // double deletion
  assert(handle.gen() == objects_[index].gen_);
```

5. If all the checks are successful, replace the stored object with an empty, default-constructed object and increment its generation. Then, place this array element at the front of the free list.

```
  objects_[index].obj_ = ImplObjectType{};
  objects_[index].gen_++;
  objects_[index].nextFree_ = freeListHead_;
  freeListHead_ = index;
  numObjects_--;
}
```

6. Dereferencing a handle is done via the `get()` method, which has both `const` and non-`const` implementations. They are identical, so we only need to check one. A mismatch in the generation helps identify access to a deleted object.

```
ImplObjectType* get(Handle<ObjectType> handle) {
  if (handle.empty()) return nullptr;
  const uint32_t index = handle.index();
  assert(index < objects_.size());
  // accessing a deleted object
  assert(handle.gen() == objects_[index].gen_);
  return &objects_[index].obj_;
}
```

7. A pool can be manually cleared, causing a destructor to be called for every object.

```
void clear() {
  objects_.clear();
  freeListHead_ = kListEndSentinel;
  numObjects_ = 0;
}
```

8. Sometimes, it is convenient to construct a `Handle<>` object knowing its index in the pool, regardless of the generation value. This is unsafe and should only be used for debugging.

```
Handle<ObjectType> getHandle(uint32_t index) const {
  assert(index < objects_.size());
  if (index >= objects_.size()) return {};
  return Handle<ObjectType>(index, objects_[index].gen_);
}
```

9. We can also check if a specific object is in the pool and obtain a handle to it. An empty handle is returned if there's no object.

```
Handle<ObjectType> findObject(const ImplObjectType* obj) {
  if (!obj) return {};
  for (size_t idx = 0; idx != objects_.size(); idx++) {
    if (objects_[idx].obj_ == *obj) {
      return Handle<ObjectType>(
        (uint32_t)idx, objects_[idx].gen_);
    }
  }
  return {};
}
```

10. The member field `numObjects_` is used to track memory leaks and prevent deallocations inside an empty pool.

```
    uint32_t numObjects() const {
      return numObjects_;
    }
    uint32_t numObjects_ = 0;
  };
```

That is how `Pools` work. The *LightweightVK* implementation in `VulkanContext` uses them to store all implementation-specific objects that are accessible by handles from the interface side. These declarations can be found in `lvk/vulkan/VulkanClasses.h`. In many situations, Vulkan objects such as `VkShaderModule` and `VkSampler` can be stored directly. However, if additional bookkeeping is required, a wrapper object is stored instead.

```
Pool<ShaderModule, ShaderModuleState> shaderModulesPool_;
Pool<RenderPipeline, RenderPipelineState> renderPipelinesPool_;
Pool<ComputePipeline, ComputePipelineState> computePipelinesPool_;
Pool<Sampler, VkSampler> samplersPool_;
Pool<Buffer, VulkanBuffer> buffersPool_;
Pool<Texture, VulkanImage> texturesPool_;
Pool<QueryPool, VkQueryPool> queriesPool_;
```

Now we know how to store various objects in `Pools` and expose access to them via `Handles`. Before concluding the topic of this chapter and finishing our introduction to Vulkan, let's take a look at how to construct bindless descriptor sets to access textures from GLSL shaders and create a simple texture mapping rendering application.

Using Vulkan descriptor indexing

Descriptor indexing became part of the Vulkan core in version 1.2 as an optional feature, and it was made mandatory in Vulkan 1.3. This feature allows applications to place all their resources into one large descriptor set and make it available to all shaders. There's no need to manage descriptor pools or construct per-shader descriptor sets. Everything is accessible to shaders at once. Shaders can access all resources in the system, and the only practical limit is performance. Let's learn how to work with descriptor sets and descriptor indexing in Vulkan by exploring the *LightweightVK* framework.

How to do it...

Let's examine the parts of the `VulkanContext` class that manage descriptors, as declared in `lvk/vulkan/VulkanClasses.h`. Integer variables `currentMaxTextures_` and `currentMaxSamplers_` define the maximum number of resources that can be stored in the descriptor set, referred to as `vkDSet_`. This descriptor set is allocated from a descriptor pool named `vkDPool_`, which is created based on the descriptor set layout specified by `vkDSL_`. Additionally, the `lastSubmitHandle_` field tracks the most recent submit operation that involved this descriptor set. This submit handle concept was discussed earlier in the previous chapter in the recipe *Using Vulkan command buffers*.

```
class VulkanContext final : public IContext {
  // ...
  uint32_t currentMaxTextures_ = 16;
  uint32_t currentMaxSamplers_ = 16;
  VkDescriptorSetLayout vkDSL_ = VK_NULL_HANDLE;
  VkDescriptorPool vkDPool_ = VK_NULL_HANDLE;
  VkDescriptorSet vkDSet_ = VK_NULL_HANDLE;
  SubmitHandle lastSubmitHandle = SubmitHandle();
  // ...
}
```

Our exploration begins with the `growDescriptorPool()` function, which is responsible for recreating Vulkan objects when necessary to handle the requested number of textures and samplers. To maintain readability, extensive error-checking code has been omitted.

1. First, ensure the number of resources is withing the hardware specific limits.

```
lvk::Result lvk::VulkanContext::growDescriptorPool(uint32_t maxTextures,
uint32_t maxSamplers, uint32_t maxAccelStructs) {
  currentMaxTextures_ = maxTextures;
  currentMaxSamplers_ = maxSamplers;
  currentMaxAccelStructs_ = maxAccelStructs;
  if (!LVK_VERIFY(maxTextures <=
      vkPhysicalDeviceVulkan12Properties_.
        maxDescriptorSetUpdateAfterBindSampledImages))
  {
    LLOGW("Max Textures exceeded: %u (max %u)",
      maxTextures,
      vkPhysicalDeviceVulkan12Properties_.
        maxDescriptorSetUpdateAfterBindSampledImages);
  }
  if (!LVK_VERIFY(maxSamplers <=
      vkPhysicalDeviceVulkan12Properties_.
        maxDescriptorSetUpdateAfterBindSamplers))
  {
    LLOGW("Max Samplers exceeded %u (max %u)",
      maxSamplers,
      vkPhysicalDeviceVulkan12Properties_.
        maxDescriptorSetUpdateAfterBindSamplers);
  }
```

2. Next, the function deallocates any existing Vulkan descriptor set layout and descriptor pool.

```
if (vkDSL_ != VK_NULL_HANDLE) {
  deferredTask(std::packaged_task<void()>(
    [device = vkDevice_, dsl = vkDSL_]() {
      vkDestroyDescriptorSetLayout(device, dsl, nullptr); }));
}
if (vkDPool_ != VK_NULL_HANDLE) {
  deferredTask(std::packaged_task<void()>(
    [device = vkDevice_, dp = vkDPool_]() {
      vkDestroyDescriptorPool(device, dp, nullptr); }));
}
```

3. The function creates a new **descriptor set layout** to be shared across all Vulkan pipelines. This layout includes bindings for all supported Vulkan resources such as sampled images, samplers, and storage images. Ray tracing-related resources are not covered here, and so are YUV images, as they are beyond the scope of the book.

```
VkShaderStageFlags stageFlags =
  VK_SHADER_STAGE_VERTEX_BIT |
  VK_SHADER_STAGE_TESSELLATION_CONTROL_BIT |
  VK_SHADER_STAGE_TESSELLATION_EVALUATION_BIT |
  VK_SHADER_STAGE_FRAGMENT_BIT |
  VK_SHADER_STAGE_COMPUTE_BIT;
const VkDescriptorSetLayoutBinding
  bindings[kBinding_NumBindings] = {
    getDSLBinding(kBinding_Textures,
      VK_DESCRIPTOR_TYPE_SAMPLED_IMAGE, maxTextures, stageFlags),
    getDSLBinding(kBinding_Samplers,
      VK_DESCRIPTOR_TYPE_SAMPLER, maxSamplers, stageFlags),
    getDSLBinding(kBinding_StorageImages,
      VK_DESCRIPTOR_TYPE_STORAGE_IMAGE, maxTextures, stageFlags),
};
```

4. The Vulkan descriptor indexing feature enables descriptor sets to be updated after they have been bound.

```
const uint32_t flags =
  VK_DESCRIPTOR_BINDING_UPDATE_AFTER_BIND_BIT |
  VK_DESCRIPTOR_BINDING_UPDATE_UNUSED_WHILE_PENDING_BIT |
  VK_DESCRIPTOR_BINDING_PARTIALLY_BOUND_BIT;
VkDescriptorBindingFlags bindingFlags[kBinding_NumBindings];
for (int i = 0; i < kBinding_NumBindings; ++i)
  bindingFlags[i] = flags;
```

5. A chain of Vulkan `Vk…CreateInfo` structures should be prepared to create a desired `VkDesciptorSetLayout` object.

```
const VkDescriptorSetLayoutBindingFlagsCreateInfo
  setLayoutBindingFlagsCI = {
    .sType = VK_STRUCTURE_TYPE_DESCRIPTOR_SET_LAYOUT_BINDING_FLAGS_
       CREATE_INFO_EXT,
    .bindingCount = kBinding_NumBindings,
    .pBindingFlags = bindingFlags,
};
const VkDescriptorSetLayoutCreateInfo dslci = {
  .sType = VK_STRUCTURE_TYPE_DESCRIPTOR_SET_LAYOUT_CREATE_INFO,
  .pNext = &setLayoutBindingFlagsCI,
  .flags =
    VK_DESCRIPTOR_SET_LAYOUT_CREATE_UPDATE_AFTER_BIND_POOL_BIT_EXT,
  .bindingCount = kBinding_NumBindings,
  .pBindings = bindings,
};
vkCreateDescriptorSetLayout(vkDevice_, &dslci, nullptr, &vkDSL_);
```

6. Using the newly created descriptor set layout, we can now create a descriptor pool. Be sure to include the flag `VK_DESCRIPTOR_POOL_CREATE_UPDATE_AFTER_BIND_BIT`, as it is necessary to support the corresponding `VK_DESCRIPTOR_BINDING_UPDATE_AFTER_BIND_BIT` flag in the descriptor set layout.

```
const VkDescriptorPoolSize poolSizes[kBinding_NumBindings]{
  VkDescriptorPoolSize{
    VK_DESCRIPTOR_TYPE_SAMPLED_IMAGE, maxTextures},
  VkDescriptorPoolSize{
    VK_DESCRIPTOR_TYPE_SAMPLER, maxSamplers},
  VkDescriptorPoolSize{
    VK_DESCRIPTOR_TYPE_STORAGE_IMAGE, maxTextures},
};
const VkDescriptorPoolCreateInfo ci = {
  .sType = VK_STRUCTURE_TYPE_DESCRIPTOR_POOL_CREATE_INFO,
  .flags = VK_DESCRIPTOR_POOL_CREATE_UPDATE_AFTER_BIND_BIT,
  .maxSets = 1,
  .poolSizeCount = kBinding_NumBindings,
  .pPoolSizes = poolSizes,
};
vkCreateDescriptorPool(vkDevice_, &ci, nullptr, &vkDPool_);
```

7. Now we can allocate a descriptor set from the `vkDPool_` descriptor pool.

```
    const VkDescriptorSetAllocateInfo ai = {
      .sType = VK_STRUCTURE_TYPE_DESCRIPTOR_SET_ALLOCATE_INFO,
      .descriptorPool = vkDPool_,
      .descriptorSetCount = 1,
      .pSetLayouts = &vkDSL_,
    };
    vkAllocateDescriptorSets(vkDevice_, &ai, &vkDSet_);
  return Result();
}
```

To create Vulkan pipelines, we need a pipeline layout. In `LightweightVK`, the pipeline layout is stored for each pipeline in the `RenderPipelineState` structure. We covered how to set this up in the previous *Chapter 2*, in the recipe *Initializing Vulkan pipelines* using the `vkDSL_` descriptor set layout created above.

The `growDescriptorPool()` function was the first step in our descriptor set management mechanism. Now, we have a descriptor set, `vkDSet_`, which must be updated before it can be used. This updating process is handled by another function, `checkAndUpdateDescriptorSets()`, which is called before dispatching Vulkan draw calls. Let's take a closer look at it. As before, some parts of the error checking have been omitted.

1. Newly created resources, such as textures and samplers, can be used immediately—just make sure they are added to descriptor sets. We covered the texture creation process earlier in the recipe *Using texture data in Vulkan*. Once a new texture is created, the `awaitingCreation_` flag is set to indicate that the descriptor set needs to be updated. If there are no new textures or samplers, no updates to the descriptor set are necessary.

```
void VulkanContext::checkAndUpdateDescriptorSets() {
  if (!awaitingCreation_) return;
```

2. As we learned in the previous recipe, *Storing Vulkan Objects*, textures and samplers are stored in pools within `VulkanContext`. Here, we grow the Vulkan descriptor pool as needed to accommodate all the textures and samplers from these pools.

```
  uint32_t newMaxTextures = currentMaxTextures_;
  uint32_t newMaxSamplers = currentMaxSamplers_;
  while (texturesPool_.objects_.size() > newMaxTextures)
    newMaxTextures *= 2;
  while (samplersPool_.objects_.size() > newMaxSamplers)
    newMaxSamplers *= 2;
  if (newMaxTextures != currentMaxTextures_ ||
      newMaxSamplers != currentMaxSamplers_) {
    growDescriptorPool(newMaxTextures, newMaxSamplers);
  }
```

3. Let's prepare the Vulkan structures to update the descriptor set with sampled and storage images. *LightweightVK* always stores a dummy texture at index `0` to prevent sparse arrays in GLSL shaders and ensure that all shaders can safely sample non-existing textures.

```
std::vector<VkDescriptorImageInfo> infoSampledImages;
std::vector<VkDescriptorImageInfo> infoStorageImages;
infoSampledImages.reserve(texturesPool_.numObjects());
infoStorageImages.reserve(texturesPool_.numObjects());
VkImageView dummyImageView =
  texturesPool_.objects_[0].obj_.imageView_;
```

4. Iterate over the texture pool and populate the `VkDescriptorImageInfo` structures based on the image properties. Multisampled images can only be accessed from shaders using `texelFetch()`, which is not supported by *LightweightVK*, so we skip them here.

```
for (const auto& obj : texturesPool_.objects_) {
  const VulkanImage& img = obj.obj_;
  const VkImageView view = obj.obj_.imageView_;
  const VkImageView storageView = obj.obj_.imageViewStorage_ ?
    obj.obj_.imageViewStorage_ : view;
  const bool isTextureAvailable = VK_SAMPLE_COUNT_1_BIT ==
    (img.vkSamples_ & VK_SAMPLE_COUNT_1_BIT);
  const bool isSampledImage =
    isTextureAvailable && img.isSampledImage();
  const bool isStorageImage =
    isTextureAvailable && img.isStorageImage();
```

5. Images are expected to be in specific layouts. Sampled images should use `VK_IMAGE_LAYOUT_SHADER_READ_ONLY_OPTIMAL`, while storage images should use `VK_IMAGE_LAYOUT_GENERAL`. *LightweightVK* automatically handles the necessary image layout conversions, which we will discuss in later chapters.

```
  infoSampledImages.push_back(VkDescriptorImageInfo{
    .sampler = VK_NULL_HANDLE,
    .imageView = isSampledImage ? view : dummyImageView,
    .imageLayout = VK_IMAGE_LAYOUT_SHADER_READ_ONLY_OPTIMAL,
  });
  infoStorageImages.push_back(VkDescriptorImageInfo{
    .sampler = VK_NULL_HANDLE,
    .imageView = isStorageImage ? storageView : dummyImageView,
    .imageLayout = VK_IMAGE_LAYOUT_GENERAL,
  });
}
```

6. Samplers are handled in a very similar manner.

```
std::vector<VkDescriptorImageInfo> infoSamplers;
infoSamplers.reserve(samplersPool_.objects_.size());
for (const auto& sampler : samplersPool_.objects_) {
  infoSamplers.push_back({
    .sampler = sampler.obj_ ?
      sampler.obj_ : samplersPool_.objects_[0].obj_,
    .imageView = VK_NULL_HANDLE,
    .imageLayout = VK_IMAGE_LAYOUT_UNDEFINED,
  });
}
```

7. The `VkWriteDescriptorSet` structure specifies the parameters for a descriptor set write operation. We need to fill in one structure for each of our three bindings, corresponding to three different descriptor types: `VK_DESCRIPTOR_TYPE_SAMPLED_IMAGE`, `VK_DESCRIPTOR_TYPE_SAMPLER`, and `VK_DESCRIPTOR_TYPE_STORAGE_IMAGE`. This code snippet is straightforward but somewhat lengthy, so we include it here in its entirety for your reference.

```
VkWriteDescriptorSet write[kBinding_NumBindings] = {};
uint32_t numWrites = 0;
if (!infoSampledImages.empty())
  write[numWrites++] = VkWriteDescriptorSet{
    .sType = VK_STRUCTURE_TYPE_WRITE_DESCRIPTOR_SET,
    .dstSet = vkDSet_,
    .dstBinding = kBinding_Textures,
    .dstArrayElement = 0,
    .descriptorCount = (uint32_t)infoSampledImages.size(),
    .descriptorType = VK_DESCRIPTOR_TYPE_SAMPLED_IMAGE,
    .pImageInfo = infoSampledImages.data(),
  };
if (!infoSamplers.empty())
  write[numWrites++] = VkWriteDescriptorSet{
    .sType = VK_STRUCTURE_TYPE_WRITE_DESCRIPTOR_SET,
    .dstSet = vkDSet_,
    .dstBinding = kBinding_Samplers,
    .dstArrayElement = 0,
    .descriptorCount = (uint32_t)infoSamplers.size(),
    .descriptorType = VK_DESCRIPTOR_TYPE_SAMPLER,
    .pImageInfo = infoSamplers.data(),
  };
if (!infoStorageImages.empty())
  write[numWrites++] = VkWriteDescriptorSet{
```

```
        .sType = VK_STRUCTURE_TYPE_WRITE_DESCRIPTOR_SET,
        .dstSet = vkDSet_,
        .dstBinding = kBinding_StorageImages,
        .dstArrayElement = 0,
        .descriptorCount = (uint32_t)infoStorageImages.size(),
        .descriptorType = VK_DESCRIPTOR_TYPE_STORAGE_IMAGE,
        .pImageInfo = infoStorageImages.data(),
      };
```

8. If we have filled in any `VkWriteDescriptorSet` structures, invoke the Vulkan function `vkUpdateDescriptorSets()` to update the descriptor set. Since we are updating the entire descriptor set, it is essential to ensure that Vulkan is not using it by waiting on a fence and calling `wait()` with the last known submit handle. This mechanism was discussed earlier in the previous chapter in the recipe *Using Vulkan command buffers*.

```
    if (numWrites) {
      immediate_->wait(immediate_->getLastSubmitHandle());
      vkUpdateDescriptorSets(vkDevice_, numWrites, write, 0, nullptr);
    }
    awaitingCreation_ = false;
  }
```

The C++ portion of the descriptor set updating process is complete. The only remaining step is to clarify how to access these descriptor sets from GLSL shaders. Let's explore how this works.

How it works...

`VulkanContext` injects helper code into GLSL shaders to simplify working with our bindless descriptor sets. Let's revisit the `VulkanContext::createShaderModuleFromGLSL()` function, which handles the injection. Here's the GLSL code it automatically adds to every fragment shader. We skip the automatically added GLSL extensions for the sake of brevity.

1. First, it declares some unbound arrays that are stored in our bindless descriptor set. You'll notice that the descriptor set ID ranges from `0` to `2`. However, don't be misled by this; it's the same descriptor set bound to three different locations. As shown earlier in this recipe, this is necessary for compatibility with MoltenVK as it does not support descriptor aliasing.

```
layout(set = 0, binding = 0) uniform texture2D kTextures2D[];
layout(set = 1, binding = 0) uniform texture3D kTextures3D[];
layout(set = 2, binding = 0) uniform textureCube kTexturesCube[];
layout(set = 3, binding = 0) uniform texture2D kTextures2DShadow[];
layout(set = 0, binding = 1) uniform sampler kSamplers[];
layout(set = 1, binding = 1) uniform samplerShadow kSamplersShadow[];
```

2. Next, some helper functions are added. These functions correspond to standard GLSL functions like `texture()`, `textureLod()`, and others. They simplify the process of working with bindless textures and descriptor indexing. We've listed just a few of them here to give you an overall picture.

```
vec4 textureBindless2D(uint textureid, uint samplerid, vec2 uv) {
  return texture(nonuniformEXT(sampler2D(
    kTextures2D[textureid], kSamplers[samplerid])), uv);
}
vec4 textureBindless2DLod(
  uint textureid, uint samplerid, vec2 uv, float lod) {
  return textureLod(nonuniformEXT(sampler2D(
    kTextures2D[textureid], kSamplers[samplerid])), uv, lod);
}
float textureBindless2DShadow(uint textureid, uint samplerid, vec3 uvw) {
  return texture(nonuniformEXT(sampler2DShadow(
    kTextures2DShadow[textureid], kSamplersShadow[samplerid])), uvw);
}
ivec2 textureBindlessSize2D(uint textureid) {
  return textureSize(nonuniformEXT(kTextures2D[textureid]), 0);
}
```

Note how `nonuniformEXT(sampler2D())` should be used here. The final argument in a function call like `texture()` determines whether the access is considered non-uniform.

3. With this, our GLSL fragment shader `Chapter03/02_STB/main.frag` can be rewritten as follows, without the need for manually declaring the lengthy data structures.

```
layout (location=0) in vec2 uv;
layout (location=0) out vec4 out_FragColor;
layout(push_constant) uniform PerFrameData {
  uniform mat4 MVP;
  uint textureId;
};
void main() {
  out_FragColor = textureBindless2D(textureId, 0, uv);
};
```

With the bindless descriptor set code in place, we can now render textured objects, like the one in the following image.

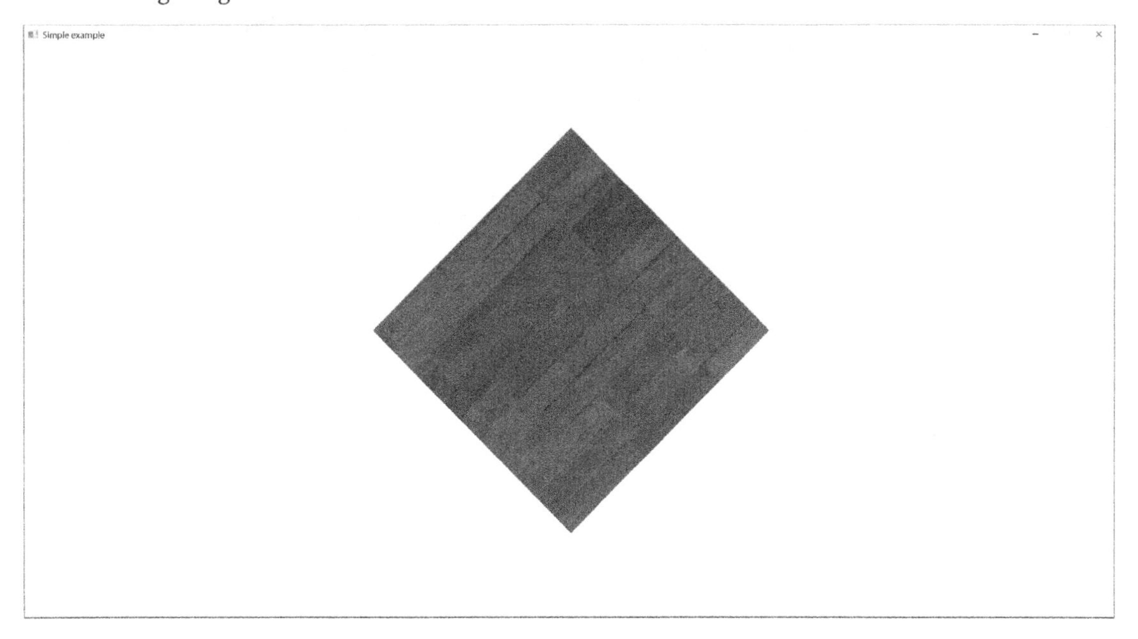

Figure 3.4: Rendering a textured quad

Now, let's move on to the next chapter and learn how to add user interaction and productivity tools to our examples.

There's more...

The topic of efficient resource management in Vulkan is vast and complex. We will revisit descriptor set management later when we cover 3D scene data management and the rendering of complex multitextured materials.

There are two major improvements that can be made to this code to prevent stalling when updating the descriptor set and boost its update performance. The first improvement is to use a couple of round-robin descriptor sets and switch between them, eliminating the need for the heavy `wait()` function call before writing descriptor sets. Another important improvement is to perform incremental updates of descriptor sets. This is crucial when creating new textures or samplers, as it allows the update to be done without any waiting by simply placing new descriptors into the available elements of the descriptor set, using the `VK_DESCRIPTOR_BINDING_UPDATE_UNUSED_WHILE_PENDING_BIT` functionality.

4

Adding User Interaction and Productivity Tools

In this chapter, we will learn how to implement basic helpers to drastically simplify the debugging of graphical applications. The demos are implemented in Vulkan using all the material from the previous three chapters. In *Chapter 3*, we demonstrated how to wrap various instances of raw Vulkan code to create and maintain basic Vulkan state and objects. In this chapter, we will show how to start implementing Vulkan rendering code in a way that is easily extensible and adaptable for different applications. Beginning with 2D user-interface rendering is the best way to learn this, as it makes things easier and allows us to focus on the rendering code without being overwhelmed by complex 3D graphics algorithms.

We will cover the following recipes:

- Rendering ImGui user interfaces
- Integrating Tracy into C++ applications
- Using Tracy GPU profiling
- Adding a frames-per-second counter
- Using cube map textures in Vulkan
- Working with a 3D camera and basic user interaction
- Adding camera animations and motion
- Implementing an immediate-mode 3D drawing canvas
- Rendering on-screen graphs with ImGui and ImPlot
- Putting it all together into a Vulkan application

Technical requirements

To run code from this chapter on your Linux or Windows PC, you will need a GPU with recent drivers that support Vulkan 1.3. The source code used in this chapter can be downloaded from `https://github.com/PacktPublishing/3D-Graphics-Rendering-Cookbook-Second-Edition/`.

Rendering ImGui user interfaces

ImGui is a popular bloat-free graphical user interface library for C++ and is essential to the interactive debugging of graphics apps. ImGui integration comes as a part of the *LightweightVK* library. In this recipe, we go step by step through the code and show how to create an example app with ImGui rendering.

Getting ready

It is recommended to revisit the *Using Vulkan descriptor indexing* recipe from *Chapter 3*, and also recall the Vulkan basics described in the other recipes of that chapter.

This recipe covers the source code of `lightweight/lvk/HelpersImGui.cpp`. The demo code for this recipe is in `Chapter04/01_ImGui`.

How to do it...

Let us start with a minimalistic ImGui demo application and take a look at how to use the ImGui Vulkan wrapper provided by *LightweightVK*:

1. First, we create an `lvk::ImGuiRenderer` object. It takes in a pointer to our `lvk::IContext`, the name of the default font, and the default font size in pixels. `ImGuiRenderer` will take care of all the low-level ImGui initialization and code:

```
unique_ptr<lvk::ImGuiRenderer> imgui =
  std::make_unique<lvk::ImGuiRenderer>(
    *ctx, "data/OpenSans-Light.ttf", 30.0f);
```

2. Let us create some GLFW callbacks that pass mouse movements and button presses into ImGui. GLFW mouse button IDs should be converted into ImGui ones:

```
glfwSetCursorPosCallback(window,
  [](auto* window, double x, double y) {
    ImGui::GetIO().MousePos = ImVec2(x, y);
  });
glfwSetMouseButtonCallback(window,
  [](auto* window, int button, int action, int mods)
  {
    double xpos, ypos;
    glfwGetCursorPos(window, &xpos, &ypos);
    const ImGuiMouseButton_ imguiButton =
```

```
    (button==GLFW_MOUSE_BUTTON_LEFT) ?
      ImGuiMouseButton_Left : (
        button == GLFW_MOUSE_BUTTON_RIGHT ?
          ImGuiMouseButton_Right :
          ImGuiMouseButton_Middle);
  ImGuiIO& io = ImGui::GetIO();
  io.MousePos = ImVec2((float)xpos, (float)ypos);
  io.MouseDown[imguiButton] = action == GLFW_PRESS;
});
```

Quick tip: Enhance your coding experience with the **AI Code Explainer** and **Quick Copy** features. Open this book in the next-gen Packt Reader. Click the **Copy** button (**1**) to quickly copy code into your coding environment, or click the **Explain** button (**2**) to get the AI assistant to explain a block of code to you.

Copy Explain

```
function calculate(a, b) {
  return {sum: a + b};
};
```

1 2

The next-gen Packt Reader is included for free with the purchase of this book. Unlock it by scanning the QR code below or visiting `https://www.packtpub.com/unlock/9781803248110`.

3. Inside our typical rendering loop, we can invoke ImGui rendering commands as follows. The `ImGuiRenderer::beginFrame()` method takes in an `lvk::Framebuffer` object so it can set up a rendering pipeline properly:

```
ICommandBuffer& buf = ctx->acquireCommandBuffer();
const lvk::Framebuffer framebuffer = {
  .color = {{ .texture = ctx->getCurrentSwapchainTexture() }}};
buf.cmdBeginRendering({ .color = { {
    .loadOp = lvk::LoadOp_Clear,
    .clearColor = {1.0f, 1.0f, 1.0f, 1.0f} } } },
  framebuffer);
imgui->beginFrame(framebuffer);
```

4. Let's draw an ImGui window with a texture. A texture index is passed into ImGui as an `ImTextureID` value so it can be used with the bindless rendering scheme we discussed in the previous chapter in the *Using Vulkan descriptor sets* recipe:

```
ImGui::Begin("Texture Viewer", nullptr,
  ImGuiWindowFlags_AlwaysAutoResize);
ImGui::Image(texture.index(), ImVec2(512, 512));
ImGui::ShowDemoWindow();
ImGui::End();
```

5. The `ImGuiRenderer::endFrame()` method is a command buffer with actual Vulkan commands to render the user interface. Then we can call `cmdEndRendering()` and submit our command buffer:

```
imgui->endFrame(buf);
buf.cmdEndRendering();
ctx->submit(buf, ctx->getCurrentSwapchainTexture());
```

This demo application should render a simple ImGui interface like that shown in the following screenshot:

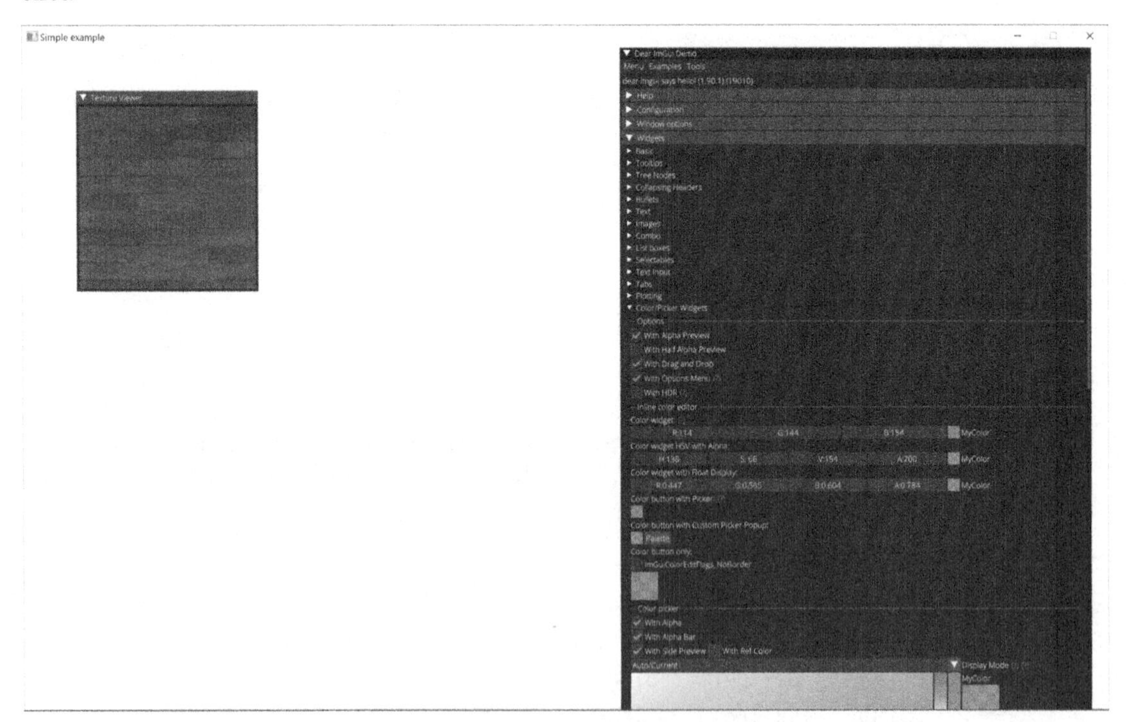

Figure 4.1: ImGui rendering

Quick tip: Need to see a high-resolution version of this image? Open this book in the next-gen Packt Reader or view it in the PDF/ePub copy.

The next-gen Packt Reader and a **free PDF/ePub copy** of this book are included with your purchase. Unlock them by scanning the QR code below or visiting `https://www.packtpub.com/unlock/9781803248110`.

Now let's take a look at the underlying low-level implementation inside *LightweightVK* that renders ImGui data.

How it works...

The `lvk::ImGuiRenderer` helper class is declared in `lvk\HelpersImGui.h`. Here is its declaration.

1. The constructor accepts a reference to `lvk::IContext`, the name of the default `.ttf` font file, and the default font size in pixels. The `updateFont()`method can be invoked at a later point to override the previously used font. This method gets called from the constructor to set the default font:

```
class ImGuiRenderer {
 public:
  explicit ImGuiRenderer(lvk::IContext& ctx,
    const char* defaultFontTTF = nullptr,
    float fontSizePixels = 24.0f);
  ~ImGuiRenderer();
  void updateFont(
    const char* defaultFontTTF, float fontSizePixels);
```

2. The `beginFrame()` and `endFrame()` methods are necessary to prepare ImGui for rendering and generate Vulkan commands from the ImGui draw data. The `setDisplayScale()` method can be used to override ImGui's `DisplayFramebufferScale` factor:

```
  void beginFrame(const lvk::Framebuffer& desc);
  void endFrame(lvk::ICommandBuffer& cmdBuffer);
  void setDisplayScale(float displayScale);
```

3. The private section of the `lvk::ImGuiRenderer` class contains a method to create a new rendering pipeline and a bunch of data necessary for rendering. There's a single set of vertex and fragment shaders, a rendering pipeline, and a texture created from the `.ttf` font file we provided at construction time:

```
private:
 Holder<RenderPipelineHandle> createNewPipelineState(
   const lvk::Framebuffer& desc);
private:
 lvk::IContext& ctx_;
 Holder<ShaderModuleHandle> vert_;
 Holder<ShaderModuleHandle> frag_;
 Holder<RenderPipelineHandle> pipeline_;
 Holder<TextureHandle> fontTexture_;
 Holder<SamplerHandle> samplerClamp_;
 float displayScale_  = 1.0f;
 uint32_t frameIndex_ = 0;
```

4. To ensure stall-free operation, *LightweightVK* uses multiple buffers to pass ImGui vertex and index data into Vulkan (`vb` and `ib` in the following code, for the vertex and index buffers respectively):

```
  struct DrawableData {
    Holder<BufferHandle> vb_;
    Holder<BufferHandle> ib_;
    uint32_t numAllocatedIndices_ = 0;
    uint32_t numAllocatedVerteices_ = 0;
  };
  static constexpr uint32_t kNumSwapchainImages = 3;
  DrawableData drawables_[kNumSwapchainImages] = {};
};
```

Now we can dive into the implementation, which resides in `lvk/HelpersImGui.cpp`.

he vertex shader uses programmable-vertex pulling, which we briefly touched on in the previous chapter in the *Dealing with buffers in Vulkan* recipe. Let's take a closer look at it.

1. ImGui provides 2D screen coordinates for each vertex, 2D texture coordinates, and an RGBA color. We declare a `Vertex` structure to hold per-vertex data and store all vertices inside the `vertices[]` array residing inside `VertexBuffer`. The `buffer_reference` GLSL layout qualifier declares a type and not an instance of a buffer, so that a reference to that buffer can be passed into the shader at a later point:

```
layout (location = 0) out vec4 out_color;
layout (location = 1) out vec2 out_uv;
```

```
layout (location = 2) out flat uint out_textureId;
struct Vertex {
  float x, y;
  float u, v;
  uint rgba;
};
layout(std430, buffer_reference) readonly buffer VertexBuffer {
  Vertex vertices[];
};
```

2. A reference to `VertexBuffer` containing our per-vertex data is passed via Vulkan push constants. Besides that, we pass a texture ID and some 2D viewport parameters represented as left, right, top, and bottom planes inside `vec4 LRTB`:

```
layout(push_constant) uniform PushConstants {
  vec4 LRTB;
  VertexBuffer vb;
  uint textureId;
} pc;
void main() {
  float L = pc.LRTB.x;
  float R = pc.LRTB.y;
  float T = pc.LRTB.z;
  float B = pc.LRTB.w;
```

Once we have the viewport parameters, we can construct an orthographic projection matrix the following way, which is similar to how `glm::ortho()` creates a projection matrix:

```
  mat4 proj = mat4(
    2.0 / (R-L),                    0.0,  0.0,  0.0,
    0.0,                2.0 / (T-B),  0.0,  0.0,
    0.0,                            0.0, -1.0,  0.0,
    (R+L) / (L-R),   (T+B) / (B-T),  0.0,  1.0);
```

3. The current vertex is extracted from the `VertexBuffer::vertices` array using the `gl_VertexIndex` built-in GLSL variable. The RGBA vertex color `v.rgba` is packed into a 32-bit unsigned integer and can be unpacked into `vec4` using the `unpackUnorm4x8()` GLSL built-in function:

```
  Vertex v = pc.vb.vertices[gl_VertexIndex];
  out_color = unpackUnorm4x8(v.rgba);
```

4. The texture coordinates and texture ID are passed into the fragment shader unchanged. The projection matrix is multiplied by the vertex position expanded into `vec4` by adding `0` as the `Z` component:

```
  out_uv = vec2(v.u, v.v);
  out_textureId = pc.textureId;
  gl_Position = proj * vec4(v.x, v.y, 0, 1);
}
```

A corresponding GLSL fragment shader is much simpler and looks as follows:

1. The input locations should match the corresponding output locations from the vertex shader:

```
layout (location = 0) in vec4 in_color;
layout (location = 1) in vec2 in_uv;
layout (location = 2) in flat uint in_textureId;
layout (location = 0) out vec4 out_color;
```

2. *LightweightVK* supports some basic sRGB framebuffer rendering. This shader constant is used to enable some rudimentary tone mapping. The texture ID is used to access a required bindless texture. The sampler is always a default sampler at index `0`. The `constant_id` GLSL modifier is used to specify specialization constants for Vulkan:

```
layout (constant_id = 0) const bool kNonLinearColorSpace = false;
void main() {
  vec4 c = in_color * texture(sampler2D(
    kTextures2D[in_textureId], kSamplers[0]), in_uv);
```

3. Here we can render our UI in linear color space into an sRGB framebuffer:

```
  out_color = kNonLinearColorSpace ?
    vec4(pow(c.rgb, vec3(2.2)), c.a) : c;
}
```

Now let's take a look at the C++ code in the `lvk::ImGuiRender` implementation. There's a private `ImGuiRenderer::createNewPipelineState()` helper function there, which is responsible for creating a new rendering pipeline for ImGui rendering. As the entire relevant Vulkan state can be dynamic in Vulkan 1.3, a single immutable pipeline is sufficient.

1. A framebuffer description is required to create a pipeline because we need information about color and depth attachment formats:

```
Holder<RenderPipelineHandle>
  ImGuiRenderer::createNewPipelineState(
    const lvk::Framebuffer& desc)
{
  const uint32_t nonLinearColorSpace =
    ctx_.getSwapChainColorSpace() == ColorSpace_SRGB_NONLINEAR ? 1:0;
```

```
  return ctx_.createRenderPipeline({
    .smVert = vert_,
    .smFrag = frag_,
```

2. The sRGB mode is enabled based on the swapchain color space and passed into the shaders as Vulkan specialization constants:

```
    .specInfo = {
      .entries = {{
        .constantId = 0,
        .size = sizeof(nonLinearColorSpace)}},
      .data = &nonLinearColorSpace,
      .dataSize = sizeof(nonLinearColorSpace) },
```

3. All ImGui elements require alpha blending to be enabled. If a depth buffer is present, it is retained unchanged but the rendering pipeline should know about it:

```
    .color = {{
      .format = ctx_.getFormat(desc.color[0].texture),
      .blendEnabled = true,
      .srcRGBBlendFactor = lvk::BlendFactor_SrcAlpha,
      .dstRGBBlendFactor = lvk::BlendFactor_OneMinusSrcAlpha,
    }},
    .depthFormat = desc.depthStencil.texture ?
      ctx_.getFormat(desc.depthStencil.texture) : lvk::Format_Invalid,
    .cullMode = lvk::CullMode_None},
    nullptr);
}
```

Another helper function `ImGuiRenderer::updateFont()` is called from the constructor. Here's how it is implemented.

1. First, it sets up ImGui font configuration parameters using the provided font size:

```
void ImGuiRenderer::updateFont(
  const char* defaultFontTTF, float fontSizePixels)
{
  ImGuiIO& io = ImGui::GetIO();
  ImFontConfig cfg = ImFontConfig();
  cfg.FontDataOwnedByAtlas = false;
  cfg.RasterizerMultiply = 1.5f;
  cfg.SizePixels = ceilf(fontSizePixels);
  cfg.PixelSnapH = true;
  cfg.OversampleH = 4;
  cfg.OversampleV = 4;
  ImFont* font = nullptr;
```

2. Then it loads the default font from a `.ttf` file:

```
    if (defaultFontTTF) {
      font = io.Fonts->AddFontFromFileTTF(
        defaultFontTTF, cfg.SizePixels, &cfg);
    }
    io.Fonts->Flags |= ImFontAtlasFlags_NoPowerOfTwoHeight;
```

3. Last but not least, the rasterized TrueType font data is retrieved from ImGui and stored as a *LightweightVK* texture. This font texture is used later for rendering via its index ID:

```
    unsigned char* pixels;
    int width, height;
    io.Fonts->GetTexDataAsRGBA32(&pixels, &width, &height);
    fontTexture_ = ctx_.createTexture({
      .type = lvk::TextureType_2D,
      .format = lvk::Format_RGBA_UN8,
      .dimensions = {(uint32_t)width, (uint32_t)height},
      .usage = lvk::TextureUsageBits_Sampled,
      .data = pixels }, nullptr);
    io.Fonts->TexID = fontTexture_.index();
    io.FontDefault = font;
  }
```

All the preparations are completed and we can now look at the constructor and destructor of `ImGuiRenderer`. Both member functions are very short.

1. The constructor initializes both ImGui and ImPlot contexts in case *LightweightVK* was compiled with optional ImPlot support. At the moment, *LightweightVK* supports only a single ImGui context:

```
  ImGuiRenderer::ImGuiRenderer(lvk::IContext& device,
    const char* defaultFontTTF,
    float fontSizePixels)
   : ctx_(device)
  {
    ImGui::CreateContext();
  #if defined(LVK_WITH_IMPLOT)
    ImPlot::CreateContext();
  #endif // LVK_WITH_IMPLOT
```

2. Here we set the `ImGuiBackendFlags_RendererHasVtxOffset` flag telling ImGui that our renderer has support for vertex offsets. It enables the output of large meshes while still using 16-bit indices, making UI rendering more efficient:

```
ImGuiIO& io = ImGui::GetIO();
io.BackendRendererName = "imgui-lvk";
io.BackendFlags |=
  ImGuiBackendFlags_RendererHasVtxOffset;
```

3. All the work to create the default font and shaders is delegated as we have just discussed:

```
  updateFont(defaultFontTTF, fontSizePixels);
  vert_ = ctx_.createShaderModule({codeVS, Stage_Vert,
    "Shader Module: imgui (vert)"});
  frag_ = ctx_.createShaderModule({codeFS, Stage_Frag,
    "Shader Module: imgui (frag)"});
  samplerClamp_ = ctx_.createSampler({
    .wrapU = lvk::SamplerWrap_Clamp,
    .wrapV = lvk::SamplerWrap_Clamp,
    .wrapW = lvk::SamplerWrap_Clamp,
  });
}
```

4. The destructor is trivial and cleans up both ImGui and the optional ImPlot:

```
ImGuiRenderer::~ImGuiRenderer() {
  ImGuiIO& io = ImGui::GetIO();
  io.Fonts->TexID = nullptr;
#if defined(LVK_WITH_IMPLOT)
  ImPlot::DestroyContext();
#endif // LVK_WITH_IMPLOT
  ImGui::DestroyContext();
}
```

There's one more simple function that we want to see before going on to the rendering: `ImGuiRenderer::beginFrame()`. It starts a new ImGui frame using the provided framebuffer. A graphics pipeline is lazily created here based on the actual framebuffer parameters because we did not have any framebuffer provided to us in the constructor:

```
void ImGuiRenderer::beginFrame(
  const lvk::Framebuffer& desc)
{
  const lvk::Dimensions dim = ctx_.getDimensions(desc.color[0].texture);
  ImGuiIO& io = ImGui::GetIO();
  io.DisplaySize = ImVec2(dim.width / displayScale_,
```

```
                          dim.height / displayScale_);
    io.DisplayFramebufferScale = ImVec2(displayScale_, displayScale_);
    io.IniFilename = nullptr;
    if (pipeline_.empty()) {
      pipeline_ = createNewPipelineState(desc);
    }
    ImGui::NewFrame();
  }
```

Now we are ready to tackle the UI rendering in the `ImGuiRenderer::endFrame()` function. This function runs every frame and populates a Vulkan command buffer. It is a bit more complicated, so let's go over it step by step to see how it works. Error checking is omitted in the following code snippets for the sake of brevity.

1. First, we should finalize ImGui frame rendering and retrieve the frame draw data:

```
void ImGuiRenderer::endFrame(lvk::ICommandBuffer& cmdBuffer) {
  ImGui::EndFrame();
  ImGui::Render();
  ImDrawData* dd = ImGui::GetDrawData();
  const float fb_width =
    dd->DisplaySize.x * dd->FramebufferScale.x;
  const float fb_height =
    dd->DisplaySize.y * dd->FramebufferScale.y;
```

2. Let's prepare the render state. We disable the depth test and depth buffer writes. A viewport is constructed based on the ImGui framebuffer size, which we set up earlier in `beginFrame()` to be equal to our *LightweightVK* framebuffer size:

```
  cmdBuffer.cmdBindDepthState({});
  cmdBuffer.cmdBindViewport({
    .x = 0.0f,
    .y = 0.0f,
    .width  = fb_width,
    .height = fb_height,
  });
```

3. The parameters of the orthographic projection matrix are prepared here. They will be passed to shaders later via Vulkan push constants inside the rendering loop together with other parameters. Clipping parameters are prepared here as well to be used inside the rendering loop:

```
  const float L = dd->DisplayPos.x;
  const float R = dd->DisplayPos.x + dd->DisplaySize.x;
  const float T = dd->DisplayPos.y;
  const float B = dd->DisplayPos.y + dd->DisplaySize.y;
```

```
const ImVec2 clipOff = dd->DisplayPos;
const ImVec2 clipScale = dd->FramebufferScale;
```

4. We have a set of separate LVK buffers per each frame. These buffers store ImGui vertex and index data for the entire frame:

```
DrawableData& drawableData = drawables_[frameIndex_];
frameIndex_ = (frameIndex_ + 1) % LVK_ARRAY_NUM_ELEMENTS(drawables_);
```

5. If there are buffers already existing from the previous frames and these buffers' sizes are insufficient to fit in the new vertex or index data, the buffers are re-created with the new size. The index buffer is created via `BufferUsageBits_Index`:

```
if (drawableData.numAllocatedIndices_ < dd->TotalIdxCount)
{
  drawableData.ib_ = ctx_.createBuffer({
    .usage = lvk::BufferUsageBits_Index,
    .storage = lvk::StorageType_HostVisible,
    .size = dd->TotalIdxCount * sizeof(ImDrawIdx),
    .debugName = "ImGui: drawableData.ib_",
  });
  drawableData.numAllocatedIndices_ = dd->TotalIdxCount;
}
```

6. The buffer to store vertices is actually a `BufferUsageBits_Storage` storage buffer, because our GLSL shaders use programmable-vertex pulling to load the vertices:

```
if (drawableData.numAllocatedVerteices_ < dd->TotalVtxCount)
{
  drawableData.vb_ = ctx_.createBuffer({
    .usage = lvk::BufferUsageBits_Storage,
    .storage = lvk::StorageType_HostVisible,
    .size = dd->TotalVtxCount * sizeof(ImDrawVert),
    .debugName = "ImGui: drawableData.vb_",
  });
  drawableData.numAllocatedVerteices_ = dd->TotalVtxCount;
}
```

7. Let's upload some data to the vertex and index buffers. The entire ImGui frame data is uploaded here. Offsets are carefully preserved so we know where every ImGui draw command data is stored:

```
ImDrawVert* vtx = (ImDrawVert*)ctx_.getMappedPtr(drawableData.vb_);
uint16_t* idx = (uint16_t*)ctx_.getMappedPtr(drawableData.ib_);
for (int n = 0; n < dd->CmdListsCount; n++) {
  const ImDrawList* cmdList = dd->CmdLists[n];
```

```
    memcpy(vtx, cmdList->VtxBuffer.Data,
      cmdList->VtxBuffer.Size * sizeof(ImDrawVert));
    memcpy(idx, cmdList->IdxBuffer.Data,
      cmdList->IdxBuffer.Size * sizeof(ImDrawIdx));
    vtx += cmdList->VtxBuffer.Size;
    idx += cmdList->IdxBuffer.Size;
  }
```

8. The host-visible memory needs to be flushed. This will allow *LightweightVK* to issue a corresponding Vulkan `vkFlushMappedMemoryRanges()` command if the memory is not host-coherent:

```
  ctx_.flushMappedMemory(drawableData.vb_, 0,
    dd->TotalVtxCount * sizeof(ImDrawVert));
  ctx_.flushMappedMemory(drawableData.ib_, 0,
    dd->TotalIdxCount * sizeof(ImDrawIdx));
```

9. Let's bind our index buffer and the rendering pipeline to a command buffer, and enter the rendering loop that iterates over all ImGui rendering commands:

```
  uint32_t idxOffset = 0;
  uint32_t vtxOffset = 0;
  cmdBuffer.cmdBindIndexBuffer(
    drawableData.ib_, lvk::IndexFormat_UI16);
  cmdBuffer.cmdBindRenderPipeline(pipeline_);
  for (int n = 0; n < dd->CmdListsCount; n++) {
    const ImDrawList* cmdList = dd->CmdLists[n];
    for (int cmd_i = 0; cmd_i < cmdList->CmdBuffer.Size; cmd_i++) {
      const ImDrawCmd& cmd = cmdList->CmdBuffer[cmd_i];
```

10. Viewport clipping is done right here on the CPU side. If the ImGui draw command is completely clipped, we should skip it:

```
      ImVec2 clipMin(
        (cmd.ClipRect.x - clipOff.x) * clipScale.x,
        (cmd.ClipRect.y - clipOff.y) * clipScale.y);
      ImVec2 clipMax(
        (cmd.ClipRect.z - clipOff.x) * clipScale.x,
        (cmd.ClipRect.w - clipOff.y) * clipScale.y);
      if (clipMin.x < 0.0f) clipMin.x = 0.0f;
      if (clipMin.y < 0.0f) clipMin.y = 0.0f;
      if (clipMax.x > fb_width ) clipMax.x = fb_width;
      if (clipMax.y > fb_height) clipMax.y = fb_height;
      if (clipMax.x <= clipMin.x ||
          clipMax.y <= clipMin.y) continue;
```

11. All the data necessary for rendering is passed into GLSL shaders via Vulkan push constants. It consists of the orthographic projection data, which is the left, right, top, and bottom planes, a reference to the vertex buffer, and a texture ID for bindless rendering:

```
        struct VulkanImguiBindData {
          float LRTB[4];
          uint64_t vb = 0;
          uint32_t textureId = 0;
          uint32_t samplerId = 0;
        } bindData = {
          .LRTB = {L, R, T, B},
          .vb = ctx_.gpuAddress(drawableData.vb_),
          .textureId = static_cast<uint32_t>(cmd.TextureId),
          .samplerId = samplerClamp_.index(),
        };
        cmdBuffer.cmdPushConstants(bindData);
```

12. Set up the scissor test so it can do precise clipping of ImGui elements:

```
        cmdBuffer.cmdBindScissorRect({
          uint32_t(clipMin.x),
          uint32_t(clipMin.y),
          uint32_t(clipMax.x - clipMin.x),
          uint32_t(clipMax.y - clipMin.y)});
```

13. The actual rendering is done via `cmdDrawIndexed()`. Here we use both the index offset and vertex offset parameters to access the correct data in our large per-frame vertex and index buffers:

```
        cmdBuffer.cmdDrawIndexed(cmd.ElemCount, 1u,
          idxOffset + cmd.IdxOffset,
          int32_t(vtxOffset + cmd.VtxOffset));
      }
      idxOffset += cmdList->IdxBuffer.Size;
      vtxOffset += cmdList->VtxBuffer.Size;
    }
  }
```

Now we have done all the ImGui rendering and can render the entire ImGui user interface using Vulkan. Let's jump to the next recipes and learn other productivity and debugging tools, such as profiling, 3D camera controls, frames-per-second counters, and a drawing canvas.

Integrating Tracy into C++ applications

In the previous chapter, *Working with Vulkan Objects*, we learned how to write small graphics applications with Vulkan and LightweightVK. In real-world applications, it is often necessary to be able to quickly get performance profiling information at runtime. In this recipe, we will show how to make use of the Tracy profiler in your 3D applications.

Getting ready

The complete source code of the demo application for this recipe is located in `Chapter04/02_TracyProfiler`.

Make sure to download a precompiled Tracy client app for your platform from `https://github.com/wolfpld/tracy`. In our book, we use Tracy version 0.11.1.

How to do it...

The Tracy profiler itself is integrated into the *LightweightVK* library. Our demo application, as well as many parts of the *LightweightVK* rendering code, is augmented with calls to profiling functions. Those calls are wrapped into a set of macros so as not to call Tracy directly. This allows turning the profiler on and off, and even switching to other profilers when necessary. Let's take a look at the demo application and then explore the underlying low-level implementation:

1. First, let's take a look at the root LightweightVK CMake configuration file `deps/src/lightweightvk/CMakeLists.txt` to see how the Tracy library is added to the project. At the beginning, we should see an option enabled by default:

   ```
   option(LVK_WITH_TRACY  "Enable Tracy profiler"  ON)
   ```

2. A few lines later in the same file, the CMake option is converted into a `TRACY_ENABLE` C++ compiler macro definition and the Tracy library is added to the project. Note that this is the `third-party/deps/src/` folder of the *LightweightVK* Git repository, which itself resides inside the `deps/src/` folder of the book repository:

   ```
   if(LVK_WITH_TRACY)
     add_definitions("-DTRACY_ENABLE=1")
     add_subdirectory(third-party/deps/src/tracy)
     lvk_set_folder(TracyClient "third-party")
   endif()
   ```

3. Let's continue scrolling the same file, `deps/src/lightweightvk/CMakeLitsts.txt`, for a few pages further. Based on the previously enabled `LVK_WITH_TRACY` CMake option, we export the `LVK_WITH_TRACY` C++ macro definition to all users of *LightweightVK*. The Tracy library is linked with the `LVKLibrary` target so that every app using *LightweightVK* has access to Tracy functions as well:

   ```
   if(LVK_WITH_TRACY)
     target_compile_definitions(
       LVKLibrary PUBLIC "LVK_WITH_TRACY=1")
   ```

```
    target_link_libraries(
      LVKLibrary PUBLIC TracyClient)
  endif()
```

4. Now let's look into `lightweightvk/lvk/LVK.h` and check out some macro definitions. The `LVK_WITH_TRACY` macro is used to enable or disable Tracy usage. Some predefined RGB colors are declared as macros to be used to mark important point-of-interest operations:

```
#if defined(LVK_WITH_TRACY)
  #include "tracy/Tracy.hpp"
  #define LVK_PROFILER_COLOR_WAIT    0xff0000
  #define LVK_PROFILER_COLOR_SUBMIT  0x0000ff
  #define LVK_PROFILER_COLOR_PRESENT 0x00ff00
  #define LVK_PROFILER_COLOR_CREATE  0xff6600
  #define LVK_PROFILER_COLOR_DESTROY 0xffa500
  #define LVK_PROFILER_COLOR_BARRIER 0xffffff
```

5. Other macros are mapped directly to Tracy functions so that we can work with Tracy zones in a non-intrusive way:

```
  #define LVK_PROFILER_FUNCTION() ZoneScoped
  #define LVK_PROFILER_FUNCTION_COLOR(color) ZoneScopedC(color)
  #define LVK_PROFILER_ZONE(name, color) { \
      ZoneScopedC(color);                \
      ZoneName(name, strlen(name))
  #define LVK_PROFILER_ZONE_END() }
```

6. The `LVK_PROFILER_THREAD` macro can be used to set the name of a C++ thread. The `LVK_PROFILER_FRAME` macro is used to mark the start of the next frame during rendering. It is used by *LightweightVK* in `lvk::VulkanSwapchain::present()` and can be helpful if you want to implement your own swapchain management code, for example, on Android using OpenXR:

```
  #define LVK_PROFILER_THREAD(name) tracy::SetThreadName(name)
  #define LVK_PROFILER_FRAME(name) FrameMarkNamed(name)
```

7. Once Tracy is disabled, all macros are defined to no-ops and zones are defined as empty C++ scopes:

```
#else
  #define LVK_PROFILER_FUNCTION()
  #define LVK_PROFILER_FUNCTION_COLOR(color)
  #define LVK_PROFILER_ZONE(name, color) {
  #define LVK_PROFILER_ZONE_END() }
  #define LVK_PROFILER_THREAD(name)
  #define LVK_PROFILER_FRAME(name)
#endif // LVK_WITH_TRACY
```

Macros LVK_PROFILER_FUNCTION and LVK_PROFILER_FUNCTION_COLOR are spread all over the *LightweightVK* code to give good profiling coverage. Let's take a look at how to use them in our own apps.

How it works...

The demo application is located in Chapter04/02_TracyProfiler/src.main.cpp. Tracy is initialized automatically together with *LightweightVK*. All we have to do now is put corresponding macros in our code. Let's take a look at how it works.

In our initialization part where we create lvk::IContext, we use LVK_PROFILER_ZONE and LVK_PROFILER_ZONE_END to mark an interesting fragment of our initialization code:

```
GLFWwindow* window = nullptr;
unique_ptr<lvk::IContext> ctx;
{
  LVK_PROFILER_ZONE("Initialization", LVK_PROFILER_COLOR_CREATE);
  int width  = -95;
  int height = -90;
  window = lvk::initWindow("Simple example", width, height);
  ctx    = lvk::createVulkanContextWithSwapchain(window, width, height, {});
  LVK_PROFILER_ZONE_END();
}
```

Inside the rendering loop, we can mark different point-of-interest code blocks the same way. The hex value is an RGB color that will be used by Tracy in the profiling window to highlight this profiling zone. Some predefined colors were mentioned earlier in this recipe:

```
lvk::ICommandBuffer& buf = ctx->acquireCommandBuffer();
LVK_PROFILER_ZONE("Fill command buffer", 0xffffff);
// ...
LVK_PROFILER_ZONE_END();
```

If we need to mark the entire function and want to have an automatic name assigned to it, we should use the LVK_PROFILER_FUNCTION macro as in the following snippet. This macro does not require closing:

```
lvk::Result lvk::compileShader(VkDevice device,
  VkShaderStageFlagBits stage, const char* code,
  VkShaderModule* outShaderModule,
  const glslang_resource_t* glslLangResource)
{
  LVK_PROFILER_FUNCTION();
  //...
  return Result();
}
```

Let's take a look at the profiler output while running this demo app. To retrieve the profiling data, you have to run a Tracy client and connect it to your graphics app. We use Tracy version 0.10, which can be downloaded from GitHub at `https://github.com/wolfpld/tracy/releases/tag/v0.10`. Here is a screenshot from a connected Tracy client showing a flame graph of our app.

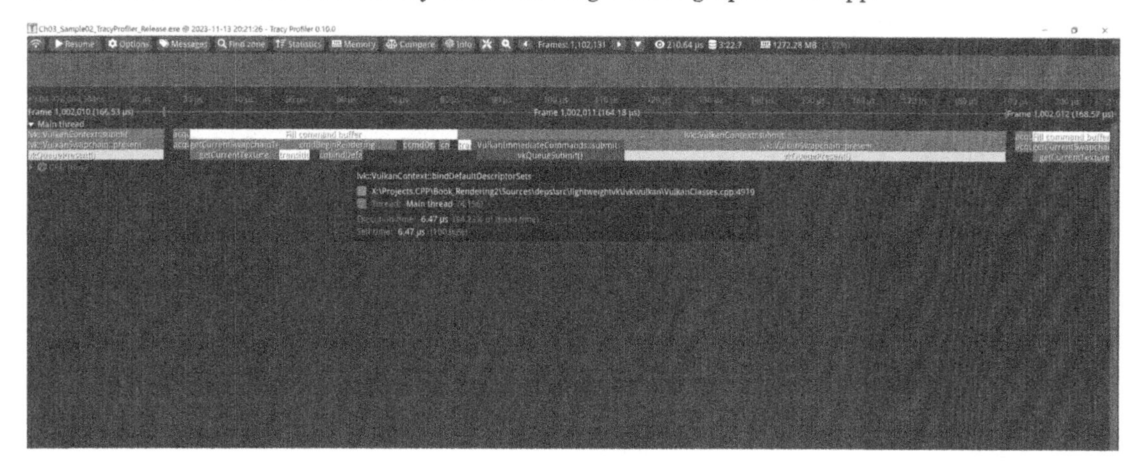

Figure 4.2: Tracy user interface

This approach allows for fully transparent enabling and disabling of the Tracy profiler at build time. Adding other profilers, such as EasyProfiler and Optick, which provide similar APIs, is mostly trivial and can be easily implemented yourself as an exercise.

Using Tracy GPU profiling

Tracy includes a feature that uses Vulkan timestamps to obtain timing information for commands executed on the GPU. If the Vulkan extension `VK_EXT_calibrated_timestamps` is supported, the GPU timing data is synchronized to the same reference frame as the CPU timing. Let's take a quick look at how to add Tracy GPU profiling to Vulkan within the `VulkanContext`.

Getting ready

The corresponding C++ source code can be found in `lvk/vulkan/VulkanClasses.cpp`.

How to do it...

Tracy GPU profiling can be integrated into existing Vulkan code as follows.

1. First, include the `TracyVulkan.hpp` header and define a helper macro to use within our rendering functions.

```
#if defined(LVK_WITH_TRACY_GPU)
  #include "tracy/TracyVulkan.hpp"
  #define LVK_PROFILER_GPU_ZONE(name, ctx, cmdBuffer, color) \
    TracyVkZoneC(ctx->pimpl_->tracyVkCtx_, cmdBuffer, name, color);
#else
```

```
  #define LVK_PROFILER_GPU_ZONE(name, ctx, cmdBuffer, color)
#endif // LVK_WITH_TRACY_GPU
```

2. Add the necessary member fields for Tracy GPU to the `VulkanContextImpl` structure.

```
struct VulkanContextImpl final {
  // ...
#if defined(LVK_WITH_TRACY_GPU)
  TracyVkCtx tracyVkCtx_ = nullptr;
  VkCommandPool tracyCommandPool_ = VK_NULL_HANDLE;
  VkCommandBuffer tracyCommandBuffer_ = VK_NULL_HANDLE;
#endif // LVK_WITH_TRACY_GPU
};
```

3. Add the following initialization code to the `VulkanContext::initContext()` function. This code checks the available Vulkan time domains and enables different Tracy GPU profiling features based on their availability. Notice how the C++ lambda is called directly in place, allowing the `hasHostQuery` variable to be declared as `const`.

```
#if defined(LVK_WITH_TRACY_GPU)
  std::vector<VkTimeDomainEXT> timeDomains;
  if (hasCalibratedTimestamps_) {
    uint32_t numTimeDomains = 0;
    vkGetPhysicalDeviceCalibrateableTimeDomainsEXT(
      vkPhysicalDevice_, &numTimeDomains, nullptr));
    timeDomains.resize(numTimeDomains);
    vkGetPhysicalDeviceCalibrateableTimeDomainsEXT(
      vkPhysicalDevice_, &numTimeDomains, timeDomains.data()));
  }
  const bool hasHostQuery = vkFeatures12_.hostQueryReset &&
    [&timeDomains]() -> bool {
      for (VkTimeDomainEXT domain : timeDomains)
        if (domain == VK_TIME_DOMAIN_CLOCK_MONOTONIC_RAW_EXT ||
            domain == VK_TIME_DOMAIN_QUERY_PERFORMANCE_COUNTER_EXT)
          return true;
      return false;
    }();
```

4. If host querying is supported, a dedicated command pool and command buffer are not required. Otherwise, we need to create them.

```
  if (hasHostQuery) {
    pimpl_->tracyVkCtx_ = TracyVkContextHostCalibrated(
      vkPhysicalDevice_, vkDevice_, vkResetQueryPool,
```

```
      vkGetPhysicalDeviceCalibrateableTimeDomainsEXT,
      vkGetCalibratedTimestampsEXT);
  } else {
    const VkCommandPoolCreateInfo ciCommandPool = {
      .sType = VK_STRUCTURE_TYPE_COMMAND_POOL_CREATE_INFO,
      .flags = VK_COMMAND_POOL_CREATE_RESET_COMMAND_BUFFER_BIT |
               VK_COMMAND_POOL_CREATE_TRANSIENT_BIT,
      .queueFamilyIndex = deviceQueues_.graphicsQueueFamilyIndex,
    };
    vkCreateCommandPool(vkDevice_,
      &ciCommandPool, nullptr, &pimpl_->tracyCommandPool_);
    lvk::setDebugObjectName(
      vkDevice_, VK_OBJECT_TYPE_COMMAND_POOL,
      (uint64_t)pimpl_->tracyCommandPool_,
      "Command Pool: VulkanContextImpl::tracyCommandPool_");
    const VkCommandBufferAllocateInfo aiCommandBuffer = {
      .sType = VK_STRUCTURE_TYPE_COMMAND_BUFFER_ALLOCATE_INFO,
      .commandPool = pimpl_->tracyCommandPool_,
      .level = VK_COMMAND_BUFFER_LEVEL_PRIMARY,
      .commandBufferCount = 1,
    };
    vkAllocateCommandBuffers(
      vkDevice_, &aiCommandBuffer, &pimpl_->tracyCommandBuffer_);
```

5. If calibrated timestamps are supported, Tracy uses a specialized code path for this scenario.

```
    if (hasCalibratedTimestamps_) {
      pimpl_->tracyVkCtx_ = TracyVkContextCalibrated(
        vkPhysicalDevice_,
        vkDevice_,
        deviceQueues_.graphicsQueue,
        pimpl_->tracyCommandBuffer_,
        vkGetPhysicalDeviceCalibrateableTimeDomainsEXT,
        vkGetCalibratedTimestampsEXT);
    } else {
      pimpl_->tracyVkCtx_ = TracyVkContext(
        vkPhysicalDevice_,
        vkDevice_,
        deviceQueues_.graphicsQueue,
        pimpl_->tracyCommandBuffer_);
    };
  }
```

```
  LVK_ASSERT(pimpl_->tracyVkCtx_);
#endif // LVK_WITH_TRACY_GPU
```

6. Add the destruction code at the very beginning of the `VulkanContext::~VulkanContext()` class destructor.

```
lvk::VulkanContext::~VulkanContext() {
  LVK_PROFILER_FUNCTION();
  VK_ASSERT(vkDeviceWaitIdle(vkDevice_));
#if defined(LVK_WITH_TRACY_GPU)
  TracyVkDestroy(pimpl_->tracyVkCtx_);
  if (pimpl_->tracyCommandPool_) {
    vkDestroyCommandPool(vkDevice_, pimpl_->tracyCommandPool_, nullptr);
  }
#endif // LVK_WITH_TRACY_GPU
  // ...
}
```

7. The final scaffolding step occurs in the `VulkanContext::submit()` function. We should collect the GPU sampling information using the `TracyVkCollect()` function before submitting the next command buffer, as shown below:

```
lvk::SubmitHandle lvk::VulkanContext::submit(
  lvk::ICommandBuffer& commandBuffer, TextureHandle present)
{
  LVK_PROFILER_FUNCTION();
  CommandBuffer* vkCmdBuffer =
    static_cast<CommandBuffer*>(&commandBuffer);
#if defined(LVK_WITH_TRACY_GPU)
  TracyVkCollect(
    pimpl_->tracyVkCtx_, vkCmdBuffer->wrapper_->cmdBuf_);
#endif // LVK_WITH_TRACY_GPU
  // ...
}
```

Now we are ready to do Vulkan GPU profiling with Tracy.

How it works...

The actual GPU profiling takes place inside the `CommandBuffer` functions, where we can now use the `LVK_PROFILER_GPU_ZONE()` macro we defined earlier. For example, here's how some of the functions will look:

```
void lvk::CommandBuffer::cmdDraw(uint32_t vertexCount,
  uint32_t instanceCount, uint32_t firstVertex, uint32_t baseInstance)
{
  LVK_PROFILER_FUNCTION();
  LVK_PROFILER_GPU_ZONE(
   "cmdDraw()", ctx_, wrapper_->cmdBuf_, LVK_PROFILER_COLOR_CMD_DRAW);
  if (vertexCount == 0) return;
  vkCmdDraw(wrapper_->cmdBuf_,
    vertexCount, instanceCount, firstVertex, baseInstance);
}
void lvk::CommandBuffer::cmdDispatchThreadGroups(
  const Dimensions& threadgroupCount, const Dependencies& deps)
{
  LVK_PROFILER_FUNCTION();
  LVK_PROFILER_GPU_ZONE("cmdDispatchThreadGroups()", ctx_,
    wrapper_->cmdBuf_, LVK_PROFILER_COLOR_CMD_DISPATCH);
  // ...
}
```

Before returning to Vulkan rendering, let's explore yet another small but useful profiling trick and learn how to implement a simple yet good frames-per-second counter.

Adding a frames-per-second counter

The **frames-per-second** (**FPS**) counter is the cornerstone of all graphical applications profiling and performance measurements. In this recipe, we will learn how to implement a simple FPS counter class and use it to roughly measure the performance of our applications.

Getting ready

The source code for this recipe can be found in `Chapter04/03_FPS`. The `FramesPerSecondCounter` class is located in `shared/UtilsFPS.h`.

How to do it...

Let's implement the FramesPerSecondCounter class containing all the machinery required to calculate the average FPS for a given time interval:

1. First, we need some member fields to store the duration of a sliding window, the number of frames rendered in the current interval, and the accumulated time of this interval. The printFPS_ Boolean field can be used to enable or disable FPS printing to the console:

```
class FramesPerSecondCounter {
public:
  float avgInterval_        = 0.5f;
  unsigned int numFrames_   = 0;
  double accumulatedTime_   = 0;
  float currentFPS_         = 0.0f;
  bool printFPS_            = true;
```

2. A single explicit constructor can override the averaging interval's default duration:

```
public:
  explicit FramesPerSecondCounter(
    float avgInterval = 0.5f)
  : avgInterval_(avgInterval)
  { assert(avgInterval > 0.0f); }
```

3. The tick() method should be called from the main loop. It accepts the time duration elapsed since the previous call and a Boolean flag, which should be set to true if a new frame has been rendered during this iteration. This flag is a convenience feature to handle situations where frame rendering can be skipped in the main loop for various reasons, such as simulation pausing. The time accumulates until it reaches the value of avgInterval_:

```
  bool tick(
    float deltaSeconds, bool frameRendered = true)
  {
    if (frameRendered) numFrames_++;
    accumulatedTime_ += deltaSeconds;
```

4. Once enough time has accumulated, we can do averaging, update the current FPS value, and print debug info to the console. We should reset the number of frames and accumulated time at this point:

```
    if (accumulatedTime_ > avgInterval_) {
      currentFPS_ = static_cast<float>(numFrames_ / accumulatedTime_);
      if (printFPS_)
        printf("FPS: %.1f\n", currentFPS_);
      numFrames_       = 0;
```

```
        accumulatedTime_ = 0;
        return true;
      }
      return false;
    }
```

5. Let's add a helper method to retrieve the current FPS value:

```
      inline float getFPS() const
      { return currentFPS_; }
    };
```

Now, let's take a look at how to use this class in our main loop. Let's augment the main loop of our demo applications to display an FPS counter in the console:

1. First, let us define a `FramesPerSecondCounter` object and a couple of variables to store the current timestamp and the delta since the last rendered frame. We have chosen to use an ad hoc 0.5-second averaging interval; feel free to experiment with different values:

```
double timeStamp    = glfwGetTime();
float deltaSeconds = 0.0f;
FramesPerSecondCounter fpsCounter(0.5f);
```

2. Within the main loop, update the current timestamp and calculate the frame duration by finding a delta between two consecutive timestamps. Then, pass this calculated delta to the `tick()` method:

```
while (!glfwWindowShouldClose(window)) {
  fpsCounter.tick(deltaSeconds);
  const double newTimeStamp = glfwGetTime();
  deltaSeconds = static_cast<float>(newTimeStamp - timeStamp);
  timeStamp    = newTimeStamp;
  // ...do the rest of your rendering here...
}
```

The console output of the running application should look similar to the following. Vertical sync is turned off:

```
FPS: 3924.7
FPS: 4322.4
FPS: 4458.9
FPS: 4445.1
FPS: 4581.4
```

The application window should look like that shown in the following screenshot, with an FPS counter rendered in the top-right corner:

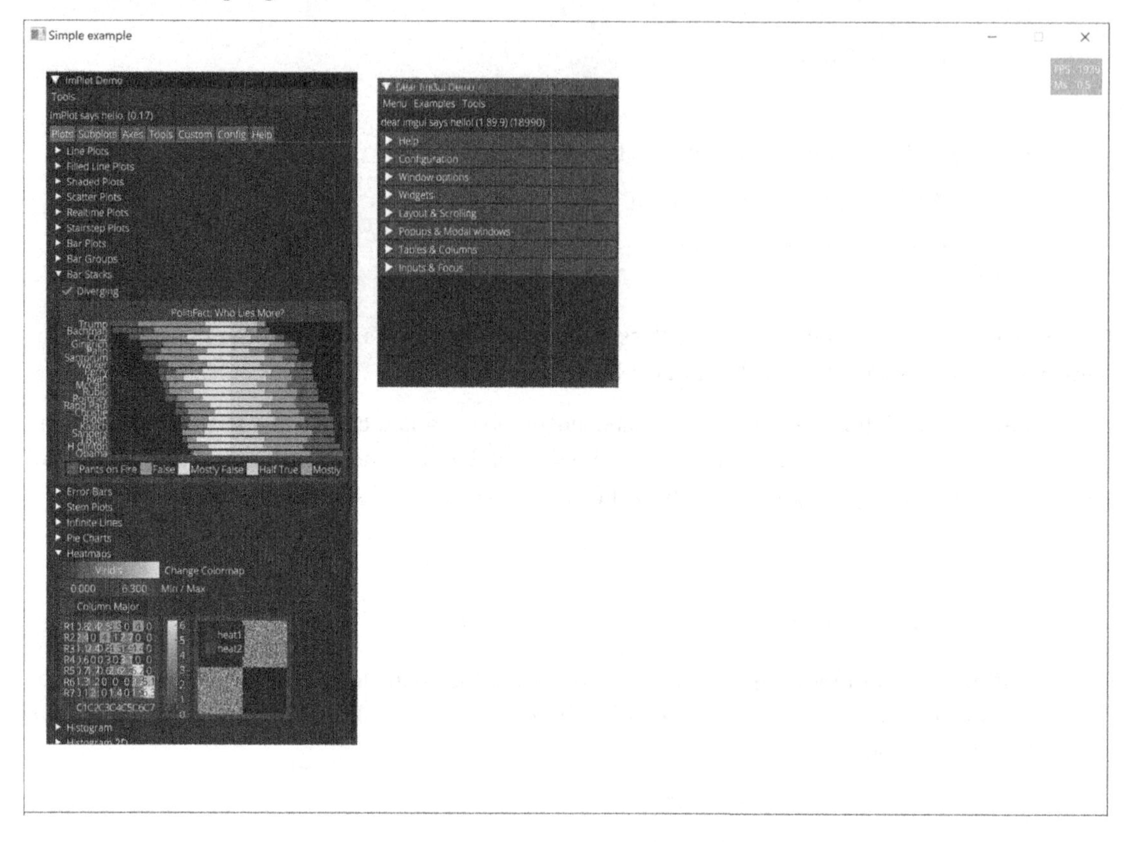

Figure 4.3: ImGui and ImPlot user interfaces with an FPS counter

Let's check the source code to learn how to add this ImGui FPS widget to your app. Here's a fragment that fills in a command buffer:

1. We set up the framebuffer parameters and start rendering. ImGui rendering is started with `imgui->beginFrame()` as we learned in the *Rendering ImGui user interface* recipe:

```
lvk::ICommandBuffer& buf = ctx->acquireCommandBuffer();
const lvk::Framebuffer framebuffer = { .color = { {
  .texture = ctx->getCurrentSwapchainTexture() } } };
buf.cmdBeginRendering(
  { .color = { { .loadOp = lvk::LoadOp_Clear,
    .clearColor = { 1.0f, 1.0f, 1.0f, 1.0f } } } },
  framebuffer);
imgui->beginFrame(framebuffer);
```

2. We get the current viewport parameters from ImGui and set the position of the next ImGui window to be aligned close to the top-right corner of the viewport work area. Sizes are hard-coded to be in the app window's pixels. The `ImGuiCond_Always` flag tells ImGui to set this position every frame:

```
if (const ImGuiViewport* v = ImGui::GetMainViewport()) {
  ImGui::SetNextWindowPos({
    v->WorkPos.x + v->WorkSize.x - 15.0f,
    v->WorkPos.y + 15.0f },
    ImGuiCond_Always, { 1.0f, 0.0f });
}
```

3. Set the next window to be transparent. We use `SetNextWindowSize()` to assign a fixed size value to the window. The width is calculated using `CalcTextSize()`. Note how that "`FPS : _______`" placeholder string is used here as a parameter to make sure the width of the window does not fluctuate based on the number of digits in the numeric FPS value:

```
ImGui::SetNextWindowBgAlpha(0.30f);
ImGui::SetNextWindowSize(
  ImVec2(ImGui::CalcTextSize("FPS : _______").x, 0));
```

4. An ImGui window that contains the FPS counter is rendered using various ImGui flags so that all unnecessary window decorations are disabled and no user interaction can happen with the window:

```
if (ImGui::Begin("##FPS", nullptr,
                 ImGuiWindowFlags_NoDecoration |
                 ImGuiWindowFlags_AlwaysAutoResize |
                 ImGuiWindowFlags_NoSavedSettings |
                 ImGuiWindowFlags_NoFocusOnAppearing |
                 ImGuiWindowFlags_NoNav |
                 ImGuiWindowFlags_NoMove))
{
  ImGui::Text("FPS : %i", (int)fpsCounter.getFPS());
  ImGui::Text("ms  : %.1f", 1000.0 / fpsCounter.getFPS());
}
ImGui::End();
```

5. After we have rendered the FPS window, let's draw ImPlot and ImGui demo windows so you can explore them. The ImPlot library will be covered a bit more in subsequent recipes:

```
ImPlot::ShowDemoWindow();
ImGui::ShowDemoWindow();
imgui->endFrame(buf);
buf.cmdEndRendering();
```

Now you know how to display a window in your apps that includes a nice FPS counter. Although this feature is straightforward, it can be tiresome to repeatedly include this code in every app. In the upcoming recipes, we will introduce a `VulkanApp` helper class that will handle various utility functions like this. But for now, let's go back to rendering and explore how to work with cube map textures.

Using cube map textures in Vulkan

A cube map is a texture that contains 6 individual 2D textures that together form 6 sides of a cube. A useful property of cube maps is that they can be sampled using a direction vector. This comes in handy when representing light coming into a scene from different directions. For example, we can store the diffuse part of the physically based lighting equation in an **irradiance cube map**, which we will touch on in *Chapter 6*.

Loading 6 faces of a cube map into *LightweightVK* is a fairly straightforward operation. However, instead of just 6 faces, cube maps are often stored as **equirectangular projections** or as vertical or horizontal crosses. The equirectangular projection is such a projection that maps longitude and latitude (vertical and horizontal lines) to straight, even lines, making it a very easy and popular way to store light probe images, as shown in *Figure 4.4* later in this recipe.

In this recipe, we will learn how to convert this cube map representation into 6 faces and render them with Vulkan.

Getting ready

There are many websites that offer high-dynamic range environment textures under various licenses. Check out `https://polyhaven.com` and `https://hdrmaps.com` for useful content.

The complete source code for this recipe can be found in the source code bundle under the name `Chapter04/04_CubeMap`.

Before we start working with cube maps, let us introduce a simple `Bitmap` helper class to work with bitmap images in 8-bit and 32-bit floating point formats. You can find it in `shared/Bitmap.h`:

1. Let us declare the interface part of the `Bitmap` class as follows:

```
class Bitmap {
public:
  Bitmap() = default;
  Bitmap(int w, int h, int comp, eBitmapFormat fmt);
  Bitmap(int w, int h, int d, int comp, eBitmapFormat fmt);
  Bitmap(int w, int h, int comp, eBitmapFormat fmt, const void* ptr);
```

2. Let's set the width, height, depth, and number of components per pixel:

```
  int w_ = 0;
  int h_ = 0;
  int d_ = 1;
  int comp_ = 3;
```

3. The type of a single component can be either unsigned byte or float. The type of this bitmap can be a 2D texture or a cube map. We store the actual pixel data of this bitmap in an `std::vector` container for simplicity:

```
  eBitmapFormat fmt_ = eBitmapFormat_UnsignedByte;
  eBitmapType type_  = eBitmapType_2D;
  std::vector<uint8_t> data_;
```

4. Next we need a helper function to get the number of bytes necessary to store one component of a specified format. This also requires a getter and setter for a two-dimensional image. We will come back to this later:

```
  static int getBytesPerComponent(eBitmapFormat fmt);
  void setPixel(int x, int y, const glm::vec4& c);
  glm::vec4 getPixel(int x, int y) const;
};
```

The implementation is also located in `shared/Bitmap.h`. Now let us use this class to build more high-level cube map conversion functions.

How to do it...

We have a cube map at `data/piazza_bologni_1k.hdr`, which is available under the CC0 license and was originally downloaded from `https://polyhaven.com/a/piazza_bologni`. The environment map image comes in an equirectangular projection and looks like this:

Figure 4.4: Equirectangular projection

Note

An equirectangular projection is very common and straightforward. In this type of projection, each horizontal position on the rectangular image corresponds to a longitude on the sphere, while each vertical position represents a latitude. Because of these relationships, it is also known as a lat/long projection.

Let us convert it into a vertical cross. In the vertical cross format, each cube map face is represented as a square inside the entire 2D image, as follows:

Figure 4.5: Vertical cross

If we use a naïve way to convert an equirectangular projection to cube map faces by iterating over its pixels, calculating the Cartesian coordinates for each pixel, and saving the pixel into a cube map face using the Cartesian coordinates, we will end up with a texture heavily damaged by a Moiré pattern caused by the insufficient sampling of the resulting cube map. A better way is to do it the other way round. That means iterating over each pixel of the resulting cube map faces, calculating source floating-point equirectangular coordinates corresponding to each pixel, and sampling the equirectangular texture using bilinear interpolation. This way the final cube map will be free of artifacts:

1. The first step is to introduce a helper function that maps integer coordinates inside a specified cube map face into floating-point normalized coordinates. This helper is handy because all faces in the vertical cross cube map have different vertical orientations:

```
vec3 faceCoordsToXYZ(int i, int j, int faceID, int faceSize) {
  const float A = 2.0f * float(i) / faceSize;
  const float B = 2.0f * float(j) / faceSize;
  if (faceID == 0) return vec3(   -1.0f, A - 1.0f,  B - 1.0f);
  if (faceID == 1) return vec3(A - 1.0f,     -1.0f,  1.0f - B);
  if (faceID == 2) return vec3(    1.0f, A - 1.0f,  1.0f - B);
  if (faceID == 3) return vec3(1.0f - A,      1.0f,  1.0f - B);
  if (faceID == 4) return vec3(B - 1.0f, A - 1.0f,  1.0f);
  if (faceID == 5) return vec3(1.0f - B, A - 1.0f, -1.0f);
  return vec3();
}
```

2. The conversion function starts as follows and calculates the face size, width, and height of the resulting bitmap. It is located in `shared/UtilsCubemap.cpp`:

```
Bitmap convertEquirectangularMapToVerticalCross(const Bitmap& b) {
  if (b.type_ != eBitmapType_2D) return Bitmap();
  const int faceSize = b.w_ / 4;
  const int w = faceSize * 3;
  const int h = faceSize * 4;
  Bitmap result(w, h, 3);
```

3. These points define the locations of individual faces inside the cross:

```
  const ivec2 kFaceOffsets[] = {
    ivec2(faceSize, faceSize * 3),
    ivec2(0, faceSize),
    ivec2(faceSize, faceSize),
    ivec2(faceSize * 2, faceSize),
    ivec2(faceSize, 0),
    ivec2(faceSize, faceSize * 2)
  };
```

4. Two constants will be necessary to clamp the texture lookup:

```
const int clampW = b.w_ - 1;
const int clampH = b.h_ - 1;
```

5. Now we can start iterating over the 6 cube map faces and each pixel inside each face:

```
for (int face = 0; face != 6; face++) {
  for (int i = 0; i != faceSize; i++) {
    for (int j = 0; j != faceSize; j++) {
```

6. We use trigonometry functions to calculate the latitude and longitude coordinates from the Cartesian cube map coordinates.

```
      const vec3  P = faceCoordsToXYZ(i, j, face, faceSize);
      const float R = hypot(P.x, P.y);
      const float theta = atan2(P.y, P.x);
      const float phi   = atan2(P.z, R);
```

To learn more about spherical coordinate systems, please follow this link: `https://en.wikipedia.org/wiki/Spherical_coordinate_system`.

7. Now we can map the latitude and longitude into floating-point coordinates inside the equirectangular image:

```
      const float Uf =
        float(2.0f * faceSize * (theta + M_PI) / M_PI);
      const float Vf =
        float(2.0f * faceSize * (M_PI / 2.0f - phi) / M_PI);
```

8. Based on these floating-point coordinates, we get two pairs of integer UV coordinates, which we will use to sample 4 texels for bilinear interpolation:

```
      const int U1 = clamp(int(floor(Uf)), 0, clampW);
      const int V1 = clamp(int(floor(Vf)), 0, clampH);
      const int U2 = clamp(U1 + 1, 0, clampW);
      const int V2 = clamp(V1 + 1, 0, clampH);
```

9. Get the fractional part for the bilinear interpolation and fetch 4 samples, `A`, `B`, `C`, and `D`, from the equirectangular map:

```
      const float s = Uf - U1;
      const float t = Vf - V1;
      const vec4 A = b.getPixel(U1, V1);
      const vec4 B = b.getPixel(U2, V1);
      const vec4 C = b.getPixel(U1, V2);
      const vec4 D = b.getPixel(U2, V2);
```

10. Do the bilinear interpolation and set the resulting pixel value in the vertical-cross cube map:

```
                const vec4 color =
                  A * (1 - s) * (1 - t) + B * (s) * (1 - t) +
                  C * (1 - s) * t + D * (s) * (t);
                result.setPixel(i + kFaceOffsets[face].x,
                                j + kFaceOffsets[face].y, color);
            }
        }
    }
    return result;
}
```

The Bitmap class takes care of the pixel format inside the image data.

Now we can write code to cut the vertical cross into tightly packed rectangular cube map faces. Here is how to do it:

1. First, let us review the layout of the vertical cross image corresponding to the Vulkan cube map faces layout.

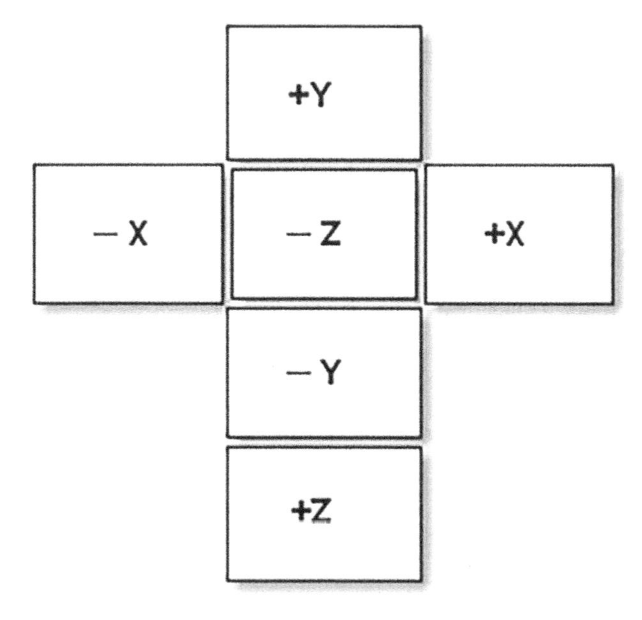

Figure 4.6: Layout of the vertical cross image

1. The layout is 3 by 4 faces, which makes it possible to calculate the dimensions of the resulting cube map as follows. The code is from `shared/UtilsCubemap.cpp`:

```
Bitmap convertVerticalCrossToCubeMapFaces(const Bitmap& b) {
  const int faceWidth  = b.w_ / 3;
  const int faceHeight = b.h_ / 4;
  Bitmap cubemap(faceWidth, faceHeight, 6, b.comp_, b.fmt_);
```

2. Let us set up pointers to read data from and write data to. This function is pixel format-agnostic so it needs to know the size of each pixel in bytes to be able to move pixels around with `memcpy()`:

```
  const uint8_t* src = b.data_.data();
  uint8_t* dst = cubemap.data_.data();
  const int pixelSize = cubemap.comp_ *
    Bitmap::getBytesPerComponent(cubemap.fmt_);
```

3. Iterate over the faces and over every pixel of each face. The order of cube map faces here corresponds to the order of Vulkan cube map faces as described in *Vulkan specification 16.5.4. Cube Map Face Selection*:

```
  for (int face = 0; face != 6; ++face) {
    for (int j = 0; j != faceHeight; ++j) {
      for (int i = 0; i != faceWidth; ++i) {
        int x = 0;
        int y = 0;
```

4. Calculate the source pixel position in the vertical cross layout based on the destination cube map face index:

```
        switch (face) {
        case 0: // +X
          x = 2 * faceWidth  + i;
          y = 1 * faceHeight + j;
          break;
        case 1: // -X
          x = i;
          y = faceHeight + j;
          break;
        case 2: // +Y
          x = 1 * faceWidth + i;
          y = j;
          break;
        case 3: // -Y
          x = 1 * faceWidth  + i;
          y = 2 * faceHeight + j;
```

```
            break;
          case 4: // +Z
            x = faceWidth  + i;
            y = faceHeight + j;
            break;
          case 5: // -Z
            x = 2 * faceWidth - (i + 1);
            y = b.h_ - (j + 1);
            break;
          }
```

5. Copy the pixel and advance to the next one:

```
          memcpy(dst, src + (y * b.w_ + x) * pixelSize, pixelSize);
          dst += pixelSize;
        }
      }
    }
    return cubemap;
  }
```

The resulting cube map contains an array of six 2D images. Let us write some more C++ code to load and convert the actual texture data and upload it into *LightweightVK*. The source code is located in `Chapter04/04_CubeMap/src/main.cpp`:

1. Use the `STB_image` floating point API to load a high dynamic range image from an `.hdr` file:

```
int w, h;
const float* img = stbi_loadf(
  "data/piazza_bologni_1k.hdr", &w, &h, nullptr, 4);
Bitmap in(w, h, 4, eBitmapFormat_Float, img);
```

2. Convert an equirectangular map to a vertical cross and save the resulting image to an `.hdr` file for further inspection:

```
Bitmap out = convertEquirectangularMapToVerticalCross(in);
stbi_image_free((void*)img);
stbi_write_hdr(".cache/screenshot.hdr", out.w_, out.h_, out.comp_,
  (const float*)out.data_.data());
```

3. Convert the loaded vertical cross image to the actual cube map faces:

```
Bitmap cubemap = convertVerticalCrossToCubeMapFaces(out);
```

4. Now, uploading texture data to LightweightVK is straightforward. We call the `IContext::createTexture()` member function to create a texture and provide a pointer to the cube map data returned by `cubemap.data_.data()`:

```
Holder<TextureHandle> cubemapTex = ctx->createTexture({
    .type       = lvk::TextureType_Cube,
    .format     = lvk::Format_RGBA_F32,
    .dimensions = {(uint32_t)cubemap.w_, (uint32_t)cubemap.h_},
    .usage      = lvk::TextureUsageBits_Sampled,
    .data       = cubemap.data_.data(),
    .debugName  = "data/piazza_bologni_1k.hdr",
});
```

Now we should take a look at how to write the GLSL shaders for this example:

1. Let us make a vertex shader `Chapter04/04_CubeMap/src/main.vert` that will take a model, view, and projection matrices as its inputs. We also need a camera position and bindless texture IDs for a mesh texture and for our cube map:

```
layout(std430, buffer_reference) readonly buffer PerFrameData {
  mat4 model;
  mat4 view;
  mat4 proj;
  vec4 cameraPos;
  uint tex;
  uint texCube;
};
```

2. A buffer reference to `PerFrameData` is passed into the shader using Vulkan **push constants** (`pc` in the following code):

```
layout(push_constant) uniform PushConstants {
  PerFrameData pc;
};
```

3. The per-vertex attributes are provided to the vertex shader. The `PerVertex` structure is used to pass parameters to a fragment shader. Normal vectors are transformed with a matrix calculated as the inverse-transpose of the model matrix:

```
struct PerVertex {
  vec2 uv;
  vec3 worldNormal;
  vec3 worldPos;
};
layout (location = 0) in vec3 pos;
```

```
layout (location = 1) in vec3 normal;
layout (location = 2) in vec2 uv;
layout (location=0) out PerVertex vtx;
void main() {
  gl_Position = pc.proj * pc.view * pc.model * vec4(pos, 1.0);
  mat4 model = pc.model;
  mat3 normalMatrix = transpose( inverse(mat3(pc.model)) );
  vtx.uv = uv;
  vtx.worldNormal = normal * normalMatrix;
  vtx.worldPos = (model * vec4(pos, 1.0)).xyz;
}
```

Now let's take a look at the fragment shader found at `Chapter04/04_CubeMap/src/main.frag`:

1. It shares the declaration of the `PerVertex` structure with the vertex shader mentioned above. The declaration is located in the file `Chapter04/04_CubeMap/src/common.sp`. We skip it here for the sake of brevity. The fragment shader uses the `textureBindlessCube()` helper function to sample the cube map using the calculated reflection vector. This function was discussed in detail in the *Using texture data in Vulkan* recipe in *Chapter 3*. The reflected direction vector is calculated using the `reflect()` GLSL built-in function:

```
layout (location=0) in PerVertex vtx;
layout (location=0) out vec4 out_FragColor;
void main() {
  vec3 n = normalize(vtx.worldNormal);
  vec3 v = normalize(pc.cameraPos.xyz - vtx.worldPos);
  vec3 reflection = -normalize(reflect(v, n));
  vec4 colorRefl = textureBindlessCube(pc.texCube, 0, reflection);
```

2. To add a more developed visual appearance, we add some diffuse lighting to our 3D model using a hardcoded light direction of `(0, 0.1, -1)`:

```
  vec4 Ka = colorRefl * 0.3;
  float NdotL = clamp(dot(n, normalize(vec3(0,0,-1))), 0.1, 1.0);
  vec4 Kd = textureBindless2D(pc.tex, 0, vtx.uv) * NdotL;
  out_FragColor = Ka + Kd;
};
```

The resulting output from the application looks as follows. Note the blown-out white areas of the sky in the reflection due to how a high dynamic range image is displayed directly onto a low dynamic range framebuffer. We will come back to this issue in *Chapter 10*, and implement a simple HDR tone-mapping example.

Figure 4.7: Reflective rubber duck

Now let's get back to improving the user interaction capabilities and learn how to implement a simple camera class to move around and debug our 3D scenes.

There's more...

In OpenGL, developers had to enable a special cube map sampling mode to ensure seamless filtering across all cube map faces. In Vulkan, all cube map texture fetches are seamless (as described under *Cube Map Edge Handling* in the Vulkan specification), except the ones with `VK_FILTER_NEAREST`, which are clamped to the face edge.

Working with a 3D camera and basic user interaction

To debug a graphical application, it is very helpful to be able to navigate and move around within a 3D scene using a keyboard or mouse. Graphics APIs themselves are not familiar with concepts of cameras and user interaction, so we have to implement a camera model that will convert user input into a view matrix usable by Vulkan. In this recipe, we will learn how to create a simple yet extensible 3D camera implementation and use it to enhance the functionality of our Vulkan examples.

Getting ready

The source code for this recipe can be found in `Chapter04/05_Camera`. The camera classes are declared and implemented in the file `shared/Camera.h`.

How to do it...

Our camera implementation will calculate a view matrix and a 3D position point based on the selected dynamic model. Let's look at the steps:

1. First, let us implement the `Camera` class, which will represent our main API to work with a 3D camera. The class stores a reference to an instance of the `CameraPositionerInterface` class, being a polymorphic implementation of the underlying camera model to allow runtime switches of camera behaviors:

```
class Camera final {
public:
  explicit Camera(CameraPositionerInterface& positioner)
    : positioner_(&positioner)  {}
  Camera(const Camera&) = default;
  Camera& operator = (const Camera&) = default;
  mat4 getViewMatrix() const {
    return positioner_->getViewMatrix();
  }
  vec3 getPosition() const {
    return positioner_->getPosition();
  }
private:
  const CameraPositionerInterface* positioner_;
};
```

The interface of `CameraPositionerInterface` contains only pure virtual methods and a default virtual destructor:

```
class CameraPositionerInterface {
public:
  virtual ~CameraPositionerInterface() = default;
  virtual mat4 getViewMatrix() const = 0;
  virtual vec3 getPosition() const = 0;
};
```

2. Now we can implement the actual camera model. We will start with a quaternion-based first-person camera that can be freely moved in space in any direction. Let us look at the `CameraPositioner_FirstPerson` class. The inner `Movement` structure contains Boolean flags that define the current motion state of our camera. This is useful to decouple keyboard and mouse inputs from the camera control logic:

```
class CameraPositioner_FirstPerson final:
  public CameraPositionerInterface
{
public:
  struct Movement {
    bool forward_   = false;
    bool backward_  = false;
```

```
    bool left_      = false;
    bool right_     = false;
    bool up_        = false;
    bool down_      = false;
    bool fastSpeed_ = false;
  } movement_;
```

3. Various numeric parameters define how responsive the camera will be to acceleration and damping. These parameters can be tweaked as you see fit:

```
  float mouseSpeed_   = 4.0f;
  float acceleration_ = 150.0f;
  float damping_      = 0.2f;
  float maxSpeed_     = 10.0f;
  float fastCoef_     = 10.0f;
```

4. We need certain private data members to control the camera state, such as the previous mouse position, the current camera position and orientation, the current movement speed, and the vector representing the "up" direction:

```
private:
  vec2 mousePos_          = vec2(0);
  vec3 cameraPosition_    = vec3(0.0f, 10.0f, 10.0f);
  quat cameraOrientation_ = quat(vec3(0));
  vec3 moveSpeed_         = vec3(0.0f);
  vec3 up_                = vec3(0.0f, 0.0f, 1.0f);
```

5. The non-default constructor takes the camera's initial position, a target position, and a vector pointing upwards. This input is similar to what one might normally use to construct a look-at viewing matrix. Indeed, we use the `glm::lookAt()` function to initialize the camera:

```
public:
  CameraPositioner_FirstPerson() = default;
  CameraPositioner_FirstPerson(const vec3& pos,
    const vec3& target, const vec3& up)
  : cameraPosition_(pos)
  , cameraOrientation_(glm::lookAt(pos, target, up))
  , up_(up)
  {}
```

6. Now, we can add some dynamics to our camera model. The `update()` method should be called every frame and take the time elapsed since the previous frame, as well as the mouse position and a mouse-button-pressed flag:

```
  void update(double deltaSeconds,
    const glm::vec2& mousePos, bool mousePressed)
```

```
{
  if (mousePressed) {
    const glm::vec2 delta = mousePos - mousePos_;
    const glm::quat deltaQuat = glm::quat(glm::vec3(
      mouseSpeed_ * delta.y, mouseSpeed_ * delta.x, 0.0f));
    cameraOrientation_ =
      glm::normalize(deltaQuat * cameraOrientation_);
    setUpVector(up_);
  }
  mousePos_ = mousePos;
```

Now, when the mouse button is pressed, we calculate a delta vector versus the previous mouse position, and use it to construct a rotation quaternion. This quaternion is used to rotate the camera. Once the camera rotation is applied, we should update the mouse position state.

7. Now we should establish the camera's coordinate system to calculate the camera movement. Let us extract the forward, right, and up vectors from the `mat4` view matrix:

```
const mat4 v = glm::mat4_cast(cameraOrientation_);
const vec3 forward = -vec3(v[0][2], v[1][2], v[2][2]);
const vec3 right   =  vec3(v[0][0], v[1][0], v[2][0]);
const vec3 up = cross(right, forward);
```

The `forward` vector corresponds to the camera's direction, which is the direction the camera is pointing at. The `right` vector corresponds to the positive X-axis of the camera space. The `up` vector is the positive Y-axis of the camera space, which is perpendicular to the first two vectors and can be calculated as their cross product.

8. The camera coordinate system has been established. Now we can apply our input state from the `Movement` structure to control the movement of our camera:

```
vec3 accel(0.0f);
if (movement_.forward_ ) accel += forward;
if (movement_.backward_) accel -= forward;
if (movement_.left_ ) accel -= right;
if (movement_.right_) accel += right;
if (movement_.up_   ) accel += up;
if (movement_.down_) accel -= up;
if (movement_.fastSpeed_) accel *= fastCoef_;
```

Instead of controlling the camera speed or position directly, we let the user input control only the acceleration vector directly. This way, the camera's behavior is much smoother, more natural, and non-jerky.

9. If, based on the input state, the calculated camera acceleration is zero, we should decelerate the camera's motion gradually, according to the `damping_` parameter. Otherwise, we should integrate the camera motion using simple Euler integration. The maximum possible speed value is clamped according to the `maxSpeed_` parameter:

```
    if (accel == vec3(0)) {
      moveSpeed_ -= moveSpeed_ * std::min((1.0f / damping_) *
        static_cast<float>(deltaSeconds), 1.0f);
    }
    else {
      moveSpeed_ += accel * acceleration_ *
        static_cast<float>(deltaSeconds);
      const float maxSpeed =
        movement_.fastSpeed_ ? maxSpeed_ * fastCoef_ : maxSpeed_;
      if (glm::length(moveSpeed_) > maxSpeed)
        moveSpeed_ = glm::normalize(moveSpeed_) * maxSpeed;
    }
    cameraPosition_ += moveSpeed_ *
      static_cast<float>(deltaSeconds);
  }
```

10. The view matrix can be calculated from the camera orientation quaternion and camera position in the following way:

```
  virtual mat4 getViewMatrix() const override {
    const mat4 t = glm::translate(mat4(1.0f), -cameraPosition_);
    const mat4 r = glm::mat4_cast(cameraOrientation_);
    return r * t;
  }
```

The translational part is inferred from the `cameraPosition_` vector and the rotational part is calculated directly from the orientation quaternion.

11. Helpful getters and setters are trivial, except for the `setUpVector()` method, which has to recalculate the camera orientation using the existing camera position and direction as follows:

```
  virtual vec3 getPosition() const override {
    return cameraPosition_;
  }
  void setPosition(const vec3& pos) {
    cameraPosition_ = pos;
  }
  void setUpVector(const vec3& up) {
    const mat4 view = getViewMatrix();
    const vec3 dir  = -vec3(view[0][2], view[1][2], view[2][2]);
```

```
        cameraOrientation_ =
          glm::lookAt(cameraPosition_, cameraPosition_ + dir, up);
      }
```

12. One additional helper function is necessary to reset the previous mouse position to prevent jerky rotation movements when, for example, the mouse cursor leaves the window:

```
      void resetMousePosition(const vec2& p) { mousePos_ = p; };
    };
```

The above class can be used in 3D applications to move the viewer around. Let us see how it works.

How it works...

The demo application is based on the cube map example from the previous *Using cube map textures in Vulkan* recipe. The updated code is located at `Chapter04/05_Camera/src/main.cpp`.

We add a mouse state and define `CameraPositioner` and `Camera`. Let them be global variables:

```
struct MouseState {
  vec2 pos         = vec2(0.0f);
  bool pressedLeft = false;
} mouseState;
const vec3 kInitialCameraPos    = vec3(0.0f, 1.0f, -1.5f);
const vec3 kInitialCameraTarget = vec3(0.0f, 0.5f,  0.0f);
CameraPositioner_FirstPerson positioner(
  kInitialCameraPos,
  kInitialCameraTarget,
  vec3(0.0f, 1.0f, 0.0f));
Camera camera(positioner);
```

The GLFW cursor position callback should update `mouseState` the following way:

```
glfwSetCursorPosCallback(
  window, [](auto* window, double x, double y) {
    int width, height;
    glfwGetFramebufferSize(window, &width, &height);
    mouseState.pos.x = static_cast<float>(x / width);
    mouseState.pos.y = 1.0f - static_cast<float>(y / height);
  }
);
```

Here, we convert window pixel coordinates into normalized `0...1` coordinates and accommodate the inverted Y-axis.

The GLFW mouse button callback passes GLFW mouse events to ImGui and sets the `pressedLeft` flag when the left mouse button is pressed:

```
glfwSetMouseButtonCallback(
  window, [](auto* window, int button, int action, int mods) {
    if (button == GLFW_MOUSE_BUTTON_LEFT)
      mouseState.pressedLeft = action == GLFW_PRESS;
    double xpos, ypos;
    glfwGetCursorPos(window, &xpos, &ypos);
    const ImGuiMouseButton_ imguiButton =
     (button == GLFW_MOUSE_BUTTON_LEFT) ?
       ImGuiMouseButton_Left :
         (button == GLFW_MOUSE_BUTTON_RIGHT ?
           ImGuiMouseButton_Right :
           ImGuiMouseButton_Middle);
    ImGuiIO& io = ImGui::GetIO();
    io.MousePos = ImVec2((float)xpos, (float)ypos);
    io.MouseDown[imguiButton] = action == GLFW_PRESS;
  });
```

To handle keyboard input for camera movement, let us write the following GLFW keyboard callback:

```
glfwSetKeyCallback(window,
  [](GLFWwindow* window, int key, int, int action, int mods) {
    const bool press = action != GLFW_RELEASE;
    if (key == GLFW_KEY_ESCAPE)
      glfwSetWindowShouldClose(window, GLFW_TRUE);
    if (key == GLFW_KEY_W) positioner.movement_.forward_ = press;
    if (key == GLFW_KEY_S) positioner.movement_.backward_= press;
    if (key == GLFW_KEY_A) positioner.movement_.left_  = press;
    if (key == GLFW_KEY_D) positioner.movement_.right_ = press;
    if (key == GLFW_KEY_1) positioner.movement_.up_    = press;
    if (key == GLFW_KEY_2) positioner.movement_.down_  = press;
    if (mods & GLFW_MOD_SHIFT)
      positioner.movement_.fastSpeed_ = press;
    if (key == GLFW_KEY_SPACE) {
      positioner.lookAt(kInitialCameraPos,
        kInitialCameraTarget, vec3(0.0f, 1.0f, 0.0f));
      positioner.setSpeed(vec3(0));
    }
  });
```

The *WSAD* keys are used to move the camera around and the *Spacebar* is used to reorient the camera up vector to the world `(0, 1, 0)` vector and reset the position camera back to the initial position. The *Shift* key is used to move the camera faster.

We can update the camera positioner from the main loop using the following statement:

```
positioner.update(
  deltaSeconds, mouseState.pos, mouseState.pressedLeft);
```

Here's a code fragment to upload matrices into a Vulkan per-frame uniform buffer, similar to how it was done with fixed values in the previous chapters:

```
const vec4 cameraPos = vec4(camera.getPosition(), 1.0f);
const mat4 p  = glm::perspective(glm::radians(60.0f), ratio, 0.1f, 1000.0f);
const mat4 m1 = glm::rotate(
  mat4(1.0f), glm::radians(-90.0f), vec3(1, 0, 0));
const mat4 m2 = glm::rotate(
  mat4(1.0f), (float)glfwGetTime(), vec3(0.0f, 1.0f, 0.0f));
const mat4 v  = glm::translate(mat4(1.0f), vec3(cameraPos));
const PerFrameData pc = {
  .model     = m2 * m1,
  .view      = camera.getViewMatrix(),
  .proj      = p,
  .cameraPos = cameraPos,
  .tex       = texture.index(),
  .texCube   = cubemapTex.index(),
};
ctx->upload(bufferPerFrame, &pc, sizeof(pc));
```

Run the demo from `Chapter04/05_Camera` to play around with the keyboard and mouse:

Figure 4.8: Camera

There is more...

This camera design approach can be extended to accommodate different motion behaviors. In the next recipe, we will learn how to implement some other useful camera positioners.

The 3D camera functionality introduced in this recipe is incredibly valuable for our book. To reduce code duplication, we've created a helper class called `VulkanApp`. This class wraps the first-person camera positioner along with other features such as the frames-per-second counter, `ImGuiRenderer`, and some others. The `VulkanApp` class will be utilized in all the subsequent recipes throughout this book. You can find it in the `shared/VulkanApp.h` and `shared/VulkanApp.cpp` files.

Adding camera animations and motion

Besides having a user-controlled first-person camera, it is convenient to be able to position and move the camera programmatically inside a 3D scene – this is helpful for debugging when we need to organize automatic screenshot tests with camera movement, for example. In this recipe, we will show how to do it and extend our minimalistic 3D camera framework from the previous recipe. We will draw a combo box using ImGui to select between two camera modes: a first-person free camera, and a fixed camera moving to a user-specified point settable from the UI.

Getting ready

The full source code for this recipe is a part of the final demo application for this chapter, and you can find it in `Chapter04/06_DemoApp`. Implementations of all new camera-related functionality are located in the `shared/Camera.h` file.

How to do it...

Let's look at how to programmatically control our 3D camera using a simple ImGui-based user interface:

1. First, we need to add a new `CameraPosition_MoveTo` camera positioner that automatically moves the camera to a specified `vec3` point. For this purpose, we have to declare a bunch of global constants and variables:

```
const vec3 kInitialCameraPos    = vec3(0.0f, 1.0f, -1.5f);
const vec3 kInitialCameraAngles = vec3(-18.5f, 180.0f, 0.0f);
CameraPositioner_MoveTo positionerMoveTo(
  kInitialCameraPos, kInitialCameraAngles);
```

2. Inside the main loop, we should update our new camera positioner. The first-person camera positioner is updated automatically inside the `VulkanApp` class mentioned in the previous recipe:

```
positioner_moveTo.update(
  deltaSeconds, mouseState.pos, mouseState.pressedLeft);
```

Now, let's draw an ImGui combo box to select which camera positioner should be used to control the camera motion:

1. First, a few more global variables will come in handy to store the current camera type, items of the combo box UI, and the new value selected in the combo box:

```
const char* cameraType = "FirstPerson";
const char* comboBoxItems[] = { "FirstPerson", "MoveTo" };
const char* currentComboBoxItem = cameraType;
```

2. To render the camera control UI with a combo box, let's write the following code. A new ImGui window starts with a call to `ImGui::Begin()`:

```
ImGui::Begin("Camera Controls", nullptr,
  ImGuiWindowFlags_AlwaysAutoResize);
{
```

3. The combo box itself is rendered via `ImGui::BeginCombo()`. The second parameter is the previewed label name to show before opening the combo box. This function will return true if the user has clicked on a label:

```
  if (ImGui::BeginCombo("##combo", currentComboBoxItem)) {
    for (int n = 0; n < IM_ARRAYSIZE(comboBoxItems); n++) {
      const bool isSelected =
        currentComboBoxItem == comboBoxItems[n];
```

4. You may set the initial focus when opening the combo box. This is useful if you want to support scrolling or keyboard navigation inside the combo box:

```
      if (ImGui::Selectable(comboBoxItems[n], isSelected))
```

```
        currentComboBoxItem = comboBoxItems[n];
      if (isSelected)
        ImGui::SetItemDefaultFocus();
    }
```

5. Finalize the ImGui combo box rendering:

```
    ImGui::EndCombo();
  }
```

6. If the `MoveTo` camera type is selected, render `vec3` input sliders to get the camera position and Euler angles from the user:

```
  if (!strcmp(cameraType, "MoveTo")) {
    if (ImGui::SliderFloat3("Position",
        glm::value_ptr(cameraPos), -10.0f, +10.0f)) {
      positionerMoveTo.setDesiredPosition(cameraPos);
    }
    if (ImGui::SliderFloat3("Pitch/Pan/Roll",
        glm::value_ptr(cameraAngles), -180.0f, +180.0f)) {
      positionerMoveTo.setDesiredAngles(cameraAngles);
    }
  }
```

7. If a new selected combo box item is different from the current camera type, print a debug message and change the active camera mode:

```
  if (currentComboBoxItem &&
      strcmp(currentComboBoxItem, cameraType)) {
    printf("Selected new camera type: %s\n", currentComboBoxItem);
    cameraType = currentComboBoxItem;
    reinitCamera(app);
  }
```

8. The resulting combo box should look as in the following screenshot:

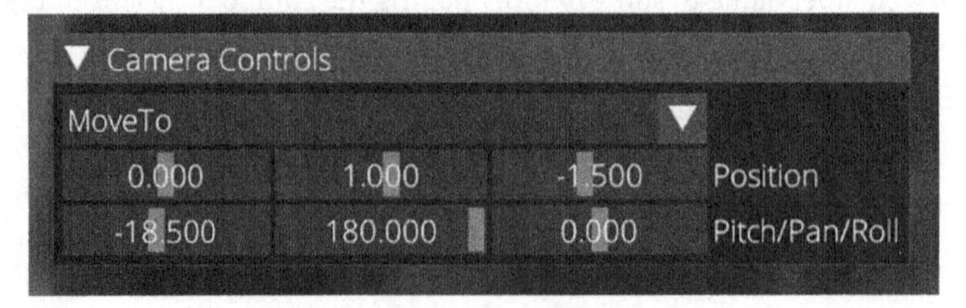

Figure 4.9: Camera controls

The preceding code is called from the main loop on every frame to draw ImGui. Check out the `Chapter04/06_DemoApp/src/main.cpp` file for the complete source code.

How it works...

Let's take a look at the implementation of the `CameraPositioner_MoveTo` class in `shared/Camera.h` we mentioned earlier in *steps 1* and *2*. In contrast to the first-person camera positioner introduced in the previous recipe, which relies on quaternions, this new positioner employs a straightforward Euler angles approach to store the camera orientation. This method is more user-friendly and intuitive for controlling the camera. The following are the steps to help us understand how this camera positioner works:

1. First, we want to have some user-configurable parameters for linear and angular damping coefficients:

```
class CameraPositioner_MoveTo final :
  public CameraPositionerInterface
{
public:
  float dampingLinear_ = 10.0f;
  vec3 dampingEulerAngles_ = vec3(5.0f, 5.0f, 5.0f);
```

2. We store the current and desired positions of the camera as well as two sets of pitch, pan, and roll Euler angles in `vec3` member fields. The current camera transformation is updated every frame and saved in a `mat4` field:

```
private:
  vec3 positionCurrent_ = vec3(0.0f);
  vec3 positionDesired_ = vec3(0.0f);
  vec3 anglesCurrent_ = vec3(0.0f); // pitch, pan, roll
  vec3 anglesDesired_ = vec3(0.0f);
  mat4 currentTransform_ = mat4(1.0f);
```

3. The constructor initializes both the current and desired data sets of the camera:

```
public:
  CameraPositioner MoveTo(const vec3& pos, const vec3& angles)
  : positionCurrent_(pos)
  , positionDesired_(pos)
  , anglesCurrent_(angles)
  , anglesDesired_(angles)
  {}
```

4. The most interesting part happens in the `update()` function. The current camera position is changed to move towards the desired camera position. The movement speed is proportional to the distance between these two positions and scaled using the linear damping coefficient:

```
void update(
  float deltaSeconds, const vec2& mousePos, bool mousePressed)
{
  positionCurrent_ += dampingLinear_ *
    deltaSeconds * (positionDesired_ - positionCurrent_);
```

5. Now, let's deal with Euler angles. We should clip them accordingly to make sure they remain within the `0...360` degrees range. This is required to prevent our camera from "spinning" around the object `2*Pi` times:

```
  anglesCurrent_ = clipAngles(anglesCurrent_);
  anglesDesired_ = clipAngles(anglesDesired_);
```

6. Similar to how we dealt with the camera position, the Euler angles are updated based on the distance between the desired and current set of angles. Before calculating the camera transformation matrix, clip the updated angles again and convert the values from degrees to radians. Note how the pitch, pan, and roll angles are swizzled before they are forwarded into `glm::yawPitchRoll()`:

```
  anglesCurrent_ -= angleDelta(anglesCurrent_, anglesDesired_)
    * dampingEulerAngles_ * deltaSeconds;
  anglesCurrent_ = clipAngles(anglesCurrent_);
  const vec3 ang = glm::radians(anglesCurrent_);
  currentTransform_ = glm::translate(
    glm::yawPitchRoll(ang.y, ang.x, ang.z), -positionCurrent_);
}
```

7. The functions for the angle clipping are straightforward and look as follows:

```
private:
  static inline float clipAngle(float d) {
    if (d < -180.0f) return d + 360.0f;
    if (d > +180.0f) return d - 360.f;
    return d;
  }
  static inline vec3 clipAngles(const vec3& angles) {
    return vec3( std::fmod(angles.x, 360.0f),
                 std::fmod(angles.y, 360.0f),
                 std::fmod(angles.z, 360.0f) );
  }
```

8. The delta between two sets of angles can be calculated in the following way:

```
static inline vec3 angleDelta( const vec3& anglesCurrent,
                               const vec3& anglesDesired )
{
  const vec3 d =
    clipAngles(anglesCurrent) - clipAngles(anglesDesired);
  return vec3(
    clipAngle(d.x), clipAngle(d.y), clipAngle(d.z));
}
};
```

Try running the demo application, `Chapter04/06_DemoApp`. Switch to the `MoveTo` camera and change the position and orientation from the ImGui user interface.

There's more...

Further camera functionality can be built on top of this example implementation. One more useful extension might be a camera that follows a spline curve defined using a set of key points for positions and targets. We will leave this as an exercise for you.

Implementing an immediate-mode 3D drawing canvas

The *Setting up Vulkan debugging capabilities* recipe from *Chapter 2*, only scratched the surface of graphical application debugging. The validation layers provided by the Vulkan API are invaluable but they do not allow you to debug logical and calculation-related errors. To see what is happening in our virtual world, we need to be able to render auxiliary graphical information such as objects' bounding boxes and plot time-varying charts of different values or plain straight lines. The Vulkan API does not provide any immediate-mode rendering facilities. All it can do is add commands to command buffers scheduled for later submission. To overcome this difficulty and add an immediate-mode rendering canvas to our applications, we have to write some additional code. Let's learn how to do it in this recipe.

Getting ready

Make sure you are proficient with all the rendering recipes from *Chapter 3*. Check the `shared/LineCanvas.h` and `shared/LineCanvas.cpp` files for a working implementation of this recipe. An example of how to use a new `LineCanvas3D` 3D line drawing class is a part of the demo app at `Chapter04/06_DemoApp`.

How to do it...

The `LineCanvas3D` class contains a CPU-accessible list of 3D lines defined by two points and a color. Each frame, the user can call the `line()` method to draw a new 3D line that should be rendered in the current frame. To render these lines into the framebuffer, we maintain a collection of Vulkan buffers to store line geometry data, which we will update every frame.

Let's take a look at the interface of this class:

1. The `LineCanvas3D` class has its internal 3D line representation as a pair of vertices for each and every line, whereas each vertex consists of a `vec4` position and a color. Each `linesBuffer` buffer holds a GPU-visible copy of the `lines_` container. We have one buffer for each swapchain image to avoid any additional Vulkan synchronization:

```
struct LineCanvas3D {
  mat4 mvp_ = mat4(1.0f);
  struct LineData {
    vec4 pos;
    vec4 color;
  };
  std::vector<LineData> lines_;
  Holder<ShaderModuleHandle> vert_;
  Holder<ShaderModuleHandle> frag_;
  Holder<RenderPipelineHandle> pipeline_;
  constexpr uint32_t kNumImages = 3;
  Holder<BufferHandle> linesBuffer_[kNumImages] = {};
  uint32_t pipelineSamples_ = 1;
  uint32_t currentBufferSize_[kNumImages] = {};
  uint32_t currentFrame_ = 0;
  void setMatrix(const mat4& mvp) { mvp_ = mvp; }
```

2. The actual drawing functionality consists of a set of functions. We want to be able to clear the canvas, render one line, and render some useful primitives, such as 3D planes, boxes, and frustums. Further utility functions can easily be built on top of the functionality provided by the `line()` member function:

```
  void clear() { lines_.clear(); }
  void line(const vec3& p1, const vec3& p2, const vec4& c);
  void plane(
    vec3& orig, const vec3& v1, const vec3& v2,
    int n1, int n2, float s1, float s2,
    const vec4& color, const vec4& outlineColor);
  void box(const mat4& m, const BoundingBox& box, const vec4& color);
  void box(const mat4& m, const vec3& size, const vec4& color);
  void frustum(
    const mat4& camView, const mat4& camProj,
    const vec4& color);
```

3. The longest method of this class is `render()`, which generates Vulkan commands into the provided command buffer to render the current contents of `LineCanvas3D`. We will look into its implementation in a few moments:

```
    void render(lvk::IContext& ctx,
      const lvk::Framebuffer& desc,
      lvk::ICommandBuffer& buf,
      uint32_t numSamples = 1);
};
```

Now, let us deal with the non-Vulkan part of the code:

1. The `line()` member function itself just adds two colored `vec3` points to the container:

```
void LineCanvas3D::line(
  const vec3& p1, const vec3& p2, const vec4& c) {
  lines_.push_back({ .pos = vec4(p1, 1.0f), .color = c });
  lines_.push_back({ .pos = vec4(p2, 1.0f), .color = c });
}
```

2. The `plane()` method uses `line()` internally to create a visual representation of a three-dimensional plane spanned by the `v1` and `v2` vectors with half-sizes `s1` and `s2`, and an origin point `o`. The `n1` and `n2` parameters specify how many lines we want to render along each coordinate direction:

```
void LineCanvas3D::plane(
  const vec3& o, const vec3& v1, const vec3& v2, int n1, int n2,
  float s1, float s2,
  const vec4& color, const vec4& outlineColor)
```

3. Draw the 4 outer lines representing a plane segment:

```
  line(o - s1 / 2.0f * v1 - s2 / 2.0f * v2,
       o - s1 / 2.0f * v1 + s2 / 2.0f * v2, outlineColor);
  line(o + s1 / 2.0f * v1 - s2 / 2.0f * v2,
       o + s1 / 2.0f * v1 + s2 / 2.0f * v2, outlineColor);
  line(o - s1 / 2.0f * v1 + s2 / 2.0f * v2,
       o + s1 / 2.0f * v1 + s2 / 2.0f * v2, outlineColor);
  line(o - s1 / 2.0f * v1 - s2 / 2.0f * v2,
       o + s1 / 2.0f * v1 - s2 / 2.0f * v2, outlineColor);
```

4. Draw `n1` horizontal lines and `n2` vertical lines inside the plane:

```
  for (int i = 1; i < n1; i++) {
    float t = ((float)i - (float)n1 / 2.0f) * s1/(float)n1;
    const vec3 o1 = o + t * v1;
    line(o1 - s2 / 2.0f * v2,
         o1 + s2 / 2.0f * v2, color);
  }
  for (int i = 1; i < n2; i++) {
```

```
        const float t = ((float)i - (float)n2 / 2.0f) * s2/(float)n2;
        const vec3 o2 = o + t * v2;
        line(o2 - s1 / 2.0f * v1,
             o2 + s1 / 2.0f * v1, color);
    }
```

5. The `box()` member function draws a colored box oriented using the provided `m` model matrix and half-size `size` along the `X`, `Y`, and `Z` axes. The idea is to create 8 corner points of the box and transform them using the `m` matrix:

```
void LineCanvas3D::box(const mat4& m,
  const vec3& size, const vec4& color)
{
  vec3 pts[8] = { vec3(+size.x, +size.y, +size.z),
                  vec3(+size.x, +size.y, -size.z),
                  vec3(+size.x, -size.y, +size.z),
                  vec3(+size.x, -size.y, -size.z),
                  vec3(-size.x, +size.y, +size.z),
                  vec3(-size.x, +size.y, -size.z),
                  vec3(-size.x, -size.y, +size.z),
                  vec3(-size.x, -size.y, -size.z) };
  for (auto& p : pts) p = vec3(m * vec4(p, 1.f));
```

6. Then render all 12 edges of the box using the `line()` function:

```
  line(pts[0], pts[1], color);
  line(pts[2], pts[3], color);
  line(pts[4], pts[5], color);
  line(pts[6], pts[7], color);
  line(pts[0], pts[2], color);
  line(pts[1], pts[3], color);
  line(pts[4], pts[6], color);
  line(pts[5], pts[7], color);
  line(pts[0], pts[4], color);
  line(pts[1], pts[5], color);
  line(pts[2], pts[6], color);
  line(pts[3], pts[7], color);
}
```

7. There's yet another overload of the `box()` function, which takes in the `BoundingBox` class declared in `shared/UtilsMath.h`. It is just a trivial wrapper over the previous variant of this function:

```
void LineCanvas3D::box(const mat4& m,
```

```
  const BoundingBox& box, const vec4& color)
{
  this->box(m * glm::translate(mat4(1.f),
    0.5f * (box.min_ + box.max_)),
    0.5f * vec3(box.max_ - box.min_), color);
}
```

8. The most interesting drawing function is `frustum()`, which renders a 3D frustum represented by a `camProj` view matrix positioned in the world using the `camView` matrix. Long story short, if you have a 3D camera somewhere in your world and its view and projection matrix are `camView` and `camProj` respectively, you can use this function to visualize that camera's viewing frustum:

Note

This code is invaluable in debugging things such as shadow maps or culling frustums. We will put it to heavy use in the final chapters of this book.

```
void LineCanvas3D::frustum(
   const mat4& camView,
   const mat4& camProj, const vec4& color)
{
```

9. The idea is somewhat similar to the `box()` function mentioned above. We create a set of corner points on a cube corresponding to 8 corners of the camera frustum (points are referred to as pp in the following code). Then, we transform each of these points with the inverse of the provided view-projection matrix, essentially warping a box into a frustum shape. Then we use `line()` to connect the dots:

```
  const vec3 corners[] = { vec3(-1, -1, -1),
                           vec3(+1, -1, -1),
                           vec3(+1, +1, -1),
                           vec3(-1, +1, -1),
                           vec3(-1, -1, +1),
                           vec3(+1, -1, +1),
                           vec3(+1, +1, +1),
                           vec3(-1, +1, +1) };
  vec3 pp[8];
  for (int i = 0; i < 8; i++) {
    glm::vec4 q = glm::inverse(camView) *
                  glm::inverse(camProj) *
                  glm::vec4(corners[i], 1.0f);
    pp[i] = glm::vec3(q.x/q.w, q.y/q.w, q.z/q.w);
  }
```

10. These are four lines representing the side edges of the camera frustum:

```
line(pp[0], pp[4], color);
line(pp[1], pp[5], color);
line(pp[2], pp[6], color);
line(pp[3], pp[7], color);
```

11. With the side edges done, we need to draw the near plane. The extra two lines are used to draw a cross inside the near plane:

```
line(pp[0], pp[1], color);
line(pp[1], pp[2], color);
line(pp[2], pp[3], color);
line(pp[3], pp[0], color);
line(pp[0], pp[2], color);
line(pp[1], pp[3], color);
```

12. Next, we do the far plane. Here again, the extra two lines are used to draw a cross to give better visual cues:

```
line(pp[4], pp[5], color);
line(pp[5], pp[6], color);
line(pp[6], pp[7], color);
line(pp[7], pp[4], color);
line(pp[4], pp[6], color);
line(pp[5], pp[7], color);
```

13. Now let's draw the sides of the frustum to give a nice perception of volume. We use a dimmed color and `100` lines on each side:

```
const vec4 gridColor = color * 0.7f;
const int gridLines  = 100;
```

14. Here are the bottom and the top sides:

```
vec3 p1 = pp[0];
vec3 p2 = pp[1];
const vec3 s1 = (pp[4]-pp[0]) / float(gridLines);
const vec3 s2 = (pp[5]-pp[1]) / float(gridLines);
for (int i = 0; i != gridLines; i++, p1 += s1, p2 += s2)
  line(p1, p2, gridColor);
p1 = pp[2];
p2 = pp[3];
const vec3 s1 = (pp[6]-pp[2]) / float(gridLines);
const vec3 s2 = (pp[7]-pp[3]) / float(gridLines);
for (int i = 0; i != gridLines; i++, p1 += s1, p2 += s2)
  line(p1, p2, gridColor);
```

15. The same should be done with the left and right sides of our frustum:

```
    p1 = pp[0];
    p2 = pp[3];
    const vec3 s1 = (pp[4]-pp[0]) / float(gridLines);
    const vec3 s2 = (pp[7]-pp[3]) / float(gridLines);
    for (int i = 0; i != gridLines; i++, p1 += s1, p2 += s2)
      line(p1, p2, gridColor);
    p1 = pp[1];
    p2 = pp[2];
    const vec3 s1 = (pp[5]-pp[1]) / float(gridLines);
    const vec3 s2 = (pp[6]-pp[2]) / float(gridLines);
    for (int i = 0; i != gridLines; i++, p1 += s1, p2 += s2)
      line(p1, p2, gridColor);
  }
```

That covers the user-facing part of our line-drawing API. Let's take a look at the actual rendering code to learn how it works in an app.

How it works...

All the rendering and graphics pipeline creation is done within a single `render()` function.

The function accepts a *LightweightVK* context, a framebuffer, and a command buffer:

```
void LineCanvas3D::render(lvk::IContext& ctx,
  const lvk::Framebuffer& desc,
  lvk::ICommandBuffer& buf,
  uint32_t numSamples)
{
  if (lines_.empty()) return;
```

The required GPU buffer size is calculated based on the current number of lines. If the current buffer capacity is not sufficient, the buffer is reallocated:

```
  const uint32_t requiredSize = lines_.size() * sizeof(LineData);
  if (currentBufferSize_[currentFrame_] < requiredSize) {
    linesBuffer_[currentFrame_] = ctx.createBuffer({
      .usage = lvk::BufferUsageBits_Storage,
      .storage = lvk::StorageType_HostVisible,
      .size = requiredSize,
      .data = lines_.data() });
    currentBufferSize_[currentFrame_] = requiredSize;
  } else {
    ctx.upload(linesBuffer_[currentFrame_], lines_.data(), requiredSize);
  }
```

If there's no rendering pipeline available, we should create a new one. We use `lvk::Topology_Line`, which matches `VK_PRIMITIVE_TOPOLOGY_LINE_LIST`. Simple alpha blending is used to render all the lines:

```
if (pipeline_.empty() || pipelineSamples_ != numSamples) {
  pipelineSamples_ = numSamples;
  vert_ = ctx.createShaderModule({
    codeVS, lvk::Stage_Vert, "Shader Module: imgui (vert)" });
  frag_ = ctx.createShaderModule({
    codeFS, lvk::Stage_Frag, "Shader Module: imgui (frag)" });
  pipeline_ = ctx.createRenderPipeline({
    .topology = lvk::Topology_Line,
    .smVert   = vert_,
    .smFrag   = frag_,
    .color    = { {
       .format = ctx.getFormat(desc.color[0].texture),
       .blendEnabled      = true,
       .srcRGBBlendFactor = lvk::BlendFactor_SrcAlpha,
       .dstRGBBlendFactor = lvk::BlendFactor_OneMinusSrcAlpha,
    } },
    .depthFormat = desc.depthStencil.texture ?
      ctx.getFormat(desc.depthStencil.texture) :
      lvk::Format_Invalid,
    .cullMode = lvk::CullMode_None,
    .samplesCount = numSamples,
  }, nullptr);
}
```

Our line-drawing vertex shader accepts the current combined model-view-projection matrix `mvp` and a GPU reference to the buffer containing the line data. Everything is updated using Vulkan push constants:

```
struct {
  mat4 mvp;
  uint64_t addr;
} pc {
  .mvp  = mvp_,
  .addr = ctx.gpuAddress(linesBuffer_[currentFrame_]),
};
buf.cmdBindRenderPipeline(pipeline_);
buf.cmdPushConstants(pc);
```

Once the Vulkan rendering state is prepared, we can render the lines and switch to the next frame to use one of the available buffers:

```
    buf.cmdDraw(lines_.size());
    currentFrame_ = (currentFrame_ + 1) % LVK_ARRAY_NUM_ELEMENTS(linesBuffer_);
  }
```

It is also worth taking a quick look at the line-drawing GLSL shaders.

The vertex shader is as follows. **Programmable vertex pulling** is used to extract line data from the provided buffer:

```
layout (location = 0) out vec4 out_color;
layout (location = 1) out vec2 out_uv;
struct Vertex {
  vec4 pos;
  vec4 rgba;
};
layout(std430, buffer_reference) readonly buffer VertexBuffer {
  Vertex vertices[];
};
layout(push_constant) uniform PushConstants {
  mat4 mvp;
  VertexBuffer vb;
};
void main() {
  Vertex v = vb.vertices[gl_VertexIndex];
  out_color = v.rgba;
  gl_Position = mvp * v.pos;
}
```

The fragment shader is trivial and simply outputs the provided color:

```
layout (location = 0) in vec4 in_color;
layout (location = 0) out vec4 out_color;
void main() {
  out_color = in_color;
}
```

That is everything regarding drawing 3D lines. For a comprehensive example showing how to use this 3D drawing canvas, check the final *Putting it all together into a Vulkan application* recipe in this chapter.

The next recipe will conclude Vulkan auxiliary rendering by showing how to render 2D lines and charts with the help of the ImGui and ImPlot libraries.

Rendering on-screen graphs with ImGui and ImPlot

In the previous recipe, we learned how to create immediate mode drawing facilities in Vulkan with basic drawing functionality. That 3D canvas was rendered on top of a 3D scene sharing a view-projection matrix with it. In this recipe, we will continue adding useful debugging features to our framework and learn how to implement pure 2D line drawing functionality. It is possible to implement such a class in a way similar to `LineCanvas3D`. However, we already use the ImGui library in our apps as described in the *Rendering ImGui user interface* recipe. Let's put it to use to render our 2D lines.

Getting ready

We recommend revisiting the *Rendering ImGui user interfaces* and *Implementing an immediate-mode 3D drawing canvas* recipes to get a better grasp of how a simple Vulkan drawing canvas can be implemented.

How to do it...

What we need at this point essentially boils down to decomposing a 2D chart or graph into a set of lines and rendering them using ImGui. Let's go through the code to see how to do it:

1. We introduce a `LineCanvas2D` class to render 2D lines. It stores a collection of 2D lines:

```
class LineCanvas2D {
public:
  void clear() { lines_.clear(); }
  void line(const vec2& p1, const vec2& p2, const vec4& c) {
    lines_.push_back({ .p1 = p1, .p2 = p2, .color = c }); }
  void render(const char* name, uint32_t width, uint32_t height);
private:
  struct LineData {
    vec2 p1, p2;
    vec4 color;
  };
  std::vector<LineData> lines_;
};
```

2. The `render()` method is quite simple. We create a new full-screen ImGui window with all decorations removed and user input disabled:

```
void LineCanvas2D::render(const char* nameImGuiWindow) {
  ImGui::SetNextWindowPos(ImVec2(0, 0));
  ImGui::SetNextWindowSize(ImGui::GetMainViewport()->Size);
  ImGui::Begin(nameImGuiWindow, nullptr,
    ImGuiWindowFlags_NoDecoration |
    ImGuiWindowFlags_AlwaysAutoResize |
    ImGuiWindowFlags_NoSavedSettings |
    ImGuiWindowFlags_NoFocusOnAppearing |
```

```
ImGuiWindowFlags_NoNav |
    ImGuiWindowFlags_NoBackground |
    ImGuiWindowFlags_NoInputs);
```

3. Then we obtain ImGui's background draw list and add all our colored lines to it one by one. The rest of the rendering will be handled as a part of the ImGui user interface rendering, as described in the *Rendering ImGui user interfaces* recipe:

```
  ImDrawList* drawList = ImGui::GetBackgroundDrawList();
  for (const LineData& l : lines_) {
    drawList->AddLine(
      ImVec2(l.p1.x, l.p1.y),
      ImVec2(l.p2.x, l.p2.y),
      ImColor(l.color.r, l.color.g, l.color.b, l.color.a));
  }
  ImGui::End();
}
```

4. Inside the `Chapter04/06_DemoApp/src/main.cpp` demo application, we can work with an instance of `LineCanvas2D` the following way:

```
canvas2d.clear();
canvas2d.line({ 100, 300 }, { 100, 400 }, vec4(1, 0, 0, 1));
canvas2d.line({ 100, 400 }, { 200, 400 }, vec4(0, 1, 0, 1));
canvas2d.line({ 200, 400 }, { 200, 300 }, vec4(0, 0, 1, 1));
canvas2d.line({ 200, 300 }, { 100, 300 }, vec4(1, 1, 0, 1));
canvas2d.render("##plane");
```

This functionality is sufficient to render 2D lines for various debugging purposes. However, there's yet another way to do rendering using the ImPlot library. Let's use it to render an FPS graph. The helper code is in `shared/Graph.h`:

1. We declare another small `LinearGraph` helper class to draw a graph of changing values, such as the number of rendered frames per second:

```
class LinearGraph {
  const char* name_ = nullptr;
  const size_t maxPoints_ = 0;
  std::deque<float> graph_;
public:
  explicit LinearGraph(const char* name,
                       size_t maxGraphPoints = 256)
  : name_(name)
  , maxPoints_(maxGraphPoints)
  {}
```

2. As we add more points to the graph, the old points are popped out, making the graph look like it is scrolling on the screen right-to-left. This is helpful to observe local fluctuations in values such as frames per second counters, and so on:

```
void addPoint(float value) {
  graph_.push_back(value);
  if (graph_.size() > maxPoints_) graph_.erase(graph_.begin());
}
```

3. The idea is to find the minimum and maximum values and normalize the graph into the `0...1` range:

```
void renderGraph(uint32_t x, uint32_t y,
  uint32_t width, uint32_t height,
  const vec4& color = vec4(1.0)) const {
  float minVal = std::numeric_limits<float>::max();
  float maxVal = std::numeric_limits<float>::min();
  for (float f : graph_) {
    if (f < minVal) minVal = f;
    if (f > maxVal) maxVal = f;
  }
  const float range = maxVal - minVal;
  float valX = 0.0;
  std::vector<float> dataX_;
  std::vector<float> dataY_;
  dataX_.reserve(graph_.size());
  dataY_.reserve(graph_.size());
  for (float f : graph_) {
    const float valY = (f - minVal) / range;
    valX += 1.0f / maxPoints_;
    dataX_.push_back(valX);
    dataY_.push_back(valY);
  }
```

4. Then we need to create an *ImGui* window to hold our graph. *ImPlot* drawing can work only inside an *ImGui* window. All decorations and user interactions are disabled:

```
ImGui::SetNextWindowPos(ImVec2(x, y));
ImGui::SetNextWindowSize(ImVec2(width, height));
ImGui::Begin(_, nullptr,
  ImGuiWindowFlags_NoDecoration |
  ImGuiWindowFlags_AlwaysAutoResize |
  ImGuiWindowFlags_NoSavedSettings |
  ImGuiWindowFlags_NoFocusOnAppearing |
```

```
    ImGuiWindowFlags_NoNav |
    ImGuiWindowFlags_NoBackground |
    ImGuiWindowFlags_NoInputs);
```

5. A new *ImPlot* plot can be started in a similar way. We disable decorations for the *ImPlot* axes and set up colors for the line drawing:

```
  if (ImPlot::BeginPlot(name_, ImVec2(width, height),
    ImPlotFlags_CanvasOnly |
    ImPlotFlags_NoFrame | ImPlotFlags_NoInputs)) {
    ImPlot::SetupAxes(nullptr, nullptr,
      ImPlotAxisFlags_NoDecorations,
      ImPlotAxisFlags_NoDecorations);
    ImPlot::PushStyleColor(ImPlotCol_Line,
      ImVec4(color.r, color.g, color.b, color.a));
    ImPlot::PushStyleColor(ImPlotCol_PlotBg,
      ImVec4(0, 0, 0, 0));
```

6. The `ImPlot::PlotLine()` function uses our collection of points' `X` and `Y` values to render a graph:

```
    ImPlot::PlotLine("#line", dataX_.data(), dataY_.data(),
      (int)graph_.size(), ImPlotLineFlags_None);
    ImPlot::PopStyleColor(2);
    ImPlot::EndPlot();
  }
  ImGui::End();
}
```

This is the entire underlying implementation code.

Let's now take a look at `Chapter04/06_DemoApp/src/main.cpp` to learn how 2D chart rendering works.

How it works...

The app at `Chapter04/06_DemoApp` makes use of `LinearGraph` to render an FPS graph, and a simple sine graph for reference. Here is how it works:

1. Both graphs are declared as global variables. They can render up to `2048` points:

```
LinearGraph fpsGraph("##fpsGraph", 2048);
LinearGraph sinGraph("##sinGraph", 2048);
```

2. Inside the main loop, we add points to both graphs like this:

```
fpsGraph.addPoint(app.fpsCounter_.getFPS());
sinGraph.addPoint(sinf(glfwGetTime() * 20.0f));
```

3. Then we render both graphs as follows:

```
sinGraph.renderGraph(0, height * 0.7f, width, height * 0.2f,
  vec4(0.0f, 1.0f, 0.0f, 1.0f));
fpsGraph.renderGraph(0, height * 0.8f, width, height * 0.2f);
```

The resulting graphs look as shown in the following screenshot.

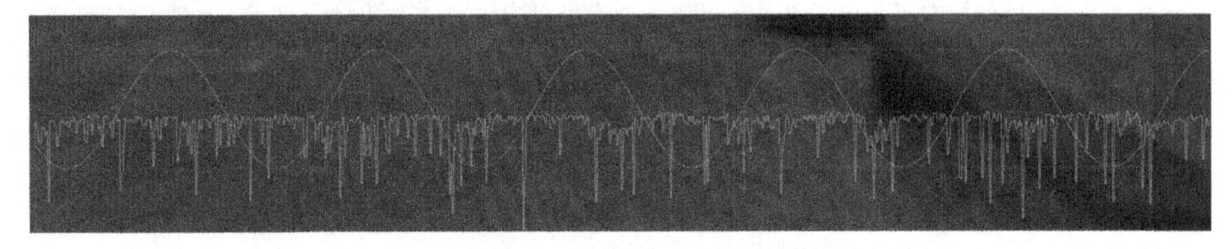

Figure 4.10: Frames-per-second and sine wave graphs

Putting it all together into a Vulkan application

In this recipe, we use all the material from previous recipes of this chapter to build a Vulkan demo application combining 3D scene rendering with 2D and 3D debug line drawing functionality.

Getting ready

This recipe is a consolidation of all the material in this chapter into a final demo app. It might be useful to revisit all the previous recipes to get to grips with the different user interaction and debugging techniques described in this chapter.

The full source code for this recipe can be found in `Chapter04/06_DemoApp`. The `VulkanApp` class used in this recipe is declared in `shared/VulkanApp.h`.

How to do it...

Let's skim through the source code to see how we can integrate the functionality from all the recipes together into a single application. We put all of the source code here so we can reference it in the subsequent chapters when necessary. All error checking is skipped again for the sake of brevity:

1. The `shared/VulkanApp.h` header provides a wrapper for *LightweightVK* context creation and GLFW window lifetime management. Check the *Initializing the Vulkan instance and graphical device* and *Initializing Vulkan swapchain* recipes in *Chapter 2* for more details:

```
#include "shared/VulkanApp.h"
#include <assimp/cimport.h>
#include <assimp/postprocess.h>
#include <assimp/scene.h>
#include "shared/LineCanvas.h"
```

2. Here we demonstrate a camera positioner for the *Adding camera animations and motion* recipe:

```
const vec3 kInitialCameraPos    = vec3(0.0f, 1.0f, -1.5f);
```

```
const vec3 kInitialCameraTarget = vec3(0.0f, 0.5f,  0.0f);
const vec3 kInitialCameraAngles = vec3(-18.5f, 180.0f, 0.0f);
CameraPositioner_MoveTo positionerMoveTo(
  kInitialCameraPos, kInitialCameraAngles);
vec3 cameraPos    = kInitialCameraPos;
vec3 cameraAngles = kInitialCameraAngles;
const char* cameraType          = "FirstPerson";
const char* comboBoxItems[]     = { "FirstPerson", "MoveTo" };
const char* currentComboBoxItem = cameraType;
```

3. The following is for the FPS graph described in the previous *Rendering on-screen graphs with ImGui and ImPlot* recipe:

```
LinearGraph fpsGraph("##fpsGraph", 2048);
LinearGraph sinGraph("##sinGraph", 2048);
```

4. The `VulkanApp` class has a built-in first-person camera as described in the *Working with a 3D camera and basic user interaction* recipe. We provide an initial camera position and target, as well as reducing the FPS-averaging interval for the purpose of drawing a nice fast-moving graph. Let's create a local variable to make the access to `lvk::IContext` stored in `VulkanApp` more convenient. We call `ctx.release()` explicitly later:

```
int main()
{
  VulkanApp app({
    .initialCameraPos = kInitialCameraPos,
    .initialCameraTarget = kInitialCameraTarget });
  app.fpsCounter_.avgInterval_ = 0.002f;
  app.fpsCounter_.printFPS_    = false;
  LineCanvas2D canvas2d;
  LineCanvas3D canvas3d;
  unique_ptr<lvk::IContext> ctx(app.ctx_.get());
```

5. All the shaders are loaded from files. The cube map rendering was described in the *Using cube map textures in Vulkan* recipe:

```
  Holder<ShaderModuleHandle> vert = loadShaderModule(
    ctx, "Chapter04/04_CubeMap/src/main.vert");
  Holder<ShaderModuleHandle> frag = loadShaderModule(
    ctx, "Chapter04/04_CubeMap/src/main.frag");
  Holder<ShaderModuleHandle> vertSkybox = loadShaderModule(
    ctx, "Chapter04/04_CubeMap/src/skybox.vert");
  Holder<ShaderModuleHandle> fragSkybox = loadShaderModule(
    ctx, "Chapter04/04_CubeMap/src/skybox.frag");
```

6. The rubber duck mesh rendering pipeline is created as follows:

```
struct VertexData {
  vec3 pos;
  vec3 n;
  vec2 tc;
};
const lvk::VertexInput vdesc = {
  .attributes    = {
    { .location = 0,
      .format = lvk::VertexFormat::Float3,
      .offset = offsetof(VertexData, pos) },
    { .location = 1,
      .format = lvk::VertexFormat::Float3,
      .offset = offsetof(VertexData, n) },
    { .location = 2,
      .format = lvk::VertexFormat::Float2,
     .offset = offsetof(VertexData, tc) }, },
  .inputBindings = { { .stride = sizeof(VertexData) } },
};
Holder<RenderPipelineHandle> pipeline =
  ctx->createRenderPipeline({
    .vertexInput = vdesc,
    .smVert      = vert,
    .smFrag      = frag,
    .color       = { {.format = ctx->getSwapchainFormat()} },
    .depthFormat = app.getDepthFormat(),
    .cullMode    = lvk::CullMode_Back,
});
```

7. The skybox rendering pipeline uses programmable-vertex pulling and has no vertex input state. See the *Using cube map textures in Vulkan* recipe for details:

```
Holder<RenderPipelineHandle> pipelineSkybox =
  ctx->createRenderPipeline({
    .smVert      = vertSkybox,
    .smFrag      = fragSkybox,
    .color       = { {.format = ctx->getSwapchainFormat()} },
    .depthFormat = app.getDepthFormat(),
});
```

8. Let's load the rubber duck from a `.gltf` file and pack it into the `vertices` and `indices` arrays:

```
const aiScene* scene = aiImportFile(
```

```
    "data/rubber_duck/scene.gltf", aiProcess_Triangulate);
  const aiMesh* mesh = scene->mMeshes[0];
  std::vector<VertexData> vertices;
  for (uint32_t i = 0; i != mesh->mNumVertices; i++) {
    const aiVector3D v = mesh->mVertices[i];
    const aiVector3D n = mesh->mNormals[i];
    const aiVector3D t = mesh->mTextureCoords[0][i];
    vertices.push_back({ .pos = vec3(v.x, v.y, v.z),
                         .n   = vec3(n.x, n.y, n.z),
                         .tc  = vec2(t.x, t.y) });
  }
  std::vector<uint32_t> indices;
  for (uint32_t i = 0; i != mesh->mNumFaces; i++)
    for (uint32_t j = 0; j != 3; j++)
      indices.push_back(mesh->mFaces[i].mIndices[j]);
  aiReleaseImport(scene);
```

9. Create two GPU buffers to hold `indices` and `vertices`:

```
  const size_t kSizeIndices  = sizeof(uint32_t) * indices.size();
  const size_t kSizeVertices = sizeof(VertexData) * vertices.size();
  Holder<BufferHandle> bufferIndices =
    ctx->createBuffer({
      .usage     = lvk::BufferUsageBits_Index,
      .storage   = lvk::StorageType_Device,
      .size      = kSizeIndices,
      .data      = indices.data(),
      .debugName = "Buffer: indices" });
  Holder<BufferHandle> bufferVertices =
    ctx->createBuffer({
      .usage     = lvk::BufferUsageBits_Vertex,
      .storage   = lvk::StorageType_Device,
      .size      = kSizeVertices,
      .data      = vertices.data(),
      .debugName = "Buffer: vertices" });
```

10. A uniform buffer is used to hold per-frame data, such as model-view-projection matrices, the camera position, and bindless IDs for both textures:

```
  struct PerFrameData {
    mat4 model;
    mat4 view;
    mat4 proj;
```

```
    vec4 cameraPos;
    uint32_t tex     = 0;
    uint32_t texCube = 0;
  };
  Holder<BufferHandle> bufferPerFrame =
    ctx->createBuffer({
      .usage     = lvk::BufferUsageBits_Uniform,
      .storage   = lvk::StorageType_Device,
      .size      = sizeof(PerFrameData),
      .debugName = "Buffer: per-frame" });
```

11. Now let's bring in a 2D texture for the rubber duck model and a cube map texture for our skybox, as described in the *Using cube map textures in Vulkan* recipe:

```
Holder<TextureHandle> texture = loadTexture(
  ctx, "data/rubber_duck/textures/Duck_baseColor.png");
Holder<TextureHandle> cubemapTex;
int w, h;
const float* img = stbi_loadf(
  "data/piazza_bologni_1k.hdr", &w, &h, nullptr, 4);
Bitmap in(w, h, 4, eBitmapFormat_Float, img);
Bitmap out = convertEquirectangularMapToVerticalCross(in);
stbi_image_free((void*)img);
stbi_write_hdr(".cache/screenshot.hdr", out.w_, out.h_,
  out.comp_, (const float*)out.data_.data());
Bitmap cubemap = convertVerticalCrossToCubeMapFaces(out);
cubemapTex = ctx->createTexture({
  .type       = lvk::TextureType_Cube,
  .format     = lvk::Format_RGBA_F32,
  .dimensions = {(uint32_t)cubemap.w_, (uint32_t)cubemap.h_},
  .usage      = lvk::TextureUsageBits_Sampled,
  .data       = cubemap.data_.data(),
  .debugName  = "data/piazza_bologni_1k.hdr" });
```

12. Run the main loop using a lambda provided by the `VulkanApp::run()` method. The camera positioner is updated as described in the *Adding camera animations and motion* recipe:

```
app.run([&](uint32_t width, uint32_t height,
  float aspectRatio, float deltaSeconds) {
  positionerMoveTo.update(deltaSeconds, app.mouseState_.pos,
    ImGui::GetIO().WantCaptureMouse ?
      false : app.mouseState_.pressedLeft);
  const mat4 p  = glm::perspective(glm::radians(60.0f),
    aspectRatio, 0.1f, 1000.0f);
```

```
const mat4 m1 = glm::rotate(mat4(1.0f),
  glm::radians(-90.0f), vec3(1, 0, 0));
const mat4 m2 = glm::rotate(mat4(1.0f),
  (float)glfwGetTime(), vec3(0.0f, 1.0f, 0.0f));
const mat4 v  = glm::translate(mat4(1.0f), app.camera_.getPosition());
const PerFrameData pc = {
  .model     = m2 * m1,
  .view      = app.camera_.getViewMatrix(),
  .proj      = p,
  .cameraPos = vec4(app.camera_.getPosition(), 1.0f),
  .tex       = texture.index(),
  .texCube   = cubemapTex.index(),
};
ctx->upload(bufferPerFrame, &pc, sizeof(pc));
```

13. To recap the details on render passes and frame buffers, check the *Dealing with buffers in Vulkan* recipe in *Chapter 3*:

```
const lvk::RenderPass renderPass = {
  .color = { { .loadOp = lvk::LoadOp_Clear,
               .clearColor = {1.0f, 1.0f, 1.0f, 1.0f} } },
  .depth = { .loadOp = lvk::LoadOp_Clear,
             .clearDepth = 1.0f } };
const lvk::Framebuffer framebuffer = {
  .color = {
    { .texture = ctx->getCurrentSwapchainTexture() } },
  .depthStencil = { .texture = app.getDepthTexture() } };
ICommandBuffer& buf = ctx->acquireCommandBuffer();
buf.cmdBeginRendering(renderPass, framebuffer);
```

14. We render the skybox as described in the *Using cube map textures in Vulkan* recipe. Note that 36 vertices are used to draw the skybox:

```
buf.cmdPushConstants(ctx->gpuAddress(bufferPerFrame));
buf.cmdPushDebugGroupLabel("Skybox", 0xff0000ff);
buf.cmdBindRenderPipeline(pipelineSkybox);
buf.cmdDraw(36);
buf.cmdPopDebugGroupLabel();
```

15. Rendering the rubber duck mesh is done as follows:

```
buf.cmdPushDebugGroupLabel("Mesh", 0xff0000ff);
buf.cmdBindVertexBuffer(0, bufferVertices);
buf.cmdBindRenderPipeline(pipeline);
buf.cmdBindDepthState({
```

```
    .compareOp = lvk::CompareOp_Less,
    .isDepthWriteEnabled = true } );
  buf.cmdBindIndexBuffer(bufferIndices, lvk::IndexFormat_UI32);
  buf.cmdDrawIndexed(indices.size());
  buf.cmdPopDebugGroupLabel();
```

16. Rendering an ImGui window with a memo for keyboard hints is done as follows:

```
app.imgui_->beginFrame(framebuffer);
ImGui::SetNextWindowPos(ImVec2(10, 10));
ImGui::Begin("Keyboard hints:", nullptr,
    ImGuiWindowFlags_AlwaysAutoResize |
    ImGuiWindowFlags_NoFocusOnAppearing |
    ImGuiWindowFlags_NoInputs |
    ImGuiWindowFlags_NoCollapse);
ImGui::Text("W/S/A/D - camera movement");
ImGui::Text("1/2 - camera up/down");
ImGui::Text("Shift - fast movement");
ImGui::Text("Space - reset view");
ImGui::End();
```

17. We render a frames-per-second counter as described in the *Adding a frames-per-second counter* recipe:

```
if (const ImGuiViewport* v = ImGui::GetMainViewport()) {
  ImGui::SetNextWindowPos({
    v->WorkPos.x + v->WorkSize.x - 15.0f,
    v->WorkPos.y + 15.0f }, ImGuiCond_Always,
    { 1.0f, 0.0f });
}
ImGui::SetNextWindowBgAlpha(0.30f);
ImGui::SetNextWindowSize(
  ImVec2(ImGui::CalcTextSize("FPS : _______").x, 0));
if (ImGui::Begin("##FPS", nullptr,
      ImGuiWindowFlags_NoDecoration |
      ImGuiWindowFlags_AlwaysAutoResize |
      ImGuiWindowFlags_NoSavedSettings |
      ImGuiWindowFlags_NoFocusOnAppearing |
      ImGuiWindowFlags_NoNav | ImGuiWindowFlags_NoMove)) {
  ImGui::Text("FPS : %i", (int)app.fpsCounter_.getFPS());
  ImGui::Text(
    "Ms  : %.1f", 1000.0 / app.fpsCounter_.getFPS());
}
ImGui::End();
```

18. Our on-screen graphs and a 2D drawing canvas are handled as shown in the *Rendering on-screen graphs with ImGui and ImPlot* recipe:

```
    sinGraph.renderGraph(0, height * 0.7f, width,
      height * 0.2f, vec4(0.0f, 1.0f, 0.0f, 1.0f));
    fpsGraph.renderGraph(0, height * 0.8f, width, height * 0.2f);
    canvas2d.clear();
    canvas2d.line({ 100, 300 }, { 100, 400 }, vec4(1, 0, 0, 1));
    canvas2d.line({ 100, 400 }, { 200, 400 }, vec4(0, 1, 0, 1));
    canvas2d.line({ 200, 400 }, { 200, 300 }, vec4(0, 0, 1, 1));
    canvas2d.line({ 200, 300 }, { 100, 300 }, vec4(1, 1, 0, 1));
    canvas2d.render("##plane");
```

19. The following code handles a 3D drawing canvas, as described in the *Implementing immediate mode 3D drawing canvas* recipe. To demonstrate the `frustum()` function, we render an ad hoc rotating frustum constructed via the `lootAt()` and `perspective()` GLM functions:

```
    canvas3d.clear();
    canvas3d.setMatrix(pc.proj * pc.view);
    canvas3d.plane(vec3(0, 0, 0), vec3(1, 0, 0), vec3(0, 0, 1),
      40, 40, 10.0f, 10.0f, vec4(1, 0, 0, 1), vec4(0, 1, 0, 1));
    canvas3d.box(mat4(1.0f), BoundingBox(vec3(-2), vec3(+2)),
      vec4(1, 1, 0, 1));
    canvas3d.frustum(
      glm::lookAt(vec3(cos(glfwGetTime()),
                  kInitialCameraPos.y, sin(glfwGetTime())),
                  kInitialCameraTarget, vec3(0.0f, 1.0f, 0.0f)),
      glm::perspective(glm::radians(60.0f), aspectRatio, 0.1f,
        30.0f), vec4(1, 1, 1, 1));
    canvas3d.render(*ctx.get(), framebuffer, buf, width, height);
```

20. Lastly, finalize the rendering, submit the command buffer to the GPU, and update the graphs:

```
    app.imgui_->endFrame(buf);
    buf.cmdEndRendering();
    ctx->submit(buf, ctx->getCurrentSwapchainTexture());
    fpsGraph.addPoint(app.fpsCounter_.getFPS());
    sinGraph.addPoint(sinf(glfwGetTime() * 20.0f));
  });
  ctx.release();
  return 0;
}
```

The following is a screenshot from the running application. The white graph displays the average FPS values and the rotating white frustum can be used to debug shadow mapping, as we will do in subsequent chapters:

Figure 4.11: Demo application

This chapter focused on combining multiple rendering aspects into one working Vulkan application. The graphical side still lacks some essential features, such as advanced lighting and materials, but we have almost everything in place to start rendering much more complex scenes. The next few chapters will cover more complicated mesh rendering techniques and physically-based lighting calculations based on the glTF2 format.

5

Working with Geometry Data

Previously, we tried different ad hoc approaches to storing and handling 3D geometry data in our graphical applications. The mesh data layout for vertex and index buffers was hardcoded in each of our demo apps. This way it was easier to focus on other important parts of the graphics pipeline. As we go into the territory of more complex graphics applications, we require additional control over the storage of different 3D meshes in system memory and GPU buffers. However, our focus still remains on guiding you through the main principles and practices rather than on pure efficiency.

In this chapter, we will learn how to store and handle mesh geometry data in a more organized way. We will cover the following recipes:

- Generating level-of-detail meshes using MeshOptimizer
- Implementing programmable vertex pulling
- Rendering instanced geometry
- Implementing instanced meshes with compute shaders
- Implementing an infinite grid GLSL shader
- Integrating tessellation into the graphics pipeline
- Organizing the mesh data storage
- Implementing automatic geometry conversion
- Indirect rendering in Vulkan
- Generating textures in Vulkan using compute shaders
- Implementing computed meshes

Technical requirements

To run the code from this chapter on your Linux or Windows PC, you'll require a GPU with up-to-date drivers that support Vulkan 1.3. The source code can be downloaded from `https://github.com/PacktPublishing/3D-Graphics-Rendering-Cookbook-Second-Edition`.

To run the demo applications from this chapter, you have to use the Amazon Lumberyard Bistro dataset from the McGuire Computer Graphics Archive `http://casual-effects.com/data/index.html`. You can download it automatically by running the `deploy_deps.py` script.

Generating level-of-detail meshes using MeshOptimizer

To get started with geometry manipulations, let's implement a mesh geometry simplification demo using the MeshOptimizer library that, besides mesh optimizations, can generate simplified meshes for real-time discrete **level-of-detail** (**LOD**) algorithms we might want to use later. Simplification is an efficient way to improve rendering performance.

For GPUs to render a mesh efficiently, all vertices in the vertex buffer should be unique and without duplicates. Solving this problem efficiently can be a complicated and computationally intensive task in any modern 3D content pipeline. MeshOptimizer is an open-source C++ library developed by Arseny Kapoulkine, which provides algorithms to help optimize meshes for modern GPU vertex and index processing pipelines. It can reindex an existing index buffer or generate an entirely new set of indices from an unindexed vertex buffer.

Let's learn how to optimize and generate simplified meshes with MeshOptimizer.

Getting ready

It is recommended that you revisit *Chapter 3*. The complete source code for this recipe can be found in `Chapter05/01_MeshOptimizer`.

How to do it...

MeshOptimizer can generate all necessary LOD meshes for a specified set of indices and vertices. Once we have our loaded mesh with Assimp, we can pass it to MeshOptimizer. Here's how to do it:

1. Let's load a mesh from a `.gltf` file using Assimp. For this demo, we need only vertex positions and indices:

```
const aiScene* scene = aiImportFile(
  "data/rubber_duck/scene.gltf", aiProcess_Triangulate);
const aiMesh* mesh = scene->mMeshes[0];
std::vector<vec3> positions;
std::vector<uint32_t> indices;
for (unsigned int i = 0; i != mesh->mNumVertices; i++) {
  const aiVector3D v = mesh->mVertices[i];
  positions.push_back(vec3(v.x, v.y, v.z));
}
for (unsigned int i = 0; i != mesh->mNumFaces; i++) {
  for (int j = 0; j != 3; j++)
    indices.push_back(mesh->mFaces[i].mIndices[j]);
}
aiReleaseImport(scene);
```

2. The LOD meshes are represented as a collection of indices that construct a new simplified mesh from the same vertices that are used for the original mesh. This way we have to store only one set of vertices and can render corresponding LODs just by switching index buffers data. As done previously, we store all indices as unsigned 32-bit integers for simplicity. Now we should generate a remap table for our existing vertex and index data:

```
std::vector<uint32_t> remap(indices.size());
const size_t vertexCount = meshopt_generateVertexRemap(
  remap.data(), indices.data(), indices.size(),
  positions.data(), indices.size(), sizeof(vec3));
```

The MeshOptimizer documentation (`https://github.com/zeux/meshoptimizer`) tells us the following:

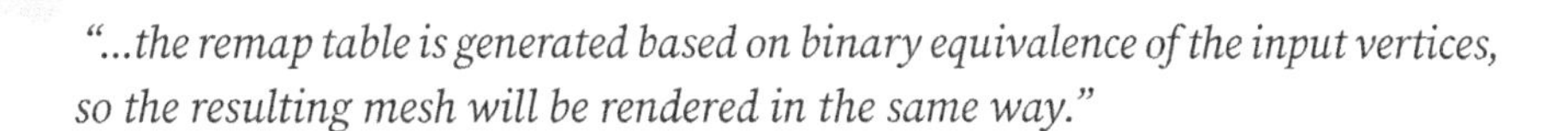

> *"...the remap table is generated based on binary equivalence of the input vertices, so the resulting mesh will be rendered in the same way."*

3. The returned `vertexCount` value corresponds to the number of unique vertices that have remained after remapping. Let's allocate space and generate new vertex and index buffers:

```
std::vector<uint32_t> remappedIndices(indices.size());
std::vector<vec3> remappedVertices(vertexCount);
meshopt_remapIndexBuffer(remappedIndices.data(),
  indices.data(), indices.size(), remap.data());
meshopt_remapVertexBuffer(remappedVertices.data(),
  positions.data(), positions.size(),
  sizeof(vec3), remap.data());
```

Now we can use other MeshOptimizer algorithms to optimize these buffers even further. The official documentation is pretty straightforward.

4. When we want to render a mesh, the GPU has to transform each vertex via a vertex shader. GPUs can reuse transformed vertices by means of a small built-in cache, usually storing between 16 and 128 vertices inside it. In order to use this small cache effectively, we need to reorder the triangles to maximize the locality of vertex references. How to do this with MeshOptimizer in place is shown next. Pay attention to how only the indices data is being touched here:

```
meshopt_optimizeVertexCache(remappedIndices.data(),
  remappedIndices.data(), indices.size(), vertexCount);
```

5. Transformed vertices form triangles that are sent for rasterization to generate fragments. Usually, each fragment is run through a depth test first, and fragments that pass the depth test get the fragment shader executed to compute the final color. As fragment shaders get more and more expensive, it becomes increasingly important to reduce the number of fragment shader invocations.

This can be achieved by reducing pixel overdraw in a mesh, and, in general, it requires the use of view-dependent algorithms. However, MeshOptimizer implements heuristics to reorder triangles and minimize overdraw from all directions. We can use it as follows:

```
meshopt_optimizeOverdraw(remappedIndices.data(),
  remappedIndices.data(), indices.size(),
  glm::value_ptr(remappedVertices[0]), vertexCount,
  sizeof(vec3), 1.05f);
```

The last parameter, `1.05`, is the threshold that determines how much the algorithm can compromise the vertex cache hit ratio. We use the recommended default value from the documentation.

6. Once we have optimized the mesh to reduce pixel overdraw, the vertex buffer access pattern can still be optimized for memory efficiency. The GPU has to fetch specified vertex attributes from the vertex buffer and pass this data into the vertex shader. To speed up this fetch, a memory cache is used, which means optimizing the locality of vertex buffer access is very important. We can use MeshOptimizer to optimize our index and vertex buffers for vertex fetch efficiency, as follows:

```
meshopt_optimizeVertexFetch(remappedVertices.data(),
  remappedIndices.data(), indices.size(),
  remappedVertices.data(), vertexCount, sizeof(vec3));
```

This function will reorder vertices in the vertex buffer and regenerate indices to match the new contents of the vertex buffer.

7. The last thing we will do in this recipe is simplify the mesh. MeshOptimizer can generate a new index buffer that uses existing vertices from the vertex buffer with a reduced number of triangles. This new index buffer can be used to render LOD meshes. The following code snippet shows you how to do this using the default threshold and target error values:

```
const float threshold          = 0.2f;
const size_t target_index_count = size_t(remappedIndices.size() * threshold);
const float target_error       = 0.01f;
std::vector<uint32_t> indicesLod;
indicesLod.resize(remappedIndices.size());
indicesLod.resize(meshopt_simplify(&indicesLod[0],
  remappedIndices.data(), remappedIndices.size(),
  &remappedVertices[0].x, vertexCount, sizeof(vec3),
  target_index_count, target_error));
indices   = remappedIndices;
positions = remappedVertices;
```

Now let's take a look at how rendering of the LOD meshes works.

How it works...

In order to render the mesh and its lower-level LOD, we need to store a vertex buffer and two index buffers – one for the mesh and one for the LOD:

1. Here's the vertex buffer for storing all vertex positions:

```
Holder<BufferHandle> vertexBuffer = ctx->createBuffer({
  .usage     = lvk::BufferUsageBits_Vertex,
  .storage   = lvk::StorageType_Device,
  .size      = sizeof(vec3) * positions.size(),
  .data      = positions.data(),
  .debugName = "Buffer: vertex" });
```

2. We use two index buffers to store both sets of indices, one for each LOD:

```
Holder<BufferHandle> indexBuffer = ctx->createBuffer({
  .usage     = lvk::BufferUsageBits_Index,
  .storage   = lvk::StorageType_Device,
  .size      = sizeof(uint32_t) * indices.size(),
  .data      = indices.data(),
  .debugName = "Buffer: index" });
Holder<BufferHandle> indexBufferLod = ctx->createBuffer({
  .usage     = lvk::BufferUsageBits_Index,
  .storage   = lvk::StorageType_Device,
  .size      = sizeof(uint32_t) * indicesLod.size(),
  .data      = indicesLod.data(),
  .debugName = "Buffer: index LOD" });
```

3. The rendering part is trivial, and the graphics pipeline setup is skipped here for the sake of brevity. We render the main mesh using the first index buffer:

```
ICommandBuffer& buf = ctx->acquireCommandBuffer();
buf.cmdBeginRendering({ ... }, { ... });
buf.cmdBindVertexBuffer(0, vertexBuffer, 0);
buf.cmdBindRenderPipeline(pipeline);
buf.cmdBindDepthState({ .compareOp = lvk::CompareOp_Less,
                       .isDepthWriteEnabled = true });
buf.cmdPushConstants(p * v1 * m);
buf.cmdBindIndexBuffer(indexBuffer, lvk::IndexFormat_UI32);
buf.cmdDrawIndexed(indices.size());
```

4. Then we render the LOD mesh using the second index buffer:

```
buf.cmdPushConstants(p * v2 * m);
buf.cmdBindIndexBuffer(indexBufferLod, lvk::IndexFormat_UI32);
buf.cmdDrawIndexed(indicesLod.size());
buf.cmdEndRendering();
ctx->submit(buf, ctx->getCurrentSwapchainTexture());
```

Here's a screenshot from the running demo application.

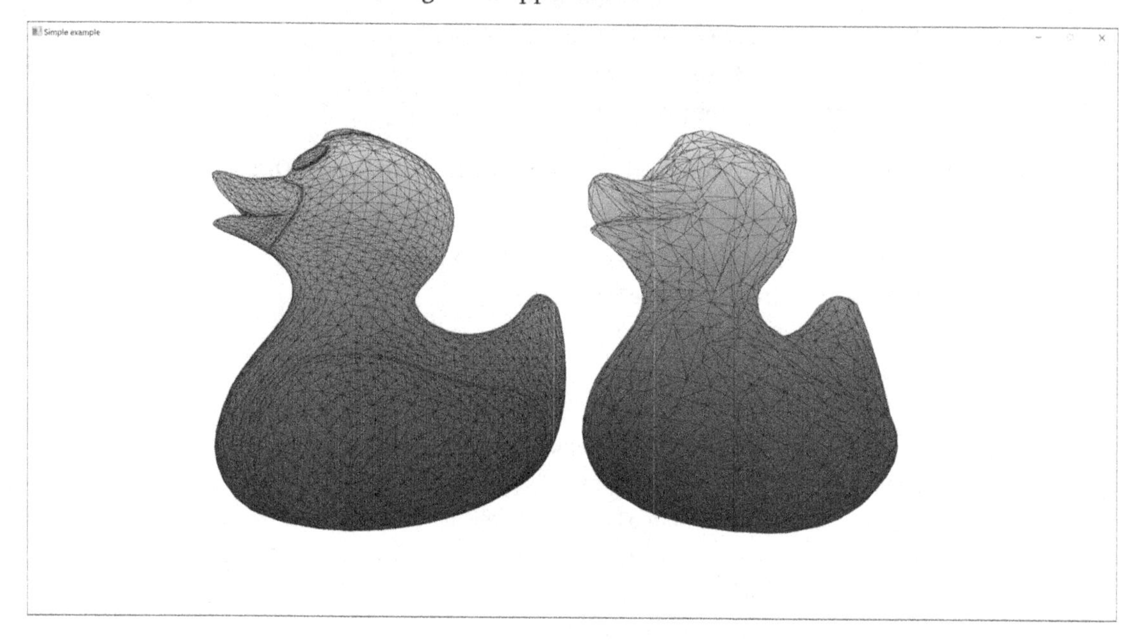

Figure 5.1: A mesh with a discrete LOD

Try changing the `threshold` parameter in the code to generate meshes with different LOD.

There's more...

The MeshOptimizer library contains many other useful algorithms, such as triangle strip generation, index and vertex buffer compression, and mesh animation data compression. All of these algorithms might be very useful for your geometry preprocessing stage depending on the kind of graphics software you are writing. Check out the official documentation and the releases page to get the latest features at `https://github.com/zeux/meshoptimizer`.

Implementing programmable vertex pulling

The concept of **programmable vertex pulling** (**PVP**) was proposed in an article called *Introducing the Programmable Vertex Pulling Rendering Pipeline* by Daniel Rákos, published in the amazing book *OpenGL Insights* in 2012. The article delves into the GPU architecture of that era and explains why this data storage approach was advantageous. Initially, vertex pulling involved storing vertex data in one-dimensional buffer textures.

Instead of setting up traditional vertex input bindings, data was accessed using `texelFetch()` and the GLSL `samplerBuffer` in the vertex shader. The built-in OpenGL GLSL variable `gl_VertexID` was used as an index to calculate texture coordinates for texel fetching. This technique emerged as a solution to CPU bottlenecks caused by numerous draw calls. By combining multiple meshes into a single buffer and rendering them with a single draw call, developers could avoid rebinding vertex arrays or buffer objects, significantly improving draw call batching.

Nowadays, buffer textures are no longer necessary, as vertex data can be fetched directly from storage or uniform buffers using offsets calculated with the built-in Vulkan GLSL variable `gl_VertexIndex`.

This technique enables merge-instancing, where multiple small meshes are combined into a larger one to be processed as part of the same batch. We will make extensive use of this approach in various examples.

In this recipe, we'll use storage buffers to implement a similar technique with Vulkan 1.3 and *LightweightVK*.

Getting ready

The complete source code for this recipe can be found in the source code bundle under the name `Chapter05/02_VertexPulling`.

How to do it...

Let's render the rubber duck 3D model `data/rubber_duck/scene.gltf` from the previous recipe. However, this time, instead of using vertex attributes, we'll use the programmable vertex-pulling technique, which requires revisiting the loading code. The approach involves allocating two buffers: one for indices and another storage buffer for vertex data. These buffers will be accessed in the vertex shader to fetch vertex positions. Here's how it can be done:

1. First, we load the 3D model via Assimp, as in the previous recipe:

   ```
   const aiScene* scene = aiImportFile(
     "data/rubber_duck/scene.gltf", aiProcess_Triangulate);
   const aiMesh* mesh = scene->mMeshes[0];
   ```

2. Convert per-vertex data into a format suitable for our GLSL shaders. We are going to use `vec3` for positions and `vec2` for texture coordinates:

   ```
   struct Vertex {
     vec3 pos;
     vec2 uv;
   };
   std::vector<Vertex> positions;
   for (unsigned int i = 0; i != mesh->mNumVertices; i++) {
     const aiVector3D v = mesh->mVertices[i];
     const aiVector3D t = mesh->mTextureCoords[0][i];
   ```

```
    positions.push_back({
      .pos = vec3(v.x, v.y, v.z), .uv = vec2(t.x, t.y) });
  }
```

3. For simplicity, we store indices as unsigned 32-bit integers. In real-world applications, consider using 16-bit indices for small meshes and be capable of switching between them:

```
std::vector<uint32_t> indices;
for (unsigned int i = 0; i != mesh->mNumFaces; i++) {
  for (int j = 0; j != 3; j++)
    indices.push_back(mesh->mFaces[i].mIndices[j]);
}
aiReleaseImport(scene);
```

4. Once the index and vertex data are ready, we can upload them into the Vulkan buffers. We should create two buffers, one for the vertices and one for the indices. Not that here, despite calling it a vertex buffer, we set the usage flag to `lvk::BufferUsageBits_Storage`:

```
Holder<BufferHandle> vertexBuffer = ctx->createBuffer({
  .usage     = lvk::BufferUsageBits_Storage,
  .storage   = lvk::StorageType_Device,
  .size      = sizeof(Vertex) * positions.size(),
  .data      = positions.data(),
  .debugName = "Buffer: vertex" });
Holder<BufferHandle> indexBuffer = ctx->createBuffer({
  .usage     = lvk::BufferUsageBits_Index,
  .storage   = lvk::StorageType_Device,
  .size      = sizeof(uint32_t) * indices.size(),
  .data      = indices.data(),
  .debugName = "Buffer: index" });
```

5. Now we can load a texture, compile the shaders, and create a render pipeline for our mesh. We'll examine the shader code shortly:

```
Holder<TextureHandle> texture =
  loadTexture(ctx, "data/rubber_duck/textures/Duck_baseColor.png");
Holder<ShaderModuleHandle> vert =
  loadShaderModule(ctx, "Chapter05/02_VertexPulling/src/main.vert");
Holder<ShaderModuleHandle> geom =
  loadShaderModule(ctx, "Chapter05/02_VertexPulling/src/main.geom");
Holder<ShaderModuleHandle> frag =
  loadShaderModule(ctx, "Chapter05/02_VertexPulling/src/main.frag");
Holder<RenderPipelineHandle> pipelineSolid =
  ctx->createRenderPipeline({
        .smVert      = vert,
```

```
        .smGeom      = geom,
        .smFrag      = frag,
        .color       = { { .format = ctx->getSwapchainFormat() } },
        .depthFormat = app.getDepthFormat(),
        .cullMode    = lvk::CullMode_Back,
    });
```

6. Before we can proceed with the actual rendering, we should pass the texture ID and storage buffer address to our GLSL shader. We can do it using Vulkan push constants. Model-view-projection matrix calculations are reused from the previous recipe.

```
const struct PushConstants {
  mat4 mvp;
  uint64_t vertices;
  uint32_t texture;
} pc {
  .mvp      = p * v * m,
  .vertices = ctx->gpuAddress(vertexBuffer),
  .texture  = texture.index(),
};
lvk::ICommandBuffer& buf = ctx->acquireCommandBuffer();
buf.cmdBeginRendering(renderPass, framebuffer);
buf.cmdPushConstants(pc);
```

7. Now, the mesh rendering can be done as follows.

```
buf.cmdBindIndexBuffer(indexBuffer, lvk::IndexFormat_UI32);
buf.cmdBindRenderPipeline(pipelineSolid);
buf.cmdBindDepthState({ .compareOp = lvk::CompareOp_Less,
                        .isDepthWriteEnabled = true });
buf.cmdDrawIndexed(indices.size());
buf.cmdEndRendering();
```

The rest of the C++ code can be found in `Chapter05/02_VertexPulling/src/main.cpp`. Now, we have to look into the GLSL vertex shader to understand how to read the vertex data from buffers. The vertex shader can be found in `Chapter05/02_VertexPulling/src/main.vert`:

1. First, we have some declarations shared between all shaders. The reason for this sharing is that our fragment shader needs to access push constants to retrieve the texture ID. Note that the `Vertex` structure does not use `vec2` and `vec3` member fields to maintain tight padding and prevent any GPU alignment issues. This structure reflects how our C++ code writes vertex data into the buffer. The buffer holds an unbounded array `in_Vertices[]`. Each element corresponds to exactly one vertex.

```
struct Vertex {
  float x, y, z;
```

```
  float u, v;
};
layout(std430, buffer_reference) readonly buffer Vertices {
  Vertex in_Vertices[];
};
layout(push_constant) uniform PerFrameData {
  mat4 MVP;
  Vertices vtx;
  uint texture;
};
```

2. Let's introduce two accessor functions to make the shader code more readable. The assemble vec3 and vec2 values are from the raw float values in the storage buffer.

```
vec3 getPosition(int i) {
  return vec3(vtx.in_Vertices[i].x,
              vtx.in_Vertices[i].y,
              vtx.in_Vertices[i].z);
}
vec2 getTexCoord(int i) {
  return vec2(vtx.in_Vertices[i].u,
              vtx.in_Vertices[i].v);
}
```

3. The rest of the shader is trivial. The previously mentioned functions are used to load the vertex positions and texture coordinates, which are passed further into the graphics pipeline.

```
layout (location=0) out vec2 uv;
void main() {
  gl_Position = MVP * vec4(getPosition(gl_VertexIndex), 1.0);
  uv = getTexCoord(gl_VertexIndex);
}
```

That's it for the PVP part. The fragment shader applies the texture and uses the barycentric coordinates trick for wireframe rendering as was described in the previous chapter. The resulting output from the program should look like the following screenshot:

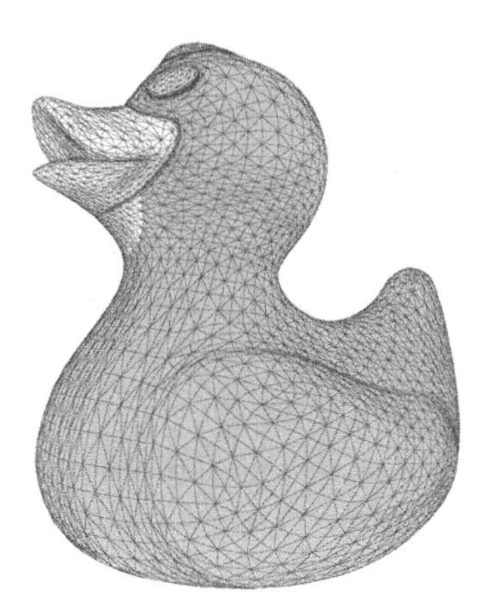

Figure 5.2: Textured mesh rendering using PVP

There's more...

PVP is a complex topic and has different performance implications. There is an open-source project that does an in-depth analysis and run-time metrics of PVP performance based on different vertex data layouts and access methods, such as storing data as an array of structures or a structure of arrays, reading data as multiple floats or a single vector type, and so on. Check it out at `https://github.com/nlguillemot/ProgrammablePulling`. It should become one of your go-to tools when designing PVP pipelines in your applications.

Rendering instanced geometry

A common task in 3D rendering is drawing multiple meshes that share the same geometry but have different transformations and materials. This can result in additional CPU overhead from generating commands for the GPU to draw each mesh individually, even though the Vulkan API already reduces CPU overhead significantly.

One solution to this issue, provided by modern graphics APIs like Vulkan, is instanced rendering. This approach allows API draw commands to include the number of instances as a parameter, while the vertex shader gains access to the current instance number via `gl_InstanceIndex`. When combined with the PVP approach from the previous recipe, this technique becomes extremely flexible. The `gl_InstanceIndex` can be used to fetch material properties, transformations, and other data directly from buffers.

Let's explore a basic instanced geometry demo to learn how to implement this in Vulkan.

Getting ready

Be sure to review the previous recipe, *Implementing programmable vertex pulling*, to understand the concept of generating vertex data within vertex shaders. The source code for this recipe is located in `Chapter05/03_MillionCubes`.

How to do it...

To demonstrate how instanced rendering works, let's render 1 million colored rotating cubes. Each cube will have its own position and distinct rotation angle around its diagonal and will be textured with one of several different colors. Let's take a look at the C++ code in `Chapter05/03_MillionCubes/src/main.cpp`:

1. First, let's generate a procedural texture for our cubes. An XOR pattern texture looks quite interesting. It's created by XOR-ing the x and y coordinates of the current texel, then applying the result to all three BGR channels using bit shifts.

```
const uint32_t texWidth  = 256;
const uint32_t texHeight = 256;
std::vector<uint32_t> pixels(texWidth * texHeight);
for (uint32_t y = 0; y != texHeight; y++) {
  for (uint32_t x = 0; x != texWidth; x++) {
    pixels[y * texWidth + x] =
      0xFF000000 + ((x^y) << 16) + ((x^y) << 8) + (x^y);
  }
}
Holder<TextureHandle> texture = ctx->createTexture({
   .type       = lvk::TextureType_2D,
   .format     = lvk::Format_BGRA_UN8,
   .dimensions = { texWidth, texHeight },
   .usage      = lvk::TextureUsageBits_Sampled,
   .data       = pixels.data(),
   .debugName  = "XOR pattern",
});
```

2. Let's create vec3 positions and `float` initial rotation angles for 1 million cubes. We can organize this data into vec4 containers and store them in an immutable storage buffer. The GLSL shader code will then perform calculations based on the elapsed time.

```
const uint32_t kNumCubes = 1024 * 1024;
std::vector<vec4> centers(kNumCubes);
for (vec4& p : centers) {
  p = vec4(glm::linearRand(-vec3(500.0f), +vec3(500.0f)),
           glm::linearRand(0.0f, 3.14159f));
}
```

```
Holder<BufferHandle> bufferPosAngle = ctx->createBuffer({
  .usage   = lvk::BufferUsageBits_Storage,
  .storage = lvk::StorageType_Device,
  .size    = sizeof(vec4) * kNumCubes,
  .data    = centers.data(),
});
```

3. We skip the usual framebuffer and pipeline creation code and jump straight into the main rendering loop. The camera motion is hardcoded to move back and forth through the swarm of cubes.

```
buf.cmdBeginRendering(renderPass, framebuffer);
const mat4 view = translate(mat4(1.0f),
  vec3(0.0f, 0.0f,
       -1000.0f + 500.0f * (1.0f - cos(-glfwGetTime() * 0.5f))));
```

4. We pass all the necessary data to the shaders using push constants. The vertex shader will need the current time to perform calculations based on the initial cube positions and rotations.

```
const struct {
  mat4 viewproj;
  uint32_t textureId;
  uint64_t bufferPosAngle;
  float time;
} pc {
  .viewproj       = proj * view,
  .textureId      = texture.index(),
  .bufferPosAngle = ctx->gpuAddress(bufferPosAngle),
  .time           = (float)glfwGetTime(),
};
```

5. Rendering is initiated with `vkCmdDraw()`, which is wrapped inside `cmdDraw()`. The first parameter, `36`, is the number of vertices required to generate a cube using triangle primitives. We'll explore how this is handled in the vertex shader shortly. The second parameter, `kNumCubes`, represents the number of instances to be rendered.

```
buf.cmdPushConstants(pc);
buf.cmdBindRenderPipeline(pipelineSolid);
buf.cmdBindDepthState({ .compareOp = lvk::CompareOp_Less,
                        .isDepthWriteEnabled = true });
buf.cmdDraw(36, kNumCubes);
buf.cmdEndRendering();
```

Now let's take a look at the GLSL code to understand how this instancing demo works under the hood.

How it works...

1. Our vertex shader starts by declaring the same `PerFrameData` structure as in our C++ code mentioned above. The shader outputs per-vertex color and texture coordinates. The storage buffer contains positions and initial angles for all the cubes.

```
layout(push_constant) uniform PerFrameData {
  mat4 viewproj;
  uint textureId;
  uvec2 bufId;
  float time;
};
layout (location=0) out vec3 color;
layout (location=1) out vec2 uv;
layout(std430, buffer_reference) readonly buffer Positions {
  vec4 pos[]; // pos: xyz, initialAngle: w
};
```

2. As you may have noticed, the C++ code in the *How to do it...* section did not provide any index data to shaders. Instead, we a going to generate vertex data in the vertex shader. Let's declare indices mapping right here. We need indices to construct `6` cube faces using triangles. Two triangles per face gives `6` points per face and `36` indices in total. That is the number passed to `vkCmdDraw()`.

```
const int indices[36] = int[36](
  0, 2, 1, 2, 3, 1, 5, 4, 1, 1, 4, 0,
  0, 4, 6, 0, 6, 2, 6, 5, 7, 6, 4, 5,
  2, 6, 3, 6, 7, 3, 7, 1, 3, 7, 5, 1);
```

3. Here are the per-instance colors for our cubes.

```
const vec3 colors[7] = vec3[7](
  vec3(1.0, 0.0, 0.0),
  vec3(0.0, 1.0, 0.0),
  vec3(0.0, 0.0, 1.0),
  vec3(1.0, 1.0, 0.0),
  vec3(0.0, 1.0, 1.0),
  vec3(1.0, 0.0, 1.0),
  vec3(1.0, 1.0, 1.0));
```

4. Since no translation or rotation matrices are passed to the vertex shader, we need to generate everything ourselves right here. Here's a GLSL function to apply a translation by a vector `v` to the current transformation `m`. This function serves as the GLSL counterpart to the C++ function `glm::translate()`.

```
mat4 translate(mat4 m, vec3 v) {
  mat4 Result = m;
  Result[3] = m[0] * v[0] + m[1] * v[1] + m[2] * v[2] + m[3];
  return Result;
}
```

5. Rotations are handled in a similar way. This is an analogue of `glm::rotate()` ported to GLSL.

```
mat4 rotate(mat4 m, float angle, vec3 v) {
  float a = angle;
  float c = cos(a);
  float s = sin(a);
  vec3 axis = normalize(v);
  vec3 temp = (float(1.0) - c) * axis;
  mat4 r;
  r[0][0] = c + temp[0] * axis[0];
  r[0][1] = temp[0] * axis[1] + s * axis[2];
  r[0][2] = temp[0] * axis[2] - s * axis[1];
  r[1][0] = temp[1] * axis[0] - s * axis[2];
  r[1][1] = c + temp[1] * axis[1];
  r[1][2] = temp[1] * axis[2] + s * axis[0];
  r[2][0] = temp[2] * axis[0] + s * axis[1];
  r[2][1] = temp[2] * axis[1] - s * axis[0];
  r[2][2] = c + temp[2] * axis[2];
  mat4 res;
  res[0] = m[0] * r[0][0] + m[1] * r[0][1] + m[2] * r[0][2];
  res[1] = m[0] * r[1][0] + m[1] * r[1][1] + m[2] * r[1][2];
  res[2] = m[0] * r[2][0] + m[1] * r[2][1] + m[2] * r[2][2];
  res[3] = m[3];
  return res;
}
```

6. With this extensive arsenal at our disposal, we can now write the `main()` function of our vertex shader. The built-in `gl_InstanceIndex` variable is used to index the storage buffer and retrieve positions and angles. Then, a model matrix for the current cube is computed using the `rotate()` and `translate()` helper functions.

```
void main() {
  vec4 center = Positions(bufId).pos[gl_InstanceIndex];
  mat4 model = rotate(translate(mat4(1.0f), center.xyz),
                 time + center.w, vec3(1.0f, 1.0f, 1.0f));
```

7. The built-in `gl_VertexIndex` variable ranges from `0` to `35`, allowing us to extract the specific index for `8` of our vertices. We then apply this formula to generate `vec3` positions for each of those `8` vertices using bit shifts.

```
uint idx = indices[gl_VertexIndex];
vec3 xyz = vec3(idx & 1, (idx & 4) >> 2, (idx & 2) >> 1);
```

8. Remap the generated `0...1` vertex coordinates to the `-0.5…+0.5` range and then scale them by the desired edge length:

```
const float edge = 1.0;
gl_Position =
  viewproj * model * vec4(edge * (xyz - vec3(0.5)), 1.0);
```

9. The UV coordinates are selected on a per-face basis, while the colors are applied per instance:

```
  int face = gl_VertexIndex / 6;
  if (face == 0 || face == 3) uv = vec2(xyz.x, xyz.z);
  if (face == 1 || face == 4) uv = vec2(xyz.x, xyz.y);
  if (face == 2 || face == 5) uv = vec2(xyz.y, xyz.z);
  color = colors[gl_InstanceIndex % 7];
}
```

10. That's all the magic that happens in the vertex shader. The fragment shader is quite simple and short:

```
layout (location=0) in vec3 color;
layout (location=1) in vec2 uv;
layout (location=0) out vec4 out_FragColor;
layout(push_constant) uniform PerFrameData {
  mat4 proj;
  uint textureId;
};
void main() {
  out_FragColor = textureBindless2D(
    textureId, 0, uv) * vec4(color, 1.0);
}
```

The running demo should look like the following screenshot, where you're flying through a swarm of 1 million cubes, moving forth and back again:

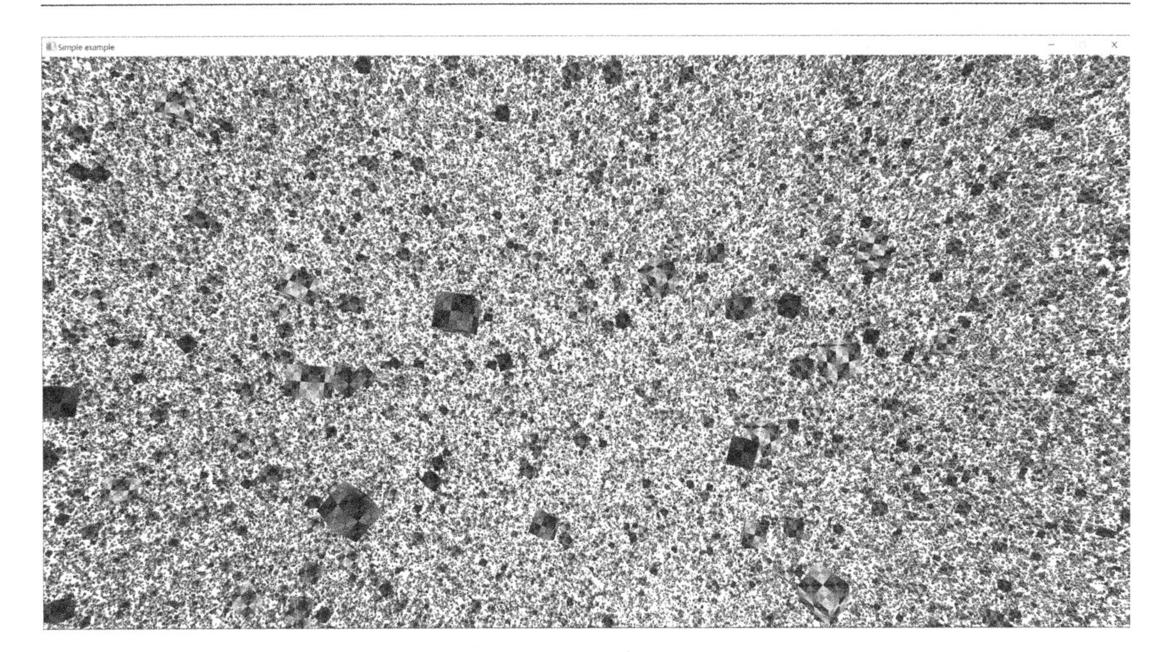

Figure 5.3: One million cubes using instanced rendering

Now, let's extend this instancing example further and draw some meshes using real mesh data.

There's more...

While this example is self-contained and much faster compared to non-instanced rendering, it can be made even faster by moving the indices out of the vertex shader and storing them in a dedicated index buffer. This allows us to take advantage of the hardware vertex cache, and model matrices can be calculated per-instance instead of per-vertex. We'll cover this in the next recipe.

This instanced rendering approach can be combined with Vulkan's indirect rendering technique for greater flexibility. We'll learn how to implement this in *Chapter 8*, in the recipe *Implementing indirect rendering with Vulkan*. The basics of indirect rendering will be covered later in this chapter in the recipe *Indirect rendering in Vulkan*.

Implementing instanced meshes with compute shaders

In the previous recipe, we learned the fundamentals of instanced rendering. While that approach covers various aspects of rendering geometry instances, such as handling model matrices and materials, it's not yet a practical implementation. Let's build on that example and show how to render instanced meshes loaded from a `.gltf` file.

To make this recipe more interesting, we'll enhance the example by precalculating per-instance model matrices using a compute shader.

Getting ready

Make sure you read the previous recipe, *Rendering instanced geometry*. The source code for this recipe can be found in `Chapter05/04_InstancedMeshes`.

How to do it...

Let's skim the C++ code to understand the big picture

1. First, we generate random positions and initial rotation angles for our meshes. We use 32,000 meshes because our GPU cannot handle 1 million meshes with this naïve brute-force approach. Pushing it to 1 million meshes is possible, and we will show some tricks to approach that number in *Chapter 11, Advanced Rendering Techniques and Optimizations.*

```
const uint32_t kNumMeshes = 32 * 1024;
std::vector<vec4> centers(kNumMeshes);
for (vec4& p : centers) {
   p = vec4(glm::linearRand(-vec3(500.0f), +vec3(500.0f)),
            glm::linearRand(0.0f, 3.14159f));
}
```

2. The center points and angles are loaded into a storage buffer in exactly the same way as in the previous recipe:

```
Holder<BufferHandle> bufferPosAngle   = ctx->createBuffer({
  .usage      = lvk::BufferUsageBits_Storage,
  .storage    = lvk::StorageType_Device,
  .size       = sizeof(vec4) * kNumMeshes,
  .data       = centers.data(),
  .debugName = "Buffer: angles & positions",
});
```

3. To store model matrices for our instances, we'll require two buffers. We'll alternate between them in a round-robin fashion during even and odd frames to skip unnecessary synchronization.

```
Holder<BufferHandle> bufferMatrices[] = {
  ctx->createBuffer({ .usage     = lvk::BufferUsageBits_Storage,
                      .storage   = lvk::StorageType_Device,
                      .size      = sizeof(mat4) * kNumMeshes,
                      .debugName = "Buffer: matrices 1" }),
  ctx->createBuffer({ .usage     = lvk::BufferUsageBits_Storage,
                      .storage   = lvk::StorageType_Device,
                      .size      = sizeof(mat4) * kNumMeshes,
                      .debugName = "Buffer: matrices 2" }),
};.
```

4. The rubber duck 3D model is loaded from `.gltf` the following way. This time, besides vertex positions and texture coordinates, we require normal vectors to do some improvized lighting. We are going to need it when we render so many meshes.

```
const aiScene* scene =
  aiImportFile("data/rubber_duck/scene.gltf", aiProcess_Triangulate);
struct Vertex {
  vec3 pos;
  vec2 uv;
  vec3 n;
};
const aiMesh* mesh = scene->mMeshes[0];
std::vector<Vertex> vertices;
std::vector<uint32_t> indices;
for (unsigned int i = 0; i != mesh->mNumVertices; i++) {
  const aiVector3D v = mesh->mVertices[i];
  const aiVector3D t = mesh->mTextureCoords[0][i];
  const aiVector3D n = mesh->mNormals[i];
  vertices.push_back({ .pos = vec3(v.x, v.y, v.z),
                       .uv  = vec2(t.x, t.y),
                       .n   = vec3(n.x, n.y, n.z) });
}
for (unsigned int i = 0; i != mesh->mNumFaces; i++) {
  for (int j = 0; j != 3; j++)
    indices.push_back(mesh->mFaces[i].mIndices[j]);
}
aiReleaseImport(scene);
```

5. The mesh data is uploaded into index and vertex buffers:

```
Holder<BufferHandle> vertexBuffer = ctx->createBuffer({
  .usage     = lvk::BufferUsageBits_Storage,
  .storage   = lvk::StorageType_Device,
  .size      = sizeof(Vertex) * vertices.size(),
  .data      = vertices.data(),
  .debugName = "Buffer: vertex" });
Holder<BufferHandle> indexBuffer = ctx->createBuffer({
  .usage     = lvk::BufferUsageBits_Index,
  .storage   = lvk::StorageType_Device,
  .size      = sizeof(uint32_t) * indices.size(),
  .data      = indices.data(),
  .debugName = "Buffer: index" });
```

6. Let's load the texture and create compute and rendering pipelines. The compute shader will generate model matrices for our instances based on the elapsed time, following the approach used in the vertex shader in the previous recipe, *Rendering instanced meshes*. However, this time, we'll do it on a per-instance basis instead of per vertex.

```
Holder<TextureHandle> texture =
  loadTexture(ctx, "data/rubber_duck/textures/Duck_baseColor.png");
Holder<ShaderModuleHandle> comp =
  loadShaderModule(ctx, "Chapter05/04_InstancedMeshes/src/main.comp");
Holder<ComputePipelineHandle> pipelineComputeMatrices =
  ctx->createComputePipeline({ smComp = comp });
Holder<ShaderModuleHandle> vert =
  loadShaderModule(ctx, "Chapter05/04_InstancedMeshes/src/main.vert");
Holder<ShaderModuleHandle> frag =
  loadShaderModule(ctx, "Chapter05/04_InstancedMeshes/src/main.frag");
Holder<RenderPipelineHandle> pipelineSolid =
  ctx->createRenderPipeline({
    .smVert      = vert,
    .smFrag      = frag,
    .color       = { { .format = ctx->getSwapchainFormat() } },
    .depthFormat = app.getDepthFormat(),
    .cullMode    = lvk::CullMode_Back });
```

7. The main loop goes like this. We use the `frameId` counter to facilitate the switching of buffers containing model matrices between even and odd frames.

```
uint32_t frameId = 0;
app.run([&](uint32_t width, uint32_t height,
            float aspectRatio, float deltaSeconds)
{
  const mat4 proj = glm::perspective(45.0f, aspectRatio, 0.2f, 1500.0f);
  const lvk::RenderPass renderPass = {
    .color = { { .loadOp = lvk::LoadOp_Clear,
                 .clearColor = { 1.0f, 1.0f, 1.0f, 1.0f } } },
    .depth = { .loadOp = lvk::LoadOp_Clear, .clearDepth = 1.0f }
  };
  const lvk::Framebuffer framebuffer = {
    .color = { { .texture = ctx->getCurrentSwapchainTexture() } },
    .depthStencil = { .texture = app.getDepthTexture() },
  };
```

8. For convenience, push constants are shared between compute and rendering pipelines:

```
lvk::ICommandBuffer& buf = ctx->acquireCommandBuffer();
const mat4 view = translate(mat4(1.0f), vec3(0.0f, 0.0f,
  -1000.0f + 500.0f * (1.0f - cos(-glfwGetTime() * 0.5f))));
const struct {
  mat4 viewproj;
  uint32_t textureId;
  uint64_t bufferPosAngle;
  uint64_t bufferMatrices;
  uint64_t bufferVertices;
  float time;
} pc {
  .viewproj       = proj * view,
  .textureId      = texture.index(),
  .bufferPosAngle = ctx->gpuAddress(bufferPosAngle),
  .bufferMatrices = ctx->gpuAddress(bufferMatrices[frameId]),
  .bufferVertices = ctx->gpuAddress(vertexBuffer),
  .time           = (float)glfwGetTime(),
};
buf.cmdPushConstants(pc);
```

9. Dispatch the compute shader. Each local workgroup handles 32 meshes – a common portable baseline supported by many GPUs:

```
buf.cmdBindComputePipeline(pipelineComputeMatrices);
buf.cmdDispatchThreadGroups({ .width = kNumMeshes / 32 } });
```

10. After the compute shader has finished updating model matrices, we can start rendering. Note that we have a non-empty dependencies parameter here, which refers to the buffer with model matrices. This is necessary to make sure a proper Vulkan buffer memory barrier is issued by *LightweightVK* to prevent race conditions between the compute shader and the vertex shader.

```
buf.cmdBeginRendering(renderPass, framebuffer,
  { .buffers = { BufferHandle(bufferMatrices[frameId]) } });
```

Note

Now let's look at the barrier. The source and destination stages, respectively, are:

```
VK_PIPELINE_STAGE_COMPUTE_SHADER_BIT
```

And:

```
VK_PIPELINE_STAGE_VERTEX_SHADER_BIT |
VK_PIPELINE_STAGE_FRAGMENT_SHADER_BIT
```

The underlying Vulkan barrier looks as follows:

```
void lvk::CommandBuffer::bufferBarrier(BufferHandle handle,
  VkPipelineStageFlags srcStage,
  VkPipelineStageFlags dstStage)
{
  lvk::VulkanBuffer* buf = ctx_->buffersPool_.get(handle);
  const VkBufferMemoryBarrier barrier = {
    .sType = VK_STRUCTURE_TYPE_BUFFER_MEMORY_BARRIER,
    .srcAccessMask = VK_ACCESS_SHADER_READ_BIT |
                     VK_ACCESS_SHADER_WRITE_BIT,
    .dstAccessMask = VK_ACCESS_SHADER_READ_BIT |
                     VK_ACCESS_SHADER_WRITE_BIT,
    .srcQueueFamilyIndex = VK_QUEUE_FAMILY_IGNORED,
    .dstQueueFamilyIndex = VK_QUEUE_FAMILY_IGNORED,
    .buffer = buf->vkBuffer_,
    .offset = 0,
    .size = VK_WHOLE_SIZE,
  };
  vkCmdPipelineBarrier(wrapper_->cmdBuf_,
    srcStage, dstStage, VkDependencyFlags{},
    0, nullptr, 1, &barrier, 0, nullptr);
}
```

11. The rest of the rendering code is pretty standard. The *LightweightVK* draw call command, `cmdDrawIndexed()`, takes the number of indices in our mesh and the number of instances `kNumMeshes`.

```
buf.cmdBindRenderPipeline(pipelineSolid);
buf.cmdBindDepthState({ .compareOp = lvk::CompareOp_Less,
                        .isDepthWriteEnabled = true });
buf.cmdBindIndexBuffer(indexBuffer, lvk::IndexFormat_UI32);
```

```
  buf.cmdDrawIndexed(indices.size(), kNumMeshes);
  buf.cmdEndRendering();
  ctx->submit(buf, ctx->getCurrentSwapchainTexture());
  frameId = (frameId + 1) & 1;
});
```

Now, let's delve into the GLSL implementation details to understand how it works internally.

How it works...

The first part is the compute shader, which prepares data for rendering. Let's take a look at `Chapter05/04_InstancedMeshes/src/main.comp`:

1. The compute shader processes 32 meshes in one local workgroup. Push constants are shared between the compute shader and the graphics pipeline. They are declared in an include file, `Chapter05/04_InstancedMeshes/src/common.sp`. We provide that file here for your convenience:

```
layout(local_size_x = 32, local_size_y = 1, local_size_z = 1) in;
// included from <Chapter05/04_InstancedMeshes/src/common.sp>
layout(push_constant) uniform PerFrameData {
  mat4 viewproj;
  uint textureId;
  uvec2 bufPosAngleId;
  uvec2 bufMatricesId;
  uvec2 bufVerticesId;
  float time;
};
layout(std430, buffer_reference) readonly buffer Positions {
  vec4 pos[]; // pos, initialAngle
};
// end of #include
```

2. The matrices buffer reference is not shared. In the compute shader, it is declared as `writeonly`, while in the vertex shader, it will be declared as `readonly`.

```
layout(std430, buffer_reference) writeonly buffer Matrices {
  mat4 mtx[];
};
```

3. The helper functions `translate()` and `rotate()` mimic the `glm::translate()` and `glm::rotate()` C++ functions in GLSL. They are reused in their entirety from the previous recipe, *Rendering instanced geometry*. Since they are quite long, we won't duplicate them here.

```
mat4 translate(mat4 m, vec3 v);
mat4 rotate(mat4 m, float angle, vec3 v);
```

4. The `main()` function reads a `vec4` value containing the `vec3` center point and initial angle, and calculates a model matrix. This is the same computation we did in the vertex shader in the previous recipe. The model matrix is then stored in a storage buffer referenced by `bufMatricesId`.

```
void main() {
  uint idx = gl_GlobalInvocationID.x;
  vec4 center = Positions(bufPosAngleId).pos[idx];
  mat4 model = rotate(translate(mat4(1.0f),
    center.xyz), time + center.w, vec3(1.0f, 1.0f, 1.0f));
  Matrices(bufMatricesId).mtx[idx] = model;
}
```

On to the rendering pipeline shaders... As we moved most of the calculations into the compute shader, the shaders for the rendering pipeline became significantly shorter.

1. The vertex shader uses the same shared declarations for push constants and buffer references:

```
#include <Chapter05/04_InstancedMeshes/src/common.sp>
layout (location=0) out vec2 uv;
layout (location=1) out vec3 normal;
layout (location=2) out vec3 color;
layout(std430, buffer_reference) readonly buffer Matrices {
  mat4 mtx[];
};
```

2. The vertex data contains normal vectors, as was declared in the C++ code earlier in this recipe:

```
struct Vertex {
  float x, y, z;
  float u, v;
  float nx, ny, nz;
};
layout(std430, buffer_reference) readonly buffer Vertices {
  Vertex in_Vertices[];
};
```

3. The vertex data is fetched from the "vertex" storage buffer, and the model matrix is retrieved from the matrices buffer, which was updated by the compute shader:

```
const vec3 colors[3] = vec3[3](vec3(1.0, 0.0, 0.0),
                               vec3(0.0, 1.0, 0.0),
                               vec3(1.0, 1.0, 1.0));
void main() {
  Vertex vtx = Vertices(bufVerticesId).in_Vertices[gl_VertexIndex];
  mat4 model = Matrices(bufMatricesId).mtx[gl_InstanceIndex];
```

4. Now we can compute the value of `gl_Position` and pass the normal vector and texture coordinates to the fragment shader:

```
  const float scale = 10.0;
  gl_Position = viewproj * model *
    vec4(scale * vtx.x, scale * vtx.y, scale * vtx.z, 1.0);
  mat3 normalMatrix = transpose( inverse(mat3(model)) );
  uv = vec2(vtx.u, vtx.v);
  normal = normalMatrix * vec3(vtx.nx, vtx.ny, vtx.nz);
  color = colors[gl_InstanceIndex % 3];
}
```

5. The fragment shader is fairly straightforward. We perform some improvised diffuse lighting calculations to enhance the visual distinctiveness of the meshes:

```
layout (location=0) in vec2 uv;
layout (location=1) in vec3 normal;
layout (location=2) in vec3 color;
layout (location=0) out vec4 out_FragColor;
layout(push_constant) uniform PerFrameData {
  mat4 viewproj;
  uint textureId;
};
void main() {
  vec3 n = normalize(normal);
  vec3 l = normalize(vec3(1.0, 0.0, 1.0));
  float NdotL = clamp(dot(n, l), 0.3, 1.0);
  out_FragColor =
    textureBindless2D(textureId, 0, uv) * NdotL * vec4(color, 1.0);
};
```

The running demo application should display a swarm of rotating ducks, as shown in the following screenshot, with the camera flying back and forth through them.

Figure 5.4: A swarm of rotating ducks using instanced rendering

There's more...

As you may have noticed, this demo uses only `32'768` instances compared to `1` million in the previous recipe. The difference is due to the cube in the previous example having just 36 indices, while the rubber duck model used here has `33'216`—nearly one thousand times more.

The naïve brute-force approach won't work for this dataset. To render `1` million duckies, we need to use additional techniques such as culling and GPU-level-of-detail management. We'll touch some of these topics in *Chapter 11*.

Now, let's switch gears and learn how to render debug grid geometry before moving on to examples of more complex mesh rendering.

Implementing an infinite grid GLSL shader

In the previous recipes of this chapter, we explored some approaches to geometry rendering. When debugging applications, it's helpful to have a visible representation of the coordinate system, allowing viewers to quickly infer the camera's orientation and position by examining the rendered image. A practical way to represent a coordinate system is by rendering an infinite grid aligned with one of the coordinate planes. Let's learn how to implement a visually appealing grid in GLSL.

Getting ready

The complete C++ source code for this recipe is available in `Chapter05/05_Grid`. The accompanying GLSL shaders, which will be reused in later recipes, are stored in the shared data folder under `data/shaders/Grid.vert` and `data/shaders/Grid.frag`.

How to do it...

To parameterize our grid, we should introduce some constants. These constants can be found and adjusted in the `data/shaders/GridParameters.h` GLSL include file. Let's take a look inside:

1. First, we need to define the size of our grid extents in world coordinates. This determines how far from the camera the grid is visible:

```
float gridSize = 100.0;
```

2. The size of one grid cell is specified in the same units as the grid size:

```
float gridCellSize = 0.025;
```

3. Let's define the colors for the grid lines. We'll use two different colors: one for regular thin lines and another for thick lines, which are rendered every tenth line. Since everything is rendered against a white background, black and 50% gray will work well.

```
vec4 gridColorThin  = vec4(0.5, 0.5, 0.5, 1.0);
vec4 gridColorThick = vec4(0.0, 0.0, 0.0, 1.0);
```

4. Our grid implementation will adjust the number of rendered lines based on the grid LOD. We will switch the LOD when the number of pixels between two adjacent grid cell lines drops below a certain threshold, which is calculated in the fragment shader:

```
const float gridMinPixelsBetweenCells = 2.0;
```

5. Let's take a look at a simple vertex shader, `data/shaders/Grid.vert`, which we use to generate and transform grid vertices. It takes in the current model-view-projection matrix, the current camera position, and the grid origin. The origin is in world space and can be used to move the grid around.

```
#version 460 core
#include <data/shaders/GridParameters.h>
layout(push_constant) uniform PerFrameData {
  mat4 MVP;
  vec4 cameraPos;
  vec4 origin;
};
layout (location=0) out vec2 uv;
layout (location=1) out vec2 out_camPos;
const vec3 pos[4] = vec3[4](
```

```
    vec3(-1.0, 0.0, -1.0),
    vec3( 1.0, 0.0, -1.0),
    vec3( 1.0, 0.0,  1.0),
    vec3(-1.0, 0.0,  1.0)
  );
  const int indices[6] = int[6]( 0, 1, 2, 2, 3, 0 );
```

6. The built-in `gl_VertexIndex` variable is used to access hardcoded quad indices and the vertices in `pos[]`. The `-1` to `+1` points are scaled by the desired grid size. The resulting vertex position is then translated by the 2D camera position in the horizontal plane and, finally, by the 3D origin position:

```
void main() {
  int idx = indices[gl_VertexIndex];
  vec3 position = pos[idx] * gridSize;
  position.x += cameraPos.x;
  position.z += cameraPos.z;
  position += origin.xyz;
  out_camPos = cameraPos.xz;
  gl_Position = MVP * vec4(position, 1.0);
  uv = position.xz;
}
```

The fragment shader, `data/shaders/Grid.frag`, is somewhat more complex. It calculates a programmatic texture that resembles a grid. The grid lines are rendered based on how quickly the UV coordinates change in screen space to avoid the Moiré pattern. For this, we'll need screen space derivatives. The screen space derivative of a variable in your shader measures how much that variable changes from one pixel to the next. The GLSL function `dFdx()` represents the horizontal change, while `dFdy()` represents the vertical change. These functions measure how fast a GLSL variable changes as you move across the screen, approximating partial derivatives in calculus. This approximation comes from relying on discrete samples at each fragment, rather than performing a full mathematical evaluation of the change:

1. First, we introduce several GLSL helper functions to assist with our calculations. These can be found in `data/shaders/GridCalculation.h`. The function names `satf()` and `satv()` stand for saturate-float and saturate-vector, respectively:

```
float log10(float x) {
  return log(x) / log(10.0);
}
float satf(float x) {
  return clamp(x, 0.0, 1.0);
}
vec2 satv(vec2 x) {
  return clamp(x, vec2(0.0), vec2(1.0));
```

```
}
float max2(vec2 v) {
  return max(v.x, v.y);
}
```

2. Let's take a look at the `gridColor()` function, which is called from `main()`, and begin by calculating the screen space length of the derivatives of the UV coordinates we previously generated in the vertex shader. We use the built-in `dFdx()` and `dFdy()` functions to calculate the necessary derivatives:

```
vec4 gridColor(vec2 uv, vec2 camPos) {
  vec2 dudv = vec2( length(vec2(dFdx(uv.x), dFdy(uv.x))),
                    length(vec2(dFdx(uv.y), dFdy(uv.y))) );
```

3. Knowing the derivatives, the current LOD of our grid can be calculated in the following way. The `gridMinPixelsBetweenCells` value controls how fast we want our LOD to increase. In this case, it is the minimum number of pixels between two adjacent cell lines of the grid:

```
  float lodLevel = max(0.0, log10((length(dudv) *
    gridMinPixelsBetweenCells) / gridCellSize) + 1.0);
  float lodFade = fract(lodLevel);
```

Besides the LOD value, we are going to need a fading factor to render smooth transitions between the adjacent levels. It can be obtained by taking a fractional part of the floating point LOD level. The logarithm base `10` is used to ensure each LOD covers `pow(10, lodLevel)` more cells than the previous size.

4. The LOD levels are blended with each other. To render them, we have to calculate the cell size for each LOD. Here, instead of calculating `pow()` three times, we can calculate it only for `lod0` and multiply each subsequent LOD cell size by `10.0`:

```
  float lod0 = gridCellSize * pow(10.0, floor(lodLevel));
  float lod1 = lod0 * 10.0;
  float lod2 = lod1 * 10.0;
```

5. To be able to draw anti-aliased lines using alpha transparency, we need to increase the screen coverage of our lines. Let's make sure each line covers up to 4 pixels. Shift grid coordinates to the centers of anti-aliased lines for subsequent alpha calculations:

```
  dudv *= 4.0;
  uv += dudv * 0.5;
```

6. Now we should get coverage alpha value corresponding to each calculated LOD level. To do that, we calculate absolute distances to cell line centers for each LOD and pick the maximum coordinate:

```
  float lod0a = max2( vec2(1.0) - abs(
    satv(mod(uv, lod0) / dudv) * 2.0 - vec2(1.0)) );
```

```
    float lod1a = max2( vec2(1.0) - abs(
      satv(mod(uv, lod1) / dudv) * 2.0 - vec2(1.0)) );
    float lod2a = max2( vec2(1.0) - abs(
      satv(mod(uv, lod2) / dudv) * 2.0 - vec2(1.0)) );
```

7. Non-zero alpha values represent non-empty transition areas of the grid. Let's blend between them using two colors to handle LOD transitions:

```
    vec4 c = lod2a > 0.0 ?
      gridColorThick :
      lod1a > 0.0 ?
        mix(gridColorThick, gridColorThin, lodFade) : gridColorThin;
```

8. Last but not least, make the grid disappear when it is far away from the camera. Use the `gridSize` value to calculate the opacity falloff:

```
    uv -= camPos;
    float opacityFalloff = (1.0 - satf(length(uv) / gridSize));
```

9. Now we can blend between the LOD level alpha values and scale the result with the opacity falloff factor. The resulting pixel color value can be stored in the framebuffer:

```
    c.a *= lod2a > 0.0 ?
      lod2a : lod1a > 0.0 ? lod1a : (lod0a * (1.0-lodFade));
    c.a *= opacityFalloff;
    return c;
  }
```

10. The shaders mentioned earlier in `data/shaders/GridCalculation.h` should be rendered using the following render pipeline state, which is created in `Chapter05/05_Grid/src/main.cpp`:

```
Holder<RenderPipelineHandle> pipeline =
  ctx->createRenderPipeline({
    .smVert       = vert,
    .smFrag       = frag,
    .color        = { {
      .format            = ctx->getSwapchainFormat(),
      .blendEnabled      = true,
      .srcRGBBlendFactor = lvk::BlendFactor_SrcAlpha,
      .dstRGBBlendFactor = lvk::BlendFactor_OneMinusSrcAlpha,
    } },
    .depthFormat = app.getDepthFormat() });
```

11. The C++ rendering code in the same file looks as follows:

```
buf.cmdBindRenderPipeline(pipeline);
buf.cmdBindDepthState({});
```

```
struct {
  mat4 mvp;
  vec4 camPos;
  vec4 origin;
} pc = {
  .mvp    = glm::perspective(
    45.0f, aspectRatio, 0.1f, 1000.0f) * app.camera_.getViewMatrix(),
  .camPos = vec4(app.camera_.getPosition(), 1.0f),
  .origin = vec4(0.0f),
};
buf.cmdPushConstants(pc);
buf.cmdDraw(6);
```

Check out the complete `Chapter05/05_Grid` for a self-contained demo app. The camera can be controlled with the WASD keys and a mouse. The resulting image should look like in the following screenshot:

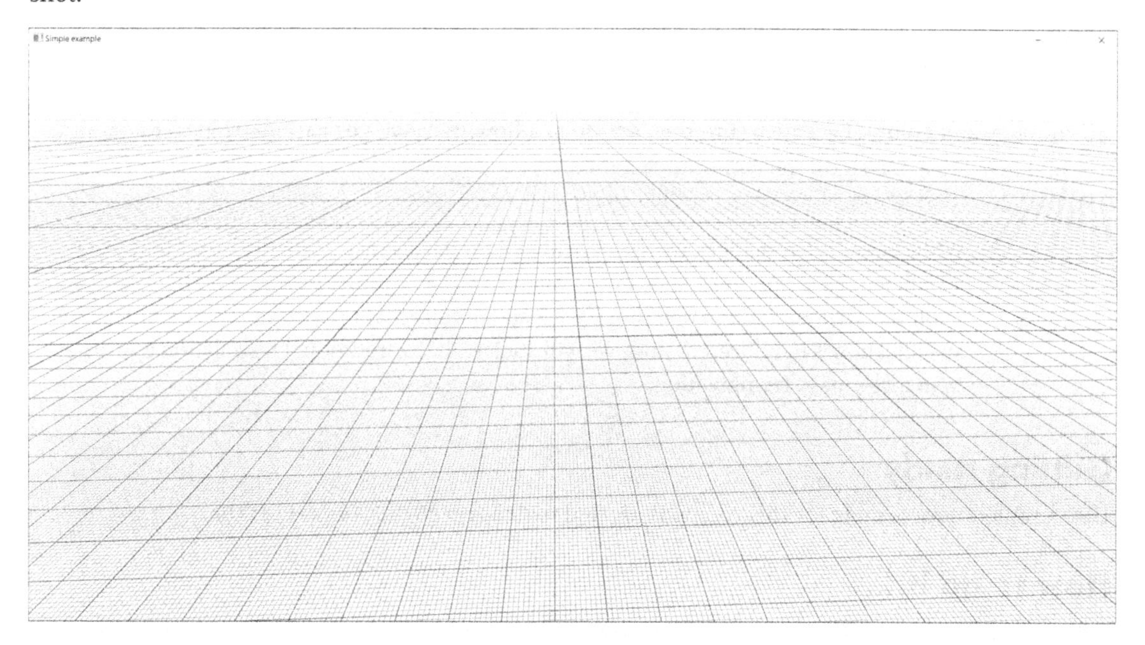

Figure 5.5: GLSL grid

There's more...

Besides considering only the distance to the camera to calculate the antialiasing falloff factor, we can use the angle between the viewing vector and the grid line. This will make the overall look and feel of the grid more visually pleasing and can be an interesting improvement if you want to implement a grid not only as an internal debugging tool but also as a part of a customer-facing product, like an editor.

This implementation was inspired by the *Our Machinery* blog. Unfortunately, it is not available anymore. However, there are some other advanced materials available on the Internet showing how to render a more complex grid suitable for customer-facing rendering. Make sure you read the blog post *The Best Darn Grid Shader (Yet)* by Ben Golus `https://bgolus.medium.com/the-best-darn-grid-shader-yet-727f9278b9d8`, which takes grid rendering a lot further.

Before moving on to the next recipe, it's worth noting that grid rendering is quite useful, and we've incorporated it into most of our subsequent demo applications. You can use the `VulkanApp::drawGrid()` function from `shared/VulkanApp.cpp` to render the grid wherever needed. This version supports MSAA multisampling, which will be used in *Chapters 10* and *11*.

Integrating tessellation into the graphics pipeline

Let's switch gears and learn how to integrate hardware tessellation into our Vulkan graphics rendering pipeline. Hardware tessellation consists of two new shader stages in the graphics pipeline. The first stage is called the **tessellation control shader** (**TCS**), and the second is the **tessellation evaluation shader** (**TES**). The tessellation control shader works with a set of vertices, known as control points, which define a geometric surface referred to as a patch. This shader can modify the control points and determine the level of tessellation required. The tessellation evaluation shader then uses the barycentric coordinates of the tessellated triangles to interpolate per-vertex attributes like texture coordinates and colors. Let's dive into the code to see how these stages can be used to dynamically triangulate a mesh based on the camera's distance.

While tessellation shaders are a powerful tool for hardware tessellation, they may not be as efficient as using mesh shaders on modern GPUs. Unfortunately, as of the time this book was written, Vulkan does not have a standardized API for mesh shaders in its core specification. Therefore, for now, we will stick with the traditional tessellation approach to manage mesh triangulation.

Getting ready

The complete source code for this recipe is located in `Chapter05/06_Tessellation`.

How to do it...

Now, we want to write GLSL shaders that calculate per-vertex tessellation levels based on the distance from the camera. This will allow us to render more geometric detail in areas closer to the viewer. To get started, we'll begin with the vertex shader in `Chapter05/06_Tessellation/src/main.vert`, which will compute the world positions of the vertices and pass them to the tessellation control shader:

1. Our per-frame data includes the usual view and projection matrices, the current camera position in world space, and the tessellation scaling factor, which the user can control via an ImGui widget. Since this data doesn't fit within `128` bytes of push constants, we store everything in a buffer. The geometry is accessed using the PVP technique and stored in the following format: vertex positions are stored as `3` floats, and texture coordinates as `2` floats:

```
// included from <Chapter05/06_Tesselation/src/common.sp>
struct Vertex {
  float x, y, z;
  float u, v;
};
layout(std430, buffer_reference) readonly buffer Vertices {
  Vertex in_Vertices[];
};
```

2. The `PerFrameData` structure is the same across all shader stages in this example.

```
layout(std430, buffer_reference) readonly buffer PerFrameData {
  mat4 model;
  mat4 view;
  mat4 proj;
  vec4 cameraPos;
  uint texture;
  float tesselationScale;
  Vertices vtx;
};
layout(push_constant) uniform PushConstants {
  PerFrameData pc;
};
```

3. Let's write some helper functions to access vertex positions and texture coordinates using the standard GLSL data types:

```
vec3 getPosition(int i) {
  return vec3(pc.vtx.in_Vertices[i].x,
              pc.vtx.in_Vertices[i].y,
              pc.vtx.in_Vertices[i].z);
}
vec2 getTexCoord(int i) {
  return vec2(pc.vtx.in_Vertices[i].u,
              pc.vtx.in_Vertices[i].v);
}
```

4. The vertex shader outputs UV texture coordinates and per-vertex world positions. The calculation is done as follows:

```
layout (location=0) out vec2 uv_in;
layout (location=1) out vec3 worldPos_in;
void main() {
  vec4 pos = vec4(getPosition(gl_VertexIndex), 1.0);
```

```
  gl_Position = pc.proj * pc.view * pc.model * pos;
  uv_in = getTexCoord(gl_VertexIndex);
  worldPos_in = (pc.model * pos).xyz;
}
```

Next, let's move on to the next shader stage and take a look at the tessellation control shader in `Chapter05/06_Tessellation/src/main.tesc`:

1. The shader operates on a group of 3 vertices, each corresponding to a single triangle in the input data. The `uv_in[]` and `worldPos_in[]` variables correspond to those in the vertex shader. Notice that here we have arrays instead of single values.

```
#include <Chapter05/06_Tessellation/src/common.sp>
layout (vertices = 3) out;
layout (location = 0) in vec2 uv_in[];
layout (location = 1) in vec3 worldPos_in[];
```

2. Let's describe the input and output data structures that correspond to each individual vertex. Besides the required vertex position, we store `vec2` texture coordinates:

```
in gl_PerVertex {
  vec4 gl_Position;
} gl_in[];
out gl_PerVertex {
  vec4 gl_Position;
} gl_out[];
struct vertex {
  vec2 uv;
};
struct vertex {
  vec2 uv;
};
layout(location = 0) out vertex Out[];
```

3. The `getTessLevel()` function calculates the desired tessellation level based on the distance of two adjacent vertices from the camera. The hardcoded distance values that are used to switch the levels are scaled using the `tessellationScale` uniform coming from the UI:

```
float getTessLevel(float distance0, float distance1) {
  const float distanceScale1 = 1.2;
  const float distanceScale2 = 1.7;
  const float avgDistance =
    (distance0 + distance1) / (2.0 * pc.tesselationScale);
  if (avgDistance <= distanceScale1) return 5.0;
  if (avgDistance <= distanceScale2) return 3.0;
```

```
    return 1.0;
  }
```

4. The `main()` function is straightforward. It simply passes the positions and UV coordinates as they are, then calculates the distance from each vertex in the triangle to the camera:

```
void main() {
  gl_out[gl_InvocationID].gl_Position =
    gl_in[gl_InvocationID].gl_Position;
  Out[gl_InvocationID].uv = uv_in[gl_InvocationID];
  vec3 c = pc.cameraPos.xyz;
  float eyeToVertexDistance0 = distance(c, worldPos_in[0]);
  float eyeToVertexDistance1 = distance(c, worldPos_in[1]);
  float eyeToVertexDistance2 = distance(c, worldPos_in[2]);
```

5. Based on these distances, we can calculate the required inner and outer tessellation levels in the following way. The inner tessellation level defines how the inner part of a triangle is subdivided into smaller triangles. The outer level defines how the outer edges of the triangle are subdivided so that they can be correctly connected to adjacent triangles:

```
  gl_TessLevelOuter[0] =
    getTessLevel(eyeToVertexDistance1, eyeToVertexDistance2);
  gl_TessLevelOuter[1] =
    getTessLevel(eyeToVertexDistance2, eyeToVertexDistance0);
  gl_TessLevelOuter[2] =
    getTessLevel(eyeToVertexDistance0, eyeToVertexDistance1);
  gl_TessLevelInner[0] = gl_TessLevelOuter[2];
};
```

Let's take a look at the tessellation evaluation shader `Chapter05/06_Tessellation/src/main.tese`:

1. We should specify triangles as the input. The `equal_spacing` spacing mode tells Vulkan that the tessellation level `n` should be clamped to the range `0...64` and rounded to the nearest integer. After that, the corresponding edge should be divided into `n` equal segments. When the tessellation primitive generator produces triangles, the orientation of triangles can be specified by an input layout declaration using the identifiers `cw` and `ccw`. We use the counter-clockwise orientation:

```
layout(triangles, equal_spacing, ccw) in;
struct vertex {
  vec2 uv;
};
in gl_PerVertex {
  vec4 gl_Position;
} gl_in[];
```

```
layout(location = 0) in vertex In[];
out gl_PerVertex {
  vec4 gl_Position;
};
layout (location=0) out vec2 uv;
```

2. These two helper functions are useful for interpolating between the `vec2` and `vec4` attribute values at the corners of the original triangle, using the barycentric coordinates of the current vertex. The built-in `gl_TessCoord` variable contains the required barycentric coordinates, ranging from `0` to `1`:

```
vec2 interpolate2(in vec2 v0, in vec2 v1, in vec2 v2) {
  return v0 * gl_TessCoord.x +
         v1 * gl_TessCoord.y +
         v2 * gl_TessCoord.z;
}
vec4 interpolate4(in vec4 v0, in vec4 v1, in vec4 v2) {
  return v0 * gl_TessCoord.x +
         v1 * gl_TessCoord.y +
         v2 * gl_TessCoord.z;
}
```

3. The actual interpolation code in `main()` is straightforward and can be written in the following way:

```
void main() {
  gl_Position = interpolate4(gl_in[0].gl_Position,
                             gl_in[1].gl_Position,
                             gl_in[2].gl_Position);
  uv = interpolate2(In[0].uv, In[1].uv, In[2].uv);
};
```

The next stage of our hardware tessellation graphics pipeline is the geometry shader `Chapter05/06_Tessellation/src/main.geom`. We use it to generate new barycentric coordinates for all the small tessellated triangles. It is used to render a nice antialiased wireframe overlay on top of our colored mesh, as we did earlier in this chapter in the *Generating level-of-detail meshes using MeshOptimizer* recipe:

1. The geometry shader processes the triangles generated by the hardware tessellator and outputs triangle strips, each containing a single triangle:

```
#version 460 core
layout(triangles) in;
layout(triangle_strip, max_vertices = 3) out;
layout(location=0) in vec2 uv[];
layout(location=0) out vec2 uvs;
layout(location=1) out vec3 barycoords;
```

2. Barycentric coordinates are assigned per vertex using these hardcoded constants:

```
void main() {
  const vec3 bc[3] = vec3[]( vec3(1.0, 0.0, 0.0),
                             vec3(0.0, 1.0, 0.0),
                             vec3(0.0, 0.0, 1.0) );
  for ( int i = 0; i < 3; i++ ) {
    gl_Position = gl_in[i].gl_Position;
    uvs = uv[i];
    barycoords = bc[i];
    EmitVertex();
  }
  EndPrimitive();
}
```

The final stage of this rendering pipeline is the fragment shader `Chapter05/06_Tessellation/src/main.frag`:

1. We take in the barycentric coordinates from the geometry shader and use them to calculate a wireframe overlay covering our mesh:

```
#include <Chapter05/06_Tessellation/src/common.sp>
layout(location=0) in vec2 uvs;
layout(location=1) in vec3 barycoords;
layout(location=0) out vec4 out_FragColor;
```

2. The `edgeFactor()` helper function returns the blending factor based on the distance to the edge and the desired thickness of the wireframe contour. Essentially, when one of the `3` barycentric coordinate values is close to `0`, it indicates that the current fragment is near one of the triangle's edges. The distance to zero controls the visible thickness of a rendered edge:

```
float edgeFactor(float thickness) {
  vec3 a3 = smoothstep( vec3(0.0),
              fwidth(barycoords) * thickness, barycoords);
  return min( min( a3.x, a3.y ), a3.z );
}
```

3. Let's sample the texture using the provided UV values and call it a day:

```
void main() {
  vec4 color = textureBindless2D(pc.texture, 0, uvs);
  out_FragColor = mix( vec4(0.1), color, edgeFactor(0.75) );
}
```

The GLSL shader part of our Vulkan hardware tessellation pipeline is over, and it is time to look into the C++ code. The source code is located in the `Chapter05/06_Tessellation/src/main.cpp` file:

1. The shaders for the tessellated mesh rendering are loaded the following way:

```
Holder<ShaderModuleHandle> vert =
  loadShaderModule(ctx, "Chapter05/06_Tessellation/src/main.vert");
Holder<ShaderModuleHandle> tesc =
  loadShaderModule(ctx, "Chapter05/06_Tessellation/src/main.tesc");
Holder<ShaderModuleHandle> geom =
  loadShaderModule(ctx, "Chapter05/06_Tessellation/src/main.geom");
Holder<ShaderModuleHandle> tese =
  loadShaderModule(ctx, "Chapter05/06_Tessellation/src/main.tese");
Holder<ShaderModuleHandle> frag =
  loadShaderModule(ctx, "Chapter05/06_Tessellation/src/main.frag");
```

2. Then, we should create a corresponding rendering pipeline:

```
Holder<RenderPipelineHandle> pipelineSolid =
  ctx->createRenderPipeline({
    .topology    = lvk::Topology_Patch,
    .smVert      = vert,
    .smTesc      = tesc,
    .smTese      = tese,
    .smGeom      = geom,
    .smFrag      = frag,
    .color       = { { .format = ctx->getSwapchainFormat() } },
    .depthFormat = app.getDepthFormat(),
    .patchControlPoints = 3,
  });;
```

The `data/rubber_duck/scene.gltf` mesh loading code is identical to that from the previous recipes, so we'll skip it here. What's more important is how we render the mesh and ImGui widget to control the tessellation scale factor. Let's take a look at the body of the rendering loop:

1. First, we calculate model-view-projection matrices for our mesh:

```
const mat4 m = glm::rotate(mat4(1.0f),
  glm::radians(-90.0f), vec3(1, 0, 0));
const mat4 v = glm::rotate(glm::translate(mat4(1.0f),
  vec3(0.0f, -0.5f, -1.5f)),
  (float)glfwGetTime(), vec3(0.0f, 1.0f, 0.0f));
const mat4 p = glm::perspective(45.0f, aspectRatio, 0.1f, 1000.0f);
```

2. Next, per-frame data is uploaded into the `bufferPerFrame`. To avoid any extra synchronization and because the data size is small, we can use the `vkCmdUpdateBuffer()` function wrapped in `cmdUpdateBuffer()`. This stores the buffer data directly in the Vulkan command buffer memory before uploading and supports updates of up to `65536` bytes.

```
const PerFrameData pc = {
  .model             = v * m,
  .view              = app.camera_.getViewMatrix(),
  .proj              = p,
  .cameraPos         = vec4(app.camera_.getPosition(), 1.0f),
  .texture           = texture.index(),
  .tessellationScale = tessellationScale,
  .vertices          = ctx->gpuAddress(vertexBuffer),
};
lvk::ICommandBuffer& buf = ctx->acquireCommandBuffer();
buf.cmdUpdateBuffer(bufferPerFrame, pc);
```

3. As usual, push constants are used to pass the address of the `bufferPerFrame` buffer into the shaders. After that, we can render the mesh using our tessellation pipeline.

```
buf.cmdBeginRendering(renderPass, framebuffer);
buf.cmdPushConstants(ctx->gpuAddress(bufferPerFrame));
buf.cmdBindIndexBuffer(indexBuffer, lvk::IndexFormat_UI32);
buf.cmdBindRenderPipeline(pipelineSolid);
buf.cmdBindDepthState({ .compareOp = lvk::CompareOp_Less,
                        .isDepthWriteEnabled = true });
buf.cmdDrawIndexed(indices.size());
```

4. We render a grid on top, as described earlier in this chapter in the *Implementing an infinite grid GLSL shader* recipe. The origin is used to position the grid just below the duck model. Inside our frame rendering loop, we can access all our ImGui rendering functionality. Here, we render a single slider with a floating-point value for the tessellation scale factor:

```
app.drawGrid(buf, p, vec3(0, -0.5f, 0));
app.imgui_->beginFrame(framebuffer);
ImGui::Begin("Camera Controls", nullptr, ImGuiWindowFlags_AlwaysAutoResize);
ImGui::SliderFloat(
  "Tessellation scale", &tessellationScale, 0.7f, 1.2f, "%.1f");
ImGui::End();
app.imgui_->endFrame(buf);
buf.cmdEndRendering();
```

Here is a screenshot from the running demo application:

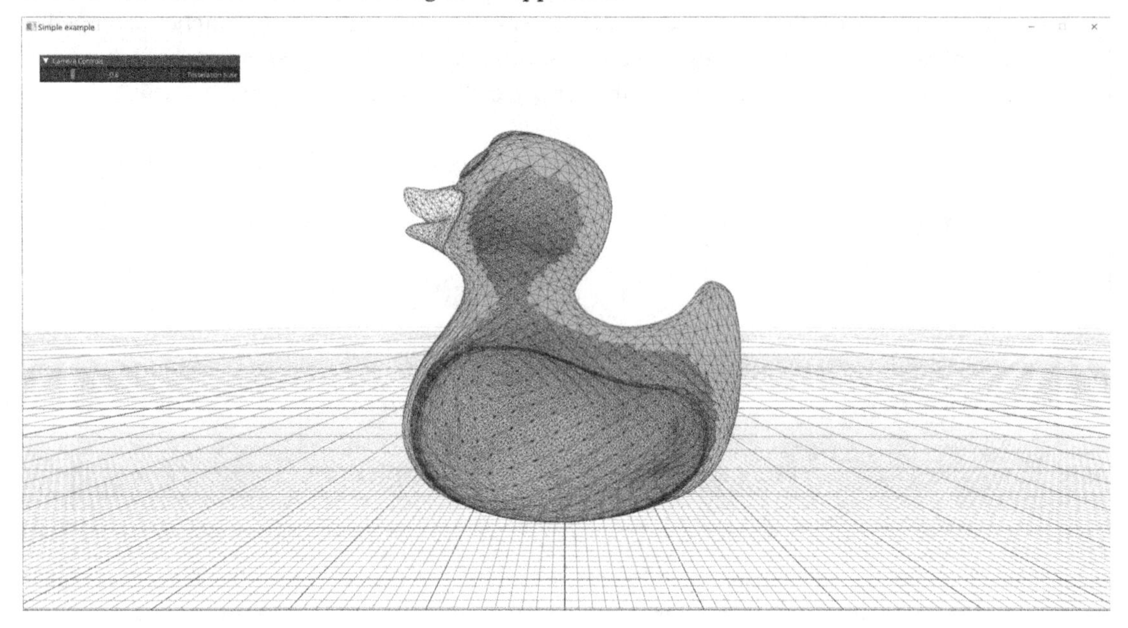

Figure 5.6: Tessellated duck

Notice how the tessellation level changes based on the distance to the camera. Try adjusting the control slider or moving the camera around to emphasize the effect.

There's more...

This recipe can be used as a cornerstone of hardware mesh tessellation techniques in your Vulkan applications. One natural step forward would be to apply a displacement map to the fine-grained tessellated vertices using the direction of normal vectors. Check out this page for further inspiration: `https://www.geeks3d.com/20100804/test-opengl-4-tessellation-with-displacement-mapping`. For those who want to go serious on adaptive tessellation of subdivision surfaces, there is a chapter in the *GPU Gems 2* book covering this advanced topic in great detail. It is now available online at `https://developer.nvidia.com/gpugems/gpugems2/part-i-geometric-complexity/chapter-7-adaptive-tessellation-subdivision-surfaces`.

Organizing mesh data storage

In the previous chapters, we used fixed, hardcoded vertex formats for our meshes, which changed between demos, and also implicitly included the material description. For example, a hardcoded texture was used to provide color information. A triangle mesh is defined by indices and vertices, with each vertex consisting of a set of attributes with distinct data formats that correspond to the `lvk::VertexInput` vertex input description. All auxiliary physical properties of an object, such as collision detection data, mass, and moments of inertia, can be represented by a mesh, while other information, like surface material properties, can be stored outside the mesh as external metadata. It's also worth noting that small 3D models, like the rubber duck we used earlier, can be loaded quite quickly.

However, larger and more intricate real-world 3D models, especially when using transmission formats like `.gltf`, may take several minutes to load. Using a runtime mesh format can address this issue by eliminating the need for parsing and instead loading flat buffers that match the internal rendering data structures. This involves using fast I/O functions like `fread()` and similar mechanisms.

Let's define a unified mesh storage format that covers most of the use cases for the rest of this book.

Getting ready

The full corresponding source code can be found in the header file `shared/Scene/VtxData.h`. Make sure you've read *Chapter 2* before going forward.

How to do it...

A vector of homogeneous **vertex attributes** stored contiguously is called a **vertex stream**. Examples of such attributes are vertex positions, texture coordinates, and normal vectors, each of the three representing one attribute. Each stream must have a format. Vertex positions are `vec3`, texture coordinates can be `vec2`, normal vectors can use the packed format `Int_2_10_10_10_REV`, and so on.

Let's define a **LOD** as an index buffer of reduced size that uses existing vertices and hence can be used directly for rendering with the original vertex buffer. We learned how to create LODs earlier in this chapter in the *Generating level-of-detail meshes using MeshOptimizer* recipe.

We define a **mesh** as a collection of all vertex data streams and a collection of all index buffers, one for each LOD. The number of elements in each vertex data stream is the same and is called the "vertex count", which we will encounter in the code below. To make things a bit simpler, we always use 32-bit offsets and indices for our data.

All the vertex data streams and LOD index buffers are packed into a single blob. This allows us to load the data with a single `fread()` call or even use memory mapping for direct data access. This simple vertex data representation also enables direct upload of meshes to the GPU. What makes it particularly useful is the ability to merge data for multiple meshes into a single file. Alternatively, this can be done by consolidating the data into two large buffers – one for indices and the other for vertex attributes. This will be especially handy later when we discuss how to implement an LOD switching technique on the GPU.

In this recipe, we will focus solely on geometrical data. The LOD creation process is covered in the recipe *Generating level-of-detail meshes using MeshOptimizer*, and the material data export process is covered in the following chapters. Let's take a look at `shared/Scene/VtxData.h` and declare the main data structures for our mesh:

1. First, we need to define an individual mesh description. We deliberately avoid using pointers, as they obscure memory allocations and complicate the saving and loading of data. Instead, we store offsets to individual data streams and LOD index buffers. These offsets function similarly to pointers but are more flexible and, importantly, more GPU-friendly. All offsets in the `Mesh` structure are given relative to the beginning of the data block. Let's declare our main data structure for a mesh. It includes the number of LODs and vertex data streams.

The LOD count, with the original mesh counting as one of the LODs, must be strictly less than `kMaxLODs`, as we don't store LOD index buffer sizes directly but calculate them from the offsets. To calculate these sizes, we store an extra empty LOD level at the end. The number of vertex data streams is stored directly, without any modifications.

```
constexpr const uint32_t kMaxLODs = 8;
struct Mesh final {
  uint32_t lodCount = 1;
  uint32_t indexOffset = 0;
  uint32_t vertexOffset = 0;
```

2. The `vertexCount` field contains the total number of vertices in this mesh. This number is described the content of the vertex buffer and can be greater than the number of vertices on any individual level of detail. We postpone the question of material data storage to the next chapters. To do this in an elegant manner, let's introduce one extra level of indirection. The `materialID` field contains an abstract identifier that allows us to reference any material data stored elsewhere:

```
  uint32_t vertexCount = 0;
  uint32_t materialID = 0;
```

3. Each mesh can potentially be displayed at different LODs. The file contains all indices for all levels of detail and offsets for the beginning of each LOD are stored in the `lodOffset` array. This array contains one extra item at the end, which serves as a marker to calculate the size of the last LOD:

```
  uint32_t lodOffset[kMaxLODs+1] = { 0 };
```

4. Instead of storing the number of indices of each LOD, we define a little helper function to calculate that number from individual offsets:

```
  uint32_t getLODIndicesCount(uint32_t lod) const {
    return lod < lodCount ? lodOffset[lod + 1] - lodOffset[lod] : 0;
  }
};
```

As you might have noticed, the `Mesh` structure is essentially an index into other buffers containing data, such as index and vertex buffers. Let's take a look at that data container:

1. The format of the vertex streams is described by the structure `lvk::VertexInput`. We already used it in *Chapter 2*. This form of vertex stream description allows very flexible storage.

```
struct MeshData {
  lvk::VertexInput streams = {};
```

2. The actual index and vertex buffer data are stored in these containers. They can accommodate multiple meshes. We use only 32-bit indices in our book for the sake of simplicity.

```
  std::vector<uint32_t> indexData;
  std::vector<uint8_t> vertexData;
```

3. Another `std::vector` stores each individual mesh description:

```
  std::vector<Mesh> meshes;
```

4. For completeness, we'll also store a bounding box for each mesh right here. Bounding boxes are extremely useful for culling, and having them pre-calculated can significantly speed up the loading process. Materials and textures are also stored here, but they'll be covered in the subsequent chapters.

```
  std::vector<BoundingBox> boxes;
  std::vector<Material> materials;
  std::vector<std::string> textureFiles;
};
```

Important note

The proposed schema supports both interleaved and non-interleaved vertex data storage formats by making use of the stride parameter in `lvk::VertexInput::VertexInputBinding`. One major drawback is that we have to account for this data reorganization in all the programmable vertex-pulling code in our shaders when we don't use the normal Vulkan vertex input. If you're developing production code, make sure to profile which storage format works faster on your target hardware before committing to one particular approach.

Before we can store these `Mesh` data structures in a file, we require some sort of a file header:

1. To ensure data integrity and to check the validity of the header, a magic hexadecimal value of `0x12345678` is stored in the first 4 bytes of the header:

```
struct MeshFileHeader {
  uint32_t magicValue = 0x12345678;
```

2. The number of individual `Mesh` descriptors in this file is stored in the `meshCount` field:

```
  uint32_t meshCount = 0;
```

3. The last two member fields store the sizes of index and vertex data in bytes, respectively. These values come in handy when checking the integrity of a mesh file as well:

```
  uint32_t indexDataSize = 0;
  uint32_t vertexDataSize = 0;
};
```

The file continues with the list of Mesh structures. After the header and a list of individual mesh descriptors, we store a large index and vertex data block, which can be loaded all at once.

How it works...

Let's go through the pseudocode from shared/Scene/VtxData.cpp for loading such a file is just a few fread() calls that look as follows. Error checks are omitted from the book text but are present in the actual code:

1. First, we read the file header with the mesh count. Error checks are skipped here in the book but are present in the bundled source code:

```
MeshFileHeader loadMeshData(const char* meshFile, MeshData& out) {
  FILE* f = fopen(meshFile, "rb");
  SCOPE_EXIT { fclose(f); };
  MeshFileHeader header;
  fread(&header, 1, sizeof(header), f)
  fread(&out.streams, 1, sizeof(out.streams), f);
```

2. Having read the header, we resize the mesh descriptors array and read in all Mesh descriptions:

```
  out.meshes.resize(header.meshCount);
  fread(out.meshes.data(), sizeof(Mesh), header.meshCount, f);
  out.boxes.resize(header.meshCount);
  fread(out.boxes.data(), sizeof(BoundingBox), header.meshCount, f);
```

3. Then we read the main geometry data blocks for this mesh, which contain the actual index and vertex data:

```
  out.indexData.resize(header.indexDataSize / sizeof(uint32_t));
  out.vertexData.resize(header.vertexDataSize);
  fread(out.indexData.data(), 1, header.indexDataSize, f);
  fread(out.vertexData.data(), 1, header.vertexDataSize, f);
  return header;
};
```

Alternatively, index and vertex buffers can be combined into a single large byte buffer. We leave this as an exercise for our readers.

Later on, the indexData and vertexData containers can be directly uploaded to the GPU. We'll revisit this idea in the subsequent recipes in this chapter.

Although you can see the result of this code in the demo app Chapter05/07_MeshRenderer for this chapter, there are some additional functionalities that need to be implemented. Let's cover a few more topics before we can run and witness the demo.

There's more...

This geometry data format is pretty simple and straightforward for the purpose of storing static mesh data. If the meshes may be changed, reloaded, or loaded asynchronously, we may store separate meshes in dedicated files.

Since it is impossible to predict all the use cases and since this book is all about the rendering and not some general gaming engine creation, it is up to the reader to make decisions on adding extra features such as mesh skinning or others. One simple example of such a decision is the addition of material data directly into the mesh file. Technically, all we need to do is add a `materialCount` field to `MeshFileHeader` and store a list of material descriptions right after the list of meshes. Even such a simple thing immediately raises more questions. Should we pack texture data in the same file? If yes, how complex should the texture format be? What material model should we use? And so on, and so forth. For now, we just leave mesh geometry data separated from material descriptions. We will come back to materials in subsequent chapters.

Implementing automatic geometry conversion

In the previous chapters, we learned how to use the Assimp library to load and render 3D models stored in different file formats. In real-world graphics applications, the process of loading a model can be tedious and multistage. Besides just loading, we might want to optimize a mesh in some specific way, such as optimizing geometry and computing multiple LOD meshes. This process might become slow for sizable meshes, so it makes perfect sense to preprocess meshes offline, before an application starts, and load them later in the app as described in the previous recipe, *Organizing mesh data storage*. Let's learn how to implement a simple framework for automatic geometry preprocessing and conversion.

In the previous edition of this book, we created a standalone tool for geometry conversion that needs to be executed before subsequent demo apps can load the converted data. It turned out to be a significant oversight on our part because many readers dove straight into running the demo apps and then reported issues when things didn't work out-of-the-box as they expected. Here, we have rectified that mistake. If an app requires converted data and cannot find it, it triggers all the necessary code to load the data from storage assets and convert it into our run-time format.

Getting ready

The source code for our geometry conversion framework is in `Chapter05/07_MeshRenderer`. Low-level loader functions are defined in `shared/Scene/VtxData.cpp`. The entire demo application is covered by multiple recipes from this chapter, including *Organizing mesh data storage*, *Implementing automatic geometry conversion*, and *Indirect rendering in Vulkan*.

How to do it...

Let us see how the Assimp library is used to export the mesh data and save it into the binary file using the data structures defined in the *Organizing mesh data storage* recipe:

1. We start by exploring a function called `convertAIMesh()`, which converts the Assimp mesh representation into our run-time format and appends it to the referenced `MeshData` parameter. Global index and vertex offsets are updated as well. The function is quite lengthy, but we will explore it in full detail here. Error checks are omitted:

```
Mesh convertAIMesh(const aiMesh* m, MeshData& meshData,
  uint32_t& indexOffset, uint32_t& vertexOffset)
{
  const bool hasTexCoords = m->HasTextureCoords(0);
```

2. The actual mesh geometry data are stored in two following arrays. We cannot output converted meshes one by one, at least not in a single-pass tool, because we do not know the total size of data in advance, so we allocate in-memory storage for all the data and then write these data blobs into the output file. We also need a reference to the global vertex buffer, where we would append new vertices from this `aiMesh`. The `outLods` container is for per-LOD index buffers. Then we just go through all vertices of `aiMesh` and convert them:

```
  std::vector<float> srcVertices;
  std::vector<uint32_t> srcIndices;
  std::vector<uint8_t>& vertices = meshData.vertexData;
  std::vector<std::vector<uint32_t>> outLods;
```

3. For this recipe, we assume there is a single LOD and that all the vertex data is stored as a continuous data stream. In other words, we have an interleaved storage of the data. We also ignore all the material information and deal exclusively with the index and vertex data for now.

```
  for (size_t i = 0; i != m->mNumVertices; i++) {
    const aiVector3D v = m->mVertices[i];
    const aiVector3D n = m->mNormals[i];
    const aiVector2D t = !hasTexCoords ? aiVector2D() : aiVector2D(
      m->mTextureCoords[0][i].x,
      m->mTextureCoords[0][i].y);
    if (g_calculateLODs) {
      srcVertices.push_back(v.x);
      srcVertices.push_back(v.y);
      srcVertices.push_back(v.z);
    }
```

4. Once we have the stream data for the vertex, we can output it into the vertex buffer. The position `v` is stored as `vec3`. The texture coordinates `uv` are stored as half-float `vec2` to save some space. The normal vector is converted into `2_10_10_10_REV`, which has the size of `uint32_t` – not bad for 3 floats.

More information

`put()` is a templated function, which mem-copies the value from its second argument into a vector of `uint8_t`:

```
template <typename T> void put(
  std::vector<uint8_t>& v, const T& value)
{
  const size_t pos = v.size();
  v.resize(v.size() + sizeof(value));
  memcpy(v.data() + pos, &value, sizeof(value));
}
```

```
    put(vertices, v);
    put(vertices, glm::packHalf2x16(vec2(t.x, t.y)));
    put(vertices, glm::packSnorm3x10_1x2(vec4(n.x, n.y, n.z, 0)));
  }
```

5. Describe the vertex streams for our demo: positions, texture coordinates, and normal vectors. The stride comes from the size of `vec3` positions, half-float `vec2` texture coordinates packed into `uint32_t`, and `2_10_10_10_REV` normals packed into `uint32_t`.

```
meshData.streams = {
  .attributes = {{ .location = 0,
                   .format = lvk::VertexFormat::Float3,
                   .offset = 0 },
                 { .location = 1,
                   .format = lvk::VertexFormat::HalfFloat2,
                   .offset = sizeof(vec3) },
                 { .location = 2,
                   .format = lvk::VertexFormat::Int_2_10_10_10_REV,
                   .offset = sizeof(vec3) + sizeof(uint32_t) } },
  .inputBindings = { { .stride =
    sizeof(vec3) + sizeof(uint32_t) + sizeof(uint32_t) } },
};
```

6. Go through all the mesh faces and create an index buffer:

```
for (unsigned int i = 0; i != m->mNumFaces; i++) {
  if (m->mFaces[i].mNumIndices != 3) continue;
  for (unsigned j = 0; j != m->mFaces[i].mNumIndices; j++)
    srcIndices.push_back(m->mFaces[i].mIndices[j]);
}
```

7. If no LOD calculation is required, we can just store `srcIndices` as LOD 0. Otherwise, we call the processLods() function, which calculates LOD levels for this mesh, as described in the *Generating LODs using MeshOptimizer* recipe.

```
if (!g_calculateLODs) {
   outLods.push_back(srcIndices);
} else {
   processLods(srcIndices, srcVertices, outLods);
}
```

8. Before updating the `indexOffset` and `vertexOffset` parameters, let's store their values in the resulting `Mesh` structure. Their values represent where all the previous index and vertex data ended before we started converting this `aiMesh`.

```
Mesh result = {
  .indexOffset  = indexOffset,
  .vertexOffset = vertexOffset,
  .vertexCount  = m->mNumVertices,
};
```

9. Stream out all the indices for all the LOD levels one after another:

```
uint32_t numIndices = 0;
for (size_t l = 0; l < outLods.size(); l++) {
  for (size_t i = 0; i < outLods[l].size(); i++) {
    meshData.indexData.push_back(outLods[l][i]);
  }
  result.lodOffset[l] = numIndices;
  numIndices += (int)outLods[l].size();
}
result.lodOffset[outLods.size()] = numIndices;
result.lodCount = (uint32_t)outLods.size();
```

10. After processing the input mesh, we increment offset counters for indices and the current starting vertex:

```
indexOffset  += numIndices;
vertexOffset += m->mNumVertices;
```

```
  return result;
}
```

Processing a 3D asset file by Assimp comprises loading the scene and converting each mesh into the internal format. Let's take a look at the `loadMeshFile()` function to see how to do it:

1. The list of flags for the `aiImportFile()` function includes options that allow further usage of imported data without any extra processing on our side. For example, all the transformation hierarchies are flattened, and the resulting transformation matrices are applied to mesh vertices.

```
void loadMeshFile(const char* fileName, MeshData& meshData)
{
  const unsigned int flags = aiProcess_JoinIdenticalVertices |
                             aiProcess_Triangulate |
                             aiProcess_GenSmoothNormals |
                             aiProcess_LimitBoneWeights |
                             aiProcess_SplitLargeMeshes |
                             aiProcess_ImproveCacheLocality |
                             aiProcess_RemoveRedundantMaterials |
                             aiProcess_FindDegenerates |
                             aiProcess_FindInvalidData |
                             aiProcess_GenUVCoords;
  const aiScene* scene = aiImportFile(fileName, flags);
```

2. After importing an Assimp scene, we resize the mesh descriptor container accordingly and call `convertAIMesh()` for each mesh in the scene. The `indexOffset` and `vertexOffset` offsets are accumulated incrementally:

```
  meshData.meshes.reserve(scene->mNumMeshes);
  meshData.boxes.reserve(scene->mNumMeshes);
  uint32_t indexOffset = 0;
  uint32_t vertexOffset = 0;
  for (unsigned int i = 0; i != scene->mNumMeshes; i++)
    meshData.meshes.push_back(
      convertAIMesh(scene->mMeshes[i], meshData,
        indexOffset, vertexOffset));
```

3. In the end, we precalculate axis-aligned bounding boxes for our mesh data. This helper function is defined in `shared/Scene/VtxData.cpp`:

```
  recalculateBoundingBoxes(meshData);
}
```

Although loading and preprocessing of data should be customizable to accommodate needs of each individual demo app, saving is pretty much standard because `MeshData` contains all the information we need. As such, the saving function `saveMeshData()` is defined in `shared/Scene/VtxData.cpp`.

Saving converted meshes into our file format is the reverse process of reading meshes from the file described in the *Organizing mesh data storage* recipe:

1. First, we fill the file header structure using the mesh number and offsets:

```
void saveMeshData(const char* fileName, const MeshData& m) {
  FILE* f = fopen(fileName, "wb");
```

2. We calculate byte sizes of index and vertex data buffers and store them in the header:

```
  const MeshFileHeader header = {
    .magicValue     = 0x12345678,
    .meshCount      = (uint32_t)m.meshes.size(),
    .indexDataSize = (uint32_t)(m.indexData.size() * sizeof(uint32_t)),
    .vertexDataSize = (uint32_t)(m.vertexData.size()),
  };
```

3. Once all the sizes are known, we save the header and the list of mesh descriptions:

```
  fwrite(&header, 1, sizeof(header), f);
  fwrite(&m.streams, 1, sizeof(m.streams), f);
  fwrite(m.meshes.data(), sizeof(Mesh), header.meshCount, f);
  fwrite(m.boxes.data(), sizeof(BoundingBox), header.meshCount, f);
```

4. After the header and other metadata, two blobs with index and vertex data are stored.

```
  fwrite(m.indexData.data(),  1, header.indexDataSize,  f);
  fwrite(m.vertexData.data(), 1, header.vertexDataSize, f);
  fclose(f);
}
```

Let's put all this code to work in our demo app.

How it works...

The mesh conversion framework is a part of our demo app. Let's take a look at that part in the `Chapter05/07_MeshRenderer/src/main.cpp` file:

1. First, we check if there's any valid cached mesh data available. The `isMeshDataValid()` function checks if the specified cached mesh file exists and does some routine sanity checks on the data sizes.

```
bool isMeshDataValid(const char* meshFile) {
  FILE* f = fopen(meshFile, "rb");
  if (!f)  false;
  SCOPE_EXIT { fclose(f); };
  MeshFileHeader header;
  if (fread(&header, 1, sizeof(header), f) != sizeof(header))
    return false;
```

```
    if (fseek(f, sizeof(Mesh) * header.meshCount, SEEK_CUR))
      return false;
    if (fseek(f, sizeof(BoundingBox) * header.meshCount, SEEK_CUR))
      return false;
    if (fseek(f, header.indexDataSize, SEEK_CUR))
      return false;
    if (fseek(f, header.vertexDataSize, SEEK_CUR))
      return false;
    return true;
  }
```

2. Then, we use the `isMeshDataValid()` function to load the Lumberyard Bistro dataset:

```
  const char* meshMeshes = ".cache/ch05_bistro.meshes";
  int main() {
    if (!isMeshDataValid(meshMeshes)) {
      printf("No cached mesh data found. Precaching...\n\n");
      MeshData meshData;
      loadMeshFile("deps/src/bistro/Exterior/exterior.obj", meshData);
      saveMeshData(meshMeshes, meshData);
    }
```

3. If the data is already precached, we just load it using the `loadMeshData()` described earlier in this recipe:

```
    MeshData meshData;
    const MeshFileHeader header = loadMeshData(meshMeshes, meshData);
```

The output mesh data is saved into the file `.cache/ch05_bistro.meshes`. Let's go through the rest of this chapter to learn how to render this mesh with Vulkan.

Indirect rendering in Vulkan

Indirect rendering is the process of issuing drawing commands to the graphics API, where most of the parameters to those commands come from GPU buffers. It is a part of many modern GPU usage paradigms and exists in all contemporary rendering APIs in some form. For example, we can do indirect rendering with Vulkan using the `vkCmdDraw*Indirect*()` family of functions. Instead of dealing with low-level Vulkan here, let's get more technical and learn how to combine indirect rendering in Vulkan with the mesh data format we introduced in the *Organizing mesh data storage* recipe.

Getting ready

In the earlier recipes, we covered building a mesh preprocessing pipeline and converting 3D meshes from transmission formats such as `.gltf2` into our run-time mesh data format. To wrap up this chapter, let's demonstrate how to render this data. To delve into something new, let's explore how to achieve this using the indirect rendering technique.

Once we have defined the mesh data structures, we also need to render them. To do so, we allocate GPU buffers for vertex and index data using the previously described functions, upload all the data to the GPU, and finally, render all the meshes.

The whole point of the previously defined `Mesh` data structure is the ability to render multiple meshes in a single Vulkan command. Since version 1.0 of the API, Vulkan supports the technique of indirect rendering. This means we do not need to issue the `vkCmdDraw()` command for each and every mesh. Instead, we create a GPU buffer and fill it with an array of `VkDrawIndirectCommand` structures, then fill these structures with appropriate offsets into our index and vertex data buffers, and finally emit a single `vkCmdDrawIndirect()` call. The `Mesh` structure described in the *Organizing mesh data storage* recipe contains the data required to fill in `VkDrawIndirectCommand`.

The full source code for this recipe is located in `Chapter05/07_MeshRenderer`. It is recommended to revisit the *Organizing mesh data storage* and *Implementing automatic geometry conversion* recipes before reading further.

How to do it...

Let's implement a simple helper class called `VKMesh`, located in `Chapter05/07_MeshRenderer/src/main.cpp`, to render our mesh using *LightweightVK*:

1. We need three buffers, an index buffer, a vertex buffer, and an indirect buffer, as well as three shaders and a rendering pipeline:

```
class VKMesh final {
  Holder<BufferHandle> bufferIndices_;
  Holder<BufferHandle> bufferVertices_;
  Holder<BufferHandle> bufferIndirect_;
  Holder<ShaderModuleHandle> vert_;
  Holder<ShaderModuleHandle> geom_;
  Holder<ShaderModuleHandle> frag_;
  Holder<RenderPipelineHandle> pipeline_;
```

2. The constructor accepts references to `MeshFileHeader` and `MeshData`, which we loaded in the previous recipe, *Implemented automatic geometry conversion*. The data buffers are used as-is and uploaded directly into the respective Vulkan buffers. The number of indices is inferred from the indices buffer size, assuming indices are stored as 32-bit unsigned integers. The depth format specification is required to create a corresponding rendering pipeline right here in this class:

```
public:
  uint32_t numIndices_ = 0;
  VKMesh(const std::unique_ptr<lvk::IContext>& ctx,
         const MeshFileHeader& header,
         const MeshData& meshData,
         lvk::Format depthFormat)
  : numIndices_(header.indexDataSize / sizeof(uint32_t)) {
```

```
const uint32_t* indices = meshData.indexData.data();
const uint8_t* vertexData = meshData.vertexData.data();
```

3. Create vertex and index buffers and upload the data into them:

```
bufferVertices_ = ctx->createBuffer({
  .usage     = lvk::BufferUsageBits_Vertex,
  .storage   = lvk::StorageType_Device,
  .size      = header.vertexDataSize,
  .data      = vertexData,
  .debugName = "Buffer: vertex" });
bufferIndices_ = ctx->createBuffer({
  .usage     = lvk::BufferUsageBits_Index,
  .storage   = lvk::StorageType_Device,
  .size      = header.indexDataSize,
  .data      = indices,
  .debugName = "Buffer: index" });
```

4. Allocate the data storage for our indirect buffer:

```
std::vector<uint8_t> drawCommands;
const uint32_t numCommands = header.meshCount;
drawCommands.resize(
  sizeof(DrawIndexedIndirectCommand) * numCommands + sizeof(uint32_t));
```

5. Store the number of draw commands at the very beginning of the indirect buffer. This approach is not used in this demo but can be useful for GPU-driven rendering when the GPU calculates the number of draw commands and stores it in a buffer.

```
memcpy(drawCommands.data(), &numCommands, sizeof(numCommands));
DrawIndexedIndirectCommand* cmd = std::launder(
  reinterpret_cast<DrawIndexedIndirectCommand*>(
    drawCommands.data() + sizeof(uint32_t)));
```

6. Fill in the content of our indirect commands buffer. Each command corresponds to a single `Mesh` structure. The indirect buffer should be allocated with the `BufferUsageBits_Indirect` usage flag.

```
for (uint32_t i = 0; i != numCommands; i++)
  *cmd++ = {
    .count         = meshData.meshes[i].getLODIndicesCount(0),
    .instanceCount = 1,
    .firstIndex    = meshData.meshes[i].indexOffset,
    .baseVertex    = meshData.meshes[i].vertexOffset,
    .baseInstance  = 0,
  };
```

More information

The `DrawIndexedIndirectCommand` structure is just our mirror-image of `VkDrawIndexedIndirectCommand` to prevent including Vulkan headers into our app. While this might be an exaggeration for a Vulkan book, this type of separation might be useful in real-world apps considering this data structure is compatible in Vulkan, Metal, DirectX, and OpenGL.

```
struct DrawIndexedIndirectCommand {
  uint32_t count;
  uint32_t instanceCount;
  uint32_t firstIndex;
   int32_t baseVertex;
  uint32_t baseInstance;
};
```

```
bufferIndirect_ = ctx->createBuffer({
  .usage     = lvk::BufferUsageBits_Indirect,
  .storage   = lvk::StorageType_Device,
  .size      = sizeof(DrawIndexedIndirectCommand) *
    numCommands + sizeof(uint32_t),
  .data      = drawCommands.data(),
  .debugName = "Buffer: indirect" });
```

7. The rendering pipeline is created using a set of vertex, geometry, and fragment shaders. The geometry shader is used for barycentric coordinates generation to render beautiful wireframes. The vertex stream descriptions from `meshData.streams` can be used directly to initialize the pipeline's vertex input:

```
vert_ = loadShaderModule(
  ctx, "Chapter05/07_MeshRenderer/src/main.vert");
geom_ = loadShaderModule(
  ctx, "Chapter05/07_MeshRenderer/src/main.geom");
frag_ = loadShaderModule(
  ctx, "Chapter05/07_MeshRenderer/src/main.frag");
pipeline_ = ctx->createRenderPipeline({
  .vertexInput = meshData.streams,
  .smVert      = vert_,
  .smGeom      = geom_,
  .smFrag      = frag_,
  .color       = { { .format = ctx->getSwapchainFormat() } },
  .depthFormat = depthFormat,
```

```
      .cullMode     = lvk::CullMode_Back,
    });
  }
```

8. The `draw()` method fills in a corresponding Vulkan command buffer to render the entire mesh. Note how `cmdDrawIndexedIndirect()` skips the first 32 bits of the indirect buffer where the number of commands is located. We will put that number to use in *Chapter 11, Advanced Rendering Techniques and Optimizations*:

```
  void draw(lvk::ICommandBuffer& buf, const MeshFileHeader& header) {
    buf.cmdBindIndexBuffer(bufferIndices_, lvk::IndexFormat_UI32);
    buf.cmdBindVertexBuffer(0, bufferVertices_);
    buf.cmdBindRenderPipeline(pipeline_);
    buf.cmdBindDepthState({ .compareOp = lvk::CompareOp_Less,
                            .isDepthWriteEnabled = true });
    buf.cmdDrawIndexedIndirect(
      bufferIndirect_, sizeof(uint32_t), header.meshCount);
  }
};
```

9. This class is used as follows after the Bistro mesh is loaded:

```
MeshData meshData;
const MeshFileHeader header = loadMeshData(meshMeshes, meshData);
const VKMesh mesh(ctx, header, meshData, app.getDepthFormat());
```

This completes the description of our initialization process. Now, let's turn to the GLSL source code:

1. The vertex shader `Chapter05/07_MeshRenderer/src/main.vert` is quite simple. We do not use programmable vertex fetching here for the sake of simplicity. Only standard per-vertex attributes are used to simplify the example.

```
layout(push_constant) uniform PerFrameData {
  mat4 MVP;
};
layout (location=0) in vec3 in_pos;
layout (location=1) in vec2 in_tc;
layout (location=2) in vec3 in_normal;
layout (location=0) out vec2 uv;
layout (location=1) out vec3 normal;
void main() {
  gl_Position = MVP * vec4(in_pos, 1.0);
  uv = in_tc;
  normal = in_normal;
};
```

2. The geometry shader, `Chapter05/07_MeshRenderer/src/main.geom`, provides the necessary barycentric coordinates, as described earlier in this chapter:

```
#version 460 core
layout (triangles) in;
layout (triangle_strip, max_vertices = 3) out;
layout (location=0) in vec2 uv[];
layout (location=1) in vec3 normal[];
layout (location=0) out vec2 uvs;
layout (location=1) out vec3 barycoords;
layout (location=2) out vec3 normals;
void main() {
  const vec3 bc[3] = vec3[](vec3(1.0, 0.0, 0.0),
                            vec3(0.0, 1.0, 0.0),
                            vec3(0.0, 0.0, 1.0) );
  for ( int i = 0; i < 3; i++ ) {
    gl_Position = gl_in[i].gl_Position;
    uvs = uv[i];
    barycoords = bc[i];
    normals = normal[i];
    EmitVertex();
  }
  EndPrimitive();
}
```

3. The fragment shader, `Chapter05/07_MeshRenderer/src/main.frag`, calculates some improvized lighting and applies a wireframe outline based on the barycentric coordinates generated by the geometry shader, as described in the *Integrating tessellation into the graphics pipeline* recipe:

```
layout (location=0) in vec2 uvs;
layout (location=1) in vec3 barycoords;
layout (location=2) in vec3 normal;
layout (location=0) out vec4 out_FragColor;
float edgeFactor(float thickness) {
  vec3 a3 = smoothstep(
    vec3( 0.0 ), fwidth(barycoords) * thickness, barycoords);
  return min( min( a3.x, a3.y ), a3.z );
}
void main() {
  float NdotL = clamp(dot(normalize(normal),
                          normalize(vec3(-1,1,-1))), 0.5, 1.0);
  vec4 color = vec4(1.0, 1.0, 1.0, 1.0) * NdotL;
```

```
    out_FragColor = mix( vec4(0.1), color, edgeFactor(1.0) );
};
```

4. The final C++ rendering code snippet is straightforward because all the heavy lifting is already done inside the `VKMesh` helper class:

```
buf.cmdBeginRendering(renderPass, framebuffer);
buf.cmdPushConstants(mvp);
mesh.draw(buf, header);
app.drawGrid(buf, p, vec3(0, -0.0f, 0));
app.imgui_->beginFrame(framebuffer);
app.drawFPS();
app.imgui_->endFrame(buf);
buf.cmdEndRendering();
```

The running application will render the following image, if the image is loaded with the Lumberyard Bistro mesh:

Figure 5.7: Lumberyard Bistro mesh geometry loaded and rendered

Generating textures in Vulkan using compute shaders

We learned how to use basic compute shaders earlier in this chapter, in the *Implementing instanced meshes with compute shaders* recipe. It is time to go through a few examples of how to use them. Let's start with some basic procedural texture generation. In this recipe, we implement a small program to display animated textures whose texel values are calculated in real time inside our custom compute shader. To add even more value to this recipe, we will port a GLSL shader from `https://www.shadertoy.com` to our Vulkan compute shader.

Getting ready

The compute pipeline creation code and Vulkan application initialization are the same as in the *Implementing instanced meshes with compute shaders* recipe. Make sure you read it before proceeding further. To use and display the generated texture, we need a textured full-screen quad renderer. Its GLSL source code can be found in `data/shaders/Quad.vert` and `data/shaders/Quad.frag`. However, the geometry used is actually a triangle covering the entire screen. We will not focus on its internals here because at this point, it should be easy for you to render a full-screen quad on your own using the material from the previous chapters.

The original shader, "Industrial Complex," that we are going to use here to generate a Vulkan texture was created by Gary "Shane" Warne (`http://rhomboid.com/`) and can be downloaded from ShaderToy: `https://www.shadertoy.com/view/MtdSWS`.

How to do it...

Let's start by discussing the process of writing a texture-generating GLSL compute shader. The simplest shader to generate an RGBA image without using any input data outputs an image by using the `gl_GlobalInvocationID` built-in variable to calculate which pixel to output. This maps directly to how ShaderToy shaders operate, thus we can transform them into a compute shader just by adding some input and output parameters and layout modifiers that are specific to compute shaders and Vulkan. Let's take a look at a minimalistic compute shader that creates a red-green gradient texture:

1. As in all other compute shaders, one mandatory line at the beginning tells the driver how to distribute the workload on the GPU. In our case, we are processing tiles of 16x16 pixels – the local workgroup size, 16x16, is supported on many GPUs, and we will use it for compatibility purposes:

```
layout (local_size_x = 16, local_size_y = 16) in;
```

2. The only buffer binding that we need to specify is the output image. This is the first time we have used the image type `image2D` in this book. Here, it means that the array `kTextures2DOut` contains an array of 2D images whose elements are nothing but pixels of a texture. The `writeonly` layout qualifier instructs the compiler to assume we will not read from this image in the shader. The binding is updated from the C++ code in the *Using Vulkan descriptor indexing* recipe in *Chapter 2*:

```
layout (set = 0, binding = 2, rgba8)
  uniform writeonly image2D kTextures2DOut[];
```

3. The GLSL compute shading language provides a set of helper functions to retrieve various image attributes. We use the built-in `imageSize()` function to determine the size of an image in pixels, and the image ID is loaded from push constants:

```
layout(push_constant) uniform uPushConstant {
  uint tex;
  float time;
```

```
} pc;
void main() {
  ivec2 dim = imageSize(kTextures2DOut[pc.tex]);
```

4. The built-in `gl_GlobalInvocationID` variable tells us which global element of our compute grid we are processing. To convert its value into 2D image coordinates, we divide it by the image dimensions. As we are dealing with 2D textures, only `x` and `y` components matter. The calling code from the C++ side executes the `vkCmdDispatch()` function and passes the output image size as the X and Y numbers of local workgroups:

```
  vec2 uv = vec2(gl_GlobalInvocationID.xy) / dim;
```

5. The actual real work we do in this shader is to call the `imageStore()` GLSL function:

```
  imageStore(kTextures2DOut[pc.tex],
    ivec2(gl_GlobalInvocationID.xy), vec4(uv, 0.0, 1.0));
}
```

Now, this example is rather limited, and all you get is a red-and-green gradient image. Let's change it a little bit to use the actual shader code from ShaderToy. The compute shader that renders a Vulkan version of the "Industrial Complex" shader from ShaderToy, available via the following URL `https://shadertoy.com/view/MtdSWS`, can be found in the `Chapter05/08_ComputeTexture/src/main.comp` file.

1. First, let's copy the entire original ShaderToy GLSL code into our new compute shader. There is a function called `mainImage()` in there, which is declared as follows:

```
void mainImage(out vec4 fragColor, in vec2 fragCoord)
```

2. We should replace it with a function that returns a `vec4` color instead of storing it in the output parameter:

```
vec4 mainImage(in vec2 fragCoord)
```

   Don't forget to add an appropriate `return` statement at the end.

3. Now, let's change the `main()` function of our compute shader to invoke `mainImage()` properly and do 5x5 accumulative antialiasing. It is a pretty neat trick:

```
void main() {
  ivec2 dim = imageSize(kTextures2DOut[pc.tex]);
  vec4 c = vec4(0.0);
  for (int dx = -2; dx != 3; dx++) {
    for (int dy = -2; dy != 3; dy++) {
      vec2 uv = vec2(gl_GlobalInvocationID.xy) / dim +
                vec2(dx, dy) / (3.0 * dim);
      c += mainImage(uv * dim);
    }
  }
```

```
  imageStore(kTextures2DOut[pc.tex],
    ivec2(gl_GlobalInvocationID.xy), c / 25.0);
}
```

4. There is still one issue that needs to be resolved before we can run this code. The ShaderToy code uses two custom input variables, `iTime` for the elapsed time and `iResolution`, which contains the size of the resulting image. To prevent any search and replace in the original GLSL code, we mimic these variables, one as a push constant and the other with a hardcoded value for simplicity.

```
layout(push_constant) uniform uPushConstant {
  uint tex;
  float time;
} pc;
vec2 iResolution = vec2( 1280.0, 720.0 );
float iTime = pc.time;
```

Important note

The GLSL `imageSize()` function can be used to obtain the `iResolution` value based on the actual size of our texture. We leave it as an exercise to the reader.

The C++ code is rather short and consists of creating a texture and invoking the previously mentioned compute shader, inserting a Vulkan pipeline barrier, and rendering a textured full-screen quad:

1. A texture is created using the `TextureUsageBits_Storage` usage flag to make it accessible to compute shaders:

```
Holder<TextureHandle> texture = ctx->createTexture({
  .type       = lvk::TextureType_2D,
  .format     = lvk::Format_RGBA_UN8,
  .dimensions = {1280, 720},
  .usage      = lvk::TextureUsageBits_Sampled |
                lvk::TextureUsageBits_Storage,
  .debugName  = "Texture: compute",
});
```

2. Compute and rendering pipelines are created the following way:

```
Holder<ShaderModuleHandle> comp = loadShaderModule(
  ctx, "Chapter05/08_ComputeTexture/src/main.comp");
Holder<ComputePipelineHandle> pipelineComputeMatrices =
  ctx->createComputePipeline({ smComp = comp });
Holder<ShaderModuleHandle> vert = loadShaderModule(
  ctx, "data/shaders/Quad.vert");
```

```
Holder<ShaderModuleHandle> frag = loadShaderModule(
  ctx, "data/shaders/Quad.frag");
Holder<RenderPipelineHandle> pipelineFullScreenQuad =
  ctx->createRenderPipeline({
    .smVert = vert,
    .smFrag = frag,
    .color  = { { .format = ctx->getSwapchainFormat() } },
});
```

3. When doing rendering, we provide the texture ID to both shaders and the current time to the compute shader using push constants:

```
lvk::ICommandBuffer& buf = ctx->acquireCommandBuffer();
const struct {
  uint32_t textureId;
  float time;
} pc {
  .textureId = texture.index(),
  .time      = (float)glfwGetTime(),
};
buf.cmdPushConstants(pc);
buf.cmdBindComputePipeline(pipelineComputeMatrices);
buf.cmdDispatchThreadGroups(
  { .width = 1280 / 16, .height = 720 / 16 });
```

4. A pipeline barrier that ensures that the compute shader finishes before texture sampling happens is created by LightweightVK when we specify a required texture dependency. Take a look at `lvk::CommandBuffer::transitionToShaderReadOnly()` for the low-level image barrier code. As our fullscreen shader uses a triangle covering the entire screen, draw 3 vertices of that triangle.

```
buf.cmdBeginRendering(renderPass, framebuffer,
  { .textures = { { TextureHandle(texture) } } });
buf.cmdBindRenderPipeline(pipelineFullScreenQuad);
buf.cmdDraw(3);
buf.cmdEndRendering();
ctx->submit(buf, ctx->getCurrentSwapchainTexture());
```

The running application should render the following image, which is similar to the output of `https://www.shadertoy.com/view/MtdSWS`:

Figure 5.8: Using compute shaders to generate textures

In the next recipe, we will continue learning the Vulkan compute pipeline and implement a mesh generation compute shader.

Implementing computed meshes

In the *Generating textures in Vulkan using compute shaders* recipe, we learned how to write pixel data into textures from compute shaders. We are going to need that data in the next chapter to implement a BRDF precomputation tool for our physically-based rendering pipeline. But before that, let's learn a few simple and interesting ways to use compute shaders in Vulkan and combine this feature with mesh geometry generation on the GPU.

We are going to run a compute shader to create the triangulated geometry of a 3D torus knot shape with different `P` and `Q` parameters.

Important note

A torus knot is a special kind of knot that lies on the surface of an unknotted torus in 3D space. Each torus knot is specified by a pair of coprime integers, p and q. To find out more, check out the Wikipedia page: `https://en.wikipedia.org/wiki/Torus_knot`.

A compute shader generates vertex data, including positions, texture coordinates, and normal vectors. This data is stored in a buffer and later utilized as a vertex buffer to render a mesh in a graphics pipeline. To make the results more visually pleasing, we will implement real-time morphing between two different torus knots that is controllable from an ImGui widget. Let's get started.

Getting ready

The source code for this example is located in `Chapter05/09_ComputeMesh`.

How to do it...

The application consists of multiple parts: the C++ part (which drives the UI and Vulkan commands), the mesh generation compute shader, the texture generations compute shader, and a rendering pipeline with simple vertex and fragment shaders. The C++ part in `Chapter05/09_ComputeMesh/src/main.cpp` is quite short, so let's tackle it first:

1. We store a queue of P-Q pairs, which defines the order of morphing. The queue always has at least two elements, which define the current and the next torus knot. We also store a floating point value, `g_MorphCoef`, which is the morphing factor `0...1` between the two adjacent P-Q pairs in the queue. The mesh is regenerated in every frame, and the morphing coefficient is increased until it reaches `1.0`. At this point, we will either stop morphing or, if there are more than two elements in the queue, remove the top element from it, reset `g_MorphCoef` to `zero`, and repeat. The `g_AnimationSpeed` value defines how fast one torus knot mesh morphs into another. The `g_UseColoredMesh` Boolean flag is used to switch between colored and textured shading of the mesh:

```
std::deque<std::pair<uint32_t, uint32_t>> morphQueue =
  { { 5, 8 }, { 5, 8 } };
float morphCoef = 0.0f;
float animationSpeed = 1.0f;
bool g_UseColoredMesh = false;
```

2. Two global constants define the tessellation level of a torus knot. Feel free to play around with them:

```
constexpr uint32_t kNumU = 1024;
constexpr uint32_t kNumV = 1024;
```

3. Regardless of the `P` and `Q` parameter values, we have a single order in which we should traverse vertices to produce torus knot triangles. The `generateIndices()` function prepares index buffer data for this purpose. Here, `6` is the number of indices generated for each rectangular grid element consisting of `2` triangles:

```
void generateIndices(uint32_t* indices) {
  for (uint32_t j = 0; j < kNumV - 1; j++) {
    for (uint32_t i = 0; i < kNumU - 1; i++) {
      uint32_t ofs = (j * (kNumU - 1) + i) * 6;
      uint32_t i1 = (j + 0) * kNumU + (i + 0);
      uint32_t i2 = (j + 0) * kNumU + (i + 1);
      uint32_t i3 = (j + 1) * kNumU + (i + 1);
      uint32_t i4 = (j + 1) * kNumU + (i + 0);
      indices[ofs + 0] = i1;
```

```
        indices[ofs + 1] = i2;
        indices[ofs + 2] = i4;
        indices[ofs + 3] = i2;
        indices[ofs + 4] = i3;
        indices[ofs + 5] = i4;
      }
    }
  }
```

Additionally, our C++ code operates an ImGui UI for selecting a configuration of a torus knot and managing various parameters. This offers insights into the code's flow, so let's examine it more closely:

1. Each torus knot is specified by a pair of coprime integers, `P` and `Q`. Here, we have preselected a few pairs that produce visually interesting results.

```
void renderGUI(TextureHandle texture) {
  static const std::vector<std::pair<uint32_t, uint32_t>> PQ = {
    {1, 1}, {2, 3}, {2, 5}, {2, 7}, {3, 4},
    {2, 9}, {3, 5}, {5, 8}, {8, 9} };
  ImGui::SetNextWindowPos(ImVec2(0, 0), ImGuiCond_Appearing);
  ImGui::Begin("Torus Knot params", nullptr,
    ImGuiWindowFlags_AlwaysAutoResize | ImGuiWindowFlags_NoCollapse);
```

2. We can control the morphing animation speed. That is how fast one mesh morphs into another. We can also switch between colored and textured shading of the mesh.

```
  ImGui::Checkbox("Use colored mesh", &g_UseColoredMesh);
  ImGui::SliderFloat(
    "Morph animation speed", &g_AnimationSpeed, 0.0f, 2.0f);
```

3. When we click a button with a different set of `P-Q` parameters, we don't regenerate the mesh right away. Instead, we let any ongoing animations complete by adding a new pair to a queue. In the main loop, after the current morphing animation concludes, we remove the front element from the queue and initiate the animation again.

```
  for (size_t i = 0; i != PQ.size(); i++) {
    const std::string title = std::to_string(PQ[i].first) + ", " +
      std::to_string(PQ[i].second);
    if (ImGui::Button(title.c_str(), ImVec2(128, 0))) {
      if (PQ[i] != g_MorphQueue.back())
        g_MorphQueue.push_back(PQ[i]);
    }
  }
```

4. The content of the morph queue is printed here for you. The current `P-Q` pair is marked with `"<---"`:

```
  ImGui::Text("Morph queue:");
  for (size_t i = 0; i != g_MorphQueue.size(); i++) {
    const bool isLastElement = (i + 1) == g_MorphQueue.size();
    ImGui::Text("  P = %u, Q = %u %s", g_MorphQueue[i].first,
      g_MorphQueue[i].second, isLastElement ? "<---" : "");
  }
  ImGui::End();
```

5. If we apply an animated texture for shading the mesh, let's also showcase it using `ImGui::Image()`:

```
  if (!g_UseColoredMesh) {
    const ImVec2 size = ImGui::GetIO().DisplaySize;
    const float  dim  = std::max(size.x, size.y);
    const ImVec2 sizeImg(0.25f * dim, 0.25f * dim);
    ImGui::SetNextWindowPos(ImVec2(size.x - sizeImg.x - 25, 0),
      ImGuiCond_Appearing);
    ImGui::Begin("Texture", nullptr,
      ImGuiWindowFlags_AlwaysAutoResize);
    ImGui::Image(texture.index(), sizeImg);
    ImGui::End();
  }
};
```

Now, let's examine the C++ code that's responsible for creating buffers and populating them with initial data:

1. The indices are immutable and are generated using the previously mentioned `generateIndices()` function. The vertex buffer size is calculated based on `12 float` elements per vertex. That is `vec4` positions, `vec4` texture coordinates, and `vec4` normal vectors. This padding is used to simplify the compute shader that writes to the vertex buffer.

```
std::vector<uint32_t> indicesGen((kNumU - 1) * (kNumV - 1) * 6);
generateIndices(indicesGen.data());
const uint32_t vertexBufferSize = 12 * sizeof(float) * kNumU * kNumV;
const uint32_t indexBufferSize  =
  sizeof(uint32_t) * (kNumU - 1) * (kNumV - 1) * 6;
Holder<BufferHandle> bufferIndex  = ctx->createBuffer({
  .usage     = lvk::BufferUsageBits_Index,
  .storage   = lvk::StorageType_Device,
  .size      = indicesGen.size() * sizeof(uint32_t),
  .data      = indicesGen.data(),
  .debugName = "Buffer: index" });
```

2. Let's create a texture that will be generated by a compute shader. This is similar to the previous recipe, *Generating textures in Vulkan using compute shaders*.

```
Holder<TextureHandle> texture = ctx->createTexture({
  .type       = lvk::TextureType_2D,
  .format     = lvk::Format_RGBA_UN8,
  .dimensions = {1024, 1024},
  .usage      = lvk::TextureUsageBits_Sampled |
                lvk::TextureUsageBits_Storage,
  .debugName  = "Texture: compute" });
```

3. The vertex buffer should be created with the `BufferUsageBits_Storage` flag to allow vertex shader usage:

```
Holder<BufferHandle> bufferVertex = ctx->createBuffer({
  .usage     = lvk::BufferUsageBits_Vertex |
               lvk::BufferUsageBits_Storage,
  .storage   = lvk::StorageType_Device,
  .size      = vertexBufferSize,
  .debugName = "Buffer: vertex" });
```

4. We create two compute pipelines: one for mesh generation and another for texture generation.

```
Holder<ShaderModuleHandle> compMesh = loadShaderModule(ctx,
  "Chapter05/09_ComputeMesh/src/main_mesh.comp");
Holder<ShaderModuleHandle> compTexture =
  loadShaderModule(ctx,
  "Chapter05/09_ComputeMesh/src/main_texture.comp");
Holder<ComputePipelineHandle> pipelineComputeMesh =
  ctx->createComputePipeline({ .smComp = compMesh });
Holder<ComputePipelineHandle> pipelineComputeTexture =
  ctx->createComputePipeline({ .smComp = compTexture });
```

5. Shader modules are loaded as usual. The geometry shader generates barycentric coordinates for wireframe outlines. Wireframe outline rendering will be enabled if you set `kNumU` and `kNumV` to 64 or lower values.

```
Holder<ShaderModuleHandle> vert =
  loadShaderModule(ctx, "Chapter05/09_ComputeMesh/src/main.vert");
Holder<ShaderModuleHandle> geom =
  loadShaderModule(ctx, "Chapter05/09_ComputeMesh/src/main.geom");
Holder<ShaderModuleHandle> frag =
  loadShaderModule(ctx, "Chapter05/09_ComputeMesh/src/main.frag");
```

6. For vertex input, we employ `vec4` to simplify padding concerns—or, to be more precise, to eliminate them entirely:

```
const lvk::VertexInput vdesc = {
  .attributes  = { { .location = 0,
                     .format = VertexFormat::Float4,
                     .offset = 0 },
                   { .location = 1,
                     .format = VertexFormat::Float4,
                     .offset = sizeof(vec4) },
                   { .location = 2,
                     .format = VertexFormat::Float4,
                     .offset = sizeof(vec4)+sizeof(vec4) } },
  .inputBindings = { { .stride = 3 * sizeof(vec4) },
};
```

7. The same shader code is specialized for textures and colored shading using specialization constants:

```
const uint32_t specColored = 1;
const uint32_t specNotColored = 0;
Holder<RenderPipelineHandle> pipelineMeshColored =
  ctx->createRenderPipeline({
    .vertexInput = vdesc,
    .smVert      = vert,
    .smGeom      = geom,
    .smFrag      = frag,
    .specInfo    = { .entries = { { .constantId = 0,
                                    .size = sizeof(uint32_t) } },
                     .data = &specColored,
                     .dataSize = sizeof(specColored) },
    .color       = { { .format = ctx->getSwapchainFormat() } },
    .depthFormat = app.getDepthFormat() });
Holder<RenderPipelineHandle> pipelineMeshTextured =
  ctx->createRenderPipeline({
    .vertexInput = vdesc,
    .smVert      = vert,
    .smGeom      = geom,
    .smFrag      = frag,
    .specInfo    = { .entries = { { .constantId = 0,
                                    .size = sizeof(uint32_t) } },
                     .data = &specNotColored,
                     .dataSize = sizeof(specNotColored) },
    .color       = { { .format = ctx->getSwapchainFormat() } },
    .depthFormat = app.getDepthFormat() });
```

The final segment of the C++ code is the rendering loop. Let's explore how to populate a command buffer:

1. First, we need to access the current `P-Q` pair from the morph queue:

```
lvk::ICommandBuffer& buf = ctx->acquireCommandBuffer();
const mat4 m = glm::translate(mat4(1.0f), vec3(0.0f, 0.0f, -18.f));
const mat4 p = glm::perspective(45.0f, aspectRatio, 0.1f, 1000.0f);
auto iter = g_MorphQueue.begin();
```

2. The number of parameters for the shaders is larger than usual, precisely `128` bytes. Push constants are shared across all shaders, compute, and graphics, for convenience purposes and to simplify the code.

```
struct PerFrame {
  mat4 mvp;
  uint64_t buffer;
  uint32_t textureId;
  float time;
  uint32_t numU, numV;
  float minU, maxU;
  float minV, maxV;
  uint32_t p1, p2;
  uint32_t q1, q2;
  float morph;
} pc = {
  .mvp       = p * m,
  .buffer    = ctx->gpuAddress(bufferVertex),
  .textureId = texture.index(),
  .time      = (float)glfwGetTime(),
  .numU      = kNumU,
  .numV      = kNumU,
  .minU      = -1.0f,
  .maxU      = +1.0f,
  .minV      = -1.0f,
  .maxV      = +1.0f,
  .p1        = iter->first,
  .p2        = (iter + 1)->first,
  .q1        = iter->second,
  .q2        = (iter + 1)->second,
  .morph     = easing(g_MorphCoef),
};
buf.cmdPushConstants(pc);
```

3. When we dispatch the mesh generation compute shader, we need to specify proper memory barriers to make sure the previous frame has finished rendering:

```
buf.cmdBindComputePipeline(pipelineComputeMesh);
buf.cmdDispatchThreadGroups(
  { .width = (kNumU * kNumV) / 2 },
  { .buffers = { { BufferHandle(bufferVertex) } } });
```

4. Do the same for the texture generation shader. We must ensure the texture regeneration is properly synchronized with rendering by issuing a Vulkan image layout transition with appropriate pipeline stages and masks. This is done by `LightweightVK` inside `lvk::CommandBuffer::cmdDispatchThreadGroups()` using a few empirical rules.

```
if (!g_UseColoredMesh) {
  buf.cmdBindComputePipeline(pipelineComputeTexture);
  buf.cmdDispatchThreadGroups(
    { .width = 1024 / 16, .height = 1024 / 16 },
    { .textures = { { TextureHandle(texture) } } });
}
```

5. When we start rendering, it's essential to specify both dependencies—the texture and the vertex buffer. The choice of the rendering pipeline depends on whether we aim to render a colored or textured mesh.

```
buf.cmdBeginRendering(
  renderPass, framebuffer,
  { .textures = { { TextureHandle(texture) } },
    .buffers  = { { BufferHandle(bufferVertex) } } });
buf.cmdBindRenderPipeline(
  g_UseColoredMesh ? pipelineMeshColored : pipelineMeshTextured);
buf.cmdBindDepthState({ .compareOp = lvk::CompareOp_Less,
   .isDepthWriteEnabled = true });
buf.cmdBindVertexBuffer(0, bufferVertex);
buf.cmdBindIndexBuffer(bufferIndex, lvk::IndexFormat_UI32);
buf.cmdDrawIndexed(indicesGen.size());
app.imgui_->beginFrame(framebuffer);
```

6. The `renderGUI()` function renders the ImGui UI, as described earlier in this recipe. In this case, we pass our generated texture to it for presentation.

```
renderGUI(texture);
app.drawFPS();
app.imgui_->endFrame(buf);
buf.cmdEndRendering();
```

That covers the C++ part. Now, let's dive into the GLSL shaders to understand how the whole demo works.

How it works...

Let's start with the compute shader, `Chapter05/09_ComputeMesh/src/main_mesh.comp`, which is responsible for generating vertex data:

1. The push constants are declared in `Chapter05/09_ComputeMesh/src/common.sp`. They are shared between all shaders, and we paste them here for convenience:

```
layout (local_size_x = 2, local_size_y = 1, local_size_z = 1) in;
// included from <Chapter05/09_ComputeMesh/src/common.sp>
layout(push_constant) uniform PerFrameData {
  mat4 MVP;
  uvec2 bufferId;
  uint textureId;
  float time;
  uint numU, numV;
  float minU, maxU, minV, maxV;
  uint P1, P2, Q1, Q2;
  float morph;
} pc;
```

2. This is the structure containing per-vertex data that we aim to generate and write into a buffer referenced by `VertexBuffer`, which is stored in `pc.bufferId` a few lines above the buffer.

```
struct VertexData {
  vec4 pos;
  vec4 tc;
  vec4 norm;
};
layout (buffer_reference) buffer VertexBuffer {
  VertexData vertices[];
} vbo;
```

3. The heart of our mesh generation algorithm is the `torusKnot()` function, which uses the following parametrization to triangulate a torus knot `https://en.wikipedia.org/wiki/Torus_knot`:

```
x = r * cos(u)
y = r * sin(u)
z = -sin(v)
```

4. The `torusKnot()` function is rather long and is implemented directly from the previously mentioned parametrization. Feel free to play with the `baseRadius`, `segmentRadius`, and `tubeRadius` values:

```
VertexData torusKnot(vec2 uv, float p, float q) {
  const float baseRadius    = 5.0;
  const float segmentRadius = 3.0;
  const float tubeRadius    = 0.5;
  float ct = cos( uv.x );
  float st = sin( uv.x );
  float qp = q / p;
  float qps = qp * segmentRadius;
  float arg = uv.x * qp;
  float sqp = sin( arg );
  float cqp = cos( arg );
  float BSQP = baseRadius + segmentRadius * cqp;
  float dxdt = -qps * sqp * ct - st * BSQP;
  float dydt = -qps * sqp * st + ct * BSQP;
  float dzdt =  qps * cqp;
  vec3 r    = vec3(BSQP * ct, BSQP * st, segmentRadius * sqp);
  vec3 drdt = vec3(dxdt, dydt, dzdt);
  vec3 v1 = normalize(cross(r, drdt));
  vec3 v2 = normalize(cross(v1, drdt));
  float cv = cos( uv.y );
  float sv = sin( uv.y );
  VertexData res;
  res.pos  = vec4(r + tubeRadius * ( v1 * sv + v2 * cv ), 1);
  res.norm = vec4(cross(v1 * cv - v2 * sv, drdt ), 0);
  return res;
}
```

5. We are running this compute shader in each frame, so, instead of generating a static set of vertices, we can actually pre-transform them to make the mesh look like it is rotating. Here is a couple of helper functions to compute the appropriate rotation matrices:

```
mat3 rotY(float angle) {
  float c = cos(angle), s = sin(angle);
  return mat3(c, 0, -s, 0, 1, 0, s, 0, c);
}
mat3 rotZ(float angle) {
  float c = cos(angle), s = sin(angle);
  return mat3(c, -s, 0, s, c, 0, 0, 0, 1);
}
```

6. Using the previously mentioned helpers, the `main()` function of our compute shader is now straightforward, and the only interesting thing worth mentioning here is the real-time morphing that blends two torus knots with different `P` and `Q` parameters. This is pretty easy because the total number of vertices always remains the same:

```
void main() {
  uint index = gl_GlobalInvocationID.x;
  vec2 numUV = vec2(pc.numU, pc.numV);
  vec2 ij = vec2(float(index / pc.numV), float(index % pc.numV));
```

7. Two sets of UV coordinates for parametrization need to be computed:

```
  const vec2 maxUV1 = 2.0 * 3.141592653 * vec2(pc.P1, 1.0);
  vec2 uv1 = ij * maxUV1 / (numUV - vec2(1));
  const vec2 maxUV2 = 2.0 * 3.141592653 * vec2(pc.P2, 1.0);
  vec2 uv2 = ij * maxUV2 / (numUV - vec2(1));
```

8. Compute the model matrix for our mesh by combining two rotation matrices:

```
  mat3 modelMatrix = rotY(0.5 * pc.time) * rotZ(0.5 * pc.time);
```

9. Compute two vertex positions for two different torus knots defined by the two sets of `P`-`Q` parameters: `P1`-`Q1` and `P2`-`Q2`.

```
  VertexData v1 = torusKnot(uv1, pc.P1, pc.Q1);
  VertexData v2 = torusKnot(uv2, pc.P2, pc.Q2);
```

10. Perform a linear blend between them using the `pc.morph` coefficient. We only need to blend the position and the normal vector. While normal vectors can be interpolated more gracefully, we leave that as another exercise for our readers:

```
  vec3 pos  = mix(v1.pos.xyz, v2.pos.xyz, pc.morph);
  vec3 norm = mix(v1.norm.xyz, v2.norm.xyz, pc.morph);
```

11. Fill in the resulting `VertexData` structure and store it in the output vertex buffer:

```
  VertexData vtx;
  vtx.pos  = vec4(modelMatrix * pos, 1);
  vtx.tc   = vec4(ij / numUV, 0, 0);
  vtx.norm = vec4(modelMatrix * norm, 0);
  VertexBuffer(pc.bufferId).vertices[index] = vtx;
}
```

The vertex shader looks as follows. Just note that the vertex attributes have proper types here:

```
layout(push_constant) uniform PerFrameData {
  mat4 MVP;
};
```

```
layout (location=0) in vec4 in_pos;
layout (location=1) in vec2 in_uv;
layout (location=2) in vec3 in_normal;
layout (location=0) out vec2 uv;
layout (location=1) out vec3 normal;
void main() {
  gl_Position = MVP * in_pos;
  uv = in_uv;
  normal = in_normal;
}
```

The geometry shader is trivial because the only thing it does is generate barycentric coordinates as described in *Integrating tessellation into the graphics pipeline*. So, let's jump straight to the fragment shader, `Chapter05/09_ComputeMesh/src/main.frag`:

1. The specialization constant is used to switch between the colored and textured versions of the shader:

```
#include <Chapter05/09_ComputeMesh/src/common.sp>
layout (location=0) in vec2 uv;
layout (location=1) in vec3 normal;
layout (location=2) in vec3 barycoords;
layout (location=0) out vec4 out_FragColor;
layout (constant_id = 0) const bool isColored = false;
```

2. Edge factor calculation for wireframe outlines as described in the *Integrating tessellation into the graphics pipeline* recipe:

```
float edgeFactor(float thickness) {
  vec3 a3 = smoothstep( vec3( 0.0 ),
    fwidth(barycoords) * thickness, barycoords);
  return min( min( a3.x, a3.y ), a3.z );
}
```

3. A function to calculate an RGB color for our mesh based on a floating point "hue" value:

```
vec3 hue2rgb(float hue) {
  float h = fract(hue);
  float r = abs(h * 6 - 3) - 1;
  float g = 2 - abs(h * 6 - 2);
  float b = 2 - abs(h * 6 - 4);
  return clamp(vec3(r,g,b), vec3(0), vec3(1));
}
void main() {
  float NdotL = dot(normalize(normal), normalize(vec3(0, 0, +1)));
```

```
float intensity = 1.0 * clamp(NdotL, 0.75, 1);
vec3 color = isColored ?
  intensity * hue2rgb(uv.x) :
  textureBindless2D(pc.textureId, 0, vec2(8,1) * uv).xyz;
out_FragColor = vec4(color, 1.0);
```

4. For high values of `numU` and `numV`, the tessellation level is so dense that no distinct wireframe edges are visible—everything collapses into a black Moiré mess. We disable wireframe overlays for values greater than 64.

```
  if (isColored && pc.numU <= 64 && pc.numV <= 64)
    out_FragColor =
      vec4( mix( vec3(0.0), color, edgeFactor(1.0) ), 1.0 );
}
```

Last but not least, there's a compute shader responsible for generating an animated texture. You can find it in `Chapter05/09_ComputeMesh/src/main_texture.comp`, and the idea is identical to the approach described in the previous recipe, *Generating textures in Vulkan using compute shaders*. We do not copy and paste that shader here.

The demo application will produce a variety of torus knots similar to the one in the following screenshot. Each time you select a new pair of `P-Q` parameters from the UI, the morphing animation will kick in and transform one knot into another. Checking the *Use colored mesh* box will apply colors to the mesh instead of a computed texture:

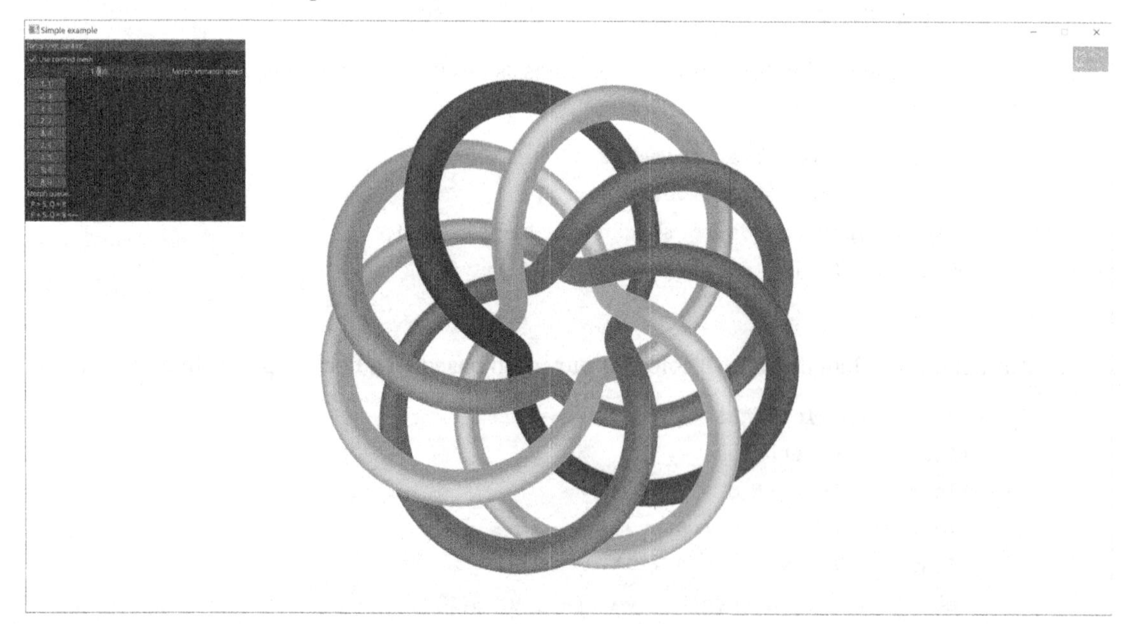

Figure 5.9: Computed mesh with real-time animation

There's more...

Refer to the Wikipedia page `https://en.wikipedia.org/wiki/Torus_knot` for additional explanation of the math details. Try setting the values of `kNumU` and `kNumV` to `32` while checking "Use colored mesh".

This recipe introduces an explicit synchronization process between two independent compute shaders and a rendering pass. The *LightweightVK* implementation is designed to make this synchronization as seamless as possible through the specification of explicit dependencies.

Since only one queue is being used, the device workload fully drains at each barrier, leaving no alternative tasks to mitigate this inefficiency. Here's a concrete example of how timeline semaphores and asynchronous compute-only queues can be utilized to accelerate a heterogeneous compute/graphics Vulkan application: `https://github.com/nvpro-samples/vk_timeline_semaphore`.

In more complex, real-world 3D applications, finer-grained synchronization mechanisms are often necessary. For valuable insights on handling synchronization effectively, refer to the Vulkan synchronization guide by Khronos: `https://github.com/KhronosGroup/Vulkan-Docs/wiki/Synchronization-Examples`.

6

Physically Based Rendering Using the glTF 2.0 Shading Model

This chapter will cover the integration of **physically based rendering** (**PBR**) into your graphics applications. We use the glTF 2.0 shading model as an example. PBR is not a single specific technique but rather a set of concepts, like using measured surface values and realistic shading models, to accurately represent real-world materials. Adding PBR to your graphics application or retrofitting an existing rendering engine with PBR might be challenging because it requires multiple big steps to be completed and work simultaneously before a correct image can be rendered.

Our goal here is to show how to implement all these steps from scratch. Some of these steps, like precomputing irradiance maps or **Bidirectional Reflectance Distribution Function** (**BRDF**) look-up tables, require additional tools to be written. We are not going to use any third-party tools here and will show how to implement the entire skeleton of a PBR pipeline from the ground up including creation of rudimental tools for the work. Some pre-calculations can be done using General-purpose Graphics Processing Unit (GPGPU) techniques and compute shaders, all of which will be covered here as well.

In this chapter, we will learn the following recipes:

- An introduction to glTF 2.0 physically based shading model
- Rendering unlit glTF 2.0 materials
- Precomputing BRDF look-up tables
- Precomputing irradiance maps and diffuse convolution
- Implementing the glTF 2.0 core metallic-roughness shading model
- Implementing the glTF 2.0 core specular-glossiness shading model

Note

In all future references to the glTF specification, we mean the glTF 2.0 Specification. glTF 1.0 is obsolete and deprecated and we do not cover it in this book.

An introduction to glTF 2.0 physically based shading model

In this section, we will learn the PBR Material basics and provide enough context for the actual implementation of some ratified glTF 2.0 PBR extensions. The actual code would be presented in the subsequent recipes and chapters. Since the topic of PBR rendering is vast, we focus on the minimalistic implementation just to guide you and get you started. In the book text right here, we focus on the GLSL shader code for the glTF 2.0 PBR shading model. Roughly speaking, rendering a physically-based image is nothing more than running a fancy pixel shader with a set of textures.

Getting ready

We assume readers already have some basic understanding of linear algebra and calculus. It is recommended to get yourself familiar with the glTF 2.0 specification, which can be found at `https://registry.khronos.org/glTF/specs/2.0/glTF-2.0.html`.

What is PBR?

Physically-based rendering (**PBR**) is a set of techniques that aim to simulate how light interacts with real-world materials. By using realistic models for light scattering and reflection, PBR materials can create much more believable and immersive visuals compared to traditional methods.

The glTF PBR material model is a standardized way of representing physically-based materials in the glTF 2.0 format. This model allows for the creation of highly realistic 3D content across diverse platforms and applications, making it a crucial tool for modern 3D development.

Light-object interactions

Let's step back and see what a ray of light is as a physical phenomenon - it's a geometric line along which light energy travels, or a beam of light. It has a starting point and a direction of propagation. Two important interactions of light with surfaces are: **reflection** and **diffusion** (also known as "specular" and "diffuse" reflection, respectively). While we intuitively understand these concepts through everyday experience, their physical characteristics may be less familiar.

When light encounters a surface, part of it bounces back in the opposite direction of the surface's normal, like how a ball rebounds at an angle off a wall. This type of reflection, occurring on smooth surfaces, creates a mirror-like effect called **specular reflection** (derived from speculum, a Latin word for "mirror").

However, not all the light reflects. Some light penetrates the surface, where it can either be absorbed, converted into heat, or scattered in various directions. The scattered light that exits the surface again is known as **diffuse light, photon diffusion,** or **subsurface scattering**. These terms all refer to the same physical phenomenon of photons movement. However, diffusion and scattering are different in how they disperse photons. Scatter involves photons being redirected in various directions, while diffusion involves photons spreading out evenly.

The way materials absorb and scatter diffuse light varies for different wavelengths of light, giving objects their distinct colors. For example, an object that absorbs most colors but scatters blue light will appear blue. This scattering is often so chaotic that it appears the same from all directions, unlike a specular mirror-like reflection.

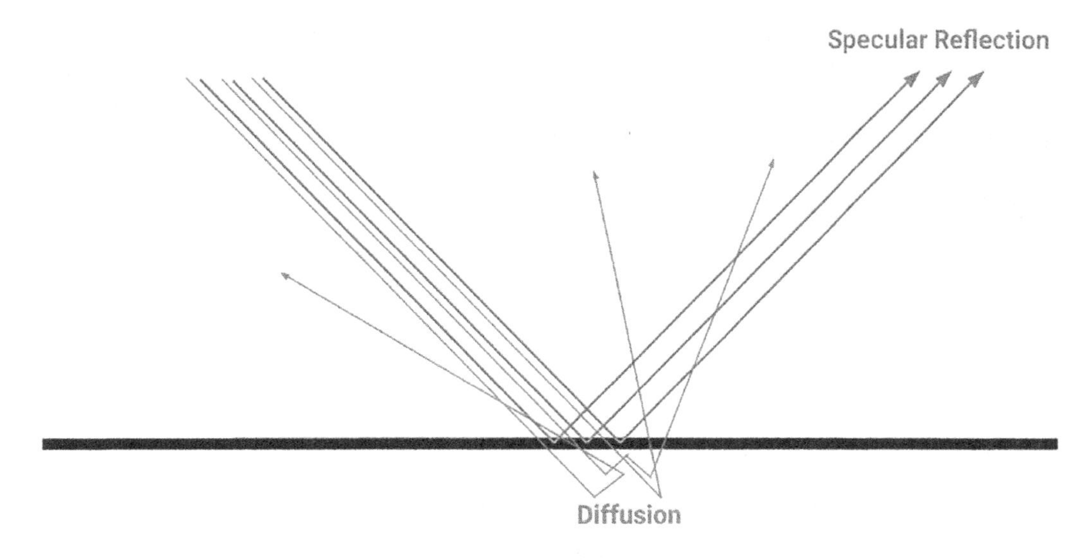

Figure 6.1: Diffuse and Specular Reflection

Simulating this behavior in computer graphics often requires only a single input: **albedo**, which represents the color defined by a mix of the fractions of various light wavelengths that scatter back out of a surface. The term **diffuse color** is often used interchangeably.

When materials have wider scattering angles, such as human skin or milk, simulating their lighting requires more complex approaches than simple light interaction with a surface. This is because the light scattering in these materials is not just limited to the surface, but also occurs within the material itself.

For thin objects, the light can even scatter out of their back side, making them **translucent**. As the scattering further decreases, like in glass, the material becomes transparent, allowing entire images to pass through it preserving their visible shape.

These unique light scattering behaviors are significantly different from the typical "close to the surface" diffusion, requiring special handling for accurate rendering.

Energy conservation

The fundamental principle in PBR revolves around the law of conservation of energy. This law asserts that within an isolated system, the overall energy remains unchanged. In the context of rendering, it signifies that the quantity of incoming light at any given location in the scene equals the combined amount of light that is reflected, transmitted, and absorbed at that location.

Enforcing energy conservation is crucial for physically-based rendering. It allows assets to adjust reflectivity and albedo values for a material without inadvertently violating the laws of physics, which often leads to unnatural- looking results. Implementing these constraints in code prevents assets from deviating too far from the reality or becoming inconsistent under varying lighting conditions.

Implementing this principle in a shading system is straightforward. We simply subtract reflected light before computing the diffuse shading. This implies that highly reflective objects will exhibit minimal to no diffuse light, as most of the light is reflected instead of penetrating the surface. Conversely, materials with strong diffusion cannot be particularly reflective.

Surface properties

In any given environment, one can readily observe various complex surfaces that exhibit distinct light interactions. These unique surface properties find representation by considering general mathematical functions known as **Bidirectional Scattering Distribution Functions (BSDFs).**

Think of a BSDF as an equation that describes how light scatters upon encountering a surface. It considers the surface's physical properties and predicts probabilities of incident light coming from one direction getting scattered in other directions.

Although the term **BSDF** might sound complex, let's break it down:

- **Bidirectional:** This refers to the two-way nature of light interaction with a surface. Incident light arrives at a surface from one direction and then scatters in various directions.
- **Scattering:** This describes how incident light can be redirected into multiple outgoing directions. This can involve reflection, transmission, or a combination of both.
- **Distribution Function:** This defines the probability of light scattering in a particular direction based on surface's characteristics. The distribution can range from perfectly uniform scattering to a concentrated reflection in a single direction.

In practice, the **BSDF** is usually split into two parts which are treated separately:

- **Bidirectional Reflectance Distribution Functions (BRDFs):** These functions specifically describe how incident light is reflected from a surface. They explain why a seemingly white light source illuminating a banana makes it appear yellow instead. The BRDF reveals that the banana primarily reflects light in the yellow part of the spectrum while absorbing or transmitting other wavelengths.
- **Bidirectional Transmittance Distribution Functions (BTDFs):** These functions specifically describe how light is transmitted through a material. This is evident in materials such as glass and plastics, where we see how incident light passes through the material.

Additionally, other types of BSDFs exist to account for more complex light interaction phenomena, such as subsurface scattering. This occurs when light enters a material and bounces around before re-emerging in a new direction at points significantly distant from the points of incidence of the incident rays.

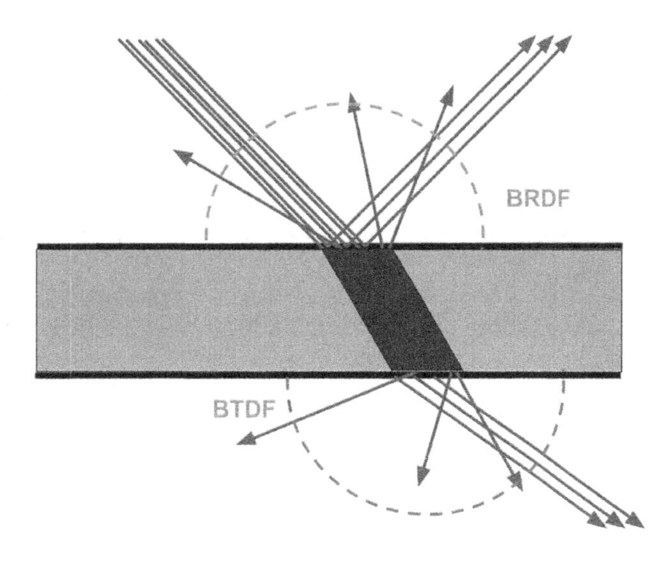

Figure 6.2: BRDF and BTDF

Types of reflection

There are 4 primary surface types characterized by their reflection distribution functions (BRDFs), which define the likelihood of light scattering in various directions:

- **Diffuse surfaces** scatter light uniformly in all directions, exemplified by the consistent color of a matte paint.
- **Glossy specular surfaces** preferentially scatter light in specific reflected directions, exhibiting blurred reflections such as specular highlights on plastic.
- **Perfect specular surfaces** scatter light precisely in a single outgoing direction, mirroring the incident light with respect to the surface normal–similar to flawless reflections seen in perfect mirrors.
- **Retro-reflective surfaces** scatter light predominantly back along the incident direction back to the light source, akin to the specular highlights observed on velvet or road signs.

However, it's improbable that a real-world surface strictly adheres to only one of these models. As a result, most materials can be modelled as intricate combinations of these surface types.

Moreover, each type of reflection— diffuse, glossy specular, perfect specular, and retro-reflective— can exhibit isotropic or anisotropic distributions:

- **Isotropic reflections** maintain a consistent amount of reflected light at a point, irrespective of the object rotation angle. This characteristic aligns with the behavior of most surfaces encountered in daily life.
- **Anisotropic reflections** vary in the amount of reflected light based on orientation of object to the light source. This occurs due to the alignment of small surface irregularities aligned predominantly in one direction, resulting in elongated and blurry reflections. Such behavior is noticeable in materials such as brushed metal and velvet.

Transmission

Reflection distribution types can be used for transmission as well, except for retro-reflection. Conversely, when light passes through a material, its path is affected by the material's properties. To illustrate this difference from reflection, consider a single light ray passing through a material, like perfect specular transmission. In perfect specular transmission, the medium's refractive index determines the direction in which light travels. This behavior adheres to **Snell's Law**, which is described using the equation, $n_1 \sin\theta_1 = n_2 \sin\theta_2$.

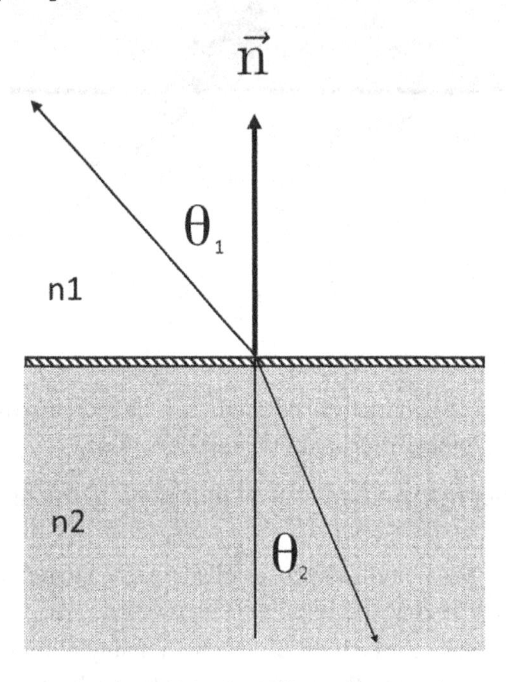

Figure 6.3: The index of refraction

Here, n represents the refractive index of the first and second media, while θ denotes the angle of the incoming light concerning the surface normal. Consequently, when both media share identical refractive indices, light proceeds in a perfectly straight path. Conversely, if the refractive indices differ, the light changes its direction upon transitioning into the next medium. A notable instance of this is when light shifts direction upon entering water from air, leading to distortions in our underwater observations. This contrasts with perfect specular reflection, where the incoming angle always equals the outgoing angle.

Fresnel equation

It is important for physically- based renderers to know how much light is reflected or transmitted on the surface. It is a combination of these effects that describes substances such as honey and stained glass that both have color and can be seen through.

These amounts are directly related to each other and described by the **Fresnel equations.**

These equations are tailored for two types of media: **conductors (metals)** and **dielectrics (nonmetals)**. Metals do not transmit light, they only reflect it entirely, or practically entirely. Dielectrics possess the property of diffuse reflection — light rays pass beneath the surface of the material and some of them are absorbed, while some are returned in the form of reflection. This is particularly evident in the specular highlight of these materials — for metals, it will be colored, while for dielectrics it appears white or, more accurately, retains the color of the incident light.

Although both conductors and dielectrics are subject to the same set of Fresnel equations, glTF 2.0 opts to develop a distinct evaluation function for dielectrics. This choice is made to leverage the notably straightforward structure that these equations assume when the refractive indices are ensured to be real numbers.

- **Nonmetals (dielectrics)**: These are materials like glass, plastic, and ceramics, which lack distinctive metallic properties.
- **Metals (conductors)**: These materials could conduct both heat and electricity to a certain extent. Examples include many metals, such as copper, silver, and gold, though not all metals exhibit this property. Unlike dielectrics, conductors do not transmit light; rather, they absorb some of the incident light, converting it into heat.

Microfacets

According to the microfacet theory, a rough surface is composed of countless microfacets or tiny surface elements, each having its own orientation with respect to the surface normal. These microfacets scatter incoming light in different directions due to their orientations, resulting in a diffused reflection rather than a perfect mirror-like reflection.

- **Blinn-Phong Model**: It was introduced by James F. Blinn in 1977 as an enhancement to the empirical Phong reflection model, devised by Bui Tuong Phong in 1973.

  This model computes the intensity of the reflected light based on the angle between the viewer's direction and the halfway vector `h=(L+V)/length(L+V)`, which is halfway between the light direction `L` and the view direction `V`. The model includes a specular term that provides a highlight on the surface, simulating the effect of a shiny surface.

- **Cook-Torrance Model**; In 1982, Robert Cook and Kenneth Torrance introduced a reflectance model that offered a more precise depiction of light reflectance in comparison to the Phong and Blinn-Phong models. The microfacet BRDF equation is as follows:

$$f_r(\omega_i, \omega_o) = \frac{F(\omega_i, h)D(h)G(\omega_i, \omega_o, h)}{4(\omega_i \cdot n)(\omega_o \cdot n)}$$

  Where:

  - $f_r(\omega_i,\omega_o)$ is the microfacet BRDF,
  - $F(\omega_i,h)$ is the Fresnel term,
  - $D(h)$ is the microfacet distribution function,
  - $G(\omega_i,\omega_o,h)$ is the geometry function,
  - ω_i is the incident light direction,

- ω_o is the outgoing light direction,
- h is the half vector,
- n is the surface normal.

The proposed approach is versatile, featuring **three interchangeable component** functions F, D, G which can be substituted with equations of your preference. Additionally, it is highly effective at accurately representing a wide range of real-world materials.

This equation represents the amount of light reflected in a specific direction ω_o given an incident light direction ω_i and the surface properties. The initial component F represents the Fresnel effect, the subsequent component `D` is a **normal distribution function** (**NDF**), and the final component accounts for the shadowing factor G referred to as the **G term**.

Of all the factors in this formulation, the NDF term typically exerts the greatest importance. The specific form of the NDF is heavily influenced by the roughness of the BRDF. To sample the microfacet BRDF model efficiently, it is customary to first sample the NDF, obtaining a random microfacet normal that conforms to the NDF, and subsequently reflecting the incident radiance along this normal to determine the outgoing direction.

The normalization requirement for the Normal Distribution Function (**NDF**) in microfacet theory ensures that the total amount of energy reflected or transmitted by a surface remains consistent across different roughness levels.

In microfacet theory, the NDF describes the statistical distribution of microfacet normals on a surface. It specifies the probability density of finding a microfacet with a particular orientation. When integrating the **BRDF** over all possible directions, the integral should yield a value representing the total reflectance or transmittance of the surface.

Normalization of the NDF guarantees that the total amount of light reflected or transmitted by the surface remains constant, regardless of the surface roughness. This ensures energy conservation, a fundamental principle in physics stating that energy cannot be created or destroyed, only transformed, or transferred.

Several NDFs are commonly used to simulate the behavior of surfaces with microscale roughness. Here are some examples: GGX, Beckmann, Blinn. In the next recipes, we will learn how to implement some of them.

What is a material?

Materials serve as high-level descriptions utilized to represent surfaces, defined by combinations of **BRDFs (Bidirectional Reflectance Distribution Functions)** and **BTDFs (Bidirectional Transmittance Distribution Functions)**. These **BSDFs** are articulated as parameters that govern the visual characteristics of the material. For instance, a matte material can be delineated by specifying a diffuse reflection value to elucidate how light interacts with the surface, along with a scalar roughness value to characterize its texture. Transitioning from a matte to a plastic material could be achieved by simply appending a glossy specular reflection value to the matte material, thus recreating the specular highlights typical of plastics.

glTF PBR specification

The glTF PBR specification approaches material representation in a manner that emphasizes realism, efficiency, and consistency across different rendering engines and applications.

One key aspect of the glTF PBR specification is its adherence to physically based principles. This means that the materials defined in glTF accurately simulate real-world behavior, such as how light interacts with surfaces. Parameters like base color (albedo), roughness, metallic, and specular are used to describe materials, aligning with physical properties like surface color, smoothness, metallicity, and specular reflectivity.

Another notable feature of the glTF PBR approach is its simplicity and ease of implementation. By standardizing the parameters used to describe materials, glTF simplifies the process of creating and exporting 3D models with PBR materials. This consistency across different applications and rendering engines streamlines the workflow for artists and developers, enabling them to work more efficiently and interchangeably.

Furthermore, the glTF PBR specification is designed for real-time rendering applications, making it well-suited for use in interactive experiences, games, and other real-time graphics applications. Its efficient representation of materials and optimized file format contribute to faster loading times and better performance in real-time rendering scenarios.

Overall, the glTF PBR specification stands out for its commitment to physical accuracy, simplicity, and efficiency, making it a preferred choice for representing materials in 3D graphics applications. Its widespread adoption and support across various platforms further cement its status as a leading standard for PBR material representation.

More information

The *Khronos 3D Formats Working Group* is continually striving to enhance PBR material capabilities by introducing new extension specifications. You can always stay updated on the status of ratified extensions by visiting Khronos GitHub page: `https://github.com/KhronosGroup/glTF/blob/main/extensions/README.md`

There's more...

For those looking to deepen their knowledge, be sure to check out the free book *Physically Based Rendering: From Theory to Implementation* by Matt Pharr, Wenzel Jakob, and Greg Humphreys available online at `http://www.pbr-book.org`. Another great reference is the book *Real Time Rendering, 4th Edition* by Tomas Akenine-Möller, Eric Haines, and Naty Hoffman.

Also, we recommend SIGGRAPH's *Physically Based Rendering* courses. For example, you can find a comprehensive collection of links on GitHub `https://github.com/neil3d/awesome-pbr`.

Besides that, the *Filament* rendering engine provides a very comprehensive explanation of PBR materials: `https://google.github.io/filament/Filament.md.html`.

Rendering unlit glTF 2.0 materials

In this recipe, we begin developing a code framework that enables us to load and render glTF 2.0 assets.

We start with the `unlit` material because it's the simplest glTF 2.0 PBR material extension, and its shader implementation is straightforward. The official name of the extension is `KHR_materials_unlit`, and you can find the specification here: `https://github.com/KhronosGroup/glTF/tree/main/extensions/2.0/Khronos/KHR_materials_unlit`.

The `unlit` material is technically not PBR-based because it may violate energy conservation principles or create artistic representations that don't follow the laws of physics. It was designed with the following motivation:

- Mobile devices with limited resources, where unlit materials offer a more performant alternative to higher-quality shading models.
- Photogrammetry, in which the lighting information is already prebaked into the texture data and no additional lighting should be applied.
- Stylized materials (such as those resembling "anime" or hand-drawn art), where lighting is intentionally excluded for aesthetic reasons.

Let's get started with the basic implementation.

Getting ready

The source code for this recipe is located in `Chapter06/01_Unlit/main.cpp`. The corresponding GLSL vertex and fragment shaders are located in `main.vert` and `main.frag`.

How to do it...

We will create a feature-rich glTF viewer in the following chapters. In this chapter, we start building a simple framework that allows us to load and render basic glTF models.

We use `VulkanApp` and the *Assimp* library from previous chapters. As usual, most of the error checking is omitted from the book text but is present in the actual source code files.

1. Let's load our `.gltf` file using *Assimp*. As you can see, the loading function is a single line of code. *Assimp* supports loading of `.gltf` and `.glb` files out of the box. We use the *DamagedHelmet* asset from the official Khronos repository `https://github.com/KhronosGroup/glTF-Sample-Assets/tree/main/Models/DamagedHelmet`. This model uses the Metallic-Roughness material, but for the demonstration purposes we apply the `unlit` shader to it

```
const aiScene* scene = aiImportFile("deps/src/glTF-Sample-
  Assets/Models/DamagedHelmet/glTF/DamagedHelmet.gltf",
    aiProcess_Triangulate);
```

2. The next step is to build the mesh geometry. The `unlit` material uses only the `baseColor` property of material in `3` different input forms: as a vertex attribute, as a static fragment shader color factor and as a base color texture input. For our vertex format, it means we need to provide the following `3` per-vertex attributes:

```
struct Vertex {
  vec3 position;
  vec4 color;
  vec2 uv;
};
```

3. To fill in these attributes, we use following code. Empty colors are filled with the white color value `(1, 1, 1, 1)` and empty texture coords are filled with zeroes `(0, 0, 0)` according to the glTF specification:

```
std::vector<Vertex> vertices;
vertices.reserve(mesh->mNumVertices);
for (unsigned int i = 0; i != mesh->mNumVertices; i++) {
  const aiVector3D v = mesh->mVertices[i];
  const aiColor4D  c = mesh->mColors[0] ?
    mesh->mColors[0][i] : aiColor4D(1, 1, 1, 1);
  const aiVector3D t = mesh->mTextureCoords[0] ?
    mesh->mTextureCoords[0][i] : aiVector3D(0, 0, 0);
  vertices.push_back({ .position = vec3(v.x, v.y, v.z),
                       .color    = vec4(c.r, c.g, c.b, c.a),
                       .uv       = vec2(t.x, 1.0f - t.y) });
}
```

If a vertex color is not presented in the mesh, then we replace it with the default white color. It is a convenient way to simplify the final shader permutation where we can just combine all 3 inputs in a simple manner. We will come back to it later in this recipe.

4. We build the index buffer using the *Assimp* mesh faces information:

```
std::vector<uint32_t> indices;
indices.reserve(3 * mesh->mNumFaces);
for (unsigned int i = 0; i != mesh->mNumFaces; i++) {
  for (int j = 0; j != 3; j++)
    indices.push_back(mesh->mFaces[i].mIndices[j]);
}
```

5. After that, we should load a diffuse, or albedo base color texture. For simplicity, we use the hardcoded file path here instead of obtaining it from the `.gltf` model:

```
Holder<TextureHandle> baseColorTexture =
  loadTexture(ctx, "deps/src/glTF-Sample-
    Assets/Models/DamagedHelmet/glTF/Default_albedo.jpg");
```

6. The vertex and index data are static and can be uploaded into corresponding Vulkan buffers:

```
Holder<BufferHandle> vertexBuffer = ctx->createBuffer({
  .usage     = lvk::BufferUsageBits_Vertex,
  .storage   = lvk::StorageType_Device,
  .size      = sizeof(Vertex) * vertices.size(),
  .data      = vertices.data(),
  .debugName = "Buffer: vertex" });
Holder<BufferHandle> indexBuffer = ctx->createBuffer({
  .usage     = lvk::BufferUsageBits_Index,
  .storage   = lvk::StorageType_Device,
  .size      = sizeof(uint32_t) * indices.size(),
  .data      = indices.data(),
  .debugName = "Buffer: index" });
```

7. To finish the mesh setup, we need to load GLSL shaders and create a render pipeline. The member fields of the `VertexInput` struct correspond to the `Vertex` struct mentioned above.

```
const lvk::VertexInput vdesc = {
  .attributes    = { { .location = 0,
                       .format = lvk::VertexFormat::Float3,
                       .offset = 0  },
                     {   .location = 1,
                       .format = lvk::VertexFormat::Float4,
                       .offset = sizeof(vec3) },
                     {   .location = 2,
                       .format = lvk::VertexFormat::Float2,
                       .offset = sizeof(vec3) + sizeof(vec4) }},
  .inputBindings = { { .stride = sizeof(Vertex) } }};
Holder<ShaderModuleHandle> vert =
  loadShaderModule(ctx, "Chapter06/01_Unlit/src/main.vert");
Holder<ShaderModuleHandle> frag =
  loadShaderModule(ctx, "Chapter06/01_Unlit/src/main.frag");
Holder<RenderPipelineHandle> pipelineSolid =
  ctx->createRenderPipeline({
    .vertexInput = vdesc,
    .smVert      = vert,
    .smFrag      = frag,
    .color       = {{ .format = ctx->getSwapchainFormat() }},
    .depthFormat = app.getDepthFormat(),
    .cullMode    = lvk::CullMode_Back });
```

This was the preparation code. Let's now look inside the application main loop.

1. Within the rendering loop, we prepare the model-view-projection matrix `mvp` and pass it into GLSL shaders using push constants and the `PerFrameData` structure together with the base color value and the albedo texture id.

```
const mat4 m1 = glm::rotate(
  mat4(1.0f), glm::radians(+90.0f), vec3(1, 0, 0));
const mat4 m2 = glm::rotate(
  mat4(1.0f), (float)glfwGetTime(), vec3(0.0f, 1.0f, 0.0f));
const mat4 v = app.camera_._.getViewMatrix();
const mat4 p = glm::perspective(45.0f, aspectRatio, 0.1f, 1000.0f);
struct PerFrameData {
  mat4 mvp;
  vec4 baseColor;
  uint32_t baseTextureId;
} perFrameData = {
  .mvp           = p * v * m2 * m1,
  .baseColor     = vec4(1, 1, 1, 1),
  .baseTextureId = baseColorTexture.index(),
};
```

2. Now the actual 3D mesh rendering is simple, so we post the code here in its entirety:

```
const lvk::RenderPass renderPass = {
  .color = { { .loadOp = lvk::LoadOp_Clear,
               .clearColor = { 1.0f, 1.0f, 1.0f, 1.0f } } },
  .depth = { .loadOp = lvk::LoadOp_Clear,
             .clearDepth = 1.0f }
};
const lvk::Framebuffer framebuffer = {
  .color  = {{ .texture = ctx->getCurrentSwapchainTexture() }},
  .depthStencil = { .texture = app.getDepthTexture() },
};
lvk::ICommandBuffer& buf = ctx->acquireCommandBuffer();
buf.cmdBeginRendering(renderPass, framebuffer);
buf.cmdBindVertexBuffer(0, vertexBuffer, 0);
buf.cmdBindIndexBuffer(indexBuffer, lvk::IndexFormat_UI32);
buf.cmdBindRenderPipeline(pipelineSolid);
buf.cmdBindDepthState({ .compareOp = lvk::CompareOp_Less,
                        .isDepthWriteEnabled = true });
buf.cmdPushConstants(perFrameData);
buf.cmdDrawIndexed(indices.size());
```

3. A few nice touches at the end to render the infinite grid as described in the recipe *Chapter 5* and the FPS counter as described in the recipe *Chapter 4*.

```
app.drawGrid(buf, p, vec3(0, -1.0f, 0));
app.imgui_->beginFrame(framebuffer);
app.drawFPS();
app.imgui_->endFrame(buf);
buf.cmdEndRendering();
ctx->submit(buf, ctx->getCurrentSwapchainTexture());
```

This was all the C++ code. Now, let's dive into the GLSL shaders for this example. They're straightforward and short.

1. The vertex shader `Chapter06/01_Unlit/src/main.vert` does the vertex transformation and pre-multiplies the per-vertex color with the provided base color value from push constants.

```
layout(push_constant) uniform PerFrameData {
  mat4 MVP;
  vec4 baseColor;
  uint textureId;
};
layout (location = 0) in vec3 pos;
layout (location = 1) in vec4 color;
layout (location = 2) in vec2 uv;
layout (location = 0) out vec2 outUV;
layout (location = 1) out vec4 outVertexColor;
void main() {
  gl_Position = MVP * vec4(pos, 1.0);
  outUV = uv;
  outVertexColor = color * baseColor;
}
```

2. The fragment shader `Chapter06/01_Unlit/src/main.frag` is very simple as well. All it does is multiply the precomputed per-vertex base color value by the color value sampled from the albedo texture. The glTF 2.0 specification mandates to provide at least one `baseColorFactor` value for the Metallic-Roughness property and this guarantees the correctness of the result as long as we keep all the remaining parameters equal to `1`.

```
layout(push_constant) uniform PerFrameData {
  mat4 MVP;
  vec4 baseColor;
  uint textureId;
};
layout (location = 0) in vec2 uv;
layout (location = 1) in vec4 vertexColor;
```

```
layout (location=0) out vec4 out_FragColor;
void main() {
  vec4 baseColorTexture = textureBindless2D(textureId, 0, uv);
  out_FragColor = textureBindless2D(textureId, 0, uv) * vertexColor;
}
```

The running application `Chapter06/01_Unlit/src/main.cpp` should look as in the following screenshot.

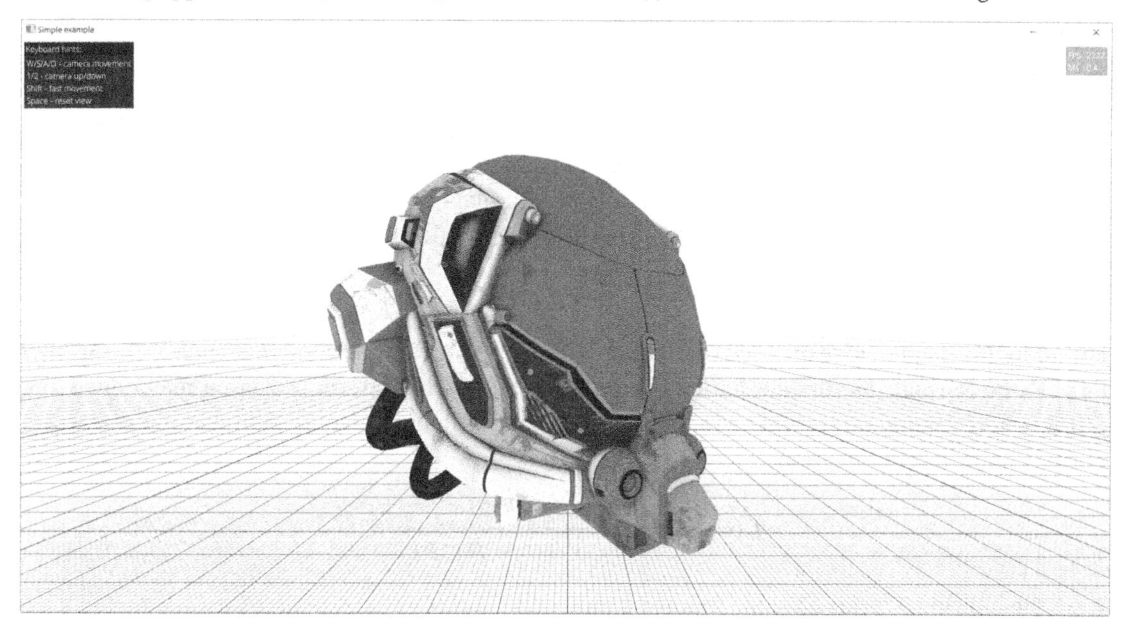

Figure 6.4: Unlit glTF2 model

In this example, we enforced the use of the albedo texture for rendering to keep the code simple for easier understanding. In the subsequent chapters, we'll fully support various combinations of material parameters as specified in the glTF 2.0 specification, providing a more accurate and complete glTF 2.0 viewer implementation.

Precomputing BRDF look-up tables

In the previous recipes, we learned the basic theory behind glTF 2.0 PBR and implemented a simple unlit glTF 2.0 renderer. Let's continue our PBR exploration and learn how to precompute the Smith GGX BRDF **look-up table** (**LUT**) for our upcoming glTF 2.0 viewer.

To render a PBR image, we have to evaluate the **BRDF** at each point on the surface being rendered, considering the surface properties and the viewing direction. This is computationally expensive and many real-time implementations, including the reference glTF-Sample-Viewer from Khronos, use precalculated tables of some sort to find the **BRDF** value based on surface roughness and the viewing direction. BRDF LUT can be stored as a two-dimensional texture.

The X-axis represents the dot product between the surface normal vector and the viewing direction, while the Y-axis represents the surface roughness values `0...1`. Each texel holds three 16-bit floating point values. The first two values represent the scale and bias to `F0`, which is the specular reflectance at normal incidence. The third value is utilized for the sheen material extension, which will be covered in the following chapter.

We are going to use Vulkan to calculate this LUT texture on the GPU and implement a compute shader to do it.

Getting ready

It is helpful to revisit the Vulkan compute pipeline creation from the *Generating textures in Vulkan using compute shaders* in *Chapter 5*. Our implementation is based on a shader from `https://github.com/KhronosGroup/glTF-Sample-Viewer`, which runs very similar computations in a fragment shader. Our GLSL compute shader can be found in `Chapter06/02_BRDF_LUT/src/main.comp`.

Why precompute?

Earlier in this chapter, we explained what BRDF is and introduced its components, such as the Fresnel term *F*, the Normal Distribution Function *NDF*, and the Geometry term *G*. As you may notice, the BRDF results depend on several factors, such as the incident and outgoing light directions, surface normal, and viewer's direction.

$$f_r(\omega_i, \omega_o) = \frac{F(\omega_i, h)D(h)G(\omega_i, \omega_o, h)}{4(\omega_i \cdot n)(\omega_o \cdot n)}$$

Were the individual terms have the following meaning.

D is the GGX NDF microfacet distribution function:

$$D(h) = \frac{\alpha^2}{\pi((n \cdot h)^2(\alpha^2 - 1) + 1)^2}$$

G accounts for mutual shadowing of microfacets and looks as follows:

$$G(\omega_i, \omega_o, h) = \frac{n \cdot v}{(n \cdot v)(1 - k) + k}$$

$$k = \frac{(Roughness + 1)^2}{8}$$

And the Fresnel F term determines how much light is reflected from the surface at a given angle of incidence:

$$F(v, h, F_0) = F_0 + (1 - F_0)(1 - (v \cdot h))^5$$

If we examine any component of the BRDF, we'll see that they are all quite complex for real-time per-pixel calculations. To simplify this, we can use an offline process to precompute certain parts of the BRDF equation.

As you can see, the *G* term and certain parts of the *F* term depend only on *v*, *h* and *Roughness* parameters. We can leverage this for precomputation. Additionally, note that we never need *n* and *v* separately, so we can always use their dot product instead.

One important question remains: how can we iterate through all possible *v* and *n* combinations? To do this, we need to integrate over all angles on a hemisphere, but we can use a simpler approximation. To keep it efficient, we make two assumptions. First, we must integrate using a limited number of samples. Second, we need to choose those samples wisely rather than randomly.

As described in the book *GPU Gems 3, Chapter 20, GPU-Based Importance Sampling* `https://developer.nvidia.com/gpugems/gpugems3/part-iii-rendering/chapter-20-gpu-based-importance-sampling`, the solution is to use Monte Carlo estimation with importance sampling. Monte Carlo estimation allows us to approximate an integral using a weighted sum of random samples. Importance sampling takes this further by prioritizing random points on the hemisphere that have a greater influence on the function being estimated.

The paper *Real Shading in Unreal Engine 4* by Brian Karis offers a detailed explanation of the mathematical aspects. We highly recommend reading it for a deeper understanding of the math behind physically-based rendering `https://blog.selfshadow.com/publications/s2013-shading-course/karis/s2013_pbs_epic_notes_v2.pdf`.

How to do it...

Before we investigate the GLSL shader code, let's implement all the necessary C++ code to process data arrays on the GPU.

To manipulate data buffers on the GPU and utilize the data effectively, we require four basic operations: loading a shader module, creating a compute pipeline, creating a buffer, and dispatching compute commands. After that, we need to transfer the data from the GPU buffer to the host memory and save it as a texture file. Let's walk through these steps by examining the code in `Chapter06/02_BRDF_LUT/src/main.cpp`.

1. The function `calculateLUT()` implements most of the described functionality. We start with the shader module loading and compute pipeline creation. The GLSL shader is specialized using the constant `kNumSamples` which defines the number of Monte Carlo trials for our LUT calculation. We store 16-bit float RGBA values in the buffer.

```
const uint32_t kBrdfW      = 256;
const uint32_t kBrdfH      = 256;
const uint32_t kNumSamples = 1024;
const uint32_t kBufferSize = 4u * sizeof(uint16_t) * kBrdfW * kBrdfH;
  void calculateLUT(const std::unique_ptr<lvk::IContext>& ctx,
  void* output, uint32_t size)
{
  Holder<ShaderModuleHandle> comp = loadShaderModule(
    ctx, "Chapter06/02_BRDF_LUT/src/main.comp");
```

```
Holder<ComputePipelineHandle> computePipelineHandle =
  ctx->createComputePipeline({
    .smComp = comp,
    .specInfo = {.entries = {{ .constantId = 0,
                               .size = sizeof(kNumSamples) }},
                 .data     = &kNumSamples,
                 .dataSize = sizeof(kNumSamples),},
});
```

2. The next step is to create a GPU storage buffer for our output data:

```
Holder<BufferHandle> dstBuffer = ctx->createBuffer({
  .usage     = lvk::BufferUsageBits_Storage,
  .storage   = lvk::StorageType_HostVisible,
  .size      = size,
  .debugName = "Compute: BRDF LUT" });
```

3. And finally, we acquire a command buffer, update push constants, and dispatch the compute commands.

```
lvk::ICommandBuffer& buf = ctx->acquireCommandBuffer();
buf.cmdBindComputePipeline(computePipelineHandle);
struct {
  uint32_t w = kBrdfW;
  uint32_t h = kBrdfH;
  uint64_t addr;
} pc {
  .addr = ctx->gpuAddress(dstBuffer),
};
buf.cmdPushConstants(pc);
buf.cmdDispatchThreadGroups({ kBrdfW / 16, kBrdfH / 16, 1 });
```

4. Before reading the generated data back to the CPU memory, we must wait for the GPU to finish processing the buffer. It can be done using the `wait()` function which waits for a command buffer to finish. We discussed it in the recipe *Chapter 2*. Once the GPU has finished working, we can copy the memory-mapped buffer back to the CPU memory referenced by the pointer `output`.

```
  ctx->wait(ctx->submit(buf));
  memcpy(output, ctx->getMappedPtr(dstBuffer), kBufferSize);
}
```

That was the C++ part. Now let's investigate the GLSL shader compute code `Chapter06/02_BRDF_LUT/src/main.comp`:

1. To break down our work into smaller pieces, we start from the shader preamble and the `main()` function of the BRDF LUT calculation shader. The preamble code sets the compute shader dispatching parameters. In our case, a 16x16 chunk of the LUT texture is calculated by one GPU work group. The number of Monte Carlo trials for numeric integration is declared as a specialization constant we can override from the C++ code:

```
layout (local_size_x=16, local_size_y=16, local_size_z=1) in;
layout (constant_id = 0) const uint NUM_SAMPLES = 1024u;
layout (std430, buffer_reference) readonly buffer Data {
  float16_t floats[];
};
```

2. We use user-provided width and height to calculate our output buffer dimensions. Last but not least, `PI` is the global "physical" constant we use in the shader:

```
layout (push_constant) uniform constants {
  uint BRDF_W;
  uint BRDF_H;
  Data data;
};
const float PI = 3.1415926536;
```

3. The `main()` function wraps the `BRDF()` function call and stores the results. First, we recalculate the worker ID to output array indices:

```
void main() {
  vec2 uv;
  uv.x = (float(gl_GlobalInvocationID.x) + 0.5) / float(BRDF_W);
  uv.y = (float(gl_GlobalInvocationID.y) + 0.5) / float(BRDF_H);
```

4. The `BRDF()` function does all the actual work. The calculated value is put into the output array:

```
  vec3 v = BRDF(uv.x, 1.0 - uv.y);
  uint offset = gl_GlobalInvocationID.y * BRDF_W +
                gl_GlobalInvocationID.x;
  data.floats[offset * 4 + 0] = float16_t(v.x);
  data.floats[offset * 4 + 1] = float16_t(v.y);
  data.floats[offset * 4 + 2] = float16_t(v.z);
}
```

5. As you can see, we use *3* channels of the texture. The *R* and *G* channels are used for GGX BRDF LUT, the third channel is used for Charlie BRDF LUT which is required for the Sheen material extension and will be covered in the next *Chapter 7*.

Now that we have described the scaffolding parts of our compute shader, we can see how the BRDF LUT values are calculated. Let's look at the steps:

1. To generate random directions in a hemisphere, we use so-called Hammersley points calculated by the following function:

```
vec2 hammersley2d(uint i, uint N) {
  uint bits = (i << 16u) | (i >> 16u);
  bits = ((bits & 0x55555555u)<<1u)|((bits & 0xAAAAAAAAu)>>1u);
  bits = ((bits & 0x33333333u)<<2u)|((bits & 0xCCCCCCCCu)>>2u);
  bits = ((bits & 0x0F0F0F0Fu)<<4u)|((bits & 0xF0F0F0F0u)>>4u);
  bits = ((bits & 0x00FF00FFu)<<8u)|((bits & 0xFF00FF00u)>>8u);
  float rdi = float(bits) * 2.3283064365386963e-10;
  return vec2(float(i) / float(N), rdi);
}
```

Important note

The code is based on the following post: `http://holger.dammertz.org:80/stuff/notes_HammersleyOnHemisphere.html`. The bit-shifting magic for this and many other applications is thoroughly described in Henry J. Warren's book called *Hacker's Delight*. Interested readers may also look up the "van der Corput sequence" to see why this can be used as a series of random directions on a hemisphere.

2. We also need some kind of a pseudorandom number generator. We use the output array indices as an input and pass them through another magic set of formulas:

```
float random(vec2 co) {
  float a  = 12.9898;
  float b  = 78.233;
  float c  = 43758.5453;
  float dt = dot( co.xy ,vec2(a,b) );
  float sn = mod(dt, 3.14);
  return fract(sin(sn) * c);
}
```

Check out this link to find some useful details about this code: `http://byteblacksmith.com/improvements-to-the-canonical-one-liner-glsl-rand-for-opengl-es-2-0`.

3. Let's look at how importance sampling is implemented according to the paper *Real Shading in Unreal Engine 4* by Brian Karis. Check out the fourth page of `https://cdn2.unrealengine.com/Resources/files/2013SiggraphPresentationsNotes-26915738.pdf`. This function maps an *i*-th 2D point *Xi* to a hemisphere with spread based on surface roughness.:

```
vec3 importanceSample_GGX(vec2 Xi, float roughness, vec3 normal)
```

```
{
  float alpha = roughness * roughness;
  float phi = 2.0 * PI * Xi.x + random(normal.xz) * 0.1;
  float cosTheta  = sqrt((1.0 - Xi.y) / (1.0 + (alpha * alpha - 1.0) * Xi.y));
  float sinTheta = sqrt(1.0 - cosTheta * cosTheta);
  vec3 H = vec3(sinTheta * cos(phi), sinTheta * sin(phi), cosTheta);
```

4. Calculations are done in tangent space defined by the vectors `up`, `tangentX`, and `tangentY` and then converted to world space.

```
  vec3 up = abs(normal.z) < 0.999 ?
    vec3(0.0, 0.0, 1.0) : vec3(1.0, 0.0, 0.0);
  vec3 tangentX = normalize(cross(up, normal));
  vec3 tangentY = normalize(cross(normal, tangentX));
  return normalize(tangentX * H.x +
                   tangentY * H.y +
                     normal * H.z);
}
```

5. Another utility function, `G_SchlicksmithGGX()`, calculates the GGX geometric shadowing factor:

```
float G_SchlicksmithGGX(
  float dotNL, float dotNV, float roughness)
{
  float k  = (roughness * roughness) / 2.0;
  float GL = dotNL / (dotNL * (1.0 - k) + k);
  float GV = dotNV / (dotNV * (1.0 - k) + k);
  return GL * GV;
}
```

6. We also precalculate LUT for Sheen material, so there are two more helper functions `V_Ashikhmin()` and `D_Charlie()`. They are based on the code from the Filament engine `https://github.com/google/filament/blob/master/shaders/src/brdf.fs#L136`.

```
float V_Ashikhmin(float NdotL, float NdotV) {
  return clamp(
    1.0 / (4.0 * (NdotL + NdotV - NdotL * NdotV)), 0.0, 1.0);
}
float D_Charlie(float sheenRoughness, float NdotH) {
  sheenRoughness = max(sheenRoughness, 0.000001);
  float invR = 1.0 / sheenRoughness;
  float cos2h = NdotH * NdotH;
  float sin2h = 1.0 - cos2h;
  return (2.0 + invR) * pow(sin2h, invR * 0.5) / (2.0 * PI);
}
```

7. Here is a corresponding sampling function `importanceSample_Charlie()` for the Sheen material which is very similar to `importanceSample_GGX()`.

```
vec3 importanceSample_Charlie(
  vec2 Xi, float roughness, vec3 normal)
{
  float alpha = roughness * roughness;
  float phi = 2.0 * PI * Xi.x;
  float sinTheta = pow(Xi.y, alpha / (2.0 *   alpha + 1.0));
  float cosTheta = sqrt(1.0 - sinTheta * sinTheta);
  vec3 H  = vec3(sinTheta * cos(phi), sinTheta * sin(phi), cosTheta);
  vec3 up = abs(normal.z) < 0.999 ?
    vec3(0.0, 0.0, 1.0) : vec3(1.0, 0.0, 0.0);
  vec3 tangentX = normalize(cross(up, normal));
  vec3 tangentY = normalize(cross(normal, tangentX));
  return normalize(tangentX * H.x +
                   tangentY * H.y +
                     normal * H.z);
}
```

8. The value of BRDF is calculated the following way, using all of the helper functions we declared above. The number of Monte Carlo trials `NUM_SAMPLES` is set earlier to be `1024`. The normal vector *N* always points along the Z-axis for the 2D lookup.

```
vec3 BRDF(float NoV, float roughness) {
  const vec3 N = vec3(0.0, 0.0, 1.0);
  vec3 V   = vec3(sqrt(1.0 - NoV*  NoV), 0.0, NoV);
  vec3 LUT = vec3(0.0);
```

9. The first loop calculates the *R* and *G* components of our LUT which correspond to the scale and bias to *F0*.

```
  for (uint i = 0u; i < NUM_SAMPLES; i++) {
    vec2 Xi = hammersley2d(i, NUM_SAMPLES);
    vec3 H = importanceSample_GGX(Xi, roughness, N);
    vec3 L = 2.0 * dot(V, H) * H - V;
    float dotNL = max(dot(N, L), 0.0);
    float dotNV = max(dot(N, V), 0.0);
    float dotVH = max(dot(V, H), 0.0);
    float dotNH = max(dot(H, N), 0.0);
    if (dotNL > 0.0) {
      float G = G_SchlicksmithGGX(dotNL, dotNV, roughness);
      float G_Vis = (G * dotVH) / (dotNH * dotNV);
      float Fc = pow(1.0 - dotVH, 5.0);
```

```
      LUT.rg += vec2((1.0 - Fc) * G_Vis, Fc * G_Vis);
    }
  }
```

10. The third component `B` used for the sheen material is calculated in another loop. We will revisit it in the next *Chapter 6*.

```
  for (uint i = 0u; i < NUM_SAMPLES; i++) {
    vec2 Xi = hammersley2d(i, NUM_SAMPLES);
    vec3 H = importanceSample_Charlie(Xi, roughness, N);
    vec3 L = 2.0 * dot(V, H) * H - V;
    float dotNL = max(dot(N, L), 0.0);
    float dotNV = max(dot(N, V), 0.0);
    float dotVH = max(dot(V, H), 0.0);
    float dotNH = max(dot(H, N), 0.0);
    if (dotNL > 0.0) {
      float sheenDistribution = D_Charlie(roughness, dotNH);
      float sheenVisibility   = V_Ashikhmin(dotNL, dotNV);
      LUT.b += sheenVisibility * sheenDistribution * dotNL * dotVH;
    }
  }
  return LUT / float(NUM_SAMPLES);
}
```

That is the entire GLSL compute shader used to precalculate the look-up table. Let's now see how it works from the C++ `main()` function.

How it works...

The `main()` function creates a KTX texture using the *KTX-Software* library so that our 16-bit RGBA LUT texture can be saved into the `.ktx` format preserving the data. Then it calls the function `calculateLUT()` we discussed above which outputs the generated LUT data into the KTX texture. The texture is saved into `data/brdfLUT.ktx`.

```
int main() {
  std::unique_ptr<lvk::IContext> ctx =
    lvk::createVulkanContextWithSwapchain(nullptr, 0, 0, {});
  ktxTextureCreateInfo createInfo = {
    .glInternalformat = GL_RGBA16F,
    .vkFormat         = VK_FORMAT_R16G16B16A16_SFLOAT,
    .baseWidth        = kBrdfW,
    .baseHeight       = kBrdfH,
    .baseDepth        = 1,
    .numDimensions    = 2,
    .numLevels        = 1,
```

```
    .numLayers        = 1,
    .numFaces         = 1,
    .generateMipmaps  = KTX_FALSE,
  };
  ktxTexture1* lutTexture = nullptr;
  ktxTexture1_Create(
    &createInfo, KTX_TEXTURE_CREATE_ALLOC_STORAGE, &lutTexture);
  calculateLUT(ctx, lutTexture->pData, kBufferSize);
  ktxTexture_WriteToNamedFile(ktxTexture(lutTexture), "data/brdfLUT.ktx");
  ktxTexture_Destroy(ktxTexture(lutTexture));
  return 0;
}
```

You can use *Pico Pixel* `https://pixelandpolygon.com` to view the generated image. It should resemble the screenshot below. The horizontal axis represents the dot product between the surface normal vector and the viewing direction, while the vertical axis represents the surface roughness values `0...1`. Each texel holds three 16-bit floating point values. The first two values represent the scale and bias to `F0`, which is the specular reflectance at normal incidence. The third value is utilized for the sheen material extension, which will be covered in the next chapter:

Figure 6.5: BRDF lookup table

This concludes the BRDF lookup table tool description. We will need yet another tool to calculate an irradiance cube map from an environment cube map which we will cover in the next recipe.

There's more...

The method described above can be used to precompute BRDF look-up tables using high -quality Monte Carlo integration and store them as textures. Dependent texture fetches can be expensive on some mobile platforms. There is an interesting run-time approximation used in Unreal Engine, which does not rely on any precomputation, as described in this post *Physically Based Shading on Mobile* by Brian Karis `https://www.unrealengine.com/en-US/blog/physically-based-shading-on-mobile`. Here is the GLSL source code:

```
vec3 EnvBRDFApprox(vec3 specularColor, float roughness, float NoV) {
  const vec4 c0 = vec4(-1, -0.0275, -0.572, 0.022);
  const vec4 c1 = vec4( 1,  0.0425,  1.04, -0.04 );
  vec4 r = roughness * c0 + c1;
  float a004 = min(r.x * r.x, exp2(-9.28 * NoV)) * r.x + r.y;
  vec2 AB = vec2(-1.04, 1.04) * a004 + r.zw;
  return specularColor * AB.x + AB.y;
}
```

Precomputing irradiance maps and diffuse convolution

As we discussed earlier in the recipe *Introduction to glTF 2.0 Physically Based Shading Model*, the second part of the split sum approximation necessary to calculate the glTF 2.0 physically-based shading model comes from the irradiance cube map which is precalculated by convolving the input environment cube map with the GGX distribution of our shading model. Our implementation is based on `https://github.com/KhronosGroup/glTF-Sample-Viewer/blob/main/source/shaders/ibl_filtering.frag`.

Image based lighting (**IBL**) is a technique for illuminating a scene using captured light information. This information can be stored as panoramic photos images. It is very hard to simulate entire real world environments, so capturing the real world and using images is a very common technique nowadays to produce realistic renders. Using **IBL** allows us to precompute the part of diffuse and specular BRDF equations and make them more runtime friendly.

If you want to dive deep into the math theory behind these computations, make sure you read Brian Karis's paper *Real Shading in Unreal Engine 4* `https://cdn2.unrealengine.com/Resources/files/2013SiggraphPresentationsNotes-26915738.pdf`.

Getting ready

Check out the source code for this recipe in `Chapter06/03_FilterEnvmap`.

How to do it...

We are doing Monte-Carlo integration inside a fragment shader which is located here: `Chapter06/03_FilterEnvmap/src/main.frag`.

The C++ source code can be found in the `Chapter06/03_FilterEnvmap/src/main.cpp` file. Let's go through the function `prefilterCubemap()` to precompute the irradiance and diffuse maps:

1. First, we need to create a cube map texture to store the results of prefiltering. We use the 32-bit RGBA floating point pixel format because most of our color-related calculations happen in linear space. Low-end mobile devices might be not performant enough and, in this case, the dynamic range can be clamped to 16-bit or even 8-bit, but that might significantly impact the visual fidelity. We use cubemap mip-levels to precompute multiple lookups for different values of material roughness `0..1`. The function takes in the `distribution` parameter which is passed to the shader to select an appropriated distribution: Lambertian, GGX, or Charlie.

```
void prefilterCubemap(
  const std::unique_ptr<lvk::IContext>& ctx,
  ktxTexture1* cube, const char* envPrefilteredCubemap,
  TextureHandle envMapCube,
  Distribution distribution,
  uint32_t sampler,
  uint32_t sampleCount)
{
  Holder<TextureHandle> prefilteredMapCube =
    ctx->createTexture({
        .type         = lvk::TextureType_Cube,
        .format       = lvk::Format_RGBA_F32,
        .dimensions   = {cube->baseWidth, cube->baseHeight, 1},
        .usage        = lvk::TextureUsageBits_Sampled |
                        lvk::TextureUsageBits_Attachment,
        .numMipLevels = (uint32_t)cube->numLevels,
        .debugName    = envPrefilteredCubemap,
      }, envPrefilteredCubemap);
```

2. We require GLSL shader modules and a rendering pipeline.

```
  Holder<ShaderModuleHandle> vert = loadShaderModule(
    ctx, "Chapter06/03_FilterEnvmap/src/main.vert");
  Holder<ShaderModuleHandle> frag = loadShaderModule(
    ctx, "Chapter06/03_FilterEnvmap/src/main.frag");
  Holder<RenderPipelineHandle> pipelineSolid =
    ctx->createRenderPipeline({
      .smVert    = vert,
      .smFrag    = frag,
      .color     = { { .format = ctx->getFormat(prefilteredMapCube) } },
      .cullMode  = lvk::CullMode_Back,
  });
```

3. Now we can start doing the actual rendering into the cube map. One command buffer is filled with all the commands necessary to render into 6 cubemap faces with all required mip-levels.

```
lvk::ICommandBuffer& buf = ctx->acquireCommandBuffer();
for (uint32_t mip = 0; mip < cube->numLevels; mip++) {
  for (uint32_t face = 0; face < 6; face++) {
```

4. We set the cube map face we want to render to and the specific mip-level we want to render.

```
    buf.cmdBeginRendering(
      { .color = { { .loadOp     = lvk::LoadOp_Clear,
                     .layer      = (uint8_t)face,
                     .level      = (uint8_t)mip,
                     .clearColor = { 1.0f, 1.0f, 1.0f, 1.0f },
                 } } },
      { .color = {{ .texture = prefilteredMapCube }} });
    buf.cmdBindRenderPipeline(pipelineSolid);
    buf.cmdBindDepthState({});
```

5. Push constants are used to pass all the data into the shaders.

```
    struct PerFrameData {
      uint32_t face;
      float roughness;
      uint32_t sampleCount;
      uint32_t width;
      uint32_t envMap;
      uint32_t distribution;
      uint32_t sampler;
    } perFrameData = {
      .face         = face,
      .roughness    = (float)(mip) / (cube->numLevels - 1),
      .sampleCount  = sampleCount,
      .width        = cube->baseWidth,
      .envMap       = envMapCube.index(),
      .distribution = uint32_t(distribution),
      .sampler      = sampler,
    };
    buf.cmdPushConstants(perFrameData);
```

6. Then, a full screen triangle is rendered to cover the entire cube face and let the fragment shader do its work. After the command buffer is filled up, we can submit it.

```
          buf.cmdDraw(3);
          buf.cmdEndRendering();
        }
      }
      ctx->submit(buf);
      // ... save results to a .ktx file
    }
```

The remaining part of the `prefilterCubemap()` function retrieves the generated cubemap data from the GPU and saves it into a `.ktx` file. Let's look into the GLSL fragment shader code which does all the heavy lifting in `Chapter06/03_FilterEnvmap/src/main.frag`.

1. To unwind the shader logic, let's start with the entry point `main()`. The code is trivial and invokes two functions. The function `uvToXYZ()` converts a cubemap face index and `vec2` coordinates into `vec3` cubemap sampling direction. The function `filterColor()` does the actual Monte Carlo sampling and we will come back to it in a moment.

```
void main() {
  vec2 newUV = uv * 2.0 - vec2(1.0);
  vec3 scan = uvToXYZ(perFrameData.face, newUV);
  vec3 direction = normalize(scan);
  out_FragColor = vec4(filterColor(direction), 1.0);
}
```

2. Here's the code of `uvToXYZ()` for your reference.

```
vec3 uvToXYZ(uint face, vec2 uv) {
  if (face == 0) return vec3(   1., uv.y,  uv.x);
  if (face == 1) return vec3(  -1., uv.y, -uv.x);
  if (face == 2) return vec3(+uv.x,   1.,  uv.y);
  if (face == 3) return vec3(+uv.x,  -1., -uv.y);
  if (face == 4) return vec3(+uv.x, uv.y,   -1.);
  if (face == 5) return vec3(-uv.x, uv.y,    1.);
}
```

The function `filterColor()` does the integration part for irradiance and Lambertian diffuse convolution. The argument `N` is the cubemap sampling direction vector. We iterate `sampleCount` samples and get the importance sampling information that includes the importance sample direction and **probability distribution function** (**PDF**) in this direction. The mathematical part is described in detail in this blog post: `https://bruop.github.io/ibl`. Here we focus on putting a minimalistic working implementation together.

```
vec3 filterColor(vec3 N) {
  vec3  color  = vec3(0.f);
  float weight = 0.0f;
  for(uint i = 0; i < perFrameData.sampleCount; i++) {
    vec4 importanceSample = getImportanceSample(i, N, perFrameData.roughness);
    vec3 H = vec3(importanceSample.xyz);
    float pdf = importanceSample.w;
```

3. Mipmap samples are filtered as described in *GPU Gems 3* §20.4. Sample Lambertian at a lower resolution to avoid too bright or "fireflies" pixels.

```
    float lod = computeLod(pdf);
    if (perFrameData.distribution == cLambertian) {
      vec3 lambertian = textureBindlessCubeLod(
        perFrameData.envMap,
        perFrameData.samplerIdx, H, lod).xyz;
      color += lambertian;
    } else if (perFrameData.distribution == cGGX ||
               perFrameData.distribution == cCharlie) {
      vec3 V = N;
      vec3 L = normalize(reflect(-V, H));
      float NdotL = dot(N, L);
      if (NdotL > 0.0) {
        if (perFrameData.roughness == 0.0) lod = 0.0;
        vec3 sampleColor = textureBindlessCubeLod(
          perFrameData.envMap,
          perFrameData.samplerIdx, L, lod).xyz;
        color  += sampleColor * NdotL;
        weight += NdotL;
      }
    }
  }
```

4. The output color value is renormalized using the sum of all `NdotL` weights, or the number of samples for Lambertian case:

```
  color /= (weight != 0.0f) ?
    weight : float(perFrameData.sampleCount);
  return color.rgb;
}
```

The importance sampling function `getImportanceSample()` returns a `vec4` value with an importance sample direction in the `.xyz` components and the **PDF** scalar value in the `.w` component. We generate a Hammersley point as we described earlier in the previous recipe *Precomputing BRDF look-up tables* and then, based on the distribution type (Lambertian, GGX, or Charlie) we generate a sample and rotate it to the normal direction. This function uses a helper structure `MicrofacetDistributionSample`.

```
struct MicrofacetDistributionSample {
  float pdf;
  float cosTheta;
  float sinTheta;
  float phi;
};
vec4 getImportanceSample(
  uint sampleIndex, vec3 N, float roughness)
{
  vec2 xi = hammersley2d(sampleIndex, perFrameData.sampleCount);
  MicrofacetDistributionSample importanceSample;
```

5. Generate points on a hemisphere with a mapping corresponding to the desired distribution. For example, Lambertian distribution uses a cosine importance.

```
  if (perFrameData.distribution == cLambertian)
    importanceSample = Lambertian(xi, roughness);
  else if (perFrameData.distribution == cGGX)
    importanceSample = GGX(xi, roughness);
  else if (perFrameData.distribution == cCharlie)
    importanceSample = Charlie(xi, roughness);
```

6. Transform the hemisphere sample point into the tangent coordinate frame. The helper function `generateTBN()` generates a tangent-bitangent-normal coordinate frame from the provided normal vector.

```
  vec3 localSpaceDirection = normalize(vec3(
    importanceSample.sinTheta * cos(importanceSample.phi),
    importanceSample.sinTheta * sin(importanceSample.phi),
    importanceSample.cosTheta));
  mat3 TBN = generateTBN(N);
  vec3 direction = TBN * localSpaceDirection;
  return vec4(direction, importanceSample.pdf);
}
```

7. We skip the details of individual distribution calculation functions `Lambertian()`, `GGX()`, and `Charlie()`. The actual GLSL shader `Chapter06/03_FilterEnvmap/src/main.frag` contains all the necessary code.

The process of importance sampling can introduce visual artifacts. One way to improve visual quality without compromising performance is by utilizing hardware-accelerated mip-mapping for swift filtering and sampling. This idea was proposed in the paper "Real-time Shading with Filtered Importance Sampling" by Mark Colbert and Jaroslav Křivánek `https://cgg.mff.cuni.cz/~jaroslav/papers/2007-sketch-fis/Final_sap_0073.pdf`. This link has a more detailed treatment of the subject: `https://developer.nvidia.com/gpugems/gpugems3/part-iii-rendering/chapter-20-gpu-based-importance-sampling`. Here we use a formula which takes a **PDF** value and calculates a proper mip-map LOD level for it:

```
float computeLod(float pdf) {
  float w = float(perFrameData.width);
  float h = float(perFrameData.height);
  float sampleCount = float(perFrameData.sampleCount);
  return 0.5 * log2( 6.0 * w * h / (sampleCount * pdf));
}
```

The rest of the code involves purely mechanical tasks, such as loading the cube map image from a file, calling the rendering functions for various distribution types (Lambertian, GGX, Charlie), and saving the result using the KTX library. Let's check out the results of prefiltering for the following input image:

Figure 6.6: Environment cube map

The convolved image should look as in the following screenshot:

Figure 6.7: Prefiltered environment cube map using diffuse convolution

Now we have all supplementary parts in place to render a PBR image. In the next recipe *Implementing the glTF 2.0 metallic-roughness shading model*, we are going to put everything together into a simple application to render a physically based glTF 2.0 3D model.

There's more...

Paul Bourke created a set of tools and a great resource explaining how to convert cube maps between different formats. Make sure to check it out: `http://paulbourke.net/panorama/cubemaps/index.html`.

Implementing the glTF 2.0 metallic-roughness shading model

This recipe will cover how to integrate physically based rendering (PBR) into your graphics pipeline. Since the topic of PBR rendering is vast, we focus on the minimalistic implementation just to guide you and get you started. In the book text right here, we focus on the metallic-roughness shading model and minimalistic C++ viewer implementation. In the following chapters we will create a more complex and feature-rich glTF viewer including advanced material extensions and geometry features.

Getting ready

It is recommended to revisit the recipe *Introduction to glTF 2.0 Physically Based Material system* before you proceed with this one. A lightweight introduction to the glTF 2.0 shading model can be found at `https://github.com/KhronosGroup/glTF-Sample-Viewer/tree/glTF-WebGL-PBR`.

The C++ source code for this recipe is in the `Chapter06/04_MetallicRoughness` folder. The GLSL shader code responsible for PBR calculations can be found in `Chapter06/04_MetallicRoughness/src/PBR.sp`.

How to do it...

Before we dive deep into the GLSL code, we'll look at how the input data is set up from the C++ side. We are going to use the *Damaged Helmet* 3D model provided by Khronos. You can find the glTF file here `deps/src/glTF-Sample-Models/2.0/DamagedHelmet/glTF/DamagedHelmet.gltf`.

Let's start with structures and helper functions first:

1. The helper struct `GLTFGlobalSamplers` contains three samplers necessary to access glTF IBL textures. It is declared in `shared/UtilsGLTF.h`.

```
struct GLTFGlobalSamplers {
  explicit GLTFGlobalSamplers(const std::unique_ptr<lvk::IContext>& ctx);
  Holder<SamplerHandle> clamp;
  Holder<SamplerHandle> wrap;
  Holder<SamplerHandle> mirror;
};
```

2. The `GLTFGlobalSamplers` constructor creates all 3 samplers the following way:

```
GLTFGlobalSamplers(const std::unique_ptr<lvk::IContext>& ctx) {
  clamp = ctx->createSampler({
    .minFilter = lvk::SamplerFilter::SamplerFilter_Linear,
    .magFilter = lvk::SamplerFilter::SamplerFilter_Linear,
    .mipMap    = lvk::SamplerMip::SamplerMip_Linear,
    .wrapU     = lvk::SamplerWrap::SamplerWrap_Clamp,
    .wrapV     = lvk::SamplerWrap::SamplerWrap_Clamp,
    .wrapW     = lvk::SamplerWrap::SamplerWrap_Clamp,
    .debugName = "Clamp Sampler" });
  wrap = ctx->createSampler({
    .minFilter = lvk::SamplerFilter::SamplerFilter_Linear,
    .magFilter = lvk::SamplerFilter::SamplerFilter_Linear,
    .mipMap    = lvk::SamplerMip::SamplerMip_Linear,
    .wrapU     = lvk::SamplerWrap::SamplerWrap_Repeat,
    .wrapV     = lvk::SamplerWrap::SamplerWrap_Repeat,
    .wrapW     = lvk::SamplerWrap::SamplerWrap_Repeat,
    .debugName = "Wrap Sampler" });
```

```
  mirror = ctx->createSampler({
    .minFilter = lvk::SamplerFilter::SamplerFilter_Linear,
    .magFilter = lvk::SamplerFilter::SamplerFilter_Linear,
    .mipMap    = lvk::SamplerMip::SamplerMip_Linear,
    .wrapU     = lvk::SamplerWrap::SamplerWrap_MirrorRepeat,
    .wrapV     = lvk::SamplerWrap::SamplerWrap_MirrorRepeat,
    .debugName = "Mirror Sampler" });
  }
```

3. The helper struct `EnvironmentMapTextures` stores all IBL environment map textures and the BRDF look up table, and provides default textures for the sake of simplicity.

```
struct EnvironmentMapTextures {
  Holder<TextureHandle> texBRDF_LUT;
  Holder<TextureHandle> envMapTexture;
  Holder<TextureHandle> envMapTextureCharlie;
  Holder<TextureHandle> envMapTextureIrradiance;
```

4. Check the previous recipe *Precomputing irradiance maps and diffuse convolution* for details of how to precalculate the IBL textures. The BRDF look up table was precalculated in the recipe *Precomputing BRDF look-up tables.*

```
  explicit EnvironmentMapTextures(
    const std::unique_ptr<lvk::IContext>& ctx)
  : EnvironmentMapTextures(ctx,
    "data/brdfLUT.ktx",
    "data/piazza_bologni_1k_prefilter.ktx",
    "data/piazza_bologni_1k_irradiance.ktx",
    "data/piazza_bologni_1k_charlie.ktx") {}
  EnvironmentMapTextures(
    const std::unique_ptr<lvk::IContext>& ctx,
    const char* brdfLUT,
    const char* prefilter,
    const char* irradiance,
    const char* prefilterCharlie = nullptr)
  {
    texBRDF_LUT = loadTexture(ctx, brdfLUT, lvk::TextureType_2D);
    envMapTexture = loadTexture(
      ctx, prefilter, lvk::TextureType_Cube);
    envMapTextureIrradiance = loadTexture(
      ctx, irradiance, lvk::TextureType_Cube);
  }
};
```

5. The structure `GLTFMaterialTextures` contains all the textures necessary to render any glTF2 model we support in our demos. It is a container for many `Holder<TextureHandle>` objects as follows:

```
struct GLTFMaterialTextures {
  // MetallicRoughness / SpecularGlossiness
  Holder<TextureHandle> baseColorTexture;
  Holder<TextureHandle> surfacePropertiesTexture;
  // Common properties
  Holder<TextureHandle> normalTexture;
  Holder<TextureHandle> occlusionTexture;
  Holder<TextureHandle> emissiveTexture;
  // Sheen
  Holder<TextureHandle> sheenColorTexture;
  Holder<TextureHandle> sheenRoughnessTexture;
  // … many other textures skipped here
}
```

6. The helper function `loadMaterialTextures()` is not shared and will be different in each app. This variant of the function loads a subset of necessary textures for our metallic-roughness demo.

```
GLTFMaterialTextures loadMaterialTextures(
  const std::unique_ptr<lvk::IContext>& ctx,
  const char* texAOFile,
  const char* texEmissiveFile,
  const char* texAlbedoFile,
  const char* texMeRFile,
  const char* texNormalFile)
{
  glTFMaterialTextures mat;
  mat.baseColorTexture = loadTexture(
    ctx, texAlbedoFile, lvk::TextureType_2D, true);
  if (mat.baseColorTexture.empty()) return {};
  mat.occlusionTexture = loadTexture(ctx, texAOFile);
  if (mat.occlusionTexture.empty()) return {};
  mat.normalTexture = loadTexture(ctx, texNormalFile);
  if (mat.normalTexture.empty()) return {};
  mat.emissiveTexture = loadTexture(
    ctx, texEmissiveFile, lvk::TextureType_2D, true);
  if (mat.emissiveTexture.empty()) return {};
  mat.surfacePropertiesTexture = loadTexture(ctx, texMeRFile);
  if (mat.surfacePropertiesTexture.empty()) return {};
```

```
  mat.wasLoaded = true;
  return mat;
}}
```

7. One important step is to load the material data and fill out the structure `MetallicRoughnessDataGPU`. We use the Assimp API to retrieve the material properties and fill the corresponding values out. The glTF specification requires well defined default values for non-optional and optional properties, so we fill them out in this snippet as well. For each texture, we read and set the data for a sampler state and a uv coordinates index:

```
struct MetallicRoughnessData {
  vec4 baseColorFactor = vec4(1.0f, 1.0f, 1.0f, 1.0f);
```

8. Here we pack `metallicFactor`, `roughnessFactor`, `normalScale`, `occlusionStrength` glTF properties into one `vec4` member field `metallicRoughnessNormalOcclusion`. We do this packing as a very basic optimization. GPU stores this data in vector registers and reading it will be more efficient if we pack all the parameters into a single `vec4` value. Another reason is avoiding any additional alignment requirements especially for `vec3` types. Similar packing is done for a `vec3` glTF property `emissiveFactor` and a `float alphaCutoff`, both of which are packed into a single `vec4` value.

```
  vec4 metallicRoughnessNormalOcclusion = vec4(1.0f, 1.0f, 1.0f, 1.0f);
  vec4 emissiveFactorAlphaCutoff = vec4(0.0f, 0.0f, 0.0f, 0.5f);
```

9. The other member fields hold texture and sampler ids for our bindless shaders. They have no default values other than `0`.

```
  uint32_t occlusionTexture        = 0;
  uint32_t occlusionTextureSampler = 0;
  uint32_t occlusionTextureUV      = 0;
  uint32_t emissiveTexture         = 0;
  uint32_t emissiveTextureSampler  = 0;
  uint32_t emissiveTextureUV       = 0;
  uint32_t baseColorTexture        = 0;
  uint32_t baseColorTextureSampler = 0;
  uint32_t baseColorTextureUV      = 0;
  uint32_t metallicRoughnessTexture        = 0;
  uint32_t metallicRoughnessTextureSampler = 0;
  uint32_t metallicRoughnessTextureUV      = 0;
  uint32_t normalTexture        = 0;
  uint32_t normalTextureSampler = 0;
  uint32_t normalTextureUV      = 0;
```

10. The `alphaMode` property defines how the alpha value is interpreted. The alpha value itself should be taken from the 4-th component of the base color for the metallic-roughness material model.

```
  uint32_t alphaMode = 0;
  enum AlphaMode {
    AlphaMode_Opaque = 0,
    AlphaMode_Mask   = 1,
    AlphaMode_Blend  = 2,
  };
};
```

11. The `MetallicRoughnessDataGPU` structure is filled out using a helper function `setupMetallicRoughnessData()`. The structure `GLTFMaterialTextures` was discussed above.

```
MetallicRoughnessDataGPU setupMetallicRoughnessData(
  const GLTFGlobalSamplers& samplers,
  const GLTFMaterialTextures& mat,
  const aiMaterial* mtlDescriptor)
{
  MetallicRoughnessDataGPU res = {
    .baseColorFactor              = vec4(1.0f, 1.0f, 1.0f, 1.0f),
    .metallicRoughnessNormalOcclusion = vec4(1.0f, 1.0f, 1.0f, 1.0f),
    .emissiveFactorAlphaCutoff    = vec4(0.0f, 0.0f, 0.0f, 0.5f),
    .occlusionTexture             = mat.occlusionTexture.index(),
    .emissiveTexture              = mat.emissiveTexture.index(),
    .baseColorTexture             = mat.baseColorTexture.index(),
    .metallicRoughnessTexture     = mat.surfacePropertiesTexture.index(),
    .normalTexture                = mat.normalTexture.index(),
  };
```

12. The rest of the function goes through reading various glTF material properties using the Assimp API. We paste just the beginning of its code here. All other material properties are loaded in a similarly repeating pattern.

```
  aiColor4D aiColor;
  if (mtlDescriptor->Get(AI_MATKEY_COLOR_DIFFUSE, aiColor) == AI_SUCCESS)
  {
    res.baseColorFactor = vec4(
      aiColor.r, aiColor.g, aiColor.b, aiColor.a);
  }
  assignUVandSampler(samplers,
    mtlDescriptor,
    aiTextureType_DIFFUSE,
    res.baseColorTextureUV,
    res.baseColorTextureSampler);
  // … many other glTF material properties are loaded here
```

13. The helper function `assignUVandSampler()` looks as follows.

```
bool assignUVandSampler(
  const GLTFGlobalSamplers& samplers,
  const aiMaterial* mtlDescriptor,
  aiTextureType textureType,
  uint32_t& uvIndex,
  uint32_t& textureSampler, int index)
{
  aiString path;
  aiTextureMapMode mapmode[3] = {
    aiTextureMapMode_Clamp,
    aiTextureMapMode_Clamp,
    aiTextureMapMode_Clamp };
  const bool res = mtlDescriptor->GetTexture(textureType, index,
    &path, 0, &uvIndex, 0, 0, mapmode) == AI_SUCCESS;
  switch (mapmode[0]) {
    case aiTextureMapMode_Clamp:
      textureSampler = samplers.clamp.index();
      break;
    case aiTextureMapMode_Wrap:
      textureSampler = samplers.wrap.index();
      break;
    case aiTextureMapMode_Mirror:
      textureSampler = samplers.mirror.index();
      break;
  }
  return res;
}
```

Now let's examine the `main()` function:

1. First, we load the glTF file using Assimp. We support only triangulated topology, hence the flag `aiProcess_Triangulate` is used to instruct Assimp to triangulate the mesh during the import.

```
const aiScene* scene = aiImportFile("deps/src/glTF-Sample-
  Assets/Models/DamagedHelmet/glTF/DamagedHelmet.gltf",
  aiProcess_Triangulate);
const aiMesh* mesh = scene->mMeshes[0];
const vec4 white = vec4(1.0f, 1.0f, 1.0f, 1.0f);
```

2. We populate the vertex data. The struct Vertex is shared across all glTF demos and is declared in `shared/UtilsGLTF.h`.

```
struct Vertex {
  vec3 position;
  vec3 normal;
  vec4 color;
  vec2 uv0;
  vec2 uv1;
};
std::vector<Vertex> vertices;
for (uint32_t i = 0; i != mesh->mNumVertices; i++) {
  const aiVector3D v   = mesh->mVertices[i];
  const aiVector3D n   = mesh->mNormals ?
    mesh->mNormals[i] : aiVector3D(0.0f, 1.0f, 0.0f);
  const aiColor4D  c   = mesh->mColors[0] ?
    mesh->mColors[0][i] : aiColor4D(1.0f, 1.0f, 1.0f, 1.0f);
```

3. A glTF model commonly uses two sets of UV texture coordinates. The first set `uv0` is used for the primary texture mapping, such as diffuse color, specular reflection, or normal mapping. These coordinates are typically used for surface details and color information. The second set `uv1` is commonly used for lightmaps or reflection maps. These maps often need separate texture coordinates to be mapped correctly onto the model, distinct from the primary texture coordinates. The glTF specification says that a viewer app should support at least 2 texture coordinate sets.

```
  const aiVector3D uv0 = mesh->mTextureCoords[0] ?
    mesh->mTextureCoords[0][i] : aiVector3D(0.0f, 0.0f, 0.0f);
  const aiVector3D uv1 = mesh->mTextureCoords[1] ?
    mesh->mTextureCoords[1][i] : aiVector3D(0.0f, 0.0f, 0.0f);
  vertices.push_back({ .position = vec3(v.x, v.y, v.z),
                       .normal   = vec3(n.x, n.y, n.z),
                       .color    = vec4(c.r, c.g, c.b, c.a),
                       .uv0      = vec2(uv0.x, 1.0f - uv0.y),
                       .uv1      = vec2(uv1.x, 1.0f - uv1.y) });
}
```

4. Let's set up indices which define our triangles and upload the resulting vertex and index data into corresponding buffers.

```
std::vector<uint32_t> indices;
for (unsigned int i = 0; i != mesh->mNumFaces; i++) {
  for (int j = 0; j != 3; j++)
    indices.push_back(mesh->mFaces[i].mIndices[j]);
```

```
}
Holder<BufferHandle> vertexBuffer = ctx->createBuffer({
  .usage     = lvk::BufferUsageBits_Vertex,
  .storage   = lvk::StorageType_Device,
  .size      = sizeof(Vertex) * vertices.size(),
  .data      = vertices.data(),
  .debugName = "Buffer: vertex" });
Holder<BufferHandle> indexBuffer = ctx->createBuffer({
  .usage     = lvk::BufferUsageBits_Index,
  .storage   = lvk::StorageType_Device,
  .size      = sizeof(uint32_t) * indices.size(),
  .data      = indices.data(),
  .debugName = "Buffer: index" });
```

5. The next step is to load all the material textures. We use the same combination of textures for most of our glTF demos, so we store them in a structure `GLTFMaterialTextures` declared in `shared/UtilsGLTF.h`.

```
std::unique_ptr<GLTFMaterialTextures> mat =
  loadMaterialTextures(ctx,
    "deps/src/glTF-Sample-Assets/Models/
      DamagedHelmet/glTF/Default_AO.jpg",
    "deps/src/glTF-Sample-Assets/Models/
      DamagedHelmet/glTF/Default_emissive.jpg",
    "deps/src/glTF-Sample-Assets/Models/
      DamagedHelmet/glTF/Default_albedo.jpg",
    "deps/src/glTF-Sample-Assets/Models/
      DamagedHelmet/glTF/Default_metalRoughness.jpg",
    "deps/src/glTF-Sample-Assets/Models/
      DamagedHelmet/glTF/Default_normal.jpg");
```

6. Before we continue with the graphics pipeline creation and rendering, we have to set up IBL samplers, textures, and BRDF look up tables. This data is shared between all our demos, so we introduced a couple of helper structs to do all this work for us. Here are the definitions within the `main()` function.

```
GLTFGlobalSamplers samplers(ctx);
EnvironmentMapTextures envMapTextures(ctx);
```

7. The next step is to create a render pipeline step for our glTF rendering. We must provide a vertex input description. Here's how to create one for our model.

```
  const lvk::VertexInput vdesc = {
    .attributes    = {
      { .location=0, .format=VertexFormat::Float3, .offset=0  },
```

```
    { .location=1, .format=VertexFormat::Float3, .offset=12 },
    { .location=2, .format=VertexFormat::Float4, .offset=24 },
    { .location=3, .format=VertexFormat::Float2, .offset=40 },
    { .location=4, .format=VertexFormat::Float2, .offset=48 }},
  .inputBindings = { { .stride = sizeof(Vertex) } },
};
```

8. A rendering pipeline should be created as follows. We will investigate the GLSL shaders in the *How it works...* section.

```
Holder<ShaderModuleHandle> vert = loadShaderModule(
  ctx, "Chapter06/04_MetallicRoughness/src/main.vert");
Holder<ShaderModuleHandle> frag = loadShaderModule(
  ctx, "Chapter06/04_MetallicRoughness/src/main.frag");
Holder<RenderPipelineHandle> pipelineSolid =
  ctx->createRenderPipeline({
    .vertexInput = vdesc,
    .smVert      = vert,
    .smFrag      = frag,
    .color       = { { .format = ctx->getSwapchainFormat() } },
    .depthFormat = app.getDepthFormat(),
    .cullMode    = lvk::CullMode_Back,
  });
```

9. We can call `setupMetallicRoughnessData()` to load all the material data from glTF and properly pack on the CPU side.

```
const aiMaterial* mtlDescriptor =
  scene->mMaterials[mesh->mMaterialIndex];
const MetallicRoughnessMaterialsPerFrame matPerFrame = {
  .materials = { setupMetallicRoughnessData(
                   samplers, mat, mtlDescriptor) },
};
```

10. We store the material data inside a dedicated Vulkan buffer and access it in GLSL shaders using a buffer device address. This address is passed into shaders through Vulkan push constants.

```
Holder<BufferHandle> matBuffer = ctx->createBuffer({
  .usage     = lvk::BufferUsageBits_Uniform,
  .storage   = lvk::StorageType_HostVisible,
  .size      = sizeof(matPerFrame),
  .data      = &matPerFrame,
  .debugName = "PerFrame materials" });
```

11. The same treatment applies to our environment textures. They should be packed for the GPU, too.

```
const EnvironmentsPerFrame envPerFrame = {
  .environments = { {
    .envMapTexture = envMapTextures.envMapTexture.index(),
    .envMapTextureSampler = samplers.clamp.index(),
    .envMapTextureIrradiance =
      envMapTextures.envMapTextureIrradiance.index(),
    .envMapTextureIrradianceSampler = samplers.clamp.index(),
    .lutBRDFTexture = envMapTextures.texBRDF_LUT.index(),
    .lutBRDFTextureSampler = samplers.clamp.index() } },
};
Holder<BufferHandle> envBuffer = ctx->createBuffer({
  .usage     = lvk::BufferUsageBits_Uniform,
  .storage   = lvk::StorageType_HostVisible,
  .size      = sizeof(envPerFrame),
  .data      = &envPerFrame,
  .debugName = "PerFrame materials" });
```

12. The maximum allowed push constant size is `128` bytes. In order to handle data exceeding this size, we set up a storage buffer.

```
struct PerDrawData {
  mat4 model;
  mat4 view;
  mat4 proj;
  vec4 cameraPos;
  uint32_t matId;
  uint32_t envId;
};
Holder<BufferHandle> perFrameBuffer = ctx->createBuffer({
  .usage     = lvk::BufferUsageBits_Storage,
  .storage   = lvk::StorageType_HostVisible,
  .size      = sizeof(PerFrameData),
  .debugName = "perFrameBuffer",
});
```

13. Everything else is just mesh rendering similar to how it was done in the previous chapters. Here is how draw commands to render a glTF mesh are generated. Framebuffer and render pass declarations are skipped for the sake of brevity.

```
const PerFrameData perFrameData = {
  .model     = m2 * m1,
  .view      = app.camera_.getViewMatrix(),
  .proj      = p,
  .cameraPos = vec4(app.camera_.getPosition(), 1.0f),
  .matId     = 0,
  .envId     = 0,
};
lvk::ICommandBuffer& buf = ctx->acquireCommandBuffer();
buf.cmdUpdateBuffer(perFrameBuffer, perFrameData);
buf.cmdBeginRendering(renderPass, framebuffer);
buf.cmdBindVertexBuffer(0, vertexBuffer, 0);
buf.cmdBindIndexBuffer(indexBuffer, lvk::IndexFormat_UI32);
buf.cmdBindRenderPipeline(pipelineSolid);
buf.cmdBindDepthState({ .compareOp = lvk::CompareOp_Less,
                        .isDepthWriteEnabled = true });
struct PerFrameData {
  uint64_t draw;
  uint64_t materials;
  uint64_t environments;
} perFrameData = {
  .draw         = ctx->gpuAddress(perFrameBuffer),
  .materials    = ctx->gpuAddress(matBuffer),
  .environments= ctx->gpuAddress(envBuffer),
};
buf.cmdPushConstants(perFrameData);
buf.cmdDrawIndexed(indices.size());
// …
```

Let's skip the rest of the C++ code, which contains trivial command buffer submission and other scaffolding, and check how GLSL shaders work.

How it works...

There are two GLSL shaders which are used to render our metallic-roughness PBR model, a vertex shader `Chapter06/04_MetallicRoughness/src/main.vert` and a fragment shader `Chapter06/04_MetallicRoughness/src/main.frag`, which include additional files for shared input declarations and our glTF PBR code GLSL library. The vertex shader uses programmable-vertex-pulling to read the vertex data from buffers. The one most important aspect of the vertex shader is that we define our own functions, such as `getModel()` or `getTexCoord()`, to hide the implementation details of vertex pulling. It allows us to be flexible when we want to change the structure of our input data. We use a similar approach for fragment shaders.

It is the fragment shader that does the actual work. Let's take a look:

1. First, we check our inputs. We use buffer references for materials and environments buffers which correspond to the C++ structures `MetallicRoughnessDataGPU` and `EnvironmentMapDataGPU`.

```
layout(std430, buffer_reference) buffer Materials;
layout(std430, buffer_reference) buffer Environments;
layout(std430, buffer_reference) buffer PerDrawData {
  mat4 model;
  mat4 view;
  mat4 proj;
  vec4 cameraPos;
  uint matId;
  uint envId;
};
```

2. We use four helper functions, `getMaterialId()`, `getMaterial()`, `getEnvironmentId()`, and `getEnvironment()` as shortcuts to access the buffer references provided in the push constants.

```
layout(push_constant) uniform PerFrameData {
  PerDrawData drawable;
  Materials materials;
  Environments environments;
} perFrame;
uint getMaterialId() {
  return perFrame.drawable.matId;
}
uint getEnvironmentId() {
  return perFrame.drawable.envId;
}
MetallicRoughnessDataGPU getMaterial(uint idx) {
  return perFrame.materials.material[idx];
}
EnvironmentMapDataGPU getEnvironment(uint idx) {
  return perFrame.environments.environment[idx];
}
```

3. Inside the file `Chapter06/04_MetallicRoughness/src/inputs.frag`, there's a bunch of helper functions – such as `sampleAO()`, `samplerEmissive()`, `sampleAlbedo()` and many others – to sample from various glTF PBR texture maps based on the material `mat`. All of them use bindless textures and samplers.

```
vec4 sampleAO(InputAttributes tc, MetallicRoughnessDataGPU mat) {
  return textureBindless2D(
    mat.occlusionTexture,
    mat.occlusionTextureSampler,
    tc.uv[mat.occlusionTextureUV]);
}
vec4 sampleEmissive(
  InputAttributes tc, MetallicRoughnessDataGPU mat) {
  return textureBindless2D(
      mat.emissiveTexture,
      mat.emissiveTextureSampler,
      tc.uv[mat.emissiveTextureUV]
    ) * vec4(mat.emissiveFactorAlphaCutoff.xyz, 1.0f);
}
vec4 sampleAlbedo(
  InputAttributes tc, MetallicRoughnessDataGPU mat) {
  return textureBindless2D(
    mat.baseColorTexture,
    mat.baseColorTextureSampler,
    tc.uv[mat.baseColorTextureUV]) * mat.baseColorFactor;
}
```

4. Within the fragment shader's `main()` function, we use these helper functions to sample the texture maps based on the material id value returned by `getMaterialId()`:

```
layout (location=0) in vec4 uv0uv1;
layout (location=1) in vec3 normal;
layout (location=2) in vec3 worldPos;
layout (location=3) in vec4 color;
layout (location=0) out vec4 out_FragColor;
void main() {
  InputAttributes tc;
  tc.uv[0] = uv0uv1.xy;
  tc.uv[1] = uv0uv1.zw;
  MetallicRoughnessDataGPU mat = getMaterial(getMaterialId());
  vec4 Kao = sampleAO(tc, mat);
  vec4 Ke  = sampleEmissive(tc, mat);
  vec4 Kd  = sampleAlbedo(tc, mat) * color;
  vec4 mrSample = sampleMetallicRoughness(tc, mat);
```

5. To calculate the proper normal mapping effect based on a provided normal map, we evaluate the normal vector per pixel. We do it in world space. The normal map is in tangent space. Hence, the function `perturbNormal()` calculates the tangent space per pixel using the derivatives of texture coordinates, which is implemented in `data/shaders/UtilsPBR.sp`, and transforms the perturbed normal to world space. Make sure you check it out. The last step here is to negate the normal for double-sided materials. We use the `gl_FrontFacing` intrinsic variable to do the check.

```
vec3 n = normalize(normal); // world-space normal
vec3 normalSample = sampleNormal(tc, getMaterialId()).xyz;
n = perturbNormal(n, worldPos, normalSample, getNormalUV(tc, mat));
if (!gl_FrontFacing) n *= -1.0f;
```

6. Now, we are ready to fill out the `PBRInfo` structure, which holds multiple inputs utilized later by the various functions in the PBR shading equation:

```
PBRInfo pbrInputs = calculatePBRInputsMetallicRoughness(
  Kd, n, perFrame.drawable.cameraPos.xyz, worldPos, mrSample);
```

7. The next step is to calculate the specular and diffuse color contributions from the IBL environment lighting. We can directly add `diffuse_color` and `specular_color` because our precalculated BRDF LUT already takes care of energy conservation:

```
vec3 specular_color = getIBLRadianceContributionGGX(pbrInputs, 1.0);
vec3 diffuse_color = getIBLRadianceLambertian(
  pbrInputs.NdotV, n, pbrInputs.perceptualRoughness,
  pbrInputs.diffuseColor, pbrInputs.reflectance0, 1.0);
vec3 color = specular_color + diffuse_color;
```

8. For this demo application, we use only one hardcoded directional light source `(0, 0, -5)`. Let's calculate its lighting contribution:

```
vec3 lightPos = vec3(0, 0, -5);
color += calculatePBRLightContribution(
  pbrInputs, normalize(lightPos - worldPos), vec3(1.0) );
```

9. Now we should multiply the color by the ambient occlusion factor. Use `1.0` in case there is no ambient occlusion texture available:

```
color = color * ( Kao.r < 0.01 ? 1.0 : Kao.r );
```

10. Finally, we apply the emissive color contribution. Before writing the framebuffer output, we convert the resulting color back into the sRGB color space using a hardcoded gamma value of `2.2`:

```
  color = pow( Ke.rgb + color, vec3(1.0/2.2) );
  out_FragColor = vec4(color, 1.0);
}
```

We mentioned a bunch of helper functions that use the `PBRInfo` structure, such as `getIBLRadianceContributionGGX()`, `getIBLRadianceLambertian()`, and `calculatePBRLightContribution()`. Let's look inside `Chapter06/04_MetallicRoughness/src/PBR.sp` to see how they work. Our implementation is based on the reference implementation of glTF 2.0 Sample Viewer from Khronos: `https://github.com/KhronosGroup/glTF-Sample-Viewer/tree/glTF-WebGL-PBR`.

1. First of all, here is the `PBRInfo` structure which holds various input parameters for our metallic-roughness glTF PBR shading model. The first values represent geometric properties of the surface at the current point.

```
struct PBRInfo {
  float NdotL; // cos angle between normal and light direction
  float NdotV; // cos angle between normal and view direction
  float NdotH; // cos angle between normal and half vector
  float LdotH; // cos angle between light dir and half vector
  float VdotH; // cos angle between view dir and half vector
  vec3 n;      // normal at surface point
  vec3 v;      // vector from surface point to camera
```

2. The following values represent material properties.

```
  float perceptualRoughness; // roughness value (input to shader)
  vec3 reflectance0;     // full reflectance color
  vec3 reflectance90;    // reflectance color at grazing angle
  float alphaRoughness; // remapped linear roughness
  vec3 diffuseColor;     // contribution from diffuse lighting
  vec3 specularColor;    // contribution from specular lighting
};
```

3. The sRGB to linear color space conversion routine is implemented this way. It is a popular rough approximation done for simplicity:

```
vec4 SRGBtoLINEAR(vec4 srgbIn) {
  vec3 linOut = pow( srgbIn.xyz,vec3(2.2) );
  return vec4(linOut, srgbIn.a);
}
```

4. Calculation of the lighting contribution from an Image-Based Light source is split into two parts – diffuse irradiance and specular radiance. First, let's start with the radiance part. We use the Lambertian diffuse term. Khronos implementation is quite complex, here we skip some of its details. For those looking into the underlying math theory, check `https://bruop.github.io/ibl/#single_scattering_results`:

```
vec3 getIBLRadianceLambertian(float NdotV, vec3 n,
  float roughness, vec3 diffuseColor, vec3 F0,
  float specularWeight)
```

```
{
  vec2 brdfSamplePoint =
    clamp(vec2(NdotV, roughness), vec2(0., 0.), vec2(1., 1.));
  EnvironmentMapDataGPU envMap =
    getEnvironment(getEnvironmentId());
  vec2 f_ab = sampleBRDF_LUT(brdfSamplePoint, envMap).rg;
  vec3 irradiance = sampleEnvMapIrradiance(n.xyz, envMap).rgb;
  vec3 Fr = max(vec3(1.0 - roughness), F0) - F0;
  vec3 k_S = F0 + Fr * pow(1.0 - NdotV, 5.0);
  vec3 FssEss = specularWeight * k_S * f_ab.x + f_ab.y;
  float Ems = (1.0 - (f_ab.x + f_ab.y));
  vec3 F_avg = specularWeight * (F0 + (1.0 - F0) / 21.0);
  vec3 FmsEms = Ems * FssEss * F_avg / (1.0 - F_avg * Ems);
  vec3 k_D = diffuseColor * (1.0 - FssEss + FmsEms);
  return (FmsEms + k_D) * irradiance;
}
```

5. contribution uses the GGX model. Please note that we use roughness as a LOD level for the precomputed mip lookup. This trick allows us to save performance to avoid excessive texture lookups and integration over them.

```
vec3 getIBLRadianceContributionGGX(
  PBRInfo pbrInputs, float specularWeight)
{
  vec3 n = pbrInputs.n;
  vec3 v =  pbrInputs.v;
  vec3 reflection = -normalize(reflect(v, n));
  EnvironmentMapDataGPU envMap = getEnvironment(getEnvironmentId());
  float mipCount = float(sampleEnvMapQueryLevels(envMap));
  float lod = pbrInputs.perceptualRoughness * (mipCount - 1);
```

6. Retrieve a scale and bias to `F0` from the BRDF lookup table:

```
  vec2 brdfSamplePoint = clamp(
    vec2(pbrInputs.NdotV, pbrInputs.perceptualRoughness),
    vec2(0.0, 0.0),
    vec2(1.0, 1.0));
  vec3 brdf = sampleBRDF_LUT(brdfSamplePoint, envMap).rgb;
```

7. Fetch values from the cube map. No conversion to the linear color space is required since HDR cube maps are already linear.

```
  vec3 specularLight =
    sampleEnvMapLod(reflection.xyz, lod, envMap).rgb;
  vec3 Fr = max(vec3(1.0 - pbrInputs.perceptualRoughness),
```

```
                    pbrInputs.reflectance0) - pbrInputs.reflectance0;
    vec3 k_S = pbrInputs.reflectance0 + Fr * pow(1.0-pbrInputs.NdotV, 5.0);
    vec3 FssEss = k_S * brdf.x + brdf.y;
    return specularWeight * specularLight * FssEss;
  }
```

Now let's go through all the helper functions which are necessary to calculate different parts of the rendering equation:

1. The `diffuseBurley()` function implements the diffuse term from the paper *Physically-Based Shading at Disney* by Brent Burley `http://blog.selfshadow.com/publications/s2012-shading-course/burley/s2012_pbs_disney_brdf_notes_v3.pdfhttp://blog.selfshadow.com/publications/s2012-shading-course/burley/s2012_pbs_disney_brdf_notes_v3.pdf`.

```
vec3 diffuseBurley(PBRInfo pbrInputs) {
  float f90 = 2.0 * pbrInputs.LdotH * pbrInputs.LdotH *
    pbrInputs.alphaRoughness - 0.5;
  return (pbrInputs.diffuseColor / M_PI) *
    (1.0 + f90 * pow((1.0 - pbrInputs.NdotL), 5.0)) *
    (1.0 + f90 * pow((1.0 - pbrInputs.NdotV), 5.0));
}
```

2. The next function models the Fresnel specular reflectance term of the rendering equation, also known as the F term:

```
vec3 specularReflection(PBRInfo pbrInputs) {
  return pbrInputs.reflectance0 +
    (pbrInputs.reflectance90 - pbrInputs.reflectance0) *
     pow(clamp(1.0 - pbrInputs.VdotH, 0.0, 1.0), 5.0);
}
```

3. The function `geometricOcclusion()` calculates the specular geometric attenuation G, where materials with a higher roughness will reflect less light back to the viewer:

```
float geometricOcclusion(PBRInfo pbrInputs) {
  float NdotL = pbrInputs.NdotL;
  float NdotV = pbrInputs.NdotV;
  float rSqr = pbrInputs.alphaRoughness * pbrInputs.alphaRoughness;
  float attenuationL = 2.0 * NdotL /
    (NdotL + sqrt(rSqr + (1.0 - rSqr) * (NdotL * NdotL)));
  float attenuationV = 2.0 * NdotV /
    (NdotV + sqrt(rSqr + (1.0 - rSqr) * (NdotV * NdotV)));
  return attenuationL * attenuationV;
}
```

4. The function `microfacetDistribution()` models the distribution D of microfacet normals across the area being drawn:

```
float microfacetDistribution(PBRInfo pbrInputs) {
  float roughnessSq = pbrInputs.alphaRoughness * pbrInputs.alphaRoughness;
  float f = (pbrInputs.NdotH * roughnessSq - pbrInputs.NdotH) *
    pbrInputs.NdotH + 1.0;
  return roughnessSq / (M_PI * f * f);
}
```

This implementation is based on *Average Irregularity Representation of a Roughened Surface for Ray Reflection* by T. S. Trowbridge, and K. P. Reitz.

5. The utility function `perturbNormal()` provides a normal in world space based on inputs. It expects a sample from the normal map `normalSample` sampled at texture coordinates `uv`, a vertex normal `n`, and a vertex position `v`.

```
vec3 perturbNormal(vec3 n, vec3 v, vec3 normalSample, vec2 uv) {
  vec3 map = normalize( 2.0 * normalSample - vec3(1.0) );
  mat3 TBN = cotangentFrame(n, v, uv);
  return normalize(TBN * map);
}
```

6. The function `cotangentFrame()` creates tangent space based on the vertex position `p`, the per-vertex normal vector `N`, and `uv` texture coordinates. This is not the best way to get the tangent basis, as it suffers from `uv` mapping discontinuities, but it's acceptable to use it in cases where per-vertex precalculated tangent basis is not provided.

```
mat3 cotangentFrame( vec3 N, vec3 p, vec2 uv ) {
  vec3 dp1 = dFdx( p );
  vec3 dp2 = dFdy( p );
  vec2 duv1 = dFdx( uv );
  vec2 duv2 = dFdy( uv );
  vec3 dp2perp = cross( dp2, N );
  vec3 dp1perp = cross( N, dp1 );
  vec3 T = dp2perp * duv1.x + dp1perp * duv2.x;
  vec3 B = dp2perp * duv1.y + dp1perp * duv2.y;
  float invmax = inversesqrt( max( dot(T,T), dot(B,B)));
```

7. Calculate handedness of the resulting cotangent frame and adjust the tangent vector if needed.

```
  float w = dot(cross(N, T), B) < 0.0 ? -1.0 : 1.0;
  T = T * w;
  return mat3( T * invmax, B * invmax, N );
}
```

There's a lot of scaffolding machinery necessary to implement the glTF PBR shading model. Before we can calculate light contribution from a light source, we should fill in the `PBRInfo` structure fields. Let's take a look at the code of `calculatePBRInputsMetallicRoughness()` to understand how this is done.

1. As it is supposed to be in glTF 2.0, roughness is stored in the `green` channel, and metallicity is stored in the `blue` channel. This layout intentionally reserves the `red` channel for optional occlusion map data. We clamp the minimal roughness value to `0.04`. It is a widely used constant for many PBR implementations. It comes from the assumption that even dielectrics have at least 4% of specular reflection:

```
PBRInfo calculatePBRInputsMetallicRoughness( vec4 albedo,
  vec3 normal, vec3 cameraPos, vec3 worldPos, vec4 mrSample)
{
  PBRInfo pbrInputs;
  MetallicRoughnessDataGPU mat = getMaterial(getMaterialId());
  float perceptualRoughness =  getRoughnessFactor(mat) * mrSample.g;
  float metallic = getMetallicFactor(mat) * mrSample.b;
  const float c_MinRoughness = 0.04;
  perceptualRoughness =
    clamp(perceptualRoughness, c_MinRoughness, 1.0);
```

2. Roughness is authored as perceptual roughness; by convention, we convert to material roughness by squaring the perceptual roughness. Perceptual roughness was introduced by Burley `https://disneyanimation.com/publications/physically-based-shading-at-disney` to make the roughness distribution more linear. The albedo value may be defined from a base texture or a flat color. Let's compute specular reflectance the following way:

```
  float alphaRoughness = perceptualRoughness *
                         perceptualRoughness;
  vec4 baseColor = albedo;
  vec3 f0 = vec3(0.04);
  vec3 diffuseColor = mix(baseColor.rgb, vec3(0), metallic);
  vec3 specularColor = mix(f0, baseColor.rgb, metallic);
  float reflectance =
    max(max(specularColor.r, specularColor.g), specularColor.b);
```

3. For a typical incident reflectance range between 4% to 100%, we should set the grazing reflectance to 100% for the typical Fresnel effect. For a very low reflectance range on highly diffuse objects, below 4%, incrementally reduce grazing reflectance to 0%:

```
  float reflectance90 = clamp(reflectance * 25.0, 0.0, 1.0);
  vec3 specularEnvironmentR0 = specularColor.rgb;
  vec3 specularEnvironmentR90 =
    vec3(1.0, 1.0, 1.0) * reflectance90;
```

```
    vec3 n = normalize(normal);
    vec3 v = normalize(cameraPos - worldPos);
```

4. Finally, we should fill in the `PBRInfo` structure with these values. It is used to calculate contributions of each individual light in the scene:

```
    pbrInputs.NdotV = clamp(abs(dot(n, v)), 0.001, 1.0);
    pbrInputs.perceptualRoughness = perceptualRoughness;
    pbrInputs.reflectance0 = specularEnvironmentR0;
    pbrInputs.reflectance90 = specularEnvironmentR90;
    pbrInputs.alphaRoughness = alphaRoughness;
    pbrInputs.diffuseColor = diffuseColor;
    pbrInputs.specularColor = specularColor;
    pbrInputs.n = n;
    pbrInputs.v = v;
    return pbrInputs;
  }
```

The lighting contribution from a single light source can be calculated in the following way using the precalculated values from `PBRInfo`:

1. Here `ld` is the vector from the surface point to the light source and `h` is the half vector between `ld` and `v`:

```
  vec3 calculatePBRLightContribution(
    inout PBRInfo pbrInputs, vec3 lightDirection, vec3 lightColor)
  {
    vec3 n = pbrInputs.n;
    vec3 v = pbrInputs.v;
    vec3 ld = normalize(lightDirection);
    vec3 h = normalize(ld +  v);
    float NdotV = pbrInputs.NdotV;
    float NdotL = clamp(dot(n, ld), 0.001, 1.0);
    float NdotH = clamp(dot(n, h), 0.0, 1.0);
    float LdotH = clamp(dot(ld, h), 0.0, 1.0);
    float VdotH = clamp(dot(v, h), 0.0, 1.0);
    vec3 color = vec3(0);
```

2. Check if the light direction is correct and calculate the shading terms `F`, `G`, `D` for the microfacet specular shading model using the helper functions described earlier in this recipe.

```
    if (NdotL > 0.0 || NdotV > 0.0) {
      pbrInputs.NdotL = NdotL;
      pbrInputs.NdotH = NdotH;
      pbrInputs.LdotH = LdotH;
```

```
    pbrInputs.VdotH = VdotH;
    vec3  F = specularReflection(pbrInputs);
    float G = geometricOcclusion(pbrInputs);
    float D = microfacetDistribution(pbrInputs);
```

3. Calculate the analytical lighting contribution. We obtain the final intensity as reflectance (BRDF) scaled by the energy of the light (cosine law).

```
    vec3 diffuseContrib = (1.0 - F) * diffuseBurley(pbrInputs);
    vec3 specContrib = F * G * D / (4.0 * NdotL * NdotV);
    color = NdotL * lightColor * (diffuseContrib + specContrib);
  }
  return color;
}
```

That is all and should be sufficient to implement the glTF metallic-roughness PBR shading model. The resulting demo application should render the following image. Also try using different glTF 2.0 files:

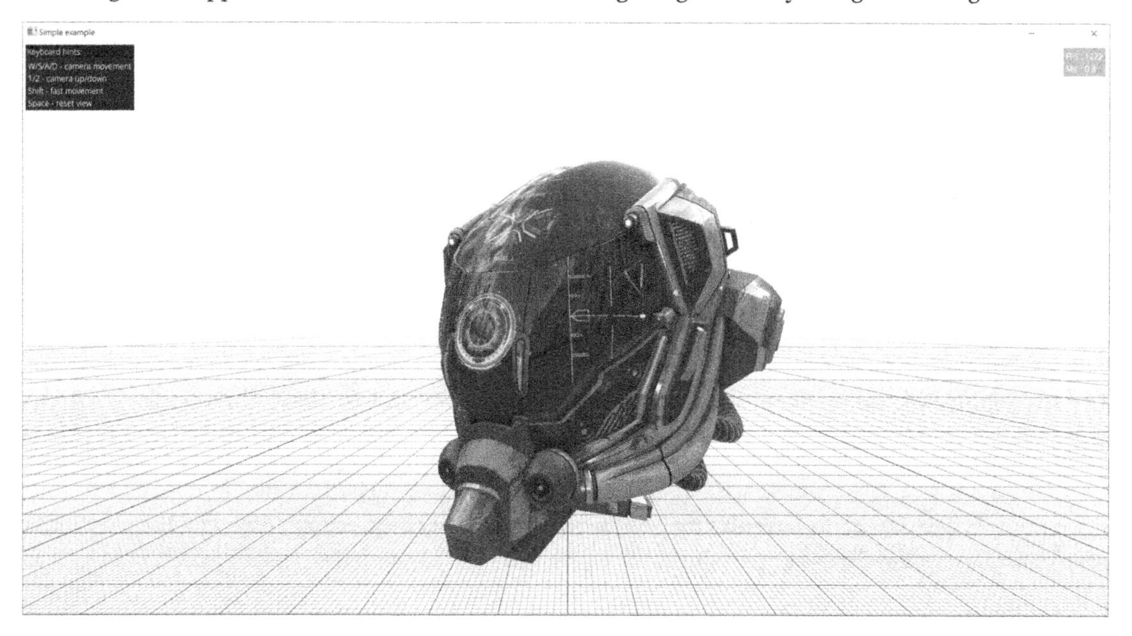

Figure 6.8: Physically based rendering of the Damaged Helmet glTF 2.0 model

There's more...

The whole area of physically based rendering is vast and given the book volume constraints, it is possible to only scratch its surface. In real life, much more complicated PBR implementations can be created, which are normally based on the requirements of content production pipelines. For an endless source of inspiration to know what can be done, we recommend looking into the Unreal Engine source code, which is available for free on GitHub: `https://github.com/EpicGames/UnrealEngine/tree/release/Engine/Shaders/Private`.

In the next recipe, we will explore one more important PBR shading model, the glTF 2.0 specular-glossiness model, and implement a demo application for it.

Implementing the glTF 2.0 specular-glossiness shading model

The specular-glossiness extension is a deprecated and archived in the official Khronos repository. However, we will demonstrate how to use it, as it remains available in many existing 3D assets. In the next chapter, we will introduce a new glTF specular extension, `KHR_materials_specular`, that replaces this older specular-glossiness shading model and show how to convert from the old model to the new extension. Originally, the specular-glossiness model was added to glTF PBR as an extension to support an artist-driven approach. For example, game development often requires greater flexibility in controlling specular effects, and the ability to adjust glossiness is essential in such scenarios.

Over time, more advanced glTF PBR extensions were introduced, making the original specular-glossiness model obsolete. To provide similar functionality within the standard metallic-roughness workflow, a new specular extension was developed. As a result, Khronos recommended discontinuing the specular-glossiness extension and archived it. Despite its deprecation, many existing 3D models were created using this shading model, so we will explore its functionality before transitioning to the newer specular extension.

What is specular-glossiness?

As you may have noticed from this chapter, PBR does not enforce a specific method for defining material models. The metallic-roughness model uses a simplified and intuitive approach, describing surfaces in terms of non-metal vs. metal and smooth vs. rough. While these parameters are not fully physically accurate, they effectively represent a wide range of real-world materials. However, in some cases, the metallic-roughness model is not sufficient to capture all variations in material appearance. The specular-glossiness model provides additional flexibility to address these limitations.

The specular-glossiness workflow is a method for defining materials based on their specular properties. This approach uses three maps: albedo, specular, and glossiness. The albedo map determines the material's base color, the specular map defines its reflectivity or specular color, and the glossiness map controls its smoothness or glossiness.

A key difference between the metallic-roughness and specular-glossiness workflows is the specular property, which defines material reflectivity using RGB channels. Each channel—red, green, and blue—represents reflectivity at different angles, allowing for a broader range of material appearances compared to the metallic-roughness model. However, this property has a significant flaw that prevents its use within the glTF PBR model. Because it does not distinguish between dielectrics and metals, most glTF PBR extensions are incompatible with the specular-glossiness model. To address this, the Khronos Group introduced the `KHR_materials_specular` extension, which adds spectral specular color support to the metallic-roughness model.

Getting ready

The C++ source code for this recipe is in the `Chapter06/05_SpecularGlossiness` folder. The GLSL shader code responsible for PBR calculations can be found in `Chapter06/05_SpecularGlossiness/src/PBR.sp`.

How to do it...

This recipe is very similar to the previous one. In fact, much of the model loading and rendering code remains unchanged, except for retrieving values for different material properties. We extend the metallic-roughness data to include specular-glossiness parameters, then apply these parameters in the shader based on the material type.

We are going to use the 3D model *SpecGlossVsMetalRough* provided by Khronos. It provides a side-by-side comparison of the same model rendered with the metallic-roughness shading and specular-glossiness shading. You can find the glTF file here: `deps/src/glTF-Sample-Assets/Models/SpecGlossVsMetalRough/glTF/`.

Let's get started. Here are C++ code changes necessary to accommodate specular-glossiness parameters:

1. We modify our material data structure `SpecularGlossinessDataGPU` by adding two new data members, `vec4 specularGlossiness` and `uint32_t materialType`. The first one will provide the necessary parameters for the specular-glossiness material and the second one will specify the exact material type. Please notice the padding member at the end of the structure. We need it to keep the binary representation of this structure aligned with GLSL shader inputs. The GLSL `st430` layout and alignment rules are not complex but might not be correctly implemented by different hardware vendors, especially on mobile devices. In this case, manual padding is just an easy and good enough way to fix compatibility between all GPUs. For further reading, we can recommend official Khronos Vulkan Guide doc: `https://github.com/KhronosGroup/Vulkan-Guide/blob/main/chapters/shader_memory_layout.adoc`.

```
struct SpecularGlossinessDataGPU {
  vec4 baseColorFactor = vec4(1.0f, 1.0f, 1.0f, 1.0f);
  vec4 metallicRoughnessNormalOcclusion = vec4(1.0f, 1.0f, 1.0f, 1.0f);
  vec4 specularGlossiness = vec4(1.0f, 1.0f, 1.0f, 1.0f);
  // … everything else remains the same
  //   as in the structure MetallicRoughnessDataGPU …
  uint32_t materialType = 0;
  uint32_t padding[3]   = {};
  enum AlphaMode : uint32_t {
    AlphaMode_Opaque = 0,
    AlphaMode_Mask   = 1,
    AlphaMode_Blend  = 2,
  };
};
```

2. This is how we identify a glTF material type. We implemented a helper function `detectMaterialType()` in `shared/UtilsGLTF.cpp`.

```
MaterialType detectMaterialType(const aiMaterial* mtl) {
  aiShadingMode shadingMode = aiShadingMode_NoShading;
  if (mtl->Get(AI_MATKEY_SHADING_MODEL, shadingMode) == AI_SUCCESS) {
    if (shadingMode == aiShadingMode_Unlit)
      return MaterialType_Unlit;
  }
  if (shadingMode == aiShadingMode_PBR_BRDF) {
    ai_real factor = 0;
    if (mtl->Get(AI_MATKEY_GLOSSINESS_FACTOR, factor) == AI_SUCCESS) {
      return MaterialType_SpecularGlossiness;
    } else if (mtl->Get(AI_MATKEY_METALLIC_FACTOR, factor) == AI_SUCCESS) {
      return MaterialType_MetallicRoughness;
    }
  }
  LLOGW("Unknown material type\n");
  return MaterialType_Invalid;
}
```

3. The next difference is how we load extra material properties. Nothing interesting here, just load and assign the values using the Assimp API.

```
SpecularGlossinessDataGPU res;
// ...
if (materialType == MaterialType_SpecularGlossiness) {
  ai_real specularFactor[3];
  if (mtlDescriptor->Get(AI_MATKEY_SPECULAR_FACTOR,
                         specularFactor) == AI_SUCCESS) {
    res.specularGlossiness.x = specularFactor[0];
    res.specularGlossiness.y = specularFactor[1];
    res.specularGlossiness.z = specularFactor[2];
  }
  assignUVandSampler(samplers, mtlDescriptor,
    aiTextureType_SPECULAR, res.surfacePropertiesTextureUV,
    res.surfacePropertiesTextureSampler);
  ai_real glossinessFactor;
  if (mtlDescriptor->Get(AI_MATKEY_GLOSSINESS_FACTOR,
                         glossinessFactor) == AI_SUCCESS) {
    res.specularGlossiness.w = glossinessFactor;
  }
}
```

The rest of the C++ code is mostly identical to the previous recipe *Implementing the glTF 2.0 metallic-roughness shading model*. The uncovered differences related to adjusting to the different material properties and providing necessary data. Let's now look at the corresponding GLSL shaders.

The differences are introduced in the code of the fragment shader `05_SpecularGlossinesss/src/main.frag` where we calculate and apply `PBRInfo` parameters:

1. First, we identify the specular-glossiness material type:

```
PBRInfo calculatePBRInputsMetallicRoughness( vec4 albedo,
  vec3 normal, vec3 cameraPos, vec3 worldPos, vec4 mrSample)
{
  PBRInfo pbrInputs;
  SpecularGlossinessDataGPU mat = getMaterial(getMaterialId());
  bool isSpecularGlossiness =
    getMaterialType(mat) == MaterialType_SpecularGlossiness;
```

2. Based on the material type, we calculate `perceptualRoughness` and `f0`, as well as the diffuse and specular color contributions. We follow the official Khronos recommendations `https://kcoley.github.io/glTF/extensions/2.0/Khronos/KHR_materials_pbrSpecularGlossiness` to convert values from specular-glossiness to metallic-roughness.

```
  float perceptualRoughness = isSpecularGlossiness ?
    getGlossinessFactor(mat):
    getRoughnessFactor(mat);
  float metallic = getMetallicFactor(mat) * mrSample.b;
  metallic = clamp(metallic, 0.0, 1.0);
  vec3 f0 = isSpecularGlossiness ?
    getSpecularFactor(mat) * mrSample.rgb :
    vec3(0.04);
  const float c_MinRoughness = 0.04;
  perceptualRoughness = isSpecularGlossiness ?
    1.0 - mrSample.a * perceptualRoughness :
    clamp(mrSample.g * perceptualRoughness, c_MinRoughness, 1.0);
  float alphaRoughness = perceptualRoughness * perceptualRoughness;
  vec4 baseColor = albedo;
  vec3 diffuseColor = isSpecularGlossiness ?
    baseColor.rgb * (1.0 - max(max(f0.r, f0.g), f0.b)) :
    mix(baseColor.rgb, vec3(0), metallic);
  vec3 specularColor = isSpecularGlossiness ?
    f0 : mix(f0, baseColor.rgb, metallic);
  // …
}
```

The remaining fragment shader code remains the same. The result image produced by this example should look as in the following screenshot:

Figure 6.9: Specular-glossiness versus metallic-roughness materials

One bottle has a metallic-roughness material and the other uses a specular-glossiness one. As you can see, the objects appear identical, and by adjusting the specular and glossiness values within a compatible range, you can achieve the exact same result with two different shading models.

There's more...

Here is an article by Don McCurdy explaining the reasons to switching over from the specular-glossiness shading model to the metallic-roughness one: `https://www.donmccurdy.com/2022/11/28/converting-gltf-pbr-materials-from-specgloss-to-metalrough`.

7

Advanced PBR Extensions

In this chapter, we will delve into advanced glTF PBR extensions that build upon the base metallic-roughness model. While the base metallic-roughness model provides a starting point, it falls short of capturing the full spectrum of real-life materials. To address this, glTF incorporates additional material layers, each with specific parameters that define their unique behaviors. Our goal here is to guide you through implementing these layers from the ground up. We will introduce the concept of layers, break down some mathematical principles behind them, and then show you how to integrate each layer into the GLSL shader code.

Most of the C++ code provided in this chapter is applicable across all the recipes we will cover here and throughout the rest of our book.

In this chapter, you will learn the following recipes:

- Introduction to glTF PBR extensions
- Implementing the `KHR_materials_clearcoat` extension
- Implementing the `KHR_materials_sheen` extension
- Implementing the `KHR_materials_transmission` extension
- Implementing the `KHR_materials_volume` extension
- Implementing the `KHR_materials_ior` extension
- Implementing the `KHR_materials_specular` extension
- Implementing the `KHR_materials_emissive_strength` extension
- Extending analytical lights support with `KHR_lights_punctual`

Our GLSL shaders code is based on the official Khronos Sample Viewer and serves as an example implementation of these extensions.

Introduction to glTF PBR extensions

In this recipe, we will explore the design approach for PBR Material extensions, offering plenty of context to help you implement various glTF PBR extensions. The actual code will be shared in subsequent recipes, with the chapter structure following the sequence in which Khronos developed these PBR extensions.

PBR specifications evolve rapidly, and the reader should be aware that some extensions may become deprecated or obsolete by the time of reading.

Getting ready

We assume our readers have some basic understanding of linear algebra and calculus. It is recommended to have the glTF 2.0 list of ratified extensions specification at hand which can be found at `https://github.com/KhronosGroup/glTF/blob/main/extensions/README.md`.

How is the glTF 2.0 PBR model designed?

In the previous chapter, we explored the core Metallic-Roughness PBR model. This model is great for depicting many types of metallic and non-metallic materials, but the real world is much more complex.

To better capture that complexity, Khronos decided not to simply extend the Metallic-Roughness model. Instead, they introduced a layered approach, much like the layers of an onion. This method lets you gradually add complexity to the PBR material, similar to how layers are built up in Adobe Standard Surface `https://github.com/Autodesk/standard-surface`.

Layering mimics real-world material structures by stacking multiple layers, each with its own light-interacting properties. To maintain physical accuracy, the first layer, called the base layer, should be either fully opaque (like metallic surfaces) or completely transparent (like glass or skin). After that, additional layers, known as dielectric slabs, can be added on top of that one by one.

When light hits the boundary between two layers, it can reflect and bounce back in the opposite direction. However, our main concern here is the light that continues to move through the material stack. As this light passes through the lower layers, it may be absorbed by the material.

The mixing operation provides a unique method for material modeling. You can think of it as a statistically weighted blend of two different materials, where you combine a certain percentage of material `A` with a certain percentage of material `B`. While this technique is great for creating new materials, it is important to remember that not all combinations are physically realistic. For example, mixing oil and water wouldn't produce a believable material.

When the mixing operation is done as a linear interpolation, it naturally follows the principle of energy conservation. This means that the total energy within the resulting material stays the same, consistent with the basic laws of physics.

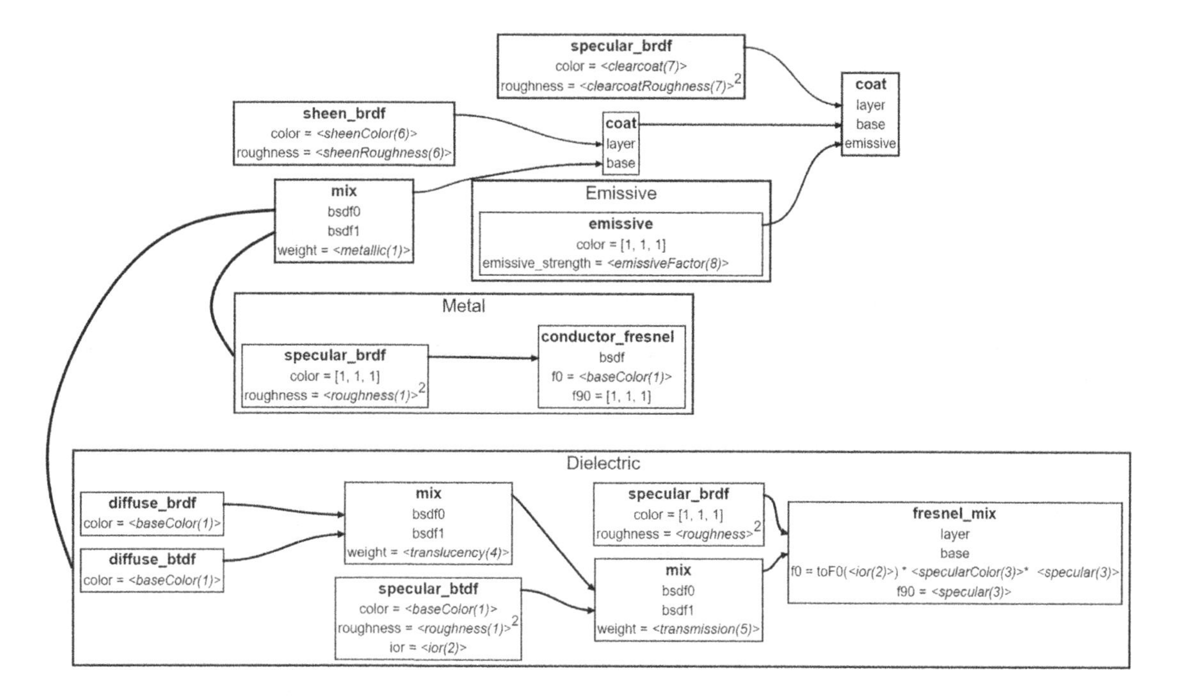

1 Core
2 KHR_materials_ior
3 KHR_materials_specular
4 KHR_materials_translucency
5 KHR_materials_transmission
6 KHR_materials_sheen
7 KHR_materials_clearcoat
8 KHR_materials_emissive_strength

Figure 7.1: glTF PBR layering and mixing

Quick tip: Need to see a high-resolution version of this image? Open this book in the next-gen Packt Reader or view it in the PDF/ePub copy.

The next-gen Packt Reader and a **free PDF/ePub copy** of this book are included with your purchase. Unlock them by scanning the QR code below or visiting `https://www.packtpub.com/unlock/9781803248110`.

In the following recipes, we will dive into several advanced material layers: specular retro-reflection (sheen), coating specular reflection, and diffuse transmission. We will also explore how these layers can be combined to create a broader range of material appearances.

Getting ready

This chapter uses a single glTF viewer sample code for all the recipes. The `main.cpp` file varies across recipes in only two ways: they use different model files to demonstrate the specific glTF PBR extensions covered and the initial camera positions are adjusted to showcase the models attractively.

The source code for the glTF Viewer itself resides in the file `shared/UtilsGLTF.cpp`. The corresponding GLSL vertex and fragment shaders are located in the `data/shaders/gltf/` folder.

Note

The structure of these GLSL shaders differs from those covered in the previous chapter. We will explore the specific implementation differences in each of the individual recipes.

How to do it...

Let's go over the main differences between our glTF Viewer implementation from the previous chapter and the newly proposed unified version.

1. We refactored the code and introduced a very basic structure in `shared/UtilsGLTF.h` to store all the necessary application data. This chapter will explain all the struct member fields.

```
struct GLTFContext {
  explicit GLTFContext(VulkanApp& app_)
  : app(app_)
  , samplers(app_.ctx_)
  , envMapTextures(app_.ctx_) {}
  GLTFDataHolder glTFDataholder;
  MaterialsPerFrame matPerFrame;
  GLTFGlobalSamplers samplers;
  EnvironmentMapTextures envMapTextures;
  GLTFFrameData frameData;
  std::vector<GLTFTransforms> transforms;
  std::vector<GLTFNode> nodesStorage;
  std::vector<GLTFMesh> meshesStorage;
  std::vector<uint32_t> opaqueNodes;
  std::vector<uint32_t> transmissionNodes;
  std::vector<uint32_t> transparentNodes;
  Holder<BufferHandle> envBuffer;
  Holder<BufferHandle> perFrameBuffer;
  Holder<BufferHandle> transformBuffer;
```

```
  Holder<RenderPipelineHandle> pipelineSolid;
  Holder<RenderPipelineHandle> pipelineTransparent;
  Holder<ShaderModuleHandle> vert;
  Holder<ShaderModuleHandle> frag;
  Holder<BufferHandle> vertexBuffer;
  Holder<BufferHandle> indexBuffer;
  Holder<BufferHandle> matBuffer;
  Holder<TextureHandle> offscreenTex[3] = {};
  uint32_t currentOffscreenTex = 0;
  GLTFNodeRef root;
  VulkanApp& app;
  bool volumetricMaterial = false;
  bool isScreenCopyRequired() const {
    return volumetricMaterial;
  }
};
```

2. Let's introduce a very basic loading and rendering API for it. The `rebuildRenderList` argument signals that model-to-world transformations of glTF nodes should be rebuilt:

```
void loadglTF(GLTFContext& context,
  const char* gltfName, const char* glTFDataPath);
void renderglTF(GLTFContext& context,
  const mat4& model, const mat4& view, const mat4& proj,
  bool rebuildRenderList = false);
```

3. We expanded the GPU data structures to include the required material properties and added a new `enum` called `MaterialType`, which allows us to provide the material ID as needed. The previous *Chapter 6* covered the old materials: unlit, metallic-roughness, and specular-glossiness. New materials will be covered here in this chapter in the subsequent recipes.

```
enum MaterialType : uint32_t {
  MaterialType_Invalid            = 0,
  MaterialType_Unlit              = 0xF,
  MaterialType_MetallicRoughness  = 0x1,
  MaterialType_SpecularGlossiness = 0x2,
  MaterialType_Sheen              = 0x4,
  MaterialType_ClearCoat          = 0x8,
  MaterialType_Specular           = 0x10,
  MaterialType_Transmission       = 0x20,
  MaterialType_Volume             = 0x40,

};
```

4. We've added `3` different vector containers to keep lists of the glTF nodes: opaque, transparent and transmission. Here is the function `buildTransformsList()` to build node transformations and collect other nodes data:

```
void buildTransformsList(GLTFContext& gltf) {
  gltf.transforms.clear();
  gltf.opaqueNodes.clear();
  gltf.transmissionNodes.clear();
  gltf.transparentNodes.clear();
```

5. The body of our recursive traversal function is declared as a local C++ lambda. It collects all the transforms into `gltf.transforms` and adds opaque, transparent, and transmissive nodes to their corresponding containers:

```
std::function<void(GLTFNodeRef gltfNode)> traverseTree =
  [&]([&](GLTFNodeRef nodeRef) {
    const GLTFNode& node = gltf.nodesStorage[nodeRef];
    for (GLTFNodeRef meshId : node.meshes) {
```

```
        const GLTFMesh& mesh = gltf.meshesStorage[meshId];
        gltf.transforms.push_back({
          .model = node.transform,
          .matId = mesh.matIdx,
          .nodeRef = nodeRef,
          .meshRef = meshId,
          .sortingType = mesh.sortingType });
```

6. Push the index of the transform that was just added to `gltf.transforms` in the code block above.

```
        uint32_t lastTransformIndex = gltf.transforms.size() - 1;
        if (mesh.sortingType == SortingType_Transparent) {
          gltf.transparentNodes.push_back(lastTransformIndex);
        } else if (mesh.sortingType==SortingType_Transmission) {
          gltf.transmissionNodes.push_back(lastTransformIndex);
        } else {
          gltf.opaqueNodes.push_back(lastTransformIndex );
        }
      }
      for (GLTFNodeRef child : node.children)
        traverseTree(child);
    };
```

7. Invoke the lambda to traverse the entire tree of glTF nodes starting from the root and store all the resulting transformations in a buffer:

```
    traverseTree(gltf.root);
    gltf.transformBuffer = gltf.app.ctx_->createBuffer({
      .usage     = lvk::BufferUsageBits_Uniform,
      .storage   = lvk::StorageType_HostVisible,
      .size      = gltf.transforms.size() * sizeof(GLTFTransforms),
      .data      = &gltf.transforms[0],
      .debugName = "Per Frame data" });
  };
```

8. Let's add a node-sorting function to correctly render glTF nodes in the right order to support transparency. We're using a very simple algorithm that sorts nodes based on the distance from the camera to the node's center. To properly render transparent nodes, they should be rendered last, from back to front.

```
  void sortTransparentNodes(
    GLTFContext& gltf, const vec3& cameraPos) {
    std::sort(
      gltf.transparentNodes.begin(),
```

```
    gltf.transparentNodes.end(),
    [&](uint32_t a, uint32_t b) {
      float sqrDistA = glm::length2(
        cameraPos-vec3(gltf.transforms[a].model[3]));
      float sqrDistB = glm::length2(
        cameraPos-vec3(gltf.transforms[b].model[3]));
      return sqrDistA < sqrDistB;
  });
}
```

Now we have to change the actual rendering function `renderGLTF()` to accommodate all the changes mentioned above.

1. First, we have to update the transforms list and sort glTF nodes based on the distance to the current camera.

```
void renderGLTF(GLTFContext& gltf,
  const mat4& model, const mat4& view, const mat4& proj,
  bool rebuildRenderList)
{
  auto& ctx = gltf.app.ctx_;
  const vec4 camPos = glm::inverse(view)[3];
  if (rebuildRenderList || gltf.transforms.empty()) {
    buildTransformsList(gltf);
  }
  sortTransparentNodes(gltf, camPos);
```

2. Store per-frame camera parameters and prepare push constants with all necessary buffers and textures:

```
  gltf.frameData = {
    .model     = model,
    .view      = view,
    .proj      = proj,
    .cameraPos = camPos,
  };
  struct PushConstants {
    uint64_t draw;
    uint64_t materials;
    uint64_t environments;
    uint64_t transforms;
    uint32_t envId;
    uint32_t transmissionFramebuffer;
    uint32_t transmissionFramebufferSampler;
```

```
} pushConstants = {
  .draw         = ctx->gpuAddress(gltf.perFrameBuffer),
  .materials    = ctx->gpuAddress(gltf.matBuffer),
  .environments = ctx->gpuAddress(gltf.envBuffer),
  .transforms   = ctx->gpuAddress(gltf.transformBuffer),
  .envId        = 0,
  .transmissionFramebuffer = 0,
  .transmissionFramebufferSampler = gltf.samplers.clamp.index(),
};
ctx->upload(
  gltf.perFrameBuffer, &gltf.frameData, sizeof(GLTFFrameData));
// ...
```

3. Let's render all opaque nodes. For this pass, no transmission framebuffer is required:

```
const lvk::RenderPass renderPass = {
  .color = { { .loadOp = lvk::LoadOp_Clear,
               .clearColor = { 1.0f, 1.0f, 1.0f, 1.0f } } },
  .depth = { .loadOp = lvk::LoadOp_Clear, .clearDepth = 1.0f },
};
const lvk::Framebuffer framebuffer = {
  .color        = { { .texture = screenCopy ?
     gltf.offscreenTex[gltf.currentOffscreenTex] :
     ctx->getCurrentSwapchainTexture() } },
  .depthStencil = { .texture = gltf.app.getDepthTexture() },
};
buf.cmdBeginRendering(renderPass, framebuffer);
buf.cmdBindVertexBuffer(0, gltf.vertexBuffer, 0);
buf.cmdBindIndexBuffer(gltf.indexBuffer, lvk::IndexFormat_UI32);
buf.cmdBindDepthState({ .compareOp = lvk::CompareOp_Less,
                        .isDepthWriteEnabled = true });
buf.cmdBindRenderPipeline(gltf.pipelineSolid);
buf.cmdPushConstants(pushConstants);
for (uint32_t transformId : gltf.opaqueNodes) {
  GLTFTransforms transform = gltf.transforms[transformId];
  buf.cmdPushDebugGroupLabel(
    gltf.nodesStorage[transform.nodeRef].name.c_str(), 0xff0000ff);
  const GLTFMesh submesh = gltf.meshesStorage[transform.meshRef];
```

4. We use the built-in GLSL variable `gl_BaseInstance` to pass the value of `transformId` into shaders. This way, we do not have to update push constants per each draw call. This is the most efficient way to do it.

Note

The `firstInstance` parameter of `vkCmdDrawIndexed()` is assigned to the built-in GLSL variable `gl_BaseInstance`. This allows you to pass an arbitrary per-draw-call `uint32_t` value into vertex shaders without involving any buffers or push constants. This is a very fast technique and should be used whenever possible.

```
vkCmdDrawIndexed(VkCommandBuffer commandBuffer,
                 uint32_t        indexCount,
                 uint32_t        instanceCount,
                 uint32_t        firstIndex,
                 int32_t         vertexOffset,
                 uint32_t        firstInstance);
```

```
    buf.cmdDrawIndexed(submesh.indexCount, 1,
      submesh.indexOffset, submesh.vertexOffset, transformId);
    buf.cmdPopDebugGroupLabel();
  }
  buf.cmdEndRendering();
  // ...
```

5. Now, we should render transparent nodes on top of opaque ones. Some transparent nodes may require a screen copy to render various effects, such as volume or index-of-refraction. Here's a very simple way to obtain it:

```
if (screenCopy) {
  buf.cmdCopyImage(
    gltf.offscreenTex[gltf.currentOffscreenTex],
    ctx->getCurrentSwapchainTexture(),
    ctx->getDimensions(ctx->getCurrentSwapchainTexture()));
  buf.cmdGenerateMipmap(
    gltf.offscreenTex[gltf.currentOffscreenTex]);
  pushConstants.transmissionFramebuffer =
    gltf.offscreenTex[gltf.currentOffscreenTex].index();
  buf.cmdPushConstants(pushConstants);
}
```

6. As we start the next render pass and use the offscreen texture, we have to synchronize it properly:

```
buf.cmdBeginRendering(renderPass, framebuffer, {
  .textures = { TextureHandle(
    gltf.offscreenTex[gltf.currentOffscreenTex]) } });
buf.cmdBindVertexBuffer(0, gltf.vertexBuffer, 0);
buf.cmdBindIndexBuffer(gltf.indexBuffer, lvk::IndexFormat_UI32);
buf.cmdBindDepthState({ .compareOp = lvk::CompareOp_Less,
                        .isDepthWriteEnabled = true });
```

7. Now we render transmission nodes:

```
buf.cmdBindRenderPipeline(gltf.pipelineSolid);
for (uint32_t transformId : gltf.transmissionNodes) {
  const GLTFTransforms transform = gltf.transforms[transformId];
  buf.cmdPushDebugGroupLabel(
    gltf.nodesStorage[transform.nodeRef].name.c_str(), 0x00FF00ff);
  const GLTFMesh submesh = gltf.meshesStorage[transform.meshRef];
  buf.cmdDrawIndexed(submesh.indexCount, 1,
    submesh.indexOffset, submesh.vertexOffset, transformId);
  buf.cmdPopDebugGroupLabel();
}
```

8. Transparent nodes come last. The same `gl_BaseInstance` trick is used to pass the value of `transformId` for each glTF mesh:

```
buf.cmdBindRenderPipeline(gltf.pipelineTransparent);
for (uint32_t transformId : gltf.transparentNodes) {
  const GLTFTransforms transform = gltf.transforms[transformId];
  buf.cmdPushDebugGroupLabel(
    gltf.nodesStorage[transform.nodeRef].name.c_str(), 0x00FF00ff);
  const GLTFMesh submesh = gltf.meshesStorage[transform.meshRef];
  buf.cmdDrawIndexed(submesh.indexCount, 1,
    submesh.indexOffset, submesh.vertexOffset, transformId);
  buf.cmdPopDebugGroupLabel();
}
```

9. Once the command buffer is filled, we can submit it and use another offscreen texture in a round-robin manner.

```
buf.cmdEndRendering();
ctx->submit(buf, ctx->getCurrentSwapchainTexture());
gltf.currentOffscreenTex = (gltf.currentOffscreenTex + 1) %
  LVK_ARRAY_NUM_ELEMENTS(gltf.offscreenTex);
```

This was a complete overview of our generic glTF rendering code. The real magic happens inside the GLSL shaders. In the next recipes, we will go through the shaders step by step to learn how to implement different glTF material extensions.

There's more...

The Khronos *3D Formats Working Group* is constantly working to improve PBR material capabilities by introducing new extension specifications. To stay up to date with the status of approved extensions, you can visit the Khronos GitHub page: `https://github.com/KhronosGroup/glTF/blob/main/extensions/README.md`

Implementing the KHR_materials_clearcoat extension

The `KHR_materials_clearcoat` extension improves glTF's core physically based rendering (PBR) model by adding a clear, reflective layer on top of another material or surface. This layer reflects light both from itself and the layers underneath. Examples of this effect include the glossy finish on car paint or the shine of a well-polished shoe.

More information

Here is a link to the Khronos glTF PBR extension: `https://github.com/KhronosGroup/glTF/blob/main/extensions/2.0/Khronos/KHR_materials_clearcoat/README.md`

Clearcoat parameters

The following parameters are provided by the `KHR_materials_clearcoat` extension:

- `clearcoatFactor` / `clearcoatTexure`: This parameter indicates the intensity of the coating. It can be set using a scalar factor or a texture. A value of `0` means no coating, while a value of `1` indicates the presence of the coating. In-between values should be used only along the boundary between coated and uncoated areas.
- `clearcoatNormalTexture`: This parameter allows a normal map to be applied to the coating layer, introducing variations and details to the coating surface.
- `clearcoatRoughnessFactor` / `clearcoatRoughnessTexture`: This parameter indicates the coating roughness. It can be set as a roughness scalar factor or a roughness texture. It works similarly to the roughness parameter of a base material but is applied to the coating layer.

Specular BRDF for the clearcoat layer

The specular BRDF for the clearcoat layer uses the specular term from the glTF 2.0 Metallic-Roughness material. However, to maintain energy conservation within the material when using a simple layering function, a slight adjustment is applied.

The microfacet Fresnel term is calculated using the `NdotV` term instead of the `VdotH` term, effectively ignoring the microscopic surface orientation within the clearcoat layer. This simplification is justified because clearcoat layers usually have very low roughness, meaning the microfacets are mostly aligned with the normal direction.

As a result, `NdotV` becomes approximately equivalent to `NdotL`. This approach ensures energy conservation within the material through a simple layering function and keeps computations efficient by omitting the `VdotH` term.

As explained in *Chapter 6*, `N` represents the normal vector at the surface point, `V` is the vector pointing from the surface to the viewer, `L` is the vector pointing from the surface point to the light source, and `H` is the half-vector that lies exactly between the directions of the light source `L` and the viewer `V`.

The provided implementation of the `clearcoat` layer within the BRDF framework makes certain assumptions that neglect some real-world material properties. Here's a breakdown of these limitations:

- **Infinitely Thin Layer**: The clearcoat layer is treated as infinitely thin, disregarding its actual thickness.
- **Neglecting Refraction**: Refraction, bending of light as it passes through the clearcoat layer, is not taken into account.
- **Independent Fresnel Terms**: The refractive indices of the clearcoat and base layers are treated as independent, with their Fresnel terms calculated separately, without accounting for any interaction between them.
- **Omitted Scattering**: The current model does not account for light scattering between the clearcoat layer and the base layer.
- **Diffraction Ignored**: Diffraction effects, the slight bending of light around the edges of microscopic facets, are not taken into account.

Despite these limitations, the clearcoat BRDF is a valuable tool for simulating clear coat effects in material modeling. It strikes a good balance between computational efficiency and producing visually plausible results, especially for clearcoat layers with low roughness.

Important note

The clearcoat extension is designed to work with glTF's core PBR shading model and is not compatible with other shading models like Unlit or Specular-glossiness. However, it can still be used alongside other PBR parameters, such as emissive materials, where the emitted light is influenced by the clearcoat layer.

Getting ready

The source code for this recipe can be found in `Chapter07/01_Clearcoat/`.

How to do it...

Let's take a look at the C++ code in `Chapter07/01_Clearcoat/src/main.cpp`.

1. First, let's load a `.gltf` file using our new glTF API:

```
VulkanApp app({
    .initialCameraPos    = vec3(0.0f, -0.2f, -1.5f),
    .initialCameraTarget = vec3(0.0f, -0.5f,  0.0f),
});
```

```
GLTFContext gltf(app);
loadGLTF(gltf,
  "deps/src/glTF-Sample-Assets/Models/ClearcoatWicker/
    glTF/ClearcoatWicker.gltf",
  "deps/src/glTF-Sample-Assets/Models/ClearcoatWicker/glTF/");
```

2. Then we render it using the `renderGLTF()` function described in the previous recipe *Introduction to glTF PBR extensions*:

```
const mat4 t = glm::translate(mat4(1.0f), vec3(0, -1, 0));
app.run([&](uint32_t width, uint32_t height,
            float aspectRatio, float deltaSeconds)
{
  const mat4 m = t * glm::rotate(
    mat4(1.0f), (float)glfwGetTime(), vec3(0.0f, 1.0f, 0.0f));
  const mat4 v = app.camera_.getViewMatrix();
  const mat4 p = glm::perspective(45.0f, aspectRatio, 0.01f, 100.0f);
  renderGLTF(gltf, m, v, p);
});
```

To load clearcoat parameters from a `.gltf` file, we have to introduce some changes to our GLTF material loader. Let's take a look at the steps required to do it.

1. First, we should add some new member fields to the `GLTFMaterialDataGPU` structure in the file `shared/UtilsGLTF.h` to store scalar values and corresponding textures:

```
struct GLTFMaterialDataGPU {
  // ...
  vec4 clearcoatTransmissionThickness = vec4(1, 1, 1, 1);
  uint32_t clearCoatTexture                  = 0;
  uint32_t clearCoatTextureSampler           = 0;
  uint32_t clearCoatTextureUV                = 0;
  uint32_t clearCoatRoughnessTexture         = 0;
  uint32_t clearCoatRoughnessTextureSampler = 0;
  uint32_t clearCoatRoughnessTextureUV       = 0;
  uint32_t clearCoatNormalTexture            = 0;
  uint32_t clearCoatNormalTextureSampler     = 0;
  uint32_t clearCoatNormalTextureUV          = 0;
  // ...
}
```

2. Let's load textures and properties using the Assimp library in `shared/UtilsGLTF.cpp`. Here we emphasize only the properties related to the clearcoat extension:

```
GLTFMaterialDataGPU setupglTFMaterialData(
  const std::unique_ptr<lvk::IContext>& ctx,
  const GLTFGlobalSamplers& samplers,
  const aiMaterial* mtlDescriptor,
  const char* assetFolder,
  GLTFDataHolder& glTFDataholder,
  bool& useVolumetric)
{
  // ...
  // clearcoat
  loadMaterialTexture(mtlDescriptor, aiTextureType_CLEARCOAT,
    assetFolder, mat.clearCoatTexture, ctx, true, 0);
  loadMaterialTexture(mtlDescriptor, aiTextureType_CLEARCOAT,
    assetFolder, mat.clearCoatRoughnessTexture, ctx, false, 1);
  loadMaterialTexture(mtlDescriptor, aiTextureType_CLEARCOAT,
    assetFolder, mat.clearCoatNormalTexture, ctx, false, 2);
  // ...
  bool useClearCoat = !mat.clearCoatTexture.empty() ||
                      !mat.clearCoatRoughnessTexture.empty() ||
                      !mat.clearCoatNormalTexture.empty();
  ai_real clearcoatFactor;
  if (mtlDescriptor->Get(AI_MATKEY_CLEARCOAT_FACTOR,
        clearcoatFactor) == AI_SUCCESS) {
    res.clearcoatTransmissionThickness.x = clearcoatFactor;
    useClearCoat = true;
  }
  ai_real clearcoatRoughnessFactor;
  if (mtlDescriptor->Get(AI_MATKEY_CLEARCOAT_ROUGHNESS_FACTOR,
        clearcoatRoughnessFactor) == AI_SUCCESS) {
    res.clearcoatTransmissionThickness.y =
      clearcoatRoughnessFactor;
    useClearCoat = true;
  }
  if (assignUVandSampler(
        samplers, mtlDescriptor, aiTextureType_CLEARCOAT,
        res.clearCoatTextureUV,
        res.clearCoatNormalTextureSampler, 0)) {
    useClearCoat = true;
  }
```

```
if (assignUVandSampler(
      samplers, mtlDescriptor, aiTextureType_CLEARCOAT,
      res.clearCoatRoughnessTextureUV,
      res.clearCoatRoughnessTextureSampler, 1)) {
  useClearCoat = true;
}
if (assignUVandSampler(
      samplers, mtlDescriptor, aiTextureType_CLEARCOAT,
      res.clearCoatNormalTextureUV,
      res.clearCoatNormalTextureSampler, 2)) {
  useClearCoat = true;
}
if (useClearCoat)
  res.materialTypeFlags |= MaterialType_ClearCoat;
```

Note

Please note that we only set the `MaterialType_ClearCoat` flag if the checks pass and the extension is present in the `.gltf` file. While it's technically possible to enable the clearcoat layer all the time – since the default settings effectively disable it – doing so is highly inefficient. The clearcoat layer adds secondary BRDF sampling, which is computationally expensive. It is better to use only the expensive features that are actually needed!

That's it for the C++ changes. Now, let's look at the GLSL shader changes, where the actual rendering work takes place:

1. Similar to the C++ changes for new parameters, we added GLSL utility functions to read clear-coat data from textures and input buffers in `data/shaders/gltf/inputs.frag`. The clearcoat factor and roughness are packed into a texture as `r` and `g` channels respectively:

```
float getClearcoatFactor(InputAttributes tc,
                         MetallicRoughnessDataGPU mat)
{
  return textureBindless2D(mat.clearCoatTexture,
    mat.clearCoatTextureSampler,
    tc.uv[mat.clearCoatTextureUV]
  ).r * mat.clearcoatTransmissionThickness.x;
}
float getClearcoatRoughnessFactor(InputAttributes tc,
                                  MetallicRoughnessDataGPU mat)
{
  return textureBindless2D(mat.clearCoatRoughnessTexture,
```

```
      mat.clearCoatRoughnessTextureSampler,
      tc.uv[mat.clearCoatRoughnessTextureUV]
    ).g * mat.clearcoatTransmissionThickness.y;
  }
```

2. We calculate the clearcoat contribution as described in the beginning of this recipe. We use the GGX BRDF to perform an additional lookup and provide clearcoat roughness, reflectance `clearcoatF0`, and normal as inputs in `data/shaders/gltf/main.frag`. Please note that we use the IOR parameter, which will be covered later in the recipe *Implementing the IOR extension*:

```
vec3 clearCoatContrib = vec3(0);
if (isClearCoat) {
  pbrInputs.clearcoatFactor = getClearcoatFactor(tc, mat);
  pbrInputs.clearcoatRoughness =
    clamp(getClearcoatRoughnessFactor(tc, mat), 0.0, 1.0);
  pbrInputs.clearcoatF0 = vec3(pow((pbrInputs.ior - 1.0) /
                          (pbrInputs.ior + 1.0), 2.0));
  pbrInputs.clearcoatF90 = vec3(1.0);
  if (mat.clearCoatNormalTextureUV > -1) {
    pbrInputs.clearcoatNormal = mat3(
      pbrInputs.t, pbrInputs.b, pbrInputs.ng) *
      sampleClearcoatNormal(tc, mat).rgb;
  } else {
    pbrInputs.clearcoatNormal = pbrInputs.ng;
  }
  clearCoatContrib = getIBLRadianceGGX(
    pbrInputs.clearcoatNormal, pbrInputs.v,
    pbrInputs.clearcoatRoughness,
    pbrInputs.clearcoatF0, 1.0, envMap);
}
```

3. We calculate the Fresnel term for the clearcoat layer using a similar approach. We apply the Schlick approximation, but with input data specific to the clearcoat:

```
vec3 clearcoatFresnel = vec3(0);
if (isClearCoat) {
  clearcoatFresnel = F_Schlick(
    pbrInputs.clearcoatF0,
    pbrInputs.clearcoatF90,
    clampedDot(pbrInputs.clearcoatNormal, pbrInputs.v));
}
```

4. Finally, at the very end of the fragment shader `data/shaders/gltf/main.frag`, we apply the clearcoat contribution on top of all layers, including emissive ones! Note the `sheenColor` value here, which will be covered in the next recipe *Implementing the sheen material extension*.

```
vec3 color =
  specularColor + diffuseColor + emissiveColor + sheenColor;
color = color * (1.0 - pbrInputs.clearcoatFactor * clearcoatFresnel) +
  clearCoatContrib;
```

This concludes all the necessary GLSL changes to implement the clearcoat extension. The demo app should look like the screenshot below:

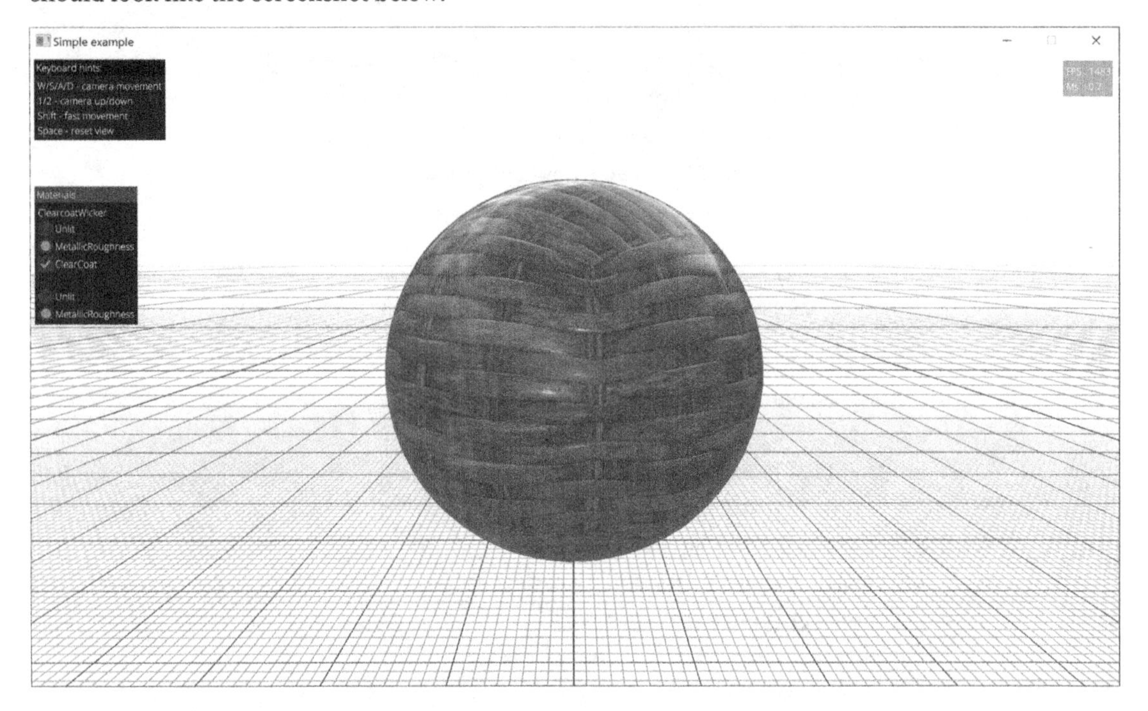

Figure 7.2: KHR_materials_clearcoat example

Notice the glossy layer on top of the ball—this is the clearcoat! Congratulations, we've completed our first advanced PBR extension.

There's more...

The Khronos glTF extensions repository includes a comprehensive list of references on clearcoat materials: `https://github.com/KhronosGroup/glTF/blob/main/extensions/2.0/Khronos/KHR_materials_clearcoat/README.md#reference`.

Implementing the KHR_materials_sheen extension

The `KHR_materials_sheen` extension improves the glTF 2.0 Metallic-Roughness material by adding a layer that simulates the sheen effect found on fabrics like satin or brushed metals. This enhancement creates more realistic and visually appealing sheen highlights.

Sheen BRDF sits on top of the glTF 2.0 Metallic-Roughness material. If the previous extension `KHR_materials_clearcoat` is also active, it is layered on top of the sheen effect.

The `sheenColorFactor` property controls the base intensity of the sheen effect, independent of the viewing angle. A value of `0` disables **sheen** entirely.

Sheen parameters

- `sheenColorTexture`/`sheenColorFactor`: If the texture is defined, the sheen color is calculated by multiplying `sheenColorFactor` and the texture's RGB value.
- `sheenRoughnessTexture`/`sheenRoughnessFactor`: If defined, the sheen roughness is calculated by multiplying with the texture's alpha channel value.

If no textures are specified, `sheenColorFactor` directly controls the sheen color, and `sheenRoughnessFactor` directly controls sheen roughness.

Simulating the sheen effect

The **sheen BRDF** simulates how light scatters off velvet-like materials. It models how light bounces off tiny fibers that are oriented perpendicular to the surface. Sheen roughness controls how much these fibers deviate from that direction:

- Lower sheen roughness: the fibers are more aligned, creating a sharper sheen highlight when light grazes the surface.
- Higher sheen roughness: the fibers are more scattered, leading to a softer sheen highlight.

The sheen BRDF is mathematically based on an exponentiated sinusoidal distribution, derived from the microfacet theory (Conty & Kulla, 2017 `https://blog.selfshadow.com/publications/s2017-shading-course/#course_content`).). The roughness is mapped using `r=sheenRoughness^2` for a more intuitive understanding of roughness changes.

Sheen roughness operates independently of the material's base roughness. This makes it possible for a material to have a rough surface texture (high base roughness) while still displaying a sharp sheen effect (low sheen roughness).

Not all incoming light interacts with the microfibers. Some light may directly reach the base layer or bounce between the fibers before doing so. The behavior of this light is governed by the underlying glTF 2.0 PBR Metallic-Roughness material properties.

Getting ready

The source code for this recipe can be found in `Chapter07/02_Sheen/`.

How to do it...

Similar to the previous recipe *Implementing the clear coat material extension*, we introduce a bunch of new material parameters. Let's take a look at the C++ code.

1. Let's load a `.gltf` file to demonstrate the effect `Chapter07/02_Sheen/src/main.cpp`:

```
GLTFContext gltf(app);
loadGLTF(gltf,
  "deps/src/glTF-Sample-Assets/Models/
    SheenChair/glTF/SheenChair.gltf",
  "deps/src/glTF-Sample-Assets/Models/SheenChair/glTF/");
```

2. In the file `shared/UtilsGLTF.h`, we added new member fields to the `GLTFMaterialDataGPU` structure in:

```
struct GLTFMaterialDataGPU {
  // ...
  vec4 sheenFactors = vec4(1.0f, 1.0f, 1.0f, 1.0f);
  uint32_t sheenColorTexture              = 0;
  uint32_t sheenColorTextureSampler       = 0;
  uint32_t sheenColorTextureUV            = 0;
  uint32_t sheenRoughnessTexture          = 0;
  uint32_t sheenRoughnessTextureSampler = 0;
  uint32_t sheenRoughnessTextureUV        = 0;
```

3. Let's load new parameters via Assimp in `shared/UtilsGLTF.cpp` and store them into the sheen material. The sheen color texture is `sRGB` and has the index `0` and the roughness texture has the index of `1`:

```
loadMaterialTexture(mtlDescriptor, aiTextureType_SHEEN,
  assetFolder, mat.sheenColorTexture, ctx, true, 0);
loadMaterialTexture(mtlDescriptor, aiTextureType_SHEEN,
  assetFolder, mat.sheenRoughnessTexture, ctx, false, 1);
bool useSheen = !mat.sheenColorTexture.empty() ||
                !mat.sheenRoughnessTexture.empty();
aiColor4D sheenColorFactor;
if (mtlDescriptor->Get(AI_MATKEY_SHEEN_COLOR_FACTOR,
    sheenColorFactor) == AI_SUCCESS) {
  res.sheenFactors = vec4(sheenColorFactor.r,
                          sheenColorFactor.g,
                          sheenColorFactor.b,
                          sheenColorFactor.a);

  useSheen = true;
}
```

```
ai_real sheenRoughnessFactor;
if (mtlDescriptor->Get(AI_MATKEY_SHEEN_ROUGHNESS_FACTOR,
    sheenRoughnessFactor) == AI_SUCCESS) {
  res.sheenFactors.w = sheenRoughnessFactor;
  useSheen = true;
}
if (assignUVandSampler(samplers, mtlDescriptor,
    aiTextureType_SHEEN, res.sheenColorTextureUV,
    res.sheenColorTextureSampler, 0)) {
  useSheen = true;
}
if (assignUVandSampler(samplers, mtlDescriptor,
    aiTextureType_SHEEN, res.sheenRoughnessTextureUV,
    res.sheenRoughnessTextureSampler, 1)) {
  useSheen = true;
}
if (useSheen) res.materialTypeFlags |= MaterialType_Sheen;
// ...
```

As you can see, we follow the same pattern as with the clearcoat extension and set the flag `MaterialType_Sheen` only when we use this extension. This is it for the main C++ code.

The Sheen extension needs a different BRDF function, which was discussed in the recipe *Precomputing BRDF look-up tables* of *Chapter 6*. We recommend reviewing that recipe to refresh your understanding of how precomputed BRDF LUTs work and revisiting the implementation details.

Now let's take a look at the GLSL shader code changes which follow a similar pattern:

1. Let's introduce some utility functions in `data/shaders/gltf/inputs.frag`. We can simplify these functions by pre-multiplying textures values by sheen factors. In the C++ code, we set `sheenColorTexture` and `sheenRoughnessTexture` to use a white 1x1 texture in case when no texture data is provided in the `.gltf` asset. In this case, it is always correct to multiply these values by identity factors. We still perform a texture lookup for this small texture, but the overhead is minimal. These small textures should always fit into the GPU's fastest cache:

```
vec4 getSheenColorFactor(
  InputAttributes tc, MetallicRoughnessDataGPU mat)
{
  return vec4(mat.sheenFactors.xyz, 1.0f) *
    textureBindless2D(mat.sheenColorTexture,
                      mat.sheenColorTextureSampler,
                      tc.uv[mat.sheenColorTextureUV]);
}
float getSheenRoughnessFactor(
```

```
  InputAttributes tc, MetallicRoughnessDataGPU mat)
{
  return mat.sheenFactors.a * textureBindless2D(
    mat.sheenRoughnessTexture,
    mat.sheenRoughnessTextureSampler,
    tc.uv[mat.sheenRoughnessTextureUV]).a;
}
```

2. The GLSL code of the sheen extension is scattered between the main fragment shader `data/shaders/gltf/main.frag` and the PBR module `data/shaders/gltf/PBR.sp`. In `main.frag`, we apply sheen parameters:

```
// ...
if (isSheen) {
  pbrInputs.sheenColorFactor = getSheenColorFactor(tc, mat).rgb;
  pbrInputs.sheenRoughnessFactor = getSheenRoughnessFactor(tc, mat);
}
// ...
```

3. In the next step during IBL calculations, we accumulate the sheen contribution calculated using the Charlie distribution:

```
vec3 sheenColor = vec3(0);
if (isSheen) {
  sheenColor += getIBLRadianceCharlie(pbrInputs, envMap);
}
```

4. Let's take a look at the implementation of `getIBLRadianceCharlie()` in the file `data/shaders/gltf/PBR.sp`. This function is similar to `getIBLRadianceGGX()` used for metallic-roughness, but is much simpler. The Sheen extension provides its own roughness value, so no perceptual adjustments are needed. All we have to do here is to multiply `sheenRoughnessFactor` by the total number of mip-levels `mipCount` to determine the correct mip-level, sample the precalculated environment map, and then multiply it by the `BRDF` and `sheenColor`.

```
vec3 getIBLRadianceCharlie(PBRInfo pbrInputs,
  EnvironmentMapDataGPU envMap) {
  float sheenRoughness = pbrInputs.sheenRoughnessFactor;
  vec3 sheenColor = pbrInputs.sheenColorFactor;
  float mipCount  = float(sampleEnvMapQueryLevels(envMap));
  float lod = sheenRoughness * float(mipCount - 1);
  vec3 reflection = normalize(reflect(-pbrInputs.v, pbrInputs.n));
  vec2 brdfSamplePoint = clamp(vec2(pbrInputs.NdotV,
    sheenRoughness), vec2(0.0, 0.0), vec2(1.0, 1.0));
  float brdf = sampleBRDF_LUT(brdfSamplePoint, envMap).b;
  vec3 sheenSample = sampleCharlieEnvMapLod(
```

```
      reflection.xyz, lod, envMap).rgb;
    return sheenSample * sheenColor * brdf;
  }
```

5. Let's go back to `data/shaders/gltf/main.frag`. We modify the sheen contribution based on the value of `occlusionStrength`. The calculation of `lights_sheen` will be covered in the last recipe *Extend analytical lights support* in this chapter. For now, assume it is just zero.

```
    vec3 lights_sheen = vec3(0);
    sheenColor = lights_sheen +
      mix(sheenColor, sheenColor * occlusion, occlusionStrength);
```

This is all the additional code required to implement the Sheen extension. The running demo app should look as in the following screenshot:

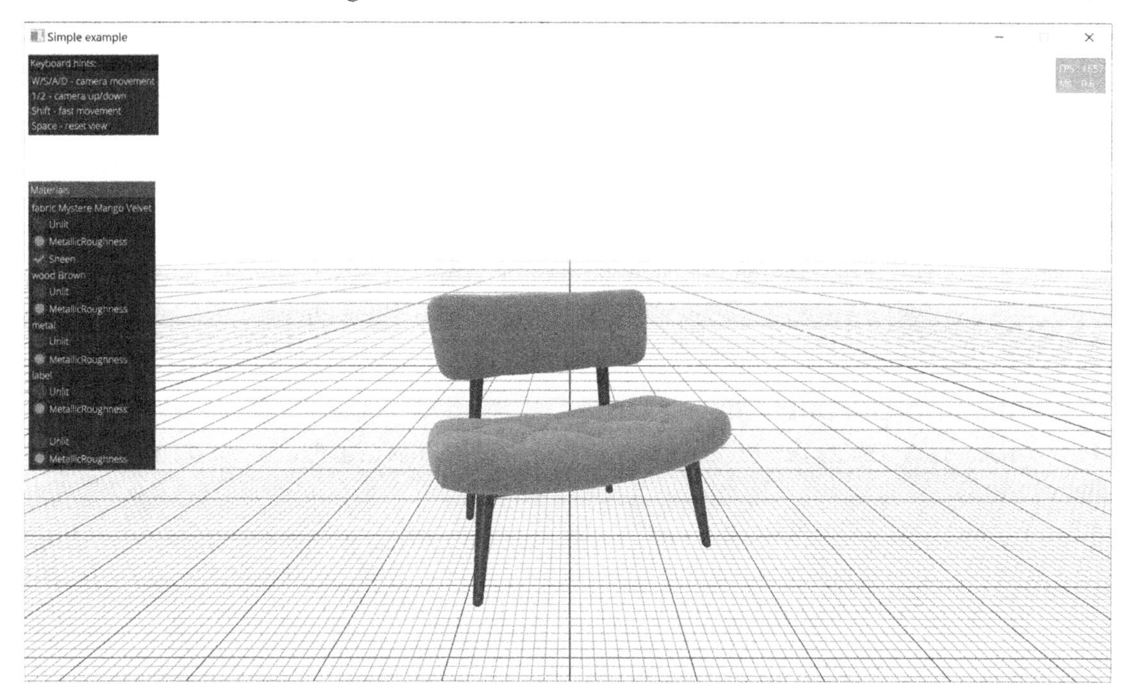

Figure 7.3: KHR_materials_sheen example

There's more...

The Khronos glTF extensions repository provides a comprehensive list of papers on different sheen materials: `https://github.com/KhronosGroup/glTF/blob/main/extensions/2.0/Khronos/KHR_materials_sheen/README.md#reference`.

Implementing the KHR_materials_transmission extension

The glTF 2.0 core specification uses a basic method for handling transparency called **alpha-as-coverage**. While this approach works well for simple materials like gauze or burlap, it doesn't do a good job of representing more complex transparent materials like glass or plastic. These materials involve complicated light interactions—such as reflection, refraction, absorption, and scattering—that alpha-as-coverage can't accurately simulate on its own.

Alpha-as-coverage basically decides if a surface is there or not. A value of `0` means nothing is visible, while a value of `1` means the surface is solid. This method works well for materials with holes or gaps that let light pass through without actually entering the material. However, for materials like glass, light interacts with the surface in more complex ways—like reflecting, refracting, or even being absorbed. Alpha-as-coverage can't handle these kinds of interactions. Additionally, it affects how intense reflections are, making more transparent materials have weaker reflections. This is the opposite of what happens with real-world transparent materials, which often have strong reflections even when they're see-through.

To overcome the limitations of alpha-as-coverage, the `KHR_materials_transmission` extension offers a more realistic way to render transparent materials in glTF. It allows for the simulation of materials that absorb, reflect, and transmit light depending on the angle of incidence and the light's wavelength. This extension is especially useful for accurately representing thin-surface materials like plastic and glass.

The `KHR_materials_transmission` extension targets the simplest cases of optical transparency: infinitely thin materials without refraction, scattering, or dispersion. This simplification enables efficient calculations of refraction and absorption.

Transmission parameters

The `KHR_materials_transmission` extension adds new properties to define material's transmission characteristics:

- `transmissionFactor`: A scalar value between `0` and `1` that represents the material's overall opacity. A value of `0` means the material is fully opaque, while a value of `1` means it is completely transparent.
- `transmissionFilter`: A color value that alters the color of the light passing through the material.

Transmission BTDF

The `KHR_materials_transmission` extension introduces a specular **BTDF** (Bidirectional Transmission Distribution Function) based on the microfacet model. It uses the same **Trowbridge-Reitz** distribution as the specular **BRDF** (Bidirectional Reflectance Distribution Function) but samples along the view vector instead of the reflection direction. This method simulates how microfacets act like tiny prisms, blurring the transmitted light.

The transmission process is modeled as two back-to-back surfaces, representing a thin material. This approach simplifies the process by avoiding the complexities of average refraction and instead focusing on refraction at the microfacet level. The roughness parameter affects both reflection and transmission since the microfacet distribution impacts both sides of the surface. Let's take a look at how to implement this glTF extension.

Getting ready

The source code for this recipe can be found in `Chapter07/03_Transmission/`.

How to do it...

This is the most complex extension discussed in this chapter, requiring changes to the C++ code to handle the complexities of transparency rendering. In addition to updating the C++ rendering code, we need to implement a specular BTDF in the GLSL shader code and incorporate the blending of two layers to accurately represent a thin material.

Let's start with C++ changes.

1. First, we should load a corresponding `.gltf` sample model in `Chapter07/03_Transmission/src/main.cpp`:

```
GLTFContext gltf(app);
loadGLTF(gltf, "deps/src/glTF-Sample-Assets/Models/
    TransmissionRoughnessTest/glTF/
    TransmissionRoughnessTest.gltf",
  "deps/src/glTF-Sample-Assets/Models/
    TransmissionRoughnessTest/glTF/");
```

2. The parameters parsing is in `shared/UtilsGLTF.cpp` and quite simple:

```
loadMaterialTexture(mtlDescriptor, aiTextureType_TRANSMISSION,
  assetFolder, mat.transmissionTexture, ctx, true, 0);
// ...
bool useTransmission = !mat.transmissionTexture.empty();
ai_real transmissionFactor = 0.0f;
if (mtlDescriptor->Get(AI_MATKEY_TRANSMISSION_FACTOR,
    transmissionFactor) == AI_SUCCESS) {
  res.clearcoatTransmissionThickness.z = transmissionFactor;
  useTransmission = true;
}
if (useTransmission) {
  res.materialTypeFlags |= MaterialType_Transmission;
  useVolumetric = true;
}
assignUVandSampler(samplers, mtlDescriptor,
```

```
aiTextureType_TRANSMISSION, res.transmissionTextureUV,
res.transmissionTextureSampler, 0);
```

Significant changes have been made to the rendering function `renderGLTF()`. We touched on some of these in the first recipe *Introduction to glTF PBR extensions*. Now, let's take a closer look at these changes. To effectively render transparent and transmission surfaces, we need to follow these steps:

1. Prepare lists of completely opaque, transmission, and transparent nodes, because these should be rendered in a particular order: opaque nodes first, then transmission, and then transparent.

Note

Keep in mind that rendering transmission nodes doesn't automatically make them transparent! Instead, we need to use the result from rendering opaque nodes. To do this, we must create a copy of the rendered surface and use it as input for the transmission nodes.

2. Pre-allocate an offscreen texture to store the rendered opaque nodes:

```
if (gltf.offscreenTex[0].empty() || isSizeChanged) {
  const lvk::Dimensions res =
    ctx->getDimensions(ctx->getCurrentSwapchainTexture());
  for (Holder<TextureHandle>& holder : gltf.offscreenTex) {
    holder = ctx->createTexture({
      .type          = lvk::TextureType_2D,
      .format        = ctx->getSwapchainFormat(),
      .dimensions    = {res.width, res.height},
      .usage         = lvk::TextureUsageBits_Attachment |
                       lvk::TextureUsageBits_Sampled,
      .numMipLevels = lvk::calcNumMipLevels(res.width, res.height),
      .debugName     = "offscreenTex" });
  }
}
```

3. Create a screen copy when necessary and pass its handle as `transmissionFramebuffer`:

```
const bool screenCopy = gltf.isScreenCopyRequired();
if (screenCopy) {
  buf.cmdCopyImage(
    gltf.offscreenTex[gltf.currentOffscreenTex],],
    ctx->getCurrentSwapchainTexture(),
    ctx->getDimensions(ctx->getCurrentSwapchainTexture()));
  buf.cmdGenerateMipmap(
    gltf.offscreenTex[gltf.currentOffscreenTex]);
```

```
    pushConstants.transmissionFramebuffer =
      gltf.offscreenTex[gltf.currentOffscreenTex].index();
    buf.cmdPushConstants(pushConstants);
  }
```

4. Now we can render transmission nodes using the screen copy as input. It's important to note that we're not using alpha blending for this pass. The nodes are still rendered as opaque, and we simulate transparency by sampling the screen copy. We also specify a texture dependency for LightweightVK here to ensure the correct Vulkan barriers are applied:

```
buf.cmdBeginRendering(renderPass, framebuffer, { .textures = {
  TextureHandle(
    gltf.offscreenTex[gltf.currentOffscreenTex]) } });
buf.cmdBindVertexBuffer(0, gltf.vertexBuffer, 0);
buf.cmdBindIndexBuffer(gltf.indexBuffer, lvk::IndexFormat_UI32);
buf.cmdBindDepthState({ .compareOp = lvk::CompareOp_Less,
                        .isDepthWriteEnabled = true });
buf.cmdBindRenderPipeline(gltf.pipelineSolid);
for (uint32_t transformId : gltf.transmissionNodes) {
  const GLTFTransforms transform = gltf.transforms[transformId];
  const GLTFMesh submesh = gltf.meshesStorage[transform.meshRef];
  buf.cmdDrawIndexed(submesh.indexCount, 1,
    submesh.indexOffset, submesh.vertexOffset, transformId);
}
```

5. The next step is to render all transparent nodes in back-to-front order. We did not change push constants and these transparent nodes use the same offscreen texture as input:

```
buf.cmdBindRenderPipeline(gltf.pipelineTransparent);
for (uint32_t transformId : gltf.transparentNodes) {
  const GLTFTransforms transform = gltf.transforms[transformId];
  const GLTFMesh submesh = gltf.meshesStorage[transform.meshRef];
  buf.cmdDrawIndexed(submesh.indexCount, 1,
    submesh.indexOffset, submesh.vertexOffset, transformId);
}}
```

This is it for the C++ changes. Let's take a look at the GLSL shader changes now.

1. First, we introduce a utility function in `data/shaders/gltf/inputs.frag` to read material inputs. The transmission factor is stored in the `r` channel of the texture:

```
float getTransmissionFactor(
  InputAttributes tc, MetallicRoughnessDataGPU mat)
{
  return mat.clearcoatTransmissionThickness.z *
```

```
    textureBindless2D(mat.transmissionTexture,
      mat.transmissionTextureSampler,
      tc.uv[mat.transmissionTextureUV]
    ).r;
}
```

2. Then, we populate inputs in `data/shaders/gltf/main.frag` if the transmission extension is enabled for this material:

```
// ...
if (isTransmission) {
  pbrInputs.transmissionFactor = getTransmissionFactor(tc, mat);
}
```

3. We calculate the transmission contribution. The volumetric part will be covered in detail in our next recipe for the `KHR_materials_volume` extension *Implementing the volume extension*. In a pure transmission implementation without the volume extension, the transmission part would be similar to GGX/Lambertian, but instead of using the reflection vector, we use the dot product `NdotV`. Implementing just the transmission extension without volume support is not practical, as the volume extension offers greater flexibility to represent effects like refraction, absorption, or scattering without adding excessive complexity on top of transmission:

```
vec3 transmission = vec3(0,0,0);
if (isTransmission) {
  transmission += getIBLVolumeRefraction(
    pbrInputs.n,
    pbrInputs.v,
    pbrInputs.perceptualRoughness,
    pbrInputs.diffuseColor,
    pbrInputs.reflectance0,
    pbrInputs.reflectance90,
    worldPos, getModel(), getViewProjection(),
    pbrInputs.ior,
    pbrInputs.thickness,
    pbrInputs.attenuation.rgb,
    pbrInputs.attenuation.a);
}
```

4. Finally, we add the calculated transmission value to the diffuse contribution scaled by `transmissionFactor`:

```
if (isTransmission) {
  diffuseColor = mix(diffuseColor, transmission,
    pbrInputs.transmissionFactor);
}
```

The resulting rendered 3D model should look as in the following screenshot:

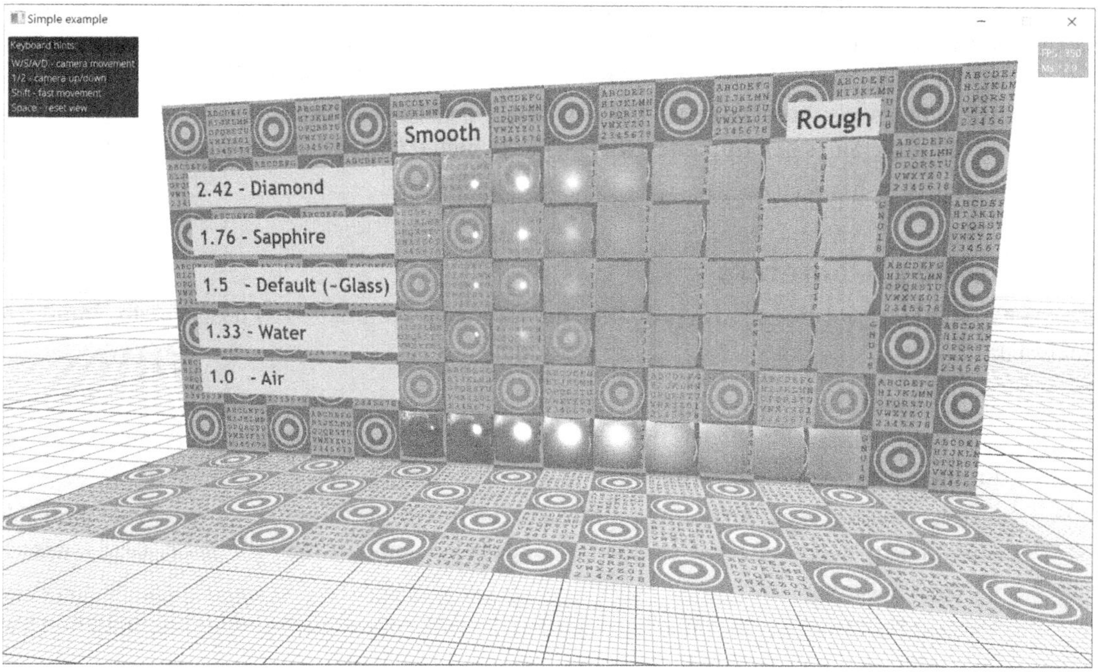

Figure 7.4: KHR_materials_transmission example

There is more...

Rendering transparent objects efficiently and accurately in real-time is challenging, particularly when dealing with overlapping transparent polygons. Issues such as order-dependent transparency and the need for separate blending operations for absorption and reflections add to the complexity. We will address some of these issues in *Chapter 11*.

The Khronos extensions repository provides comprehensive reference materials for the transmission extension: `https://github.com/KhronosGroup/glTF/blob/main/extensions/2.0/Khronos/KHR_materials_transmission/README.md#reference`.

Implementing the KHR_materials_volume extension

The `KHR_materials_volume` extension adds volumetric effects to the glTF 2.0 ecosystem, enabling the creation of materials with depth and internal structure. It's crucial for accurately rendering materials such as smoke, fog, clouds, and translucent objects.

Volumetric effects are different from surface-based materials. While surface-based materials focus on how light interacts with a surface, volumetric materials describe how light moves through a medium. This includes simulating how light scatters and gets absorbed as it passes through the volume.

To create realistic volumetric effects, the `KHR_materials_volume` extension needs to work alongside other extensions that define light interactions with the material's surface. The `KHR_materials_transmission` extension is key here, as it lets light rays pass through the surface and enter the volume. Once inside, the light's interaction with the material is no longer affected by the surface properties. Instead, the light travels through the volume, undergoing refraction and attenuation. When it exits the volume, its direction is determined by the angle at which it leaves the volume boundary.

Let's explore how to add this extension to our glTF renderer.

Volumetric parameters

The `KHR_materials_volume` extension defines the following parameters to describe a volumetric material:

- `thicknessFactor`: A scalar floating-point value that represents the base thickness of the volume. This value is multiplied by the thickness texture value (if available) to determine the final thickness at any point on the surface.
- `attenuationDistance`: A floating-point value that indicates the distance over which the volume's density decreases. This parameter controls how quickly the volume's opacity fades as light passes through it.
- `attenuationColor`: A color value representing the base color of the volume's attenuation. This color influences how light is absorbed as it travels through the volume.
- `thicknessTexture`: An optional texture that adds extra detail about the volume's thickness. The values in the texture are multiplied by the `thicknessFactor` to determine the final thickness at each point on the surface.

Here's how these parameters work together:

- **Thickness:** The `thicknessFactor` and `thicknessTexture` are multiplied to define the volume's depth. A higher thickness value results in a thicker volume.
- **Attenuation:** The `attenuationDistance` and `attenuationColor` control how light is absorbed as it travels through the volume. A smaller value of `attenuationDistance` leads to quicker attenuation. The value of `attenuationColor` determines the color change due to absorption.

Note

The `KHR_materials_volume` extension currently assumes a homogeneous volume, where the material properties are uniform throughout. Future extensions may add support for heterogeneous volumes with varying properties.

Getting ready

The source code for this recipe can be found in `Chapter07/04_Volume/`. Please review the previous recipe *Implementing the transmission extension*, where we covered the transmission and volumetric C++ rendering flow.

How to do it...

This extension requires only a handful of changes across our existing C++ and GLSL shader code. This extension requires `KHR_materials_transmission` support and works only in conjunction with it.

Let's explore how to create advanced volumetric effects such as refraction, absorption, or scattering, starting with the C++.++. code

1. First, we load a corresponding `.gltf` model in `Chapter07/04_Volume/main.cpp`:

```
GLTFContext gltf(app);
loadGLTF(gltf, "deps/src/glTF-Sample-Assets/Models/
    DragonAttenuation/glTF/DragonAttenuation.gltf",
 "deps/src/glTF-Sample-Assets/Models/DragonAttenuation/glTF/");
```

2. Parsing the volume parameters using Assimp in `shared/UtilsGLTF.cpp`:

```
// …
loadMaterialTexture(mtlDescriptor, aiTextureType_TRANSMISSION,
  assetFolder, mat.thicknessTexture, ctx, true, 1);
bool useVolume = !mat.thicknessTexture.empty();
ai_real thicknessFactor = 0.0f;
if (mtlDescriptor->Get(AI_MATKEY_VOLUME_THICKNESS_FACTOR,
    thicknessFactor) == AI_SUCCESS) {
  res.clearcoatTransmissionThickness.w = thicknessFactor;
  useVolume = true;
}
ai_real attenuationDistance = 0.0f;
if (mtlDescriptor->Get(AI_MATKEY_VOLUME_ATTENUATION_DISTANCE,
    attenuationDistance) == AI_SUCCESS) {
  res.attenuation.w = attenuationDistance;
  useVolume = true;
}
aiColor4D volumeAttenuationColor;
if (mtlDescriptor->Get(AI_MATKEY_VOLUME_ATTENUATION_COLOR,
    volumeAttenuationColor) == AI_SUCCESS) {
  res.attenuation.x = volumeAttenuationColor.r;
  res.attenuation.y = volumeAttenuationColor.g;
  res.attenuation.z = volumeAttenuationColor.b;
  useVolume         = true;
}
if (useVolume) {
  res.materialTypeFlags |= MaterialType_Transmission |
                           MaterialType_Volume;
```

```
    useVolumetric = true;
  }
  assignUVandSampler(samplers, mtlDescriptor,
    aiTextureType_TRANSMISSION, res.thicknessTextureUV,
    res.thicknessTextureSampler, 1);
  // ...
```

Most of the magic is hidden in GLSL shaders. Let's step inside `data/shaders/gltf/PBR.sp` and check important helper functions.

1. The function `getVolumeTransmissionRay()` calculates the direction of refracted light `refractionVector` and uses the `modelScale` factor to get an actual lookup vector inside a volume. Note that the `thickness` factor is designed to be normalized to the actual scale of the mesh.

```
vec3 getVolumeTransmissionRay(
  vec3 n, vec3 v, float thickness, float ior, mat4 modelMatrix)
{
  vec3 refractionVector = refract(-v, n, 1.0 / ior);
```

2. Compute rotation-independent scaling of the model matrix. The `thickness` factor is specified in local space:

```
  vec3 modelScale = vec3(length(modelMatrix[0].xyz),
                         length(modelMatrix[1].xyz),
                         length(modelMatrix[2].xyz));
  return
    normalize(refractionVector) * thickness * modelScale.xyz;
}
```

Another helper function is `getIBLVolumeRefraction()`. This function has several important steps:

1. The first step is to get a transmission ray `transmissionRay` and calculate the final refraction position:

```
vec3 getIBLVolumeRefraction(vec3 n, vec3 v,
  float perceptualRoughness, vec3 baseColor, vec3 f0, vec3 f90,
  vec3 position, mat4 modelMatrix, mat4 viewProjMatrix,
  float ior, float thickness, vec3 attenuationColor,
  float attenuationDistance)
{
  vec3 transmissionRay =
    getVolumeTransmissionRay(n, v, thickness, ior, modelMatrix);
  vec3 refractedRayExit = position + transmissionRay;
```

2. We project the refracted vector onto the framebuffer, convert it to normalized device coordinates, and sample the color of the pixel where the refracted ray intersects the framebuffer. The refracted framebuffer coordinates should be transformed from the `-1...+1` range into `0...1` and then flipped vertically:

```
vec4 ndcPos = viewProjMatrix * vec4(refractedRayExit, 1.0);
vec2 refractionCoords = ndcPos.xy / ndcPos.w;
refractionCoords += 1.0;
refractionCoords /= 2.0;
refractionCoords.y = 1.0 - refractionCoords.y;
vec3 transmittedLight = getTransmissionSample(
  refractionCoords, perceptualRoughness, ior);
```

3. After that, we apply volume attenuation and sample GGX BRDF to get the specular component and modulate it by `baseColor` and `attenuatedColor`:

```
  vec3 attenuatedColor = applyVolumeAttenuation(transmittedLight,
    length(transmissionRay), attenuationColor, attenuationDistance);
  float NdotV = clampedDot(n, v);
  vec2 brdfSamplePoint = clamp(vec2(NdotV, perceptualRoughness),
    vec2(0.0, 0.0), vec2(1.0, 1.0));
  vec2 brdf = sampleBRDF_LUT(brdfSamplePoint,
    getEnvironmentMap(getEnvironmentId())).rg;
  vec3 specularColor = f0 * brdf.x + f90 * brdf.y;
  return (1.0 - specularColor) * attenuatedColor * baseColor;
}
```

4. Here's the function `getTransmissionSample()`. We use a copy of the framebuffer as we explained in the previous recipe *Implementing the transmission extension*:

```
vec3 getTransmissionSample(
  vec2 fragCoord, float roughness, float ior)
{
  const ivec2 size =
    textureBindlessSize2D(perFrame.transmissionFramebuffer);
  const vec2 uv = fragCoord;
  float framebufferLod =
    log2(float(size.x)) * applyIorToRoughness(roughness, ior);
  vec3 transmittedLight = textureBindless2DLod(
    perFrame.transmissionFramebuffer,
    perFrame.transmissionFramebufferSampler,
    uv, framebufferLod).rgb;
  return transmittedLight;
}
```

5. The helper function `applyVolumeAttenuation()` looks as follows. The attenuation distance of `0` means the transmitted color is not attenuated at all. Light attenuation is computed using Beer-Lambert law https://en.wikipedia.org/wiki/Beer%E2%80%93Lambert_law:

```
vec3 applyVolumeAttenuation(vec3 radiance,
  float transmissionDistance,
  vec3 attenuationColor,
  float attenuationDistance)
{
  if (attenuationDistance == 0.0) return radiance;
  vec3 attenuationCoefficient =
    -log(attenuationColor) / attenuationDistance;
  vec3 transmittance =
    exp(-attenuationCoefficient * transmissionDistance);
  return transmittance * radiance;
}
```

6. Now we can go back to `data/shaders/gltf/main.frag` and use `getIBLVolumeRefraction()`, together with other helper functions as described in the previous recipe *Implementing the transmission extension*:

```
// ...
vec3 transmission = vec3(0,0,0);
if (isTransmission) {
  transmission += getIBLVolumeRefraction(
    pbrInputs.n, pbrInputs.v,
    pbrInputs.perceptualRoughness,
    pbrInputs.diffuseColor, pbrInputs.reflectance0,
    pbrInputs.reflectance90,
    worldPos, getModel(), getViewProjection(),
    pbrInputs.ior, pbrInputs.thickness,
    pbrInputs.attenuation.rgb, pbrInputs.attenuation.w);
}
```

The resulting demo application should render a translucent dragon, similar to the screenshot below. You can move the camera around the dragon to observe how light interacts with the volume from different angles. This will allow you to see how the light passes through the medium and interacts with the volumetric material.

Figure 7.5: KHR_materials_volume example

After we explored a range of complex PBR glTF extensions, it is time to switch gears. Let's take a look at something a bit more straightforward: implementing the **Index-of-Refraction** extension. This simpler extension is a great way to continue building your understanding while giving you a break from the more complex topics we have covered.

There is more...

An alternative way to implement this extension could involve volume ray-casting or ray-tracing. We leave that as an exercise for our readers.

The Khronos extensions repository provides comprehensive reference materials for this extension: `https://github.com/KhronosGroup/glTF/blob/main/extensions/2.0/Khronos/KHR_materials_volume/README.md#overview`.

Implementing the KHR_materials_ior extension

The `KHR_materials_ior` extension for glTF 2.0 adds the concept of **Index-of-Refraction** (**IOR**) to materials, allowing for more accurate and realistic simulations of transparent objects. IOR is a key material property that dictates how light bends when it passes through a substance `https://en.wikipedia.org/wiki/Refractive_index`.

The **IOR** is a dimensionless number that shows the ratio of the speed of light in a vacuum to its speed in a particular medium. Different materials have different IOR values, which influence how light bends when it enters or exits the material. A higher IOR means more refraction. For instance, the IOR of air is nearly `1`, water has an IOR of about `1.33`, and glass has an IOR of around `1.5`.

IOR parameters

The `KHR_materials_ior` extension adds a single property to the glTF material definition:

`ior`: A floating-point value representing the index-of-refraction of the material.

This value is used in conjunction with the `KHR_materials_transmission` extension to calculate the refraction direction of light rays when passing through the material.

Getting ready

The source code for this recipe can be found in `Chapter07/05_IOR/`. Check out the recipe *Implementing the transmission extension* to recap how `KHR_materials_transmission` is implemented.

How to do it...

This extension requires only a handful of changes across C++ and GLSL shader code. Let's start with C++.

1. In `Chapter07/05_IOR/main.cpp`, load a corresponding `.gltf` model:

   ```
   GLTFContext gltf(app);
   loadGLTF(gltf,
     "deps/src/glTF-Sample-Assets/Models/
       MosquitoInAmber/glTF/MosquitoInAmber.gltf",
     "deps/src/glTF-Sample-Assets/Models/MosquitoInAmber/glTF/");
   ```

2. Here is the code in `shared/UtilsGLTF.cpp` for parsing the IOR material parameter with *Assimp*. You will notice that we don't set any material flag, it's not needed. IOR is just a value and does not alter the functionality of shaders.

   ```
   ai_real ior;
   if (mtlDescriptor->Get(AI_MATKEY_REFRACTI, ior) == AI_SUCCESS) {
     res.ior = ior;
   }
   ```

Here goes the GLSL shader code. We need to modify a few values:

1. The first value is in the PBR module `data/shaders/gltf/PBR.sp` in the function `calculatePBRInputsMetallicRoughness()`. The default index of refraction value, `ior = 1.5`, results in the `f0` term being calculated as `0.04`:

   ```
   PBRInfo calculatePBRInputsMetallicRoughness(InputAttributes tc,
     vec4 albedo, vec4 mrSample, MetallicRoughnessDataGPU mat) {
     PBRInfo pbrInputs;
     // ...
     vec3 f0 = isSpecularGlossiness ?
       getSpecularFactor(mat) * mrSample.rgb :
       vec3(pow((pbrInputs.ior - 1)/( pbrInputs.ior + 1), 2));
   ```

2. The second part is in `data/shaders/gltf/main.frag` and modifies the clearcoat reflectance value:

    ```
    if (isClearCoat) {
      // ...
      pbrInputs.clearcoatF0 = vec3(
        pow((pbrInputs.ior - 1.0) / (pbrInputs.ior + 1.0), 2.0));
      // ...
    }
    ```

That's it! The application should now render a mosquito encased in a piece of amber, as shown in the following screenshot:

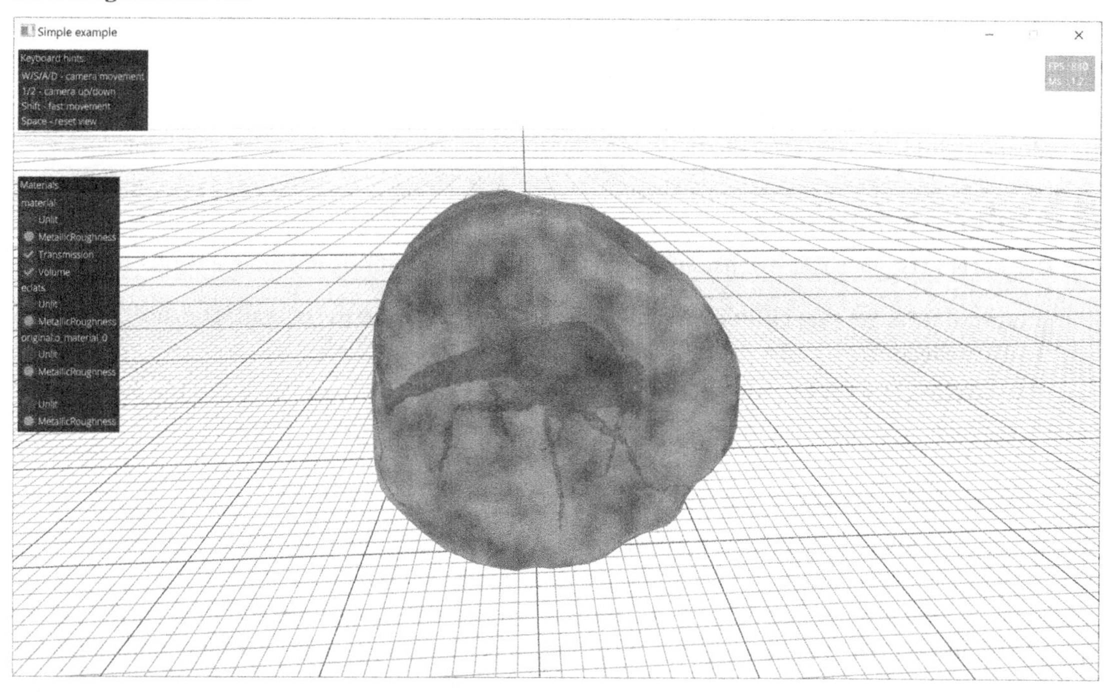

Figure 7.6: KHR_materials_ior example

There is more...

The official extension specification includes a normative section that explains how this glTF extension `KHR_material_ior` interacts with other extensions: `https://github.com/KhronosGroup/glTF/blob/main/extensions/2.0/Khronos/KHR_materials_ior/README.md#interaction-with-other-extensions`.

Implementing the KHR_materials_specular extension

In the previous chapter, we discussed the **Specular-Glossiness** PBR model. One of its main issues is the lack of compatibility with most other extensions. This is because it introduces non-physically based material properties, including an unclear distinction between dielectrics and metals in specular-glossiness mode, which makes it impossible to combine with the metallic-roughness model or other extension properties.

As an alternative, Khronos proposed the `KHR_materials_specular` extension, which addresses these issues and offers the functionality of `KHR_materials_pbrSpecularGlossiness` without compromising the physical accuracy of the **Metallic-Roughness** PBR model. This makes it compatible with most glTF PBR extensions. At the time of writing, the `KHR_materials_specular` extension is only incompatible with the `KHR_materials_pbrSpecularGlossiness` and `KHR_materials_unlit` extensions.

The `KHR_materials_specular` extension allows for more precise control over specular reflections in glTF materials. While the core glTF specification includes a basic specular BRDF, this extension introduces additional parameters to better fine-tune the appearance of specular highlights.

Specular parameters

The `KHR_materials_specular` extension introduces several parameters to enhance specular reflections:

- `specularFactor/specularTexture`: A scalar value that scales the overall intensity of the specular reflection.
- `specularColorFactor/specularColorTexture`: A color value that modifies the color of the specular reflection.

Specular-Glossiness conversion

You can convert **Specular-Glossiness** materials to the **Metallic-Roughness** workflow using the `KHR_materials_ior` extension described in the previous recipe *Implementing the index-of-refraction extension*. By setting the `IOR` parameter to `0`, the material is treated as a dielectric with the maximum specular reflection. The `IOR` parameter controls the upper limit of the specular reflection's strength, and setting it to `0` maximizes this strength, allowing full control over specular reflection through the `specularColorFactor`. This method eliminates the need to classify materials as either dielectric or metallic. It's important to note that materials using the `KHR_materials_volume` extension are incompatible with this conversion due to their non-zero `IOR` value. For new materials, it's often better to use the **Metallic-Roughness** model directly.

Getting ready

The source code for this recipe can be found in `Chapter07/06_Specular/`. Reread two previous recipes *Implementing the index-of-refraction extension* and *Implementing the volume extension* as this extension interacts with them.

How to do it...

This extension doesn't require any major changes to the C++ code, aside from reading the additional material properties via Assimp:

1. Let's load a new `.gltf` model in `Chapter07/06_Specular/src/main.cpp`:

```
VulkanApp app({
  .initialCameraPos    = vec3(0.0f, -0.5f, -1.0f),
  .initialCameraTarget = vec3(0.0f, -1.0f,  0.0f),
});
GLTFContext gltf(app);
loadGLTF(gltf, "deps/src/glTF-Sample-Assets/Models/
    SpecularSilkPouf/glTF/SpecularSilkPouf.gltf",
  "deps/src/glTF-Sample-Assets/Models/SpecularSilkPouf/glTF/");
```

2. To make this example more interesting, we added rotation to our 3D model:

```
const bool rotateModel = true;
const mat4 t = glm::translate(mat4(1.0f), vec3(0, -1, 0));
app.run([&](uint32_t width, uint32_t height,
  float aspectRatio, float deltaSeconds)
{
  const mat4 m = t * glm::rotate(mat4(1.0f),
    rotateModel ? (float)glfwGetTime() : 0.0f,
    vec3(0.0f, 1.0f, 0.0f));
  const mat4 p = glm::perspective(45.0f, aspectRatio, 0.01f, 100.0f);
  renderGLTF(gltf, m, app.camera_.getViewMatrix(), p);
});
```

3. Now, let's load material properties in `shared/UtilsGLTF.cpp`:

```
loadMaterialTexture(mtlDescriptor, aiTextureType_SPECULAR,
  assetFolder, mat.specularTexture, ctx, true, 0);
loadMaterialTexture(mtlDescriptor, aiTextureType_SPECULAR,
  assetFolder, mat.specularColorTexture, ctx, true, 1);
// ...
bool useSpecular = !mat.specularColorTexture.empty() ||
                   !mat.specularTexture.empty();
ai_real specularFactor;
if (mtlDescriptor->Get(AI_MATKEY_SPECULAR_FACTOR,
    specularFactor) == AI_SUCCESS) {
  res.specularFactors.w = specularFactor;
  useSpecular = true;
}
```

```
  assignUVandSampler(samplers, mtlDescriptor,
    aiTextureType_SPECULAR, res.specularTextureUV,
    res.specularTextureSampler, 0);
  aiColor4D specularColorFactor;
  if (mtlDescriptor->Get(AI_MATKEY_COLOR_SPECULAR,
      specularColorFactor) == AI_SUCCESS) {
    res.specularFactors = vec4(specularColorFactor.r,
                               specularColorFactor.g,
                               specularColorFactor.b,
                               res.specularFactors.w);
    useSpecular = true;
  }
  assignUVandSampler(samplers, mtlDescriptor,
    aiTextureType_SPECULAR, res.specularColorTextureUV,
    res.specularColorTextureSampler, 1);
  if (useSpecular)
    res.materialTypeFlags |= MaterialType_Specular;
}
```

GLSL shader changes are a bit more complex. The `specularColor` parameter introduces color variations into the specular reflection. It is integrated into the Fresnel term, influencing the specular reflectance at different viewing angles. At normal incidence, the specular color directly scales the base reflectance (`F0`), while at grazing angles, the reflectance approaches `1.0` regardless of the specular color. To maintain energy conservation, the maximum component of the specular color is used to calculate the scaling factor for the Fresnel term, preventing excessive energy in the specular reflection.

1. First, we introduce some utility functions in `data/shaders/gltf/inputs.frag`:

```
vec3 getSpecularColorFactor(
  InputAttributes tc, MetallicRoughnessDataGPU mat)
{
  return mat.specularFactors.rgb *
    textureBindless2D(mat.specularColorTexture,
                      mat.specularColorTextureSampler,
                      tc.uv[mat.specularColorTextureUV]).rgb;
}
float getSpecularFactor(
  InputAttributes tc, MetallicRoughnessDataGPU mat)
{
  return mat.specularFactors.a *
    textureBindless2D(mat.specularTexture,
                      mat.specularTextureSampler,
                      tc.uv[mat.specularTextureUV]).a;
}
```

2. We add a new field `specularWeight` to the `PBRInfo` structure in `data/shaders/gltf/PBR.sp`.

```
struct PBRInfo {
  // ...
  float specularWeight;
};
```

3. We modify the accessor function for obtaining the `F0` reflectance and populate the `specularWeight` field in the `calculatePBRInputsMetallicRoughness()` function:

```
PBRInfo calculatePBRInputsMetallicRoughness(InputAttributes tc,
  vec4 albedo, vec4 mrSample, MetallicRoughnessDataGPU mat) {
  // ...
  if (isSpecular) {
    vec3 dielectricSpecularF0 =
      min(f0 * getSpecularColorFactor(tc, mat), vec3(1.0));
    f0 = mix(dielectricSpecularF0, pbrInputs.baseColor.rgb, metallic);
    pbrInputs.specularWeight = getSpecularFactor(tc, mat);
  }
  // ...
  vec3 specularColor = isSpecularGlossiness ?
    f0 : mix(f0, pbrInputs.baseColor.rgb, metallic);
  float reflectance = max(
    max(specularColor.r, specularColor.g), specularColor.b);
  // ...
```

4. Now we can use these parameters in `data/shaders/gltf/main.frag` to calculate the specular and diffuse components of the IBL contribution:

```
  vec3 specularColor = getIBLRadianceContributionGGX(
    pbrInputs, pbrInputs.specularWeight, envMap);
  vec3 diffuseColor = getIBLRadianceLambertian(pbrInputs.NdotV,
    n, pbrInputs.perceptualRoughness, pbrInputs.diffuseColor,
    pbrInputs.reflectance0, pbrInputs.specularWeight, envMap);
```

5. Here is how the specular contribution is calculated. Note that we multiply by `specularWeight` at the end of the `getIBLRadianceContributionGGX()` function:

```
vec3 getIBLRadianceContributionGGX(PBRInfo pbrInputs,
  float specularWeight, EnvironmentMapDataGPU envMap) {
  vec3 n = pbrInputs.n;
  vec3 v = pbrInputs.v;
  vec3 reflection = normalize(reflect(-v, n));
  float mipCount = float(sampleEnvMapQueryLevels(envMap));
```

```
  float lod = pbrInputs.perceptualRoughness * (mipCount - 1);
  vec2 brdfSamplePoint = clamp(
    vec2(pbrInputs.NdotV, pbrInputs.perceptualRoughness),
    vec2(0.0, 0.0), vec2(1.0, 1.0));
  vec3 brdf = sampleBRDF_LUT(brdfSamplePoint, envMap).rgb;
  vec3 specularLight = sampleEnvMapLod(reflection.xyz, lod, envMap).rgb;
  vec3 Fr = max(vec3(1.0 - pbrInputs.perceptualRoughness),
                pbrInputs.reflectance0) - pbrInputs.reflectance0;
  vec3 k_S = pbrInputs.reflectance0 + Fr * pow(1.0-pbrInputs.NdotV, 5.0);
  vec3 FssEss = k_S * brdf.x + brdf.y;
  return specularWeight * specularLight * FssEss;
}
```

6. The diffuse contribution looks like this. We scale the Fresnel term and replace it with a `vec3` RGB value to incorporate the `specularColor` contribution into `F0`:

```
vec3 getIBLRadianceLambertian(float NdotV, vec3 n,
  float roughness, vec3 diffuseColor, vec3 F0,
  float specularWeight, EnvironmentMapDataGPU envMap) {
  vec2 brdfSamplePoint =
    clamp(vec2(NdotV, roughness), vec2(0., 0.), vec2(1., 1.));
  vec2 f_ab = sampleBRDF_LUT(brdfSamplePoint, envMap).rg;
  vec3 irradiance = sampleEnvMapIrradiance(n.xyz, envMap).rgb;
  vec3 Fr  = max(vec3(1.0 - roughness), F0) - F0;
  vec3 k_S = F0 + Fr * pow(1.0 - NdotV, 5.0);
  vec3 FssEss = specularWeight * k_S * f_ab.x + f_ab.y;
  float Ems   = (1.0 - (f_ab.x + f_ab.y));
  vec3 F_avg  = specularWeight * (F0 + (1.0 - F0) / 21.0);
  vec3 FmsEms = Ems * FssEss * F_avg / (1.0 - F_avg * Ems);
  vec3 k_D    = diffuseColor * (1.0 - FssEss + FmsEms);
  return (FmsEms + k_D) * irradiance;
}
```

7. Those are all the changes required to implement the `KHR_materials_specular` extension in our glTF renderer. The demo application should render a rotating torus as in the following screenshot:

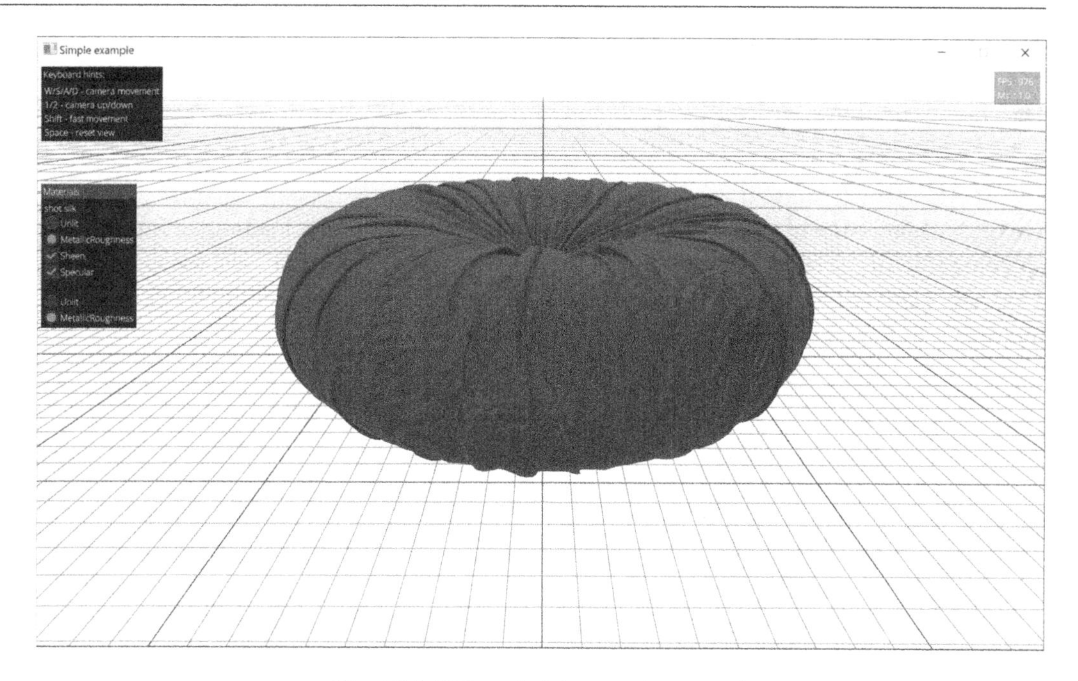

Figure 7.7: KHR_materials_specular example

There is more...

For further details on the motivation behind this approach, please refer to the Khronos specification `https://github.com/KhronosGroup/glTF/blob/main/extensions/2.0/Khronos/KHR_materials_specular/README.md`. This page provides detailed information about the extension, including explanations of the BRDF and insights into converting between different PBR models.

Implementing the KHR_materials_emissive_strength extension

The metallic-roughness core model supports light emission, but before the introduction of the `KHR_materials_emissive_strength` extension, it was difficult to control the intensity of a material's light emission. This made it challenging to create realistic glowing objects or materials that function as light sources in a scene.

The `KHR_materials_emissive_strength` extension overcomes this limitation by introducing a new property called `emissiveStrength`. This property allows for precise control over the intensity of a material's emitted light. With values ranging from `0.0` for no emission to higher values for increased intensity, artists and designers gain more control over the lighting in their scenes.

Getting ready

The source code for this recipe can be found in `Chapter07/07_EmissiveStrength/`.

How to do it...

This extension is one of the simplest to implement. All it requires is loading an intensity value and applying it to the existing emissive value. Essentially, you only need to take the `emissiveStrength` property from *Assimp*, which determines how intense the material's emitted light should be, and multiply it by the emissive color.

1. Let's load a new 3D model in `Chapter07/07_EmissiveStrength/src/main.cpp` to demonstrate this extension.

```
VulkanApp app({
  .initialCameraPos = vec3(0.0f, 5.0f, -10.0f),
});
GLTFContext gltf(app);
loadGLTF(gltf, "deps/src/glTF-Sample-Assets/Models/
    EmissiveStrengthTest/glTF/EmissiveStrengthTest.gltf",
  "deps/src/glTF-Sample-Assets/Models/
    EmissiveStrengthTest/glTF/");
```

2. Here's the C++ code in `shared/UtilsGLTF.cpp` to retrieve the material properties from *Assimp*:

```
if (mtlDescriptor->Get(AI_MATKEY_COLOR_EMISSIVE,
     aiColor) == AI_SUCCESS) {
  res.emissiveFactorAlphaCutoff = vec4(aiColor.r,
                                       aiColor.g,
                                       aiColor.b, 0.5f);
}
assignUVandSampler(samplers, mtlDescriptor,
  aiTextureType_EMISSIVE,
  res.emissiveTextureUV,
  res.emissiveTextureSampler);
ai_real emissiveStrength = 1.0f;
if (mtlDescriptor->Get(AI_MATKEY_EMISSIVE_INTENSITY,
    emissiveStrength) == AI_SUCCESS) {
  res.emissiveFactorAlphaCutoff *= vec4(
    emissiveStrength, emissiveStrength, emissiveStrength, 1.0);
}
```

The resulting demo app should render a set of five glowing cubes, as shown in the following screenshot:

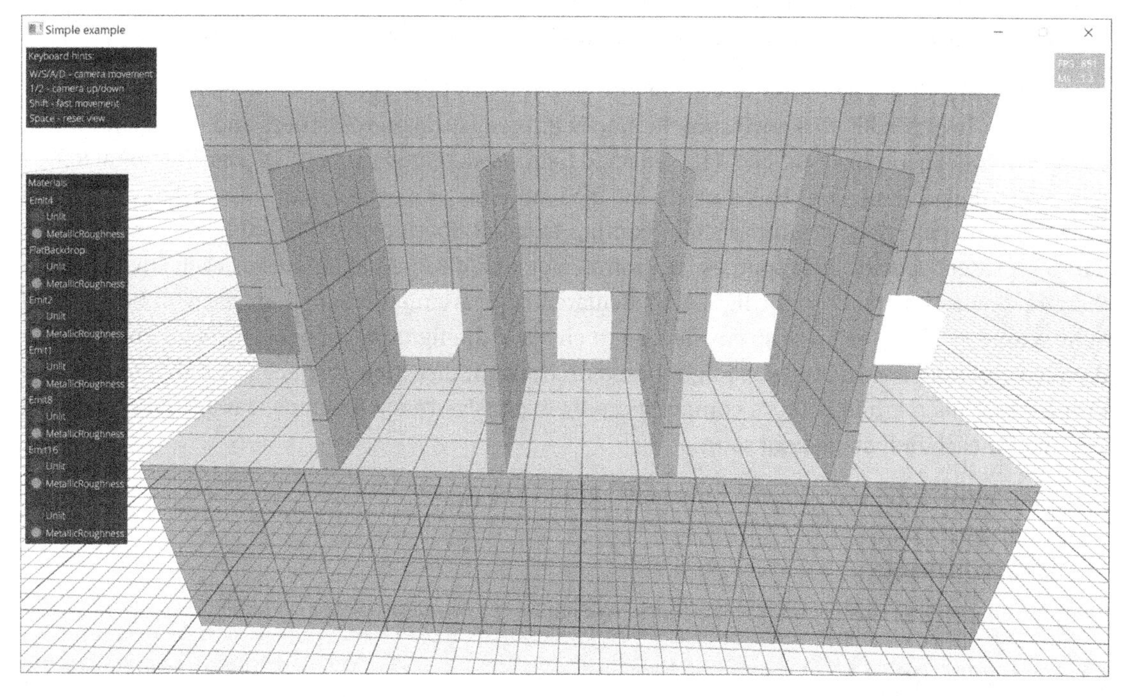

Figure 7.8: KHR_materials_emissive_strength example

Now, let's jump to the final recipe of this chapter, where we will dive into implementing support for glTF analytical lights.

Extend analytical lights support with KHR_lights_punctual

This is the final recipe in this chapter, and we will add support for analytical light sources to our glTF viewer. In the next chapter, we will cover the `KHR_lights_punctual` extension, which will allow us to load lighting information directly from glTF assets. In this recipe, we will only be dealing with shader changes.

In the context of glTF PBR, the terms "analytical" and "punctual" lights are often used interchangeably to describe the same type of light source:

- **Analytical light**: This refers to light sources defined by mathematical equations, enabling precise calculations of its effect on lighting.
- **Punctual light**: This describes light sources that are infinitely small points that emit light in specific directions and intensities.

We will explore these concepts in more detail in the next chapter. In this recipe, fFor simplicity, we'll use both terms interchangeably.

Image-Based Lighting vs Punctual lights

Let's review the difference between Image-Based Lighting (**IBL**) and punctual lights.

IBL simulates indirect lighting from the environment using pre-captured or pre-computed environment maps. In glTF PBR, this environment map is filtered based on roughness and normal direction to approximate incoming radiance. The reflected light is calculated using **BRDF** (Bidirectional Reflectance Distribution Function) based on the surface material properties, with integration performed over the hemisphere to account for light coming from all directions. Punctual lights, on the other hand, represent specific light sources like point, spot, and directional lights. For each surface point, the direction and distance to the light are calculated, with attenuation applied based on how far the light source is. Shadows are also considered to check if the light reaches the surface. The BRDF is then used to calculate the reflected light based on the light direction, surface normal, and material properties. This method is more computationally expensive than IBL since it requires calculating lighting for each individual light source.

Let's take a look at how to add glTF punctual lights to our viewer.

Getting ready

The source code for this recipe can be found in `Chapter07/08_AnalyticalLight`.

How to do it...

C++ code changes are pretty small and straightforward. We introduce additional structures to provide light information data.

1. First, let's load a corresponding `.gltf` model in `Chapter07/08_AnalyticalLight/src/main.cpp`.

```
VulkanApp app({
  .initialCameraPos    = vec3(0.0f, 3.5f, -5.0f),
  .initialCameraTarget = vec3(0.0f, 2.0f,  0.0f),
});
GLTFContext gltf(app);
loadGLTF(gltf, "deps/src/glTF-Sample-Assets/Models/
    LightsPunctualLamp/glTF/LightsPunctualLamp.gltf",
  "deps/src/glTF-Sample-Assets/Models/
    LightsPunctualLamp/glTF/");
```

2. Declare an enumeration for different light types in `shared/UtilsGLTF.h`:

```
enum LightType : uint32_t {
  LightType_Directional = 0,
  LightType_Point       = 1,
  LightType_Spot        = 2,
};
```

3. Here's a structure called `LightDataGPU` to store light information in GPU buffers. It has default values defining a dummy directional light:

```
struct LightDataGPU {
  vec3 direction      = vec3(0, 0, 1);
  float range         = 10000.0;
  vec3 color          = vec3(1, 1, 1);
  float intensity     = 1.0;
  vec3 position       = vec3(0, 0, -5);
  float innerConeCos = 0.0;
  float outerConeCos = 0.78;
  LightType type      = LightType_Directional;
  int padding[2];
};
struct EnvironmentsPerFrame {
  EnvironmentMapDataGPU environments[kMaxEnvironments];
  LightDataGPU lights[kMaxLights];
  uint32_t lightCount;
};
```

4. We set up the light sources in `shared/UtilsGLTF.cpp` as a part of our per-frame constants:

```
const EnvironmentsPerFrame envPerFrame = {
  .environments = { { … } },
  .lights      = { LightDataGPU() },
  .lightCount = 1,
};
```

The changes to the GLSL shader code are substantial. We need to reimplement the specular and diffuse contributions for Metallic-Roughness and other extensions, applying these calculations for each light source individually. In this recipe, we will not go too deep into the implementation details, but we strongly recommend reviewing the actual shaders and the reference materials provided in the comments to fully understand this topic. Here's a brief glimpse of the changes.

1. Let's introduce a couple of utility functions in `data/shaders/gltf/inputs.frag` to conveniently access the light data:

```
uint getLightsCount() {
  return perFrame.environments.lightsCount;
}
Light getLight(uint i) {
  return perFrame.environments.lights[i];
}
```

2. In `data/shaders/gltf/main.frag`, we introduce accumulation variables for each individual contribution component:

```
vec3 lights_diffuse      = vec3(0);
vec3 lights_specular     = vec3(0);
vec3 lights_sheen        = vec3(0);
vec3 lights_clearcoat    = vec3(0);
vec3 lights_transmission = vec3(0);
float albedoSheenScaling = 1.0;
```

3. We iterate over all light sources, calculating the necessary terms for each one and checking if the light source is visible from the rendering point. Then we calculate the light intensity:

```
for (int i = 0; i < getLightsCount(); ++i) {
  Light light = getLight(i);
  vec3 l = normalize(pointToLight);
  vec3 h = normalize(l + v);
  float NdotL = clampedDot(n, l);
  float NdotV = clampedDot(n, v);
  float NdotH = clampedDot(n, h);
  float LdotH = clampedDot(l, h);
  float VdotH = clampedDot(v, h);
  if (NdotL > 0.0 || NdotV > 0.0) {
    vec3 intensity = getLightIntensity(light, pointToLight);
```

Note

Evaluating all lights for every object can be quite costly. Alternatives like clustered or deferred shading can help improve performance in this situation.

4. Then we calculate diffuse and specular contributions for this light:

```
    lights_diffuse += intensity * NdotL *
      getBRDFLambertian(pbrInputs.reflectance0,
        pbrInputs.reflectance90, pbrInputs.diffuseColor,
        pbrInputs.specularWeight, VdotH);
    lights_specular += intensity * NdotL *
      getBRDFSpecularGGX(pbrInputs.reflectance0,
        pbrInputs.reflectance90, pbrInputs.alphaRoughness,
        pbrInputs.specularWeight, VdotH, NdotL, NdotV, NdotH);
```

5. The sheen contribution is now calculated as follows:

```
        if (isSheen) {
          lights_sheen += intensity *
            getPunctualRadianceSheen(pbrInputs.sheenColorFactor,
              pbrInputs.sheenRoughnessFactor, NdotL, NdotV, NdotH);
          albedoSheenScaling =
            min(1.0 - max3(pbrInputs.sheenColorFactor) *
              albedoSheenScalingFactor(NdotV,
                pbrInputs.sheenRoughnessFactor),
                1.0 - max3(pbrInputs.sheenColorFactor) *
             albedoSheenScalingFactor(NdotL, pbrInputs.sheenRoughnessFactor));
        }
```

6. The new clearcoat contribution is calculated in a similar way. Transmission and volume contributions are skipped here for the sake of brevity:

```
        if (isClearCoat) {
          lights_clearcoat += intensity *
            getPunctualRadianceClearCoat(
              pbrInputs.clearcoatNormal, v, l, h, VdotH,
          pbrInputs.clearcoatF0, pbrInputs.clearcoatF90,
            pbrInputs.clearcoatRoughness);
        }
        // … transmission & volume effects are skipped for brevity
      }
    }
```

7. We used a helper function `getLightIntensity()` which is declared in `data/shaders/gltf/PBR.sp`:

```
vec3 getLightIntensity(Light light, vec3 pointToLight) {
  float rangeAttenuation = 1.0;
  float spotAttenuation = 1.0;
  if (light.type != LightType_Directional) {
    rangeAttenuation =
      getRangeAttenuation(light.range, length(pointToLight));
  }
  if (light.type == LightType_Spot) {
    spotAttenuation = getSpotAttenuation(pointToLight,
      light.direction, light.outerConeCos, light.innerConeCos);
  }
  return rangeAttenuation * spotAttenuation * light.intensity * light.color;
}
```

Other helper functions, such as `getBRDFLambertian()`, `getBRDFSpecularGGX()`, `getBRDFSpecularSheen()`, `getPunctualRadianceSheen()`, and many others mentioned in the GLSL code earlier, are defined in `data/shaders/gltf/PBR.sp`. These functions contain the math for calculating the specific BRDF terms. For brevity, we do not include them here.

The running application should render a mesh illuminated by an analytical directional light, as shown in the screenshot below:

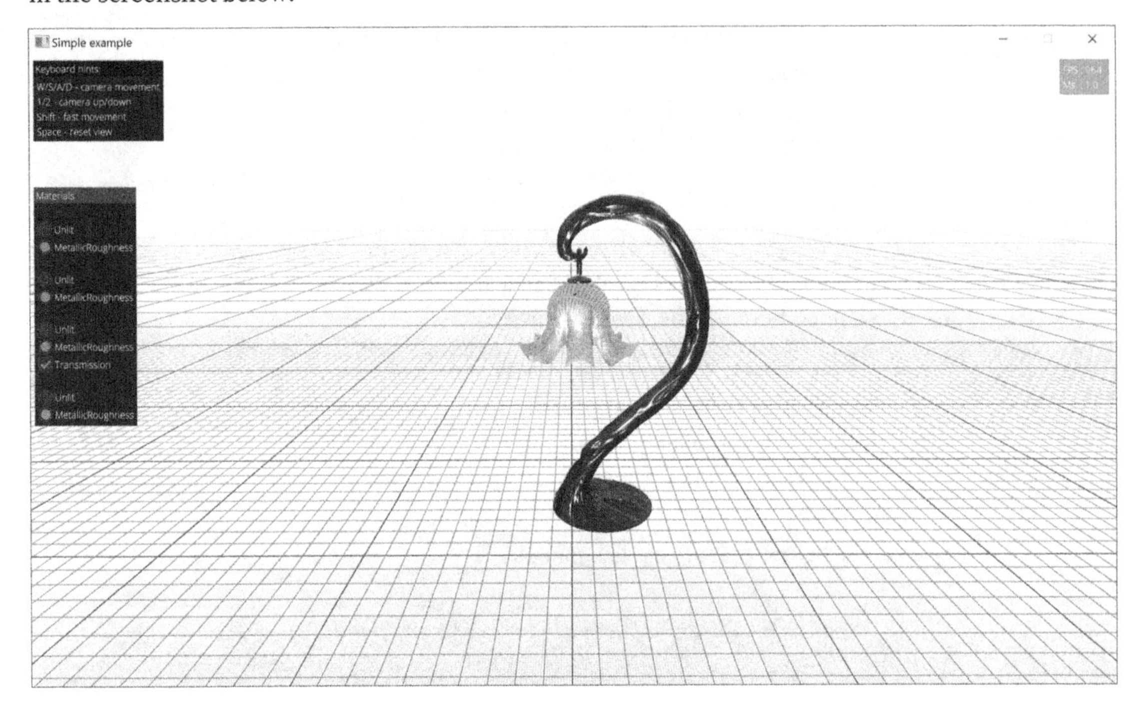

Figure 7.9: Analytical lights example

With this example, we wrap up our chapter on advanced glTF PBR extensions. We have delved into complex topics and extended our understanding of how to work with various lighting models and extensions. In the next chapter, *Graphics Rendering Pipeline*, we will shift our focus to the broader organization of 3D scenes. We will explore various data structures and strategies needed to efficiently manage and render multiple 3D models. This will involve a detailed look at how to structure and optimize data for rendering, ensuring smooth and effective visualization of complex scenes.

8

Graphics Rendering Pipeline

In this chapter, we will learn how to build a hierarchical scene representation and set up a rendering pipeline that combines the rendering of geometry and materials, as discussed in earlier chapters. Instead of implementing a naïve object-oriented scene graph, where each node is an object stored on the heap, we will explore a data-oriented design approach. This will simplify the memory layout of our scene and greatly improve the performance of scene graph manipulations. This approach also serves as an introduction to data-oriented design principles and how to apply them in practice. The scene graph and material representations we cover are compatible with glTF 2.0. We will also look at how to organize the rendering process for complex scenes with multiple materials. We will cover the following recipes:

- How not to do a scene graph
- Using data-oriented design for a scene graph
- Loading and saving a scene graph
- Implementing transformation trees
- Implementing a material system
- Implementing automatic material conversion
- Using descriptor indexing and arrays of textures in Vulkan
- Implementing indirect rendering with Vulkan
- Putting it all together into a scene editing application
- Deleting nodes and merging scene graphs
- Rendering large scenes

How not to do a scene graph

Numerous hobby 3D engines take a straightforward and naïve class-based approach to implementing a scene graph. It is always tempting to define a structure like the one in the following code; please do not do that though:

```
struct SceneNode {
  SceneNode* parent;
```

```
  vector<SceneNode*> children;
  mat4 localTransform;
  mat4 globalTransform;
  Mesh* mesh;
  Material* material;
  void render();
};
```

On top of this structure, one might define various recursive traversal methods, such as the dreaded `render()` function. For example, let's say we have a root object:

```
SceneNode* root;
```

Then, rendering the scene graph could be as simple as:

```
root->render();
```

In this approach, the `render()` method performs several tasks. First, it calculates the global transform for the current node based on the local transform, which is relative to the parent node. Then, depending on the rendering API being used, it sends the mesh geometry and material information to the rendering pipeline. Finally, it makes a recursive call to render all the child nodes. Things can get tricky when you need to handle transparent nodes, and your results may vary:

```
void SceneNode::render() {
  globalTransform =
    (parent ? parent->globalTransform : mat4(1)) * localTransform;
  apiSpecificRenderCalls();
  for (auto& c: this->children)
    c->render();
}
```

While this is a simple and "canonical" object-oriented approach, it comes with several significant drawbacks:

- Non-locality of data due to the use of pointers, unless a custom memory allocator is implemented.
- Performance issues arise due to explicit recursion, which can become especially problematic with deep recursion.
- Risk of memory leaks and crashes when using raw pointers, or a slowdown from atomic operations if smart pointers are used instead.
- Challenges with circular references, requiring the use of weak pointers or other tricks to avoid memory leaks when using smart pointers.

- Recursive loading and saving of the structure is complicated, slow, and prone to errors. We've also seen the visitor pattern used in this context, which only adds to the complexity without solving any performance issues.
- Difficulty implementing extensions, as it requires continually adding more and more fields to the `SceneNode` structure.

As the 3D engine grows and the requirements for the scene graph increase, more fields, arrays, callbacks, and pointers need to be added and managed within the `SceneNode` structure. This makes the approach (besides being slow) fragile and difficult to maintain over time.

Let's take a step back and reconsider how we can preserve the relative structure of the scene without relying on large, monolithic classes filled with heavy dynamic containers.

Using data-oriented design for a scene graph

To represent complex nested visual objects like robotic arms, planetary systems, or intricately branched animated trees, one effective approach is to break the object into smaller parts and track the hierarchical relationships between them. This is known as a scene graph, which is a directed graph showing the parent-child relationships among various objects in a scene. We avoid using the term "acyclic graph" because, for convenience, it's sometimes useful to have controlled circular references between nodes. Many 3D graphics tutorials for hobbyists take a straightforward but less optimal approach, as we discussed in the previous recipe, *How not to do a scene graph*. Now, let's go deeper into the rabbit hole and explore how to use data-oriented design to build a more efficient scene graph.

In this recipe, we will learn how to get started with designing a reasonably efficient scene graph, with a focus on fixed hierarchies. In *Chapter 11*, we will take a closer look at this topic, discussing how to handle runtime topology changes and other scene graph operations.

Getting ready

The source code for this recipe is a part of the scene rendering apps `Chapter08/02_SceneGraph` and `Chapter08/03_LargeScene`. In the previous edition of our book, we included this code in a separate scene preprocessing tool that had to be run manually before the demo apps could use the converted data. This method turned out to be cumbersome and confusing for many readers. This time, all scene conversion and preprocessing tasks will be handled automatically. Let's see how to do it.

How to do it...

Let's break down and refactor the `SceneNode` struct from the previous recipe *How not to do a scene graph*, step by step. It makes sense to store a linear array of individual nodes and replace all the "external" pointers, like `Mesh*` and `Material*`, with appropriately-sized integer handles that simply act as indices into corresponding arrays. We will keep the array of child nodes and references to parent nodes separate. This approach not only significantly improves cache locality but also reduces memory usage, as pointers are 64-bit, while indices can be just 32-bit or, in many scenarios, even 16-bit.

The local and global transforms are also stored in separate arrays, which can, by the way, be easily mapped to a GPU buffer without conversion, making it directly accessible from GLSL shaders. Let's look at the implementation:

1. We have a new simplified scene node declaration and our new scene is a composite of arrays:

```
struct SceneNode {
  int mesh_;
  int material_;
};
struct Scene {
  vector<SceneNode> nodes_;
  vector<mat4> local_;
  vector<mat4> global_;
};
```

2. One question remains: how do we store the hierarchy where each node can have multiple children? A well-known solution is the **Left Child, Right Sibling** tree representation. Since a scene graph is really a tree, at least in theory, without any optimization-related circular references, we can convert any tree with more than two children into a binary tree by "tilting" its branches, as illustrated in the following figure:

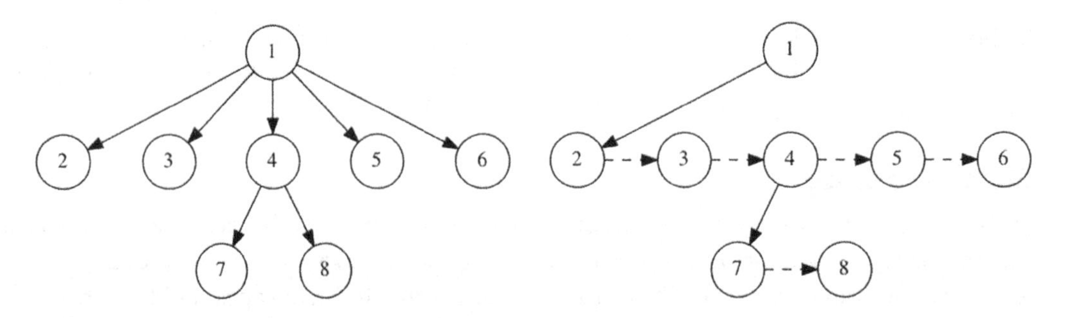

Figure 8.1: Tree representations

More information

For more information on the "Left Child, Right Sibling" representation, you can check out the additional reading available on Wikipedia at `https://en.wikipedia.org/wiki/Left-child_right-sibling_binary_tree`.

1. In this figure *8.1*, the left image depicts a standard tree where each node can have a variable number of children, while the right image illustrates a new structure that stores only a single reference to the first child and another reference to the next "sibling." Here, "sibling node" refers to a child node of the same parent. This transformation eliminates the need to store a `std::vector` in each scene node. If we then "tilt" the right image, we arrive at a familiar binary tree structure, where solid left-pointing arrows represent the "First Child" reference and dashed right-pointing arrows indicate the "Next Sibling" reference:

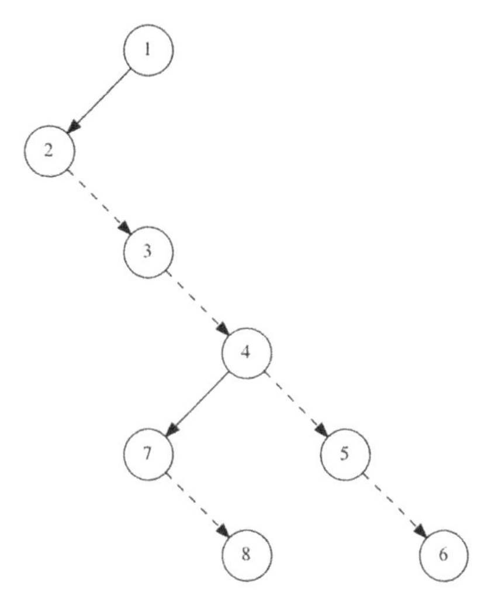

Figure 8.2: A tilted tree

2. Let's incorporate indices into the `SceneNode` structure to reflect this storage schema. In addition to the mesh and material indices for each node, we will include a reference to the parent, an index for the first child (or a negative value if there are no child nodes), and an index for the next sibling scene node:

```
struct SceneNode {
  int mesh_, material_;
  int parent_;
  int firstChild_;
  int rightSibling_;
};
```

What we have now is a compact constant-sized object that is **plain old data**. While the tree traversal and modification routines may seem unconventional, they are essentially just iterations over a linked list. It's important to note a significant drawback: random access to a child node is now slower on average since we must traverse each node in the list. However, this isn't a major issue for our purposes, as we typically traverse either all the children or none of them.

3. Before we move on to the implementation, let's make another unconventional transformation to the new `SceneNode` structure. It currently contains indices for the mesh and material, along with hierarchical information, but the local and global transformations are stored separately. This indicates that we can declare the following structure `shared/Scene/Scene.h` to manage our scene hierarchy:

```
struct Hierarchy {
  int parent = -1;
  int firstChild = -1;
  int nextSibling = -1;
  int lastSibling = -1;
  int level = 0;
};
```

The `lastSibling` member field is a way to speed up the process of adding new child nodes to a parent. We've renamed "left" to "first" and "right" to "next" because relative positions of the tree nodes are not relevant in this context. The `level` field stores a cached depth of the node from the root of the scene graph. The root node has a level of zero, and each child node's level is one greater than that of its parent.

Note

Additionally, the `Mesh` and `Material` objects for each node can be stored in separate arrays. However, if not all nodes have a mesh or material, we can utilize a sparse representation, such as a hash table, to map nodes to their corresponding meshes and materials. The absence of such mapping simply indicates that a scene node is used solely for storing transformations or hierarchical relationships. While hash tables are not as "linear" as arrays, they can easily be converted to and from arrays of `{key, value}` pairs of some sort. We use `std::unordered_map` for the sake of simplicity. Try choosing a faster hash table for your production code. If the majority of your scene nodes contain meshes and materials, you might want to prefer using dense storage data structures, such as arrays, right from the get-go.

4. Now, we can declare a new `Scene` structure in `shared/Scene/Scene.h` with logical "compartments," which we will later refer to as **components**:

```
struct Scene {
  vector<mat4> localTransforms;
  vector<mat4> globalTransforms;
  vector<Hierarchy> hierarchy;
  // Meshes for scene nodes (Node -> Mesh)
  unordered_map<uint32_t, uint32_t> meshForNode;
  // Materials for scene nodes (Node -> Material)
  unordered_map<uint32_t, uint32_t> materialForNode;
```

5. The following components are not strictly necessary but they help a lot while debugging scene graph manipulation routines or while implementing an interactive scene editor, where the ability to see some human-readable node identifiers is crucial:

```
    // Node name component: (Node -> name)
    unordered_map<uint32_t, uint32_t> nameForNode;
    // Collection of scene node names
    vector<std::string> nodeNames;
    // Collection of debug material names
    vector<std::string> materialNames;
  };
```

   One thing that's missing is the SceneNode structure itself, which is now represented by integer indices in the arrays of the Scene container. It might feel quite amusing and unconventional for someone with an object-oriented mindset to discuss SceneNode without actually having or needing a scene node class.

The routine for converting Assimp's aiScene into our scene format is implemented in the shared header file Chapter08/SceneUtils.h. This process uses a top-down recursive traversal, creating implicit scene node objects within the Scene container. Let's walk through the steps needed to traverse a scene stored in this format.

1. The traversal starts from some aiNode and a given parent node ID, passed as a parameter. A new node identifier is returned by the addNode() routine described below. If aiNode contains a name, we store it in the Scene::nodeNames array:

```
void traverse(const aiScene* sourceScene,
  Scene& scene, aiNode* node,
  int parent, int atLevel)
{
  const int newNodeID = addNode(scene, parent, atLevel);
  if (node->mName.C_Str()) {
    uint32_t stringID = (uint32_t)scene.nodeNames.size();
    scene.nodeNames.push_back(std::string(node->mName.C_Str()));
    scene.nameForNode[newNodeID] = stringID;
  }
```

2. If the aiNode object has meshes attached to it, we create a subnode for each of those meshes. To make debugging easier, we also add the name of each new mesh subnode:

```
  for (size_t i = 0; i < node->mNumMeshes; i++) {
    const int newSubNodeID = addNode(scene, newNode, atLevel+1);
    const uint32_t stringID = (uint32_t)scene.nodeNames.size();
    scene.nodeNames.push_back(
      std::string(node->mName.C_Str()) + "_Mesh_" + std::to_string(i));
    scene.nameForNode[newSubNodeID] = stringID;
```

3. Each of the meshes is assigned to the newly created subnode. Assimp guarantees a mesh has its material assigned, so we assign that material to our node. Since we use subnodes only for attaching meshes, we set local and global transformations to identity:

```
    const int mesh = (int)node->mMeshes[i];
    scene.meshForNode[newSubNodeID] = mesh;
    scene.materialForNode[newSubNodeID] =
      sourceScene->mMeshes[mesh]->mMaterialIndex;
    scene.globalTransform[newSubNode] = mat4(1.0f);
    scene.localTransform[newSubNode]  = mat4(1.0f);
  }
```

4. The global transformation is set to identity at the beginning of node conversion. It will be recalculated at the first frame or if the node is marked as changed. See the *Implementing transformation trees* recipe in this chapter for the implementation details. The local transformation is fetched from `aiNode` and converted into a `glm::mat4` object:

```
  scene.globalTransform[newNode] = mat4(1.0f);
  scene.localTransform[newNode] = toMat4(N->mTransformation);
```

5. At the end, we recursively traverse the children of this `aiNode` object:

```
  for (unsigned int n = 0; n < N->mNumChildren; n++)
    traverse(sourceScene, scene,
      N->mChildren[n], newNode, atLevel+1);
}
```

The most complex part of this code that deals with the `Scene` data structure is the `addNode()` routine, which allocates a new scene node and adds it to the scene hierarchy. Let's take a look at how to implement it in `shared/Scene/Scene.cpp`.

1. First, the addition process obtains a new node identifier, which corresponds to the current size of the hierarchy array. New identity transformations are then added to the local and global transform arrays. The hierarchy for the newly added node consists solely of the reference to its parent:

```
int addNode(Scene& scene, int parent, int level) {
  const int node = (int)scene.hierarchy.size();
  scene.localTransform.push_back(mat4(1.0f));
  scene.globalTransform.push_back(mat4(1.0f));
  scene.hierarchy.push_back({ .parent = parent });
```

2. If we have a parent node, we update its first child reference and, potentially, the next sibling reference of another node. If the parent node has no children, we simply set its `firstChild` field. Otherwise, we need to traverse the siblings of this child to find the appropriate place to add the next sibling:

```
if (parent > -1) {
  const int s = scene.hierarchy[parent].firstChild;
  if (s == -1) {
    scene.hierarchy[parent].firstChild  = node;
    scene.hierarchy[node].lastSibling = node;
  } else {
    int dest = scene.hierarchy[s].lastSibling;
    if (dest <= -1) {
      for (dest = s;
           scene.hierarchy[dest].nextSibling !=-1;
           dest = scene.hierarchy[dest].nextSibling);
    }
    scene.hierarchy[dest].nextSibling = node;
    scene.hierarchy[s].lastSibling = node;
  }
}
```

After the `for` loop, we set our new node as the next sibling of the last child. It's worth noting that this linear traversal of the siblings isn't necessary if we store the index of the last added child node. Later, in the *Implementing transformation trees* recipe, we'll demonstrate how to modify `addNode()` to eliminate the loop.

3. The level of this node is stored for correct global transformation updating. To keep the structure valid, we store negative indices for the newly added node:

```
    scene.hierarchy[node].level = level;
    scene.hierarchy[node].nextSibling = -1;
    scene.hierarchy[node].firstChild  = -1;
    return node;
  }
```

We use this `traverse()` routine to load mesh data with Assimp in our sample applications using a helper function, `loadMeshFile()`.

There's more...

Data-oriented design (**DOD**) is a broad field, and we've only covered a few techniques here. We recommend checking out the online book *Data-Oriented Design* by Richard Fabian to become more familiar with additional DOD concepts: `https://www.dataorienteddesign.com/dodbook`.

The demo application Chapter08/02_SceneGraph implements some basic scene graph editing capabilities using ImGui, which can help you get started with integrating scene graphs into your productivity tools and writing an editor for your 3D applications. The following renderSceneTreeUI() recursive function from Chapter08/02_SceneGraph/src/main.cpp is responsible for the rendering of the scene graph tree hierarchy in the UI and selecting a node for editing:

```
int renderSceneTreeUI(
  const Scene& scene, int node, int selectedNode)
{
  const std::string name = getNodeName(scene, node);
  const std::string label = name.empty() ?
    (std::string("Node") + std::to_string(node)) : name;
  const bool isLeaf = scene.hierarchy[node].firstChild<0;
  ImGuiTreeNodeFlags flags = isLeaf ?
    ImGuiTreeNodeFlags_Leaf |
    ImGuiTreeNodeFlags_Bullet : 0;
  if (node == selectedNode)
    flags |= ImGuiTreeNodeFlags_Selected;
  ImVec4 color = isLeaf ? ImVec4(0, 1, 0, 1) : ImVec4(1, 1, 1, 1);
  ImGui::PushStyleColor(ImGuiCol_Text, color);
  const bool isOpened = ImGui::TreeNodeEx(
    &scene.hierarchy[node], flags, "%s", label.c_str());
  ImGui::PopStyleColor();
  ImGui::PushID(node);
  if (ImGui::IsItemHovered() && isLeaf)
    selectedNode = node;
  if (isOpened) {
    for (int ch = scene.hierarchy[node].firstChild;
         ch != -1;
         ch = scene.hierarchy[ch].nextSibling) {
      if (int subNode =
          renderSceneTreeUI(scene, ch, selectedNode);
          subNode > -1) {
        selectedNode = subNode;
      }
    }
    ImGui::TreePop();
  }
  ImGui::PopID();
  return selectedNode;
}
```

This function can be used as a basis for building various editing functionality for nodes, materials, and other scene graph content. Take a look at `editNodeUI()` and `editMaterialUI()` for some more examples.

Loading and saving a scene graph

To quote Frederick Brooks, "Show me your data structures, and I don't need to see your code." By now, you should have a pretty good idea of how to implement basic operations on the scene graph. However, the rest of this chapter will walk you through all the necessary routines in detail. In this section, we'll provide an overview of how to load and save our scene graph structure.

Getting ready

Make sure you read the previous recipe, *Using data-oriented design for a scene graph*, before proceeding any further.

How to do it...

The loading process consists of a series of `fread()` calls, followed by two `loadMap()` calls, which are handled within the `loadScene()` helper function located in `shared/Scene/Scene.cpp`. As with other examples in this book, we've left out error handling in the text, but the accompanying source code includes the necessary checks.

1. After opening the file, we read the number of scene nodes stored. Then, we resize the data containers to match:

```
void loadScene(const char* fileName, Scene& scene) {
  FILE* f = fopen(fileName, "rb");
  uint32_t sz = 0;
  fread(&sz, sizeof(sz), 1, f);
  scene.hierarchy.resize(sz);
  scene.globalTransform.resize(sz);
  scene.localTransform.resize(sz);
```

2. A series of `fread()` calls is sufficient to read the transformations and hierarchical data for all the scene nodes:

```
  fread(scene.localTransform.data(), sizeof(glm::mat4), sz, f);
  fread(scene.globalTransform.data(), sizeof(glm::mat4), sz, f);
  fread(scene.hierarchy.data(), sizeof(Hierarchy), sz, f);
```

3. The node-to-material and node-to-mesh mappings are loaded using the `loadMap()` helper function. We will look into it in a moment:

```
  loadMap(f, scene.materialForNode);
  loadMap(f, scene.meshForNode);
```

4. If any data remains, we proceed to read the scene node names and material names:

```
    if (!feof(f)) {
      loadMap(f, scene.nameForNode);
      loadStringList(f, scene.nodeNames);
      loadStringList(f, scene.materialNames);
    }
    fclose(f);
}
```

Saving the scene data is essentially the reverse of the `loadScene()` routine. Let's take a look at how it's implemented:

1. At the start of the file, we write the number of scene nodes:

```
void saveScene(
    const char* fileName, const Scene& scene)
{
    FILE* f = fopen(fileName, "wb");
    const uint32_t sz = (uint32_t)scene.hierarchy.size();
    fwrite(&sz, sizeof(sz), 1, f);
```

2. Three `fwrite()` calls save the local and global transformations, followed by the hierarchical information:

```
    fwrite(scene.localTransform.data(), sizeof(glm::mat4), sz, f);
    fwrite(scene.globalTransform.data(), sizeof(glm::mat4), sz, f);
    fwrite(scene.hierarchy.data(), sizeof(Hierarchy), sz, f);
```

3. Two `saveMap()` calls store the node-to-material and node-to-mesh mappings:

```
    saveMap(f, scene.materialForNode);
    saveMap(f, scene.meshForNode);
```

4. If the scene node and material names are not empty, we also save these mappings:

```
    if (!scene.nodeNames.empty() &&
        !scene.nameForNode.empty()) {
      saveMap(f, scene.nameForNode);
      saveStringList(f, scene.nodeNames);
      saveStringList(f, scene.materialNames);
    }
    fclose(f);
}
```

Let's briefly describe the helper routines `loadMap()` and `saveMap()`, which handle loading and saving our `std::unordered_map` containers, respectively. Loading a `std::unordered_map` is done in three steps:

1. First, we read the number of integer values from the file. Each of our `{key, value}` pairs consists of two unsigned integers, which are stored sequentially:

```
void loadMap(
  FILE* f, unordered_map<uint32_t, uint32_t>& map)
{
  vector<uint32_t> ms;
  uint32_t sz = 0;
  fread(&sz, 1, sizeof(sz), f);
```

2. Then, all the key and value pairs are loaded with a single `fread()` call:

```
  ms.resize(sz);
  fread(ms.data(), sizeof(uint32_t), sz, f);
```

3. Finally, all the pairs from the array are inserted into a hash table:

```
  for (size_t i = 0; i < (sz / 2); i++)
    map[ms[i * 2 + 0]] = ms[i * 2 + 1];
}
```

The saving routine for `std::unordered_map` is implemented by reversing the `loadMap()` process, line by line:

1. First, we allocate a temporary vector to hold the `{key, value}` pairs:

```
void saveMap(FILE* f,
  const std::unordered_map<uint32_t, uint32_t>& map)
{
  vector<uint32_t> ms;
  ms.reserve(map.size() * 2);
```

2. All the values from `std::unordered_map` are copied into this vector:

```
  for (const auto& m : map) {
    ms.push_back(m.first);
    ms.push_back(m.second);
  }
```

3. The number of stored integer values is written to the file, followed by the actual integer values that represent our `{key, value}` pairs. This is done using `fwrite()` calls:

```
  const uint32_t sz = static_cast<uint32_t>(ms.size());
  fwrite(&sz, sizeof(sz), 1, f);
  fwrite(ms.data(), sizeof(uint32_t), ms.size(), f);
}
```

There's more...

Changes to the topology of nodes in our scene graph present a (solvable) management problem. The relevant source code is discussed in the recipe *Deleting nodes and merging scene graphs* in *Chapter 11*. We just need to keep all the mesh geometries in a single large GPU buffer, and we will explain how to implement this later in the chapter.

The material conversion routines will be implemented in the *Implementing a material system* recipe later in this chapter, along with the complete scene loading and saving code that includes the scene mesh geometry. Now, let's continue exploring our scene graph management code.

Implementing transformation trees

The typical application of a scene graph is to represent spatial relationships. For rendering, we need to compute a global 3D affine transformation for each node in the scene graph. This recipe explains how to calculate global transformations from local ones, avoiding unnecessary computations.

Getting ready

Using the previously defined `Scene` structure, we demonstrate how to recalculate global transformations. It's helpful to review the *Using data-oriented design for a scene graph* recipe before continuing. To begin, let's revisit a risky but tempting idea we had earlier – a recursive global transformation calculator within the non-existent `SceneNode::render()` method:

```
SceneNode::render() {
  mat4 parentTransform = parent ? parent->globalTransform : mat4(1);
  this->globalTransform = parentTransform * localTransform;
  // ... rendering and recursion
}
```

This approach results in data being scattered across memory, causing cache thrashing and severely reducing performance.

More information

For more information about caches and performance, check out the CppCon 2014 presentation *Data-Oriented Design and C++* by Mike Acton.

It is always better to separate operations like rendering, scene traversal, and transform calculation, and to store related data close together in memory. Additionally, processing similar tasks in large batches helps. This separation becomes even more critical as the number of nodes grows.

We've already seen how to render a large number of static meshes in a single GPU draw call by combining indirect rendering with programmable vertex pulling. Here, we will focus on how to minimize global transformation recalculations.

How to do it...

It is always best to avoid unnecessary calculations. For global transformations of scene nodes, we need a way to flag specific nodes whose transforms have changed in the current frame. Since changes to a node can affect its children, we also need to mark those children as changed. To handle this, let's introduce a data structure to track "dirty" nodes that have been updated since the last frame. The full source code is located in `shared/Scene/Scene.cpp`.

1. In the `Scene` structure, we should declare an array of `std::vectors` called `changedAtThisFrame` to quickly track any node that has changed at a specific depth level in the scene graph. This is why we needed to store each node's depth level value:

```
struct Scene {
  // ...
  vector<int> changedAtThisFrame[MAX_NODE_LEVEL];
};
```

2. The `markAsChanged()` function starts with a given node ID and recursively traverses all its child nodes, adding them to the `changedAtThisFrame` array. The process begins by marking the node itself as changed:

```
void markAsChanged(Scene& scene, int node) {
  const int level = scene.hierarchy_[node].level;
  scene.changedAtThisFrame_[level].push_back(node);
```

3. Next, we move to the first child and proceed to the next sibling, continuing to descend the hierarchy:

```
  for (int s = scene.hierarchy[node].firstChild;
       s != -1;
       s = scene.hierarchy[s].nextSibling) {
    markAsChanged(scene, s);
  }
}
```

Read more

A pedantic reader might point out that with a slightly more sophisticated dirty flag tracking, we could avoid making `markAsChanged()` recursive and instead implement it as a pure `O(1)` call. This is correct, and those interested in that approach are encouraged to check out the presentation *Handling Massive Transform Updates in a SceneGraph* by Markus Tavenrath (`https://x.com/CorporateShark/status/1840797491140444290`) and his `TransformTree` data structure. However, we will stick to a simpler method here for the sake of easier teaching.

To recalculate all the global transformations for the changed nodes, we implement the following `recalculateGlobalTransforms()` function. If no local transformations have been updated and the scene is essentially static, no processing will occur.

1. We begin with the list of changed scene nodes at depth level `0`, assuming there is only one root node. The root node is unique because it has no parent, and its global transform is identical to its local transform. This design helps us avoid unnecessary branching by not requiring a parent check for all other nodes. After the update, we clear the list of changed nodes depth at level `0`:

```
void recalculateGlobalTransforms(Scene& scene) {
  if (!scene.changedAtThisFrame_[0].empty()) {
    const int c = scene.changedAtThisFrame_[0][0];
    scene.globalTransform_[c] = scene.localTransform_[c];
    scene.changedAtThisFrame_[0].clear();
  }
```

2. For all the higher depth levels, we can be certain that there are parents, so the loop runs in a linear fashion without any conditions inside. We begin at the depth level `1` since the root level has already been processed. The exit condition for the loop is when the current level's list is empty. Additionally, we ensure we don't descend deeper than `MAX_NODE_LEVEL`:

```
  for (int i = 1 ; i < MAX_NODE_LEVEL &&
       (!scene.changedAtThisFrame[i].empty()); i++ )
  {
```

3. We iterate through all the changed nodes at the current depth level. For each node we process, we retrieve the parent global transform and multiply it by the local node transform:

```
    for (int c : scene.changedAtThisFrame[i]) {
      int p = scene.hierarchy[c].parent;
      scene.globalTransform[c] =
        scene.globalTransform[p] *
        scene.localTransform[c];
    }
```

4. At the end of the node iteration, we should clear the list of changed nodes for this depth level:

```
    scene.changedAtThisFrame[i].clear();
  }
}
```

The core of this approach is that we avoid recalculating any global transformations multiple times. By starting from the root layer of the scene graph tree, all the changed layers below the root obtain a valid global transformation from their parents.

Note

Depending on how often local transformations are updated, it may be more efficient to forgo the list of recently updated nodes and simply perform a full update every time. Be sure to profile your actual code before making a decision.

There's more...

As an advanced exercise, you can offload the computation of changed node transformations to the GPU. This is relatively straightforward to implement, given that all the data is stored in arrays and we have compute shaders and buffer management code already in place. Those interested in pursuing a GPU version are encouraged to check out the presentation *Handling Massive Transform Updates in a SceneGraph* by Markus Tavenrath.

Implementing a material system

Chapter 6 provided an overview of the PBR shading model and included all the necessary GLSL shaders for rendering a single 3D object with multiple textures. In this chapter, we focus on how to organize scene rendering with multiple objects, each having different materials and properties. Our scene material system uses the same `GLTFMaterialDataGPU` structure from `shared/UtilsGLTF.h` and is compatible with our glTF 2.0 rendering code. However, to simplify the understanding of the code flow, we will focus on the scene representation and won't delve into parsing all the material properties again.

Getting ready

In the previous chapters, we focused on rendering individual objects and applying a PBR shading model to them. In the recipe *Using data-oriented design for a scene graph*, we learned the overall structure for organizing a scene and used opaque integers as material handles. Here, we will define a data structure for storing material parameters for the entire scene and demonstrate how this structure can be utilized in GLSL shaders. The process of converting material parameters from those loaded by Assimp will be explained later in this chapter in the recipe *Implementing automatic material conversion*.

Be sure to review the recipe *Organizing mesh data storage* in *Chapter 5* to refresh your understanding of how scene geometry data is organized.

How to do it...

We need data structures to represent our materials in both CPU memory, for quick loading and saving to a file, and in a GPU buffer. We already have `GLTFMaterialDataGPU` for the latter purpose. Now, let's introduce a similar structure for the CPU.

1. One reason for using a custom data structure is to allow for loading and storing material data without the need to parse complex glTF files. Another reason is to include additional data for our rendering needs. Let's introduce a set of convenience flags for our materials:

```
enum MaterialFlags {
```

```
  sMaterialFlags_CastShadow     = 0x1,
  sMaterialFlags_ReceiveShadow = 0x2,
  sMaterialFlags_Transparent    = 0x4,
};
```

2. The `Material` struct includes both the numeric values that define the basic lighting properties of the material and a set of texture indices. The first two fields store the emissive and diffuse colors for our simple shading model:

```
struct Material {
  vec4 emissiveFactor        = vec4(0, 0, 0, 0);
  vec4 baseColorFactor       = vec4(1, 1, 1, 1);
  float roughness            = 1.0f;
  float transparencyFactor = 1.0f;
  float alphaTest            = 0.0f;
  float metallicFactor       = 0.0f;
```

3. The textures are stored as indices in a `MeshData::textureFiles` array, which we will discuss in the next step. This approach is essential for deduplicating textures, as different materials may utilize the same texture that we want to reuse. It is important to note that the indices here are signed integers, where a value of `-1` indicates the absence of a texture. In contrast, the GPU version of this structure uses unsigned values and dummy textures. To customize our rendering pipeline, we might want to use flags that vary from material to material or from object to object. In the demos for this book, we don't require this level of flexibility, as we render all objects with a single shader. However, we do have a placeholder for storing these flags:

```
  int baseColorTexture = -1;
  int emissiveTexture  = -1;
  int normalTexture    = -1;
  int opacityTexture   = -1;
  uint32_t flags = sMaterialFlags_CastShadow |
                   sMaterialFlags_ReceiveShadow;
};
```

4. Before we can save and load materials, we need to revisit another data structure, `MeshData`. In *Chapter 5*, we introduced a similar struct called `MeshData` that contained only geometry data. Now, let's add two new member fields – a vector of `Material` structs (`materials`) and a vector of texture file names (`textureFiles`):

```
struct MeshData {
  lvk::VertexInput streams = {};
  vector<uint32_t> indexData;
  vector<uint8_t> vertexData;
  vector<Mesh> meshes;
  vector<BoundingBox> boxes;
```

```
    vector<Material> materials;
    vector<std::string> textureFiles;
  };
```

5. Once we have these data structures established, let's examine the code for loading and saving scene materials in `shared/Scene/VtxData.cpp`. The following `loadMeshDataMaterials()` function is responsible for reading a list of materials from a file. We will omit all the error checks in this discussion for brevity, but the actual code is much more detailed:

```
void loadMeshDataMaterials(const char* fileName, MeshData& out) {
  FILE* f = fopen(fileName, "rb");
  uint64_t numMaterials  = 0;
  uint64_t materialsSize = 0;
  fread(&numMaterials, 1, sizeof(numMaterials), f);
  fread(&materialsSize, 1, sizeof(materialsSize), f);
  out.materials.resize(numMaterials);
  fread(out.materials.data(), 1, materialsSize, f);
  loadStringList(f, out.textureFiles);
  fclose(f);
}
```

6. The saving function `saveMeshDataMaterials()` is quite similar:

```
void saveMeshDataMaterials(const char* fileName,
                           const MeshData& m) {
  FILE* f = fopen(fileName, "wb");
  uint64_t numMaterials  = m.materials.size();
  uint64_t materialsSize = m.materials.size() * sizeof(Material);
  fwrite(&numMaterials, 1, sizeof(numMaterials), f);
  fwrite(&materialsSize, 1, sizeof(materialsSize), f);
  fwrite(m.materials.data(), sizeof(Material), numMaterials, f);
  saveStringList(f, m.textureFiles);
  fclose(f);
}
```

The code uses the helper functions `saveStringList()`, which appends a list of strings to an opened binary file, and `loadStringList()`. They are located in `shared/Utils.cpp`.

Now we can take a look at the demo application and explore how it works.

How it works...

The demo application is located at `Chapter08/02_SceneGraph/src/main.cpp`.

1. At the start of the program, we check whether our scene is precached. If it isn't, we load a `.gltf` file into our `MeshData` and `Scene` data structures and save them for future use. The `isMesh...Valid()` helper functions check for the existence and validity of the cache files and can be found in `shared/Scene/VtxData.cpp`:

```
const char* fileNameCachedMeshes = ".cache/ch08_orrery.meshes";
const char* fileNameCachedMaterials = ".cache/ch08_orrery.materials";
const char* fileNameCachedHierarchy = ".cache/ch08_orrery.scene";
if (!isMeshDataValid(fileNameCachedMeshes) ||
  !isMeshHierarchyValid(fileNameCachedHierarchy)||
  !isMeshMaterialsValid(fileNameCachedMaterials))
{
  MeshData meshData;
  Scene ourScene;
```

2. The `loadMeshFile()` function is similar to the one described in the recipe *Implementing automatic geometry conversion* in *Chapter 5*. This new version implemented in `Chapter08/SceneUtils.h` is capable of loading materials and textures. We will explore it further later in this chapter in the recipe *Implementing automatic material conversion*:

```
  loadMeshFile("data/meshes/orrery/scene.gltf",
    meshData, ourScene, true);
  saveMeshData(fileNameCachedMeshes, meshData);
  saveMeshDataMaterials(
    fileNameCachedMaterials, meshData);
  saveScene(fileNameCachedHierarchy, ourScene);
}
```

3. Once the scene is loaded from a `.gltf` file and precached into our compact binary data formats, we can load it easily and proceed with rendering. The helper function `loadMeshData()` was discussed in *Chapter 5*, and the `loadScene()` function was covered in the recipe *Loading and saving a scene graph* earlier in this chapter:

```
MeshData meshData;
const MeshFileHeader header =
  loadMeshData(fileNameCachedMeshes, meshData);
loadMeshDataMaterials(fileNameCachedMaterials, meshData);
Scene scene;
loadScene(fileNameCachedHierarchy, scene);
```

4. Before we enter the application rendering loop, we create a `VKMesh` object, which is responsible for the rendering of our `MeshData` and `Scene`. We will discuss it in the recipe *Putting it all together into a scene editing application*:

```
VulkanApp app({
  .initialCameraPos = vec3(-0.88f, 1.26f, 1.07f),
  .initialCameraTarget = vec3(0, -0.6f, 0) });
unique_ptr<lvk::IContext> ctx(app.ctx_.get());
const VKMesh mesh(ctx, header, meshData, scene, app.getDepthFormat());
```

Now let's take a look at how to import the materials from Assimp. The next recipe shows how to extract and pack the values from the Assimp library's `aiMaterial` structure to our `Material` structure.

Implementing automatic material conversion

In the previous recipe, we learned how to create a runtime data storage format for scene materials and how to save and load it from disk. In the recipe *Implementing the glTF 2.0 metallic-roughness shading model* in *Chapter 6*, we explored how to load PBR material properties. Now, let's do another quick intermezzo and learn how to extract material properties from Assimp data structures and convert textures for the demo applications in this chapter.

Getting ready

Refer to the previous recipe, *Implementing a material system*, where we learned how to load and store multiple meshes with different materials. Now, it's time to explain how to import material data from popular 3D asset formats.

How to do it...

Let's take a look at the `convertAIMaterial()` function from `Chapter08/VKMesh08.h`. It retrieves all the required parameters from Assimp's `aiMaterial` structure and returns a `Material` object suitable for storage.

1. Each texture is later addressed by an integer identifier. We store a list of texture filenames in the `files` parameter. The `opacityMap` parameter contains a list of textures that need to be combined with opacity/transparency maps:

```
Material convertAIMaterial(
  const aiMaterial* M,
  vector<std::string>& files,
  vector<std::string>& opacityMaps)
{
  Material D;
  aiColor4D Color;
```

2. The Assimp API provides **getter** functions that we use to extract individual color parameters. These functions allow us to read all the required color data. Since most of the code for handling different color values is quite similar, we won't include it all here to keep things concise:

```
if (aiGetMaterialColor(M, AI_MATKEY_COLOR_AMBIENT,
    &Color) == AI_SUCCESS)
{
  D.emissiveFactor = {
    Color.r, Color.g, Color.b, Color.a };
  if (D.emissiveFactor.w > 1.0f)
    D.emissiveFactor.w = 1.0f;
}
// ...
```

3. After reading the colors and various scalar factors, we move on to handling textures. All textures for our materials are stored in external files, and the filenames are extracted using the `aiGetMaterialTexture()` function:

```
aiString path;
aiTextureMapping mapping;
unsigned int uvIndex  = 0;
float blend           = 1.0f;
aiTextureOp textureOp = aiTextureOp_Add;
aiTextureMapMode textureMapMode[2] = {
  aiTextureMapMode_Wrap,
  aiTextureMapMode_Wrap };
unsigned int textureFlags = 0;
```

4. This function requires several parameters, but for simplicity, we mostly ignore them in our converter. The first texture is an emissive map, and we use the `addUnique()` function to add the texture file to our `files` list:

```
if (aiGetMaterialTexture(M,
      aiTextureType_EMISSIVE, 0, &path,
      &mapping, &uvIndex, &blend, &textureOp,
      textureMapMode,
      &textureFlags) == AI_SUCCESS)
{
  D.emissiveTexture = addUnique(files, path.C_Str());
}
```

5. The diffuse map is stored in the baseColorTexture field of our material structure. Here, we apply some ad hoc heuristics based on the material name to mark the material as transparent. This is necessary to improve the appearance of the Bistro model by customizing some materials:

```
if (aiGetMaterialTexture(M, aiTextureType_DIFFUSE,
    0, &path, &mapping, &uvIndex, &blend,
    &textureOp, textureMapMode,
    &textureFlags) == AI_SUCCESS)
{
  D.baseColorTexture = addUnique(files, path.C_Str());
  const string albedoMap = string(path.C_Str());
  if (albedoMap.find("grey_30") != albedoMap.npos)
    D.flags |= sMaterialFlags_Transparent;
}
```

6. The normal map can be extracted from either the aiTextureType_NORMALS property or aiTextureType_HEIGHT in aiMaterial. We check for the presence of an aiTextureType_NORMALS texture map and store the texture index in the normalTexture field:

```
if (aiGetMaterialTexture(M, aiTextureType_NORMALS,
    0, &path, &mapping, &uvIndex, &blend,
    &textureOp, textureMapMode,
    &textureFlags) == AI_SUCCESS)
{
  D.normalTexture = addUnique(files, path.C_Str());
}
```

7. If there is no tangent space normal map, we should check if a heightmap texture is present, which can be converted to a normal map at a later stage of the conversion process:

```
if ( (D.normalTexture == -1) &&
     (aiGetMaterialTexture(M,
        aiTextureType_HEIGHT, 0, &path,
        &mapping, &uvIndex, &blend, &textureOp,
        textureMapMode,
        &textureFlags) == AI_SUCCESS) )
{
    D.normalTexture = addUnique(files, path.C_Str());
}
```

8. The final map we use is the opacity map, with each one stored in the output `opacityMaps` array. We pack the opacity maps into the alpha channel of our albedo textures:

```
  if (aiGetMaterialTexture(M, aiTextureType_OPACITY,
        0, &path, &mapping, &uvIndex, &blend,
        &textureOp, textureMapMode,
        &textureFlags) == AI_SUCCESS)
  {
    D.opacityTexture = addUnique(opacityMaps, path.C_Str());
    D.alphaTest      = 0.5f;
  }
```

9. The final part of the material conversion process uses heuristics to infer material properties based on the material's name. In our largest test scene, we only check for glass-like materials, but other common names like "gold" or "silver" can also be used to assign metallic coefficients and albedo colors. This simple trick helps improve the appearance of the test scene. Once complete, the `Material` object is returned for further processing:

```
  aiString Name;
  std::string materialName;
  auto name = [&materialName](
    const char* substr) -> bool {
    return materialName.find(substr) != std::string::npos;
  };
  if (name("MASTER_Glass_Clean") ||
      name("MenuSign_02_Glass") ||
      name("Vespa_Headlight")) {
    D.alphaTest         = 0.75f;
    D.transparencyFactor = 0.2f;
    D.flags |= sMaterialFlags_Transparent;
  } else if (name("MASTER_Glass_Exterior") ||
             name("MASTER_Focus_Glass")) {
    D.alphaTest         = 0.75f;
    D.transparencyFactor = 0.3f;
    D.flags |= sMaterialFlags_Transparent;
  } else if (name("MASTER_Frosted_Glass") {
    // ...
  }
  return D;
}
```

10. One last thing to note is the `addUnique()` function, which populates the list of texture files. It checks if a filename is already in the collection. If the filename isn't there, it's added, and its index is returned. Otherwise, the index of the previously added texture file is returned:

```
int addUnique(
  vector<std::string>& files, const std::string& file)
{
  if (file.empty()) return -1;
  const auto i = std::find(std::begin(files), std::end(files), file);
  if (i != files.end())
    return (int)std::distance(files.begin(), i);
  files.push_back(file);
  return (int)files.size() - 1;
}
```

Note

The `addUnique()` function has a complexity of `O(n²)`. This is acceptable in this case because it's only used during the material conversion stage, where the number of texture files is relatively small.

We have now read all the material properties from Assimp and built a list of texture filenames needed to render our scene. Since the 3D models come from the internet, these textures can vary significantly in size and be stored in various uncompressed image formats. For runtime rendering, we want consistent texture sizes, and we also want to compress them into a GPU-friendly format, as we did in the recipe *Compressing textures into the BC7 format* in *Chapter 1*. This conversion is handled by the `convertAndDownscaleAllTextures()` function. Let's take a look at how it works.

How it works...

The helper function `convertAndDownscaleAllTextures()` is defined in `Chapter08/SceneUtils.h`. It iterates through all the texture files in a multi-threaded manner and calls the `convertTexture()` function for each texture file. This file defines two C++ macros: `DEMO_TEXTURE_MAX_SIZE` and `DEMO_TEXTURE_CACHE_FOLDER`. These specify the maximum texture dimensions for rescaling and the output cache folder, respectively. We'll redefine these macros in later chapters to reuse this code with different texture resolutions.

1. This routine takes several parameters – a list of material descriptions, an output directory for the converted textures `basePath`, and containers for all the texture filenames and opacity maps:

```
void convertAndDownscaleAllTextures(
  const vector<Material>& materials,
  const std::string& basePath,
  vector<std::string>& files,
  vector<std::string>& opacityMaps)
{
```

2. Each opacity map is combined with the base color map. To maintain the correspondence between the opacity map list and the global texture indices, we use a standard C++ hash table. We iterate through all the materials and check if they have both an opacity map and an albedo map. If both maps are present, we associate the opacity map with the albedo map:

```
std::unordered_map<std::string, uint32_t>
  opacityMapIndices(files.size());
for (const auto& m : materials) {
  if (m.opacityTexture   != -1 &&
      m.baseColorTexture != -1)
    opacityMapIndices[files[m.baseColorTexture]] =
      (uint32_t)m.opacityTexture;
}
```

3. The following lambda function takes a source texture filename from the 3D model and returns a modified cached texture filename. It also performs the conversion of the texture data by calling the `convertTexture()` function, which we will explore in a moment:

```
auto converter = [&](const std::string& s) -> std::string {
  return convertTexture(
    s, basePath, opacityMapIndices, opacityMaps);
};
```

4. Now, we can use the `std::transform()` algorithm with the `std::execution::par` argument to convert all the texture files in a multi-threaded manner:

```
  std::transform(std::execution::par,
    std::begin(files), std::end(files),
    std::begin(files), converter);
}
```

The `std::execution::par` parameter is a C++20 feature that enables parallel processing of the container. Since converting the texture data can be a lengthy process, this simple parallelization may significantly reduce our processing time.

A single texture map is converted to our runtime data format using the `convertTexture()` routine and is stored as a `.ktx` file. Let's take a look at how it works:

1. The function takes a source texture filename from the 3D model, a `basePath` folder containing the 3D model file, and a collection of mappings for the opacity map indices:

```
std::string convertTexture(
  const std::string& file,
  const std::string& basePath,
  unordered_map<std::string, uint32_t>& opacityMapIndices,
  const vector<std::string>& opacityMaps)
{
```

2. All of our output textures will have a maximum size of 512x512 pixels, controlled by the macro `DEMO_TEXTURE_MAX_SIZE`, to make sure our demos run fast. We also create a destination folder to store the cached textures:

```
const int maxNewWidth  = DEMO_TEXTURE_MAX_SIZE;
const int maxNewHeight = DEMO_TEXTURE_MAX_SIZE;
namespace fs = std::filesystem;
if (!fs::exists(DEMO_TEXTURE_CACHE_FOLDER)) {
  fs::create_directories(DEMO_TEXTURE_CACHE_FOLDER);
}
```

3. The output filename is created by concatenating a fixed output directory with the source filename, replacing all path separators with double underscores. To ensure compatibility across Windows, Linux, and macOS, we replace all path separators with the / symbol:

```
const string srcFile = replaceAll(basePath + file, "\\", "/");
const string newFile = std::string(
  DEMO_TEXTURE_CACHE_FOLDER) +
    lowercaseString(replaceAll(
      replaceAll(srcFile, "..", "__"), "/", "__")+std::string("__rescaled")
) + std::string(".ktx");
```

4. As we did in the previous chapters, we use the `stb_image` library to load textures. We enforce that the loaded image is in RGBA format, even if there is no opacity information. This is a convenient shortcut that simplifies our texture handling code significantly:

```
int origWidth, origHeight, texChannels;
stbi_uc* pixels =
  stbi_load(fixTextureFile(srcFile).c_str(),
  &origWidth, &origHeight, &texChannels,
  STBI_rgb_alpha);
uint8_t* src = pixels;
texChannels  = STBI_rgb_alpha;
uint8_t* src = pixels;
texChannels  = STBI_rgb_alpha;
```

Note

The `fixTextureFile()` function addresses situations where the 3D model material data references texture files with an incorrect case in their filenames. For instance, the `.mtl` file might contain `map_Ka Texture01.png`, while the actual filename on the filesystem is in lowercase, `texture01.png`. This function helps us resolve naming inconsistencies in the Bistro scene on Linux.

5. If the texture fails to load, we set our empty temporary array as the input data to prevent having to exit at this point:

```
std::vector<uint8_t> tmpImage(maxNewWidth * maxNewHeight * 4);
if (!src) {
  printf("Failed to load [%s] texture\n", srcFile.c_str());
  origWidth   = maxNewWidth;
  origHeight  = maxNewHeight;
  texChannels = STBI_rgb_alpha;
  src         = tmpImage.data();
}
```

6. If the texture has an associated opacity map stored in the hash table, we load that opacity map and combine its contents with the albedo map. As with the source texture file, we replace path separators to ensure cross-platform compatibility. The opacity map is loaded as a simple grayscale image:

```
if (opacityMapIndices.count(file) > 0) {
  const std::string opacityMapFile =
    replaceAll(basePath + opacityMaps[opacityMapIndices[file]], "\\", "/");
  int opacityWidth, opacityHeight;
  stbi_uc* opacityPixels = stbi_load(
    fixTextureFile(opacityMapFile).c_str(),
    &opacityWidth, &opacityHeight, nullptr, 1);
```

7. After successfully loading the opacity map with the correct dimensions, we store the opacity values in the alpha component of the albedo texture. The `stb_image` library uses explicit memory management, so we free the loaded opacity map manually:

```
  assert(opacityPixels);
  assert(origWidth == opacityWidth);
  assert(origHeight == opacityHeight);
  if (opacityPixels) {
    for (int y = 0; y != opacityHeight; y++)
      for (int x = 0; x != opacityWidth; x++) {
        int idx = y * opacityWidth + x;
        src[idx * texChannels + 3] = opacityPixels[idx];
      }
  }
  stbi_image_free(opacityPixels);
}
```

8. Once the texture is combined with the optional opacity map, we can downscale it, generate all the mip levels, and compress it into the BC7 format. We will generate all the mip levels ourselves:

```
const int newW = std::min(origWidth,  maxNewWidth);
const int newH = std::min(origHeight, maxNewHeight);
const uint32_t numMipLevels = lvk::calcNumMipLevels(newW, newH);
ktxTextureCreateInfo createInfoKTX2 = {
  .glInternalformat = GL_RGBA8,
  .vkFormat         = VK_FORMAT_R8G8B8A8_UNORM,
  .baseWidth        = (uint32_t)newW,
  .baseHeight       = (uint32_t)newH,
  .baseDepth        = 1u,
  .numDimensions    = 2u,
  .numLevels        = numMipLevels,
  .numLayers        = 1u,
  .numFaces         = 1u,
  .generateMipmaps  = KTX_FALSE,
};
ktxTexture2* textureKTX2 = nullptr;
ktxTexture2_Create(&createInfoKTX2,
  KTX_TEXTURE_CREATE_ALLOC_STORAGE, &textureKTX2);
```

9. Here's the loop that generates all the mip levels from the full-sized texture. The `stb_image_resize` library offers a simple `stbir_resize_uint8_linear()` function to rescale an image without significant loss of quality:

```
int w = newW;
int h = newH;
for (uint32_t i = 0; i != numMipLevels; ++i) {
  size_t offset = 0;
  ktxTexture_GetImageOffset(
    ktxTexture(textureKTX2), i, 0, 0, &offset);
  stbir_resize_uint8_linear( (const uint8_t*)src,
    origWidth, origHeight, 0,
    ktxTexture_GetData(ktxTexture(textureKTX2)) + offset,
    w, h, 0, STBIR_RGBA);
  h = h > 1 ? h >> 1 : 1;
  w = w > 1 ? w >> 1 : 1;
}
```

10. Then, we compress the entire image into the BC7 format using the KTX-Software library. This involves two steps – first, compressing the image into the Basis format, and then transcoding it into BC7:

```
ktxTexture2_CompressBasis(textureKTX2, 255);
ktxTexture2_TranscodeBasis(textureKTX2, KTX_TTF_BC7_RGBA, 0);
ktxTextureCreateInfo createInfoKTX1 = {
```

```
    .glInternalformat = GL_COMPRESSED_RGBA_BPTC_UNORM,
    .vkFormat         = VK_FORMAT_BC7_UNORM_BLOCK,
    .baseWidth        = (uint32_t)newW,
    .baseHeight       = (uint32_t)newH,
    .baseDepth        = 1u,
    .numDimensions    = 2u,
    .numLevels        = numMipLevels,
    .numLayers        = 1u,
    .numFaces         = 1u,
    .generateMipmaps  = KTX_FALSE,
  };
  ktxTexture1* textureKTX1 = nullptr;
  ktxTexture1_Create(&createInfoKTX1,
    KTX_TEXTURE_CREATE_ALLOC_STORAGE, &textureKTX1);
  for (uint32_t i = 0; i != numMipLevels; ++i) {
    size_t offset1 = 0;
    ktxTexture_GetImageOffset(
      ktxTexture(textureKTX1), i, 0, 0, &offset1);
    size_t offset2 = 0;
    ktxTexture_GetImageOffset(
      ktxTexture(textureKTX2), i, 0, 0, &offset2);
    memcpy( ktxTexture_GetData(
        ktxTexture(textureKTX1)) + offset1,
      ktxTexture_GetData(
        ktxTexture(textureKTX2)) + offset2,
      ktxTexture_GetImageSize(ktxTexture(textureKTX1), i) );
  }
```

11. Finally, we save the resulting `.ktx` file and release all resources. Regardless of the conversion outcome, we return the new texture filename:

```
  ktxTexture_WriteToNamedFile(
    ktxTexture(textureKTX1), newFile.c_str());
  ktxTexture_Destroy(ktxTexture(textureKTX1));
  ktxTexture_Destroy(ktxTexture(textureKTX2));
  return newFile;
}
```

This ensures that even if some textures from the original 3D model are missing, the converted dataset remains valid and requires significantly fewer runtime checks.

There's more...

This relatively lengthy recipe has detailed all the necessary routines for retrieving and precaching material and texture data from external 3D assets. This will greatly enhance loading speeds for large 3D models, such as the Bistro model, which is essential for interactive debugging of complex graphics applications. Now, let's switch back to Vulkan and explore how to render our scene data with *LightweightVK*.

Using descriptor indexing and arrays of textures in Vulkan

Before we can use the material system described earlier in this chapter on the GPU, as discussed in the recipe *Implementing a material system*, let's first explore some Vulkan functionality that is essential for conveniently handling numerous materials with multiple textures on the GPU. Descriptor indexing is a highly useful feature that became mandatory in Vulkan 1.3 and is now widely supported on both desktop and mobile devices. We briefly discussed this in *Chapter 3*, in the recipe *Using Vulkan descriptor indexing*, where we explored how LightweightVK manages descriptor sets. Descriptor indexing allows for the creation of unbounded descriptor sets and the use of non-uniform dynamic indexing to access textures within them. This enables materials to be stored in shader storage buffers, with each one referencing the necessary textures using integer IDs. These IDs can be retrieved from buffers and used directly to index into the appropriate descriptor set containing all the textures needed by the application. Now, let's take a closer look at how to use this feature.

Our example app uses three different explosion animations released by Unity Technologies under the liberal CC0 license: `https://blogs.unity3d.com/2016/11/28/free-vfx-image-sequences-flipbooks`.

Getting ready

Make sure to revisit the recipe *Using Vulkan descriptor indexing* in *Chapter 3* to understand the underlying Vulkan implementation.

The source code for this recipe is located at `Chapter08/01_DescriptorIndexing`. All the textures we use are stored in the `deps/src/explosion` folder.

How to do it...

Let's build an app that renders animated explosions using the flipbook technique. We will store 100 images representing the explosion frames, load them into GPU memory as textures, and apply the appropriate texture based on the current animation time. Since no binding is needed, we just need to update the texture ID at the right time and pass it to our GLSL shaders.

The application renders an animated explosion every time the user clicks anywhere inside the window. Multiple explosions, each using a different flipbook, can be rendered at the same time. Let's explore the C++ code from `Chapter08/01_DescriptorIndexing/src/main.cpp` to learn how to do it.

1. First, we need to define some constants and structures to represent our animations. Here, `position` represents the screen location of an explosion, `startTime` marks the timestamp when the animation started, `time` indicates the current animation time, `textureIndex` is the index of the current texture in the flipbook, and `firstFrame` is the index of the first frame of the current flipbook in the texture array:

```
const double   kAnimationFPS = 50.0;
const uint32_t kNumFlipbooks = 3;
const uint32_t kNumFlipbookFrames = 100;
struct AnimationState {
  vec2 position          = vec2(0);
  double startTime       = 0;
  float time             = 0;
  uint32_t textureIndex = 0;
  uint32_t firstFrame    = 0;
};
```

2. The `g_Animations` vector holds the current state of animations for all visible explosions and is updated every frame. The `g_AnimationsKeyframe` vector saves the animation state of all explosions when we want to pause the simulation and manually adjust the current animation by offsetting it from stored keyframes using the ImGui UI. We'll explain this further in a moment. The `timelineOffset` parameter is used to offset the time from the keyframe values "locked" in `g_AnimationsKeyframe`:

```
std::vector<AnimationState> g_Animations;
std::vector<AnimationState> g_AnimationsKeyframe;
float timelineOffset = 0.0f;
bool showTimeline = false;
```

3. Here's the animation update logic placed into a separate function, `updateAnimations()`. The current texture index is updated for each animation based on its start time. As we go through all the animations, we can safely remove the ones that are finished. Instead of using the swap-and-pop trick to remove an element from the vector, which can cause distracting Z-fighting where animations suddenly pop in front of one another, we use a simple, straightforward removal with `erase()`:

```
void updateAnimations(float deltaSeconds) {
  for (size_t i = 0; i < g_Animations.size();) {
    g_Animations[i].time += deltaSeconds;
    g_Animations[i].textureIndex =
      g_Animations[i].firstFrame +
```

```
        (uint32_t)(kAnimationFPS * g_Animations[i].time);
      if (g_Animations[i].textureIndex >=
          kNumFlipbookFrames + g_Animations[i].firstFrame)
        g_Animations.erase(g_Animations.begin() + i);
      else i++;
    }
  }
```

4. The `setAnimationsOffset()` function takes a time `offset` and applies it to the values in `g_AnimationsKeyframe` to compute the new values for `g_Animations`:

```
void setAnimationsOffset(float offset) {
  for (size_t i = 0; i < g_Animations.size(); i++) {
    g_Animations[i].time = std::max(
      g_AnimationsKeyframe[i].time + offset, 0.0f);
    g_Animations[i].textureIndex =
      g_Animations[i].firstFrame + (uint32_t)
       (kAnimationFPS * g_Animations[i].time);
    g_Animations[i].textureIndex = std::min(
      g_Animations[i].textureIndex,
      kNumFlipbookFrames + g_Animations[i].firstFrame - 1);
  }
}
```

Now that all the utilities are in place, we're ready to dive into the `main()` function.

1. First, we create our app and load all the textures:

```
int main() {
  VulkanApp app;
  unique_ptr<lvk::IContext> ctx(app.ctx_.get());
```

2. We have 3 different explosion types, each with 100 frames defined as `kNumFlipbooks` and `kNumFlipbookFrames`, respectively. All the texture files are loaded from the `deps/src/explosion/` folder and its subfolders – `explosion/`, `explosion1/`, and `explosion2/`:

```
  std::vector<Holder<TextureHandle>> textures;
  textures.reserve(kNumFlipbooks * kNumFlipbookFrames);
  for (uint32_t book = 0;
       book != kNumFlipbooks; book++) {
    for (uint32_t frame = 0;
         frame != kNumFlipbookFrames; frame++) {
      char fname[1024];
      snprintf(fname, sizeof(fname),
        "deps/src/explosion/explosion%01u/explosion%02u-frame%03u.tga",
```

```
        book, book, frame + 1);
      textures.emplace_back(
        loadTexture(ctx, fname));
    }
  }
```

3. Next, we load the vertex and fragment shaders, which we'll examine shortly.

```
Holder<ShaderModuleHandle> vert =
  loadShaderModule(ctx,
  "Chapter08/01_DescriptorIndexing/src/main.vert");
Holder<ShaderModuleHandle> frag =
  loadShaderModule(ctx,
  "Chapter08/01_DescriptorIndexing/src/main.frag");
```

4. The rendering pipeline is set up using alpha blending mode with `BlendFactor_SrcAlpha` and `BlendFactor_OneMinusSrcAlpha`, allowing us to render transparent texels correctly:

```
Holder<RenderPipelineHandle> pipelineQuad =
  ctx->createRenderPipeline({
    .topology = lvk::Topology_TriangleStrip,
    .smVert   = vert,
    .smFrag   = frag,
    .color    = { {
      .format = ctx->getSwapchainFormat(),
      .blendEnabled = true,
      .srcRGBBlendFactor = lvk::BlendFactor_SrcAlpha,
      .dstRGBBlendFactor = lvk::BlendFactor_OneMinusSrcAlpha,
    } },
});
```

5. Let's put the first explosion in the center of the screen:

```
g_Animations.push_back(AnimationState{
  .position     = vec2(0.5f, 0.5f),
  .startTime    = glfwGetTime(),
  .textureIndex = 0,
  .firstFrame   = kNumFlipbookFrames * (uint32_t)(rand() % 3),
});
```

6. Before entering the rendering loop, we'll set up the GLFW mouse and key callbacks. The mouse callback will allow us to add a new explosion to the screen with a mouse click. The flipbook starting offset `firstFrame` is selected randomly for each explosion:

```
app.addMouseButtonCallback([](
  GLFWwindow* window,
```

```
int button, int action, int mods)
{
  VulkanApp* app = (VulkanApp*)
    glfwGetWindowUserPointer(window); ImGuiIO& io = ImGui::GetIO();
  if (button == GLFW_MOUSE_BUTTON_LEFT &&
      action == GLFW_PRESS &&
      !io.WantCaptureMouse && !showTimeline) {
    g_Animations.push_back(AnimationState{
      .position     = app->mouseState_.pos,
      .startTime    = glfwGetTime(),
      .textureIndex = 0,
      .firstFrame   = kNumFlipbookFrames *
        (uint32_t)(rand() % kNumFlipbooks),
    });
  }
});
```

7. The key callback is used to toggle the timeline UI and "lock" the current state of all explosion animations as a keyframe into the `g_AnimationKeyframe` vector:

```
app.addKeyCallback([](GLFWwindow* window,
  int key, int scancode, int action, int mods) {
  ImGuiIO& io          = ImGui::GetIO();
  const bool pressed = action != GLFW_RELEASE;
  if (key == GLFW_KEY_SPACE && pressed &&
      !io.WantCaptureKeyboard)
    showTimeline = !showTimeline;
```

8. Save the current animation state as a keyframe. If there are no animations playing, disable the timeline UI:

```
  if (showTimeline) {
    timelineOffset        = 0.0f;
    g_AnimationsKeyframe = g_Animations;
  }
  if (g_Animations.empty())
    showTimeline = false;
});
```

Now that everything is set up, let's dive into the rendering loop and see how it all comes together to render the animated explosions.

How it works...

The application's main loop looks as follows:

1. When the timeline UI is active, we pause the animation updates:

```
app.run([&](uint32_t width, uint32_t height,
  float aspectRatio, float deltaSeconds) {
  if (!showTimeline) updateAnimations(deltaSeconds);
```

2. Let's introduce a helper lambda to calculate transparency at the start and end of animations using the `smoothstep()` function. This way, our explosions will appear and disappear smoothly, without any distractions. Here, `p1` represents the appearance transition for the first 10% of the animation, and `p2` represents the disappearance transition for the last 20% of the animation:

```
auto easing = [](float t) -> float {
  const float p1 = 0.1f;
  const float p2 = 0.8f;
  if (t <= p1)
    return glm::smoothstep(0.0f, 1.0f, t / p1);
  if (t >= p2)
    return glm::smoothstep(1.0f, 0.0f, (t - p2) / (1.0f - p2));
  return 1.0f;
};
```

3. The render passes and framebuffer configurations are straightforward. As usual, we acquire a command buffer for rendering:

```
const lvk::RenderPass renderPass = {
  .color = {{ .loadOp = lvk::LoadOp_Clear,
              .clearColor = { 1, 1, 1, 1 } } },
};
const lvk::Framebuffer framebuffer = {
  .color = {{ .texture = ctx->getCurrentSwapchainTexture() }},
};
ICommandBuffer& buf = ctx->acquireCommandBuffer();
buf.cmdBeginRendering(renderPass, framebuffer);
buf.cmdBindRenderPipeline(pipelineQuad);
```

4. We render each explosion as a quad, and GLSL shader parameters for each explosion are passed using Vulkan push constants. The time in seconds (`s.time`) is converted to normalized time, `t`, ranging from `0` to `1`. The current `textureIndex` is converted to the LightweightVK texture index:

```
for (const AnimationState& s : g_Animations) {
  const float t = s.time / (kNumFlipbookFrames / kAnimationFPS);
  const struct {
```

```
    mat4 proj;
    uint32_t textureId;
    vec2 pos;
    vec2 size;
    float alphaScale;
  } pc {
    .proj = glm::ortho(
      0.0f, float(width), 0.0f, float(height)),
    .textureId  = textures[s.textureIndex].index(),
    .pos  = s.position * vec2(width, height),
    .size = vec2(height * 0.5f),
    .alphaScale = easing(t),
  };
  buf.cmdPushConstants(pc);
  buf.cmdDraw(4);
}
```

Note

Curious readers may notice several inefficiencies in this code. A faster approach would be to store all the explosion data in a buffer and then call `cmdDraw()` with the number of instances equal to the number of explosions. This is true. However, it was a deliberate choice. Since the number of active explosions in this demo is very low – capping at a few dozen at most – the performance gain would be modest, but it would introduce extra complexity, such as managing multiple round-robin uniform buffers and ensuring proper synchronization. We leave this as an exercise for you.

5. After rendering all the explosions, we can render our ImGui UI, which includes the FPS counter and hints:

```
app.imgui_->beginFrame(framebuffer);
app.drawFPS();
ImGui::SetNextWindowPos(ImVec2(10, 10));
ImGui::Begin("Hints:", nullptr,
  ImGuiWindowFlags_AlwaysAutoResize |
  ImGuiWindowFlags_NoFocusOnAppearing |
  ImGuiWindowFlags_NoInputs |
  ImGuiWindowFlags_NoCollapse);
if (showTimeline) {
  ImGui::Text("SPACE - toggle timeline");
} else {
```

```
    ImGui::Text("SPACE - toggle timeline");
    ImGui::Text("Left click - set an explosion");
  }
  ImGui::End();
```

6. The timeline editing UI can be rendered in a similar way:

```
  if (showTimeline) {
    const ImGuiViewport* v = ImGui::GetMainViewport();
    ImGui::SetNextWindowContentSize({ v->Size.x - 520, 0 });
    ImGui::SetNextWindowPos(ImVec2(350, 10), ImGuiCond_Always);
    ImGui::Begin("Timeline:", nullptr,
      ImGuiWindowFlags_NoCollapse |
      ImGuiWindowFlags_NoResize);
```

7. Each time we move the slider, the offset value is used to calculate the new animation state based on the "locked" keyframe, which we stored earlier in `g_AnimationsKeyframe` when toggling the timeline editing UI:

```
    if (ImGui::SliderFloat("Time offset", &timelineOffset, -2.0f, +2.0f))
    {
      setAnimationsOffset(timelineOffset);
    }
    ImGui::End();
  }
  app.imgui_->endFrame(buf);
  buf.cmdEndRendering();
  ctx->submit(
    buf, ctx->getCurrentSwapchainTexture());
});
```

That covers the C++ part. Now, let's take a look at the GLSL shaders for our application.

The vertex shader `Chapter08/01_DescriptorIndexing/src/main.vert` to render our textured quads looks the following way:

1. The `PerFrameData` structure corresponds to the C++ push constants we passed into the shaders and is shared between the vertex and fragment shaders. It is declared in the `Chapter08/01_DescriptorIndexing/src/common.sp` file:

```
#version 460
// #include <Chapter08/01_DescriptorIndexing/src/common.sp>
layout(push_constant) uniform PerFrameData {
  mat4 proj;
  uint textureId;
  float x;
```

```
  float y;
  float width;
  float height;
  float alphaScale;
} pc;
```

2. We use the programmable vertex pulling technique to calculate quad vertices using the value of `gl_VertexIndex`:

```
layout (location=0) out vec2 uv;
const vec2 pos[4] = vec2[4](
  vec2( 0.5, -0.5),
  vec2( 0.5,  0.5),
  vec2(-0.5, -0.5),
  vec2(-0.5,  0.5)
);
void main() {
  uv = pos[gl_VertexIndex] + vec2(0.5);
```

3. The clip-space position is calculated by scaling and offsetting the unit quad stored in pos using the values of `width`, `height`, and the xy screen position of the explosion:

```
  vec2 p = vec2(pc.x, pc.y) +
    pos[gl_VertexIndex] * vec2(pc.width, pc.height);
  gl_Position = pc.proj * vec4(p, 0.0, 1.0);
}
```

4. The fragment shader `Chapter08/01_DescriptorIndexing/src/main.frag` is straightforward. All we need to do here is scale the texture's alpha value by the `alphaScale` value that we obtained in C++ using the easing function:

```
#include <Chapter08/01_DescriptorIndexing/src/common.sp>
layout (location=0) in vec2 uv;
layout (location=0) out vec4 out_FragColor;
void main() {
  out_FragColor = vec4(vec3(1), pc.alphaScale) *
    textureBindless2D(pc.textureId, 0, uv);
}
```

Now we can run our application. Click a few times in the window and press the Spacebar to see something similar to the screenshot below:

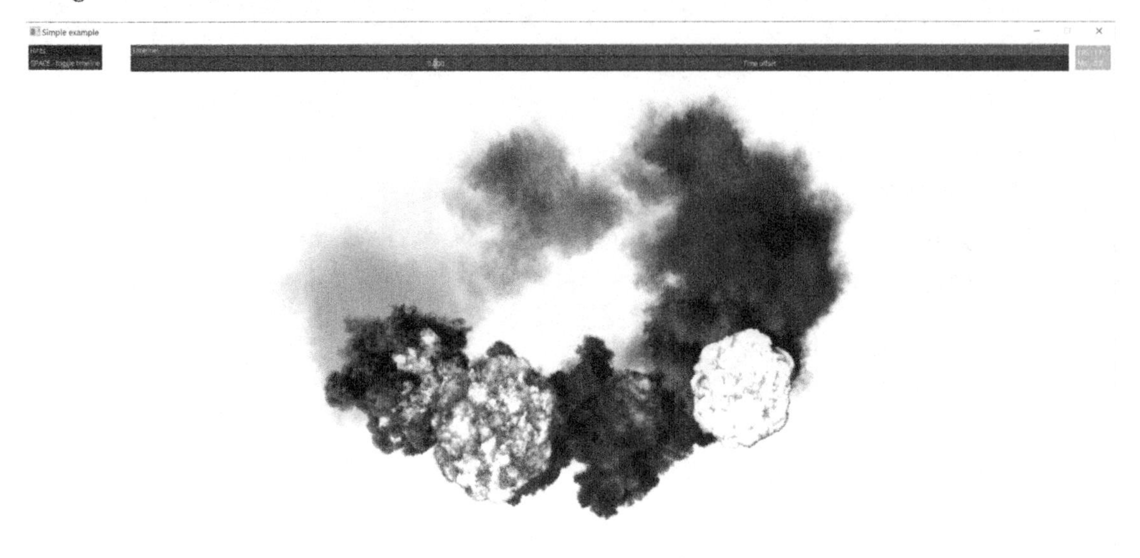

Figure 8.3: Animated explosions using descriptor indexing and arrays of textures

You can interact with the timeline UI by moving the slider, allowing you to control the animation's progress. By dragging the slider, you can rewind and fast-forward the animation, giving you precise control over its playback.

Now, let's move on to the next recipe and explore how to use descriptor indexing to render our scene using the material system we described in *Implementing a material system*.

Implementing indirect rendering with Vulkan

In the previous chapters, we covered geometry rendering with basic texturing. In the previous recipe *Using descriptor indexing and arrays of textures in Vulkan*, we learned how to wrangle multiple textures. This recipe explains how to use Vulkan's indirect rendering feature to render the scene data discussed earlier in the chapter, specifically in the recipes *Loading and saving a scene graph* and *Implementing a material system*. While this recipe doesn't have a standalone example app, all the helper code provided here will be used in the upcoming recipes.

Indirect rendering allows a significant part of scene traversal to be precalculated and offloaded from the CPU to the GPU, which can greatly boost performance. The key idea is to organize all scene data into arrays that shaders can access using integer IDs, eliminating the need for any API state changes during rendering. The indirect buffer generated by the application contains an array of fixed-size structs, with each struct holding the parameters for a single draw command. Data representation is the key here. Let's explore how to implement this approach for our scene.

Getting ready

Be sure to revisit the recipes on our scene graph data structures, especially *Loading and saving a scene graph* and *Implementing a material system*.

This recipe explains the source code for the `VKMesh` helper class, which is defined in `Chapter08/VKMesh08.h`.

How to do it...

In the recipe *Implementing a material system*, we learned how to manage materials for the entire scene, including textures. The data structures introduced there are intended for CPU-based scene processing. Now, let's introduce a helper function, `convertToGPUMaterial()`, which converts the CPU material representation into a GPU-friendly format.

1. We have simplified The `convertToGPUMaterial()` function by leaving out the full glTF 2.0 PBR material properties. Instead, we'll focus on managing textures to keep things concise. The `TextureCache` container holds the loaded textures, while the `TextureFiles` container stores the filenames. Both containers are indexed by a `textureId` and were set up in the recipe *Implementing a material system*:

```
using TextureCache = std::vector<Holder<TextureHandle>>;
using TextureFiles = std::vector<std::string>;
GLTFMaterialDataGPU convertToGPUMaterial(
  const std::unique_ptr<lvk::IContext>& ctx,
  const Material& mat,
  const TextureFiles& files,
  TextureCache& cache)
{
  GLTFMaterialDataGPU result = {
    .baseColorFactor = mat.baseColorFactor,
    .metallicRoughnessNormalOcclusion =
      vec4(mat.metallicFactor, mat.roughness, 1, 1),
    .emissiveFactorAlphaCutoff =
      vec4(vec3(mat.emissiveFactor), mat.alphaTest),
  };
```

2. Let's introduce a local lambda function to retrieve a texture index from the cache. If the texture with this index isn't in the cache yet, it will be loaded and added to the cache. This helps prevent duplicate textures, as different materials that use the same textures will store the same texture IDs:

```
  auto getTextureFromCache = [&cache, &ctx, &files](
    int textureId) -> uint32_t {
    if (textureId == -1) return 0;
    if (cache.size() <= textureId)
```

```
      cache.resize(textureId + 1);
    if (cache[textureId].empty()) {
      cache[textureId] =
        loadTexture(ctx, files[textureId].c_str());
    }
    return cache[textureId].index();
  };
```

3. Now we can retrieve all the necessary material textures using the cache:

```
  result.baseColorTexture =
    getTextureFromCache(mat.baseColorTexture);
  result.emissiveTexture =
    getTextureFromCache(mat.emissiveTexture);
  result.normalTexture =
    getTextureFromCache(mat.normalTexture);
  result.transmissionTexture =
    getTextureFromCache(mat.opacityTexture);
  return result;
}
```

Once this utility function is in place, we can move on to the `VKMesh` class, which handles all aspects of scene rendering. It is responsible for storing GPU buffers for geometry, transformations, materials, and other auxiliary data. Despite its name, this class can manage the entire scene, and we'll see how it works shortly.

1. First, we need to store the number of indices in a mesh, as this class handles only indexed meshes. It also needs buffers for both indices and vertices:

```
class VKMesh final {
public:
  const unique_ptr<lvk::IContext>& ctx;
  uint32_t numIndices_ = 0;
  uint32_t numMeshes_  = 0;
  Holder<BufferHandle> bufferIndices_;
  Holder<BufferHandle> bufferVertices_;
```

2. To make things really fast, we use indirect rendering for our scene. The indirect buffer will contain an array of `VkDrawIndexedIndirectCommand` structures, one for each individual mesh in the scene. The transforms buffer contains an array of `mat4` model-to-world matrices for the entire scene. These matrices were generated in the recipe *Implementing transformation trees*:

Note

Here's a quick look at the `VkDrawIndexedIndirectCommand` declaration, which corresponds to the parameters of one `vkCmdDrawIndexedIndirect()` draw call:

```
struct VkDrawIndexedIndirectCommand {
  uint32_t indexCount;
  uint32_t instanceCount;
  uint32_t firstIndex;
  int32_t  vertexOffset;
  uint32_t firstInstance;
};
```

We've already used the `firstInstance` parameter to pass an integer index into GLSL shaders through the `gl_BaseInstance` built-in variable. We'll use the same technique again, but this time, for a series of draw calls to render the entire scene efficiently.

```
Holder<BufferHandle> bufferIndirect_;
Holder<BufferHandle> bufferTransforms_;
```

3. For each mesh, we need to store its transform index and material index. Let's introduce a `DrawData` struct and store an array of these in the `bufferDrawData_` buffer. This array can be accessed from GLSL shaders using the `gl_BaseInstance` built-in variable to retrieve the appropriate indices. We'll explore how this works when we begin populating the `bufferIndirect_` buffer:

```
struct DrawData {
  uint32_t transformId;
  uint32_t materialId;
};
Holder<BufferHandle> bufferDrawData_;
```

4. The materials buffer holds an array of `GLTFMaterialDataGPU` structs, which we discussed earlier. The remaining member fields contain vertex and fragment shaders, as well as two rendering pipelines – one for standard rendering and the other for wireframe rendering. We also store the texture cache data here:

```
Holder<BufferHandle> bufferMaterials_;
Holder<ShaderModuleHandle> vert_;
Holder<ShaderModuleHandle> frag_;
Holder<RenderPipelineHandle> pipeline_;
Holder<RenderPipelineHandle> pipelineWireframe_;
TextureFiles textureFiles_;
mutable TextureCache textureCache_;
```

5. The constructor of the `VKMesh` class takes parameters representing the entire scene, which were loaded in the recipe *Loading and saving a scene graph*. The depth-buffer format is application-dependent and is passed in from outside the class. The `numSamples` parameter is used for MSAA and will be explained in the *Implementing MSAA in Vulkan* recipe in *Chapter 10*. For now, we will use the default value, `1`:

```
VKMesh(const std::unique_ptr<lvk::IContext>& ctx,
       const MeshData& meshData,
       const Scene& scene,
       lvk::Format colorFormat,
       lvk::Format depthFormat,
       uint32_t numSamples = 1)
: ctx(ctx)
, numIndices_((uint32_t)meshData.indexData.size())
, numMeshes_((uint32_t)meshData.meshes.size())
, textureFiles_(meshData.textureFiles) {
```

6. The `MeshFileHeader` and `MeshData` structs already contain the packed geometry vertex and index data. All that is left to do is upload them into GPU buffers. Since the geometry is static, we don't need to worry about it anymore. To refresh your memory on how this data was prepared, refer to *Chapter 5*:

```
  const MeshFileHeader header = meshData.getMeshFileHeader();
  const uint32_t* indices = meshData.indexData.data();
  const uint8_t* vertexData = meshData.vertexData.data();
  bufferVertices_ = ctx->createBuffer({
    .usage     = lvk::BufferUsageBits_Vertex,
    .storage   = lvk::StorageType_Device,
    .size      = header.vertexDataSize,
    .data      = vertexData });
  bufferIndices_ = ctx->createBuffer({
    .usage     = lvk::BufferUsageBits_Index,
    .storage   = lvk::StorageType_Device,
    .size      = header.indexDataSize,
    .data      = indices });
```

7. The model-to-world transformation `mat4` matrices are copied from the `Scene::globalTransforms` vector. To refresh your memory on how this was done, refer to the recipe *Implementing transformation trees*:

```
  bufferTransforms_ = ctx->createBuffer({
    .usage   = lvk::BufferUsageBits_Storage,
    .storage = lvk::StorageType_Device,
    .size    = scene.globalTransform.size() * sizeof(glm::mat4),
    .data    = scene.globalTransform.data() });
```

8. Now, let's pack all the materials into a GPU buffer. We iterate through all the materials in `MeshData::materials` and use the helper function `convertToGPUMaterial()`, which we introduced earlier in this recipe:

```
std::vector<GLTFMaterialDataGPU> materials;
materials.reserve(meshData.materials.size());
for (const auto& mat : meshData.materials)
  materials.push_back( convertToGPUMaterial(
    ctx, mat, textureFiles_, textureCache_) );
bufferMaterials_ = ctx->createBuffer({
  .usage   = lvk::BufferUsageBits_Storage,
  .storage = lvk::StorageType_Device,
  .size    = materials.size() *
    sizeof(decltype(materials)::value_type),
  .data    = materials.data() });
```

9. So far, we've uploaded the geometry, transforms, and material data for the entire scene into GPU buffers, but we haven't actually rendered anything yet. For example, our vertex and index buffers contain the geometry for all the meshes in the scene, but there's no information about how many of them we actually want to render (if any). Now, we'll go through all the meshes in the scene and create draw commands to render them:

```
vector<DrawIndexedIndirectCommand> drawCommands;
vector<DrawData> drawData;
const uint32_t numCommands = header.meshCount;
drawCommands.resize(numCommands);
drawData.resize(numCommands);
DrawIndexedIndirectCommand* cmd = drawCommands.data();
DrawData* dd = drawData.data();
uint32_t ddIndex = 0;
```

10. Populate both the indirect commands buffer and the draw data buffer. Note that the `baseInstance` parameter holds the index of a `DrawData` instance, and the corresponding `DrawData` instance contains the transform ID and material ID for that mesh:

```
for (auto& i : scene.meshForNode) {
  const Mesh& mesh = meshData.meshes[i.second];
  *cmd++ = { .count         = mesh.getLODIndicesCount(0),
             .instanceCount = 1,
             .firstIndex    = mesh.indexOffset,
             .baseVertex    = mesh.vertexOffset,
             .baseInstance  = ddIndex++ };
  *dd++ = { .transformId = i.first,
            .materialId  = mesh.materialID };
}
```

11. The data for both buffers is now ready, and we can upload it to the GPU:

```
bufferIndirect_ = ctx->createBuffer({
  .usage   = lvk::BufferUsageBits_Indirect,
  .storage = lvk::StorageType_Device,
  .size    = sizeof(DrawIndexedIndirectCommand)*numCommands,
  .data    = drawCommands.data() });
bufferDrawData_ = ctx->createBuffer({
  .usage   = lvk::BufferUsageBits_Storage,
  .storage = lvk::StorageType_Device,
  .size    = sizeof(DrawData) * numCommands,
  .data    = drawData.data() });
```

12. The rest of the `VKMesh` constructor code handles the creation of rendering pipelines for both normal and wireframe rendering. We'll include it here without any additional explanations. The `samplesCount` and `minSampleShading` parameters are used for MSAA and will be explained in the *Implementing MSAA in Vulkan* recipe in *Chapter 10*:

```
  vert_ = loadShaderModule(
    ctx, "Chapter08/02_SceneGraph/src/main.vert");
  frag_ = loadShaderModule(
    ctx, "Chapter08/02_SceneGraph/src/main.frag");
  pipeline_ = ctx->createRenderPipeline({
    .vertexInput  = meshData.streams,
    .smVert       = vert_,
    .smFrag       = frag_,
    .color        = { { .format = colorFormat } },
    .depthFormat  = depthFormat,
    .cullMode     = lvk::CullMode_None,
    .samplesCount = numSamples,
    .minSampleShading = numSamples>1 ? 0.25f : 0 });
  pipelineWireframe_ = ctx->createRenderPipeline({
    .vertexInput  = meshData.streams,
    .smVert       = vert_,
    .smFrag       = frag_,
    .color        = { { .format = colorFormat } },
    .depthFormat  = depthFormat,
    .cullMode     = lvk::CullMode_None,
    .polygonMode  = lvk::PolygonMode_Line,
    .samplesCount = numSamples });
}
```

This approach to data storage makes it a useful building block for a rendering engine, where individual mesh instances can be added or removed dynamically at runtime. Now let's take a look at how actual rendering works.

How it works...

The actual rendering takes place in the `VKMesh::draw()` method, and it is fairly straightforward.

1. First, we bind the index and vertex buffers for rendering the entire scene, along with the appropriate rendering pipeline and the depth state:

```
  void draw(
    lvk::ICommandBuffer& buf,
    const mat4& view,
    const mat4& proj,
    TextureHandle texSkyboxIrradiance = {},
    bool wireframe = false) const
  {
    buf.cmdBindIndexBuffer(bufferIndices_, lvk::IndexFormat_UI32);
    buf.cmdBindVertexBuffer(0, bufferVertices_);
    buf.cmdBindRenderPipeline(
      wireframe ? pipelineWireframe_ : pipeline_);
    buf.cmdBindDepthState({
      .compareOp = lvk::CompareOp_Less,
      .isDepthWriteEnabled = true });
```

2. Next, we prepare the push constant data for our GLSL shaders. Unlike the Explosions example from the previous recipe, *Using descriptor indexing and arrays of textures in Vulkan*, this approach uses a single push constant value for the entire scene. This is done for simplicity and performance reasons, as we aim to render a scene with thousands of meshes in one draw call:

```
    const struct {
      mat4 viewProj;
      uint64_t bufferTransforms;
      uint64_t bufferDrawData;
      uint64_t bufferMaterials;
      uint32_t texSkyboxIrradiance;
    } pc = {
      .viewProj         = proj * view,
      .bufferTransforms = ctx->gpuAddress(bufferTransforms_),
      .bufferDrawData   = ctx->gpuAddress(bufferDrawData_),
      .bufferMaterials = ctx->gpuAddress(bufferMaterials_),
      .texSkyboxIrradiance = texSkyboxIrradiance.index(),
    };
    buf.cmdPushConstants(pc);
```

3. Rendering can now be done in a single draw call, `vkCmdDrawIndexedIndirect()`. We can render an entire scene with thousands of meshes, each using a distinct material, all with just one indirect draw call:

```
    buf.cmdDrawIndexedIndirect(
      bufferIndirect_, 0, numMeshes_);
  }
```

4. A few more important things to note here. Since we want our scene to be dynamic, we might need to update the model-to-world transformations every frame. The member function `updateGlobalTransforms()` takes a vector of `mat4` matrices and uploads it to the `bufferTransforms_` buffer:

```
  void updateGlobalTransforms(
    const mat4* data, size_t numMatrices) const
  {
    ctx->upload(bufferTransforms_, data, numMatrices * sizeof(mat4));
  }
```

5. Similarly, with materials, we may need to update a material occasionally. The `updateMaterialIndex()` function is a member function designed to handle this. We will demonstrate how to use it in the next recipe:

```
  void updateMaterial(
    const Material* materials,
    int updateMaterialIndex) const
  {
    if (updateMaterialIndex < 0) return;
    const GLTFMaterialDataGPU mat =
      convertToGPUMaterial(ctx,
        materials[updateMaterialIndex],
        textureFiles_, textureCache_);
    ctx->upload(bufferMaterials_, &mat, sizeof(mat),
      sizeof(mat) * updateMaterialIndex);
  }
};
```

With the helper class `VKMesh` in place, we now have the foundation to render our scene data using Vulkan and LightweightVK. This class handles all the complexities of managing buffers, materials, transformations, and indirect draw commands, allowing us to efficiently render an entire scene. Now that we've covered the setup, let's move on to the next recipe, where we will see how all of this comes together in our first complete scene rendering application.

Putting it all together into a scene editing application

In the previous recipes, we introduced a significant amount of scaffolding code to manage the scene hierarchy, handle materials, and prepare GPU buffers for rendering. Now, let's put all of that together by implementing a complete scene-rendering application based on those code fragments. To make the application even more engaging, we will add some scene graph editing capabilities, allowing us to interactively modify and update the scene in real time.

Getting ready

Before proceeding with this recipe, be sure to revisit the recipe *Implementing indirect rendering with Vulkan* along with all the related recipes to review the data structures we use for storing and managing our scene. This will ensure you have a solid understanding of how everything fits together before diving into the next steps.

The source code for this recipe is located at `Chapter08/02_SceneGraph`.

How to do it...

Let's go through the C++ part of the application `Chapter08/02_SceneGraph/src/main.cpp`, starting from the `main()` function.

1. First, we load and precache our scene data using the method described earlier in this chapter in the recipe *Loading and saving a scene graph*:

```
const char* fileNameCachedMeshes = ".cache/ch08_orrery.meshes";
const char* fileNameCachedMaterials = ".cache/ch08_orrery.materials";
const char* fileNameCachedHierarchy = ".cache/ch08_orrery.scene";
int main() {
  if (!isMeshDataValid(fileNameCachedMeshes)||
      !isMeshHierarchyValid(fileNameCachedHierarchy)||
      !isMeshMaterialsValid(fileNameCachedMaterials))
 {
    printf("No cached data found. Precaching...\n\n");
    MeshData meshData;
    Scene ourScene;
    loadMeshFile("data/meshes/orrery/scene.gltf",
      meshData, ourScene, true);
    saveMeshData(fileNameCachedMeshes, meshData);
    saveMeshDataMaterials(fileNameCachedMaterials, meshData);
    saveScene(fileNameCachedHierarchy, ourScene);
  }
```

2. Next, we load the precached scene, which significantly speeds up subsequent runs of our application after the initial caching is complete. This will be especially important in the final recipe of this chapter, where we load the complex Lumberyard Bistro mesh. For our `VKMesh` class, we will need instances of `MeshData`, `MeshFileHeader`, and `Scene`:

```
MeshData meshData;
const MeshFileHeader header =
  loadMeshData(fileNameCachedMeshes, meshData);
loadMeshDataMaterials(fileNameCachedMaterials, meshData);
Scene scene;
loadScene(fileNameCachedHierarchy, scene);
```

3. Let's create a `VulkanApp` and a 3D drawing canvas, following the steps outlined in the recipe *Implementing immediate mode 3D drawing canvas* in *Chapter 4*. We will use the canvas to render the bounding boxes of the meshes in our scene. The `drawWireframe` flag toggles the wireframe rendering mode, while `selectedNode` stores the ID of the node we want to edit in the UI. We will demonstrate how this works shortly:

```
VulkanApp app({
  .initialCameraPos = vec3(-0.88f, 1.26f, 1.07f),
  .initialCameraTarget = vec3(0, -0.6f, 0),
});
LineCanvas3D canvas3d;
unique_ptr<lvk::IContext> ctx(app.ctx_.get());
bool drawWireframe = false;
int selectedNode = -1;
```

4. Before entering the rendering loop, let's create our `VKMesh` instance using the loaded scene data and the depth buffer format obtained from `VulkanApp`. Since there's only one rendering pass, the setup for `RenderPass` and `FrameBuffer` is trivial:

```
const VKMesh mesh(ctx, meshData, scene,
  ctx->getSwapchainFormat(),
  app.getDepthFormat());
app.run([&](uint32_t width, uint32_t height,
            float aspectRatio, float deltaSeconds)
{
  const lvk::RenderPass renderPass = {
    .color = {
      { .loadOp = lvk::LoadOp_Clear,
        .clearColor = { 1.f, 1.f, 1.f, 1.f } } },
    .depth = { .loadOp = lvk::LoadOp_Clear,
               .clearDepth = 1.f }
```

```
};
const lvk::Framebuffer framebuffer = {
  .color = {{.texture = ctx->getCurrentSwapchainTexture()}},
  .depthStencil = { .texture = app.getDepthTexture() },
};
```

5. The view and projection matrices remain unchanged for the entire frame. The `updateMaterialIndex` variable is used to update a single material from the UI:

```
const mat4 proj = glm::perspective(45.0f,
  aspectRatio, 0.01f, 100.0f);
const mat4 view = app.camera_.getViewMatrix();
int updateMaterialIndex = -1;
```

6. Now, let's render the scene. All the work has already been done in `VKMesh::draw()` from the previous recipe, *Implementing indirect rendering with Vulkan*. We simply need to call it here with the appropriate parameters:

```
ICommandBuffer& buf = ctx->acquireCommandBuffer();
buf.cmdBeginRendering(renderPass, framebuffer);
buf.cmdPushDebugGroupLabel("Mesh", 0xff0000ff);
mesh.draw(buf, view, proj, {}, drawWireframe);
buf.cmdPopDebugGroupLabel();
```

7. On top of the mesh, we render a grid, as explained in the recipe *Implementing an infinite grid GLSL shader* in *Chapter 5*. This is accompanied by an FPS counter and a default keyboard helpers memo.

```
app.drawGrid(buf, proj, vec3(0, -1.0f, 0));
app.imgui_->beginFrame(framebuffer);
app.drawFPS();
app.drawMemo();
```

8. Next, we render the bounding boxes for all meshes in our scene using the 3D canvas from the recipe *Implementing immediate mode 3D drawing canvas* in *Chapter 4*:

```
canvas3d.clear();
canvas3d.setMatrix(proj * view);
for (auto& p : scene.meshForNode) {
  const BoundingBox box = meshData.boxes[p.second];
  canvas3d.box(scene.globalTransform[p.first],
    box, vec4(1, 0, 0, 1));
}
```

9. The extra step is to draw the scene editing UI itself using the **ImGui** library. The UI displays the entire scene graph tree with all the nodes and a checkbox to toggle the wireframe mode for each node.

```
const ImGuiViewport* v = ImGui::GetMainViewport();
ImGui::SetNextWindowPos(ImVec2(10, 200));
ImGui::SetNextWindowSize(
  ImVec2(v->WorkSize.x / 6,
         v->WorkSize.y - 210));
ImGui::Begin("Scene graph", nullptr,
  ImGuiWindowFlags_NoFocusOnAppearing |
  ImGuiWindowFlags_NoCollapse |
  ImGuiWindowFlags_NoResize);
ImGui::Checkbox("Draw wireframe", &drawWireframe);
ImGui::Separator();
```

10. The scene graph tree is rendered recursively using the `renderSceneTreeUI()` function, which we'll examine in more detail shortly, in the *How it works...* section. Our UI allows selecting a scene node by clicking on a node name in the tree. Once a node is selected, a new UI window appears, enabling the editing of the selected node's properties. This functionality is handled by the `editNodeUI()` function, which we will explore shortly:

```
const int node = renderSceneTreeUI(scene, 0, selectedNode);
if (node > -1) selectedNode = node;
ImGui::End();
editNodeUI(scene, meshData, view, proj,
  selectedNode, updateMaterialIndex,
  mesh.textureCache_);
```

11. If a node is selected, we render its bounding box in green. After that, we can finalize both the 3D canvas and ImGui rendering, and then submit our command buffer:

```
if (selectedNode > -1 && scene.hierarchy[
      selectedNode].firstChild < 0)
{
  const uint32_t meshId = scene.meshForNode[selectedNode];
  const BoundingBox box = meshData.boxes[meshId];
  canvas3d.box(
    scene.globalTransform[selectedNode],
    box, vec4(0, 1, 0, 1));
}
canvas3d.render(*ctx.get(), framebuffer, buf);
app.imgui_->endFrame(buf);
```

```
    buf.cmdEndRendering();
    ctx->submit(buf, ctx->getCurrentSwapchainTexture());
```

12. After rendering, we need to update the model-to-world transformation matrices, as described earlier in the recipe *Implementing transformation trees*. If any transformations were changed, we must update the GPU buffer by calling `VKMesh::updateGlobalTransforms()`, which we implemented in the recipe *Implementing indirect rendering with Vulkan*:

```
    if (recalculateGlobalTransforms(scene))
      mesh.updateGlobalTransforms(
        scene.globalTransform.data(),
        scene.globalTransform.size());
```

13. The same applies to materials. If a material has been modified through the UI, we need to update the GPU buffer data for that material:

```
    if (updateMaterialIndex > -1)
      mesh.updateMaterial(meshData.materials.data(),
        updateMaterialIndex);
  });
  ctx.release();
  return 0;
}
```

This concludes the high-level flow of our sample application. Now, let's dive deeper into how the scene editing helper functions work. We'll explore the details of the functions `renderSceneTreeUI()` and `editNodeUI()`, as well as how they interact with the rest of the application to allow us to manipulate and edit the scene in real time.

How it works...

One helper function is `renderSceneTreeUI()`, which recursively renders the scene graph. This function is central to rendering the scene graph UI, so let's take a look at it in detail.

Note

Here's the `getNodeName()` helper function for your convenience:

```
std::string getNodeName(const Scene& scene, int node) {
  const int strID = scene.nameForNode.contains(node) ?
    scene.nameForNode.at(node) : -1;
  return strID > -1 ?
    scene.nodeNames[strID] : std::string();
}
```

1. The function takes three arguments: the `Scene` we want to render, the ID of the node we want to display in the UI, and the ID of the selected node. Since the UI includes node names, we can use `getNodeName()` from `shared/Scene/Scene.h` to retrieve the names of the nodes:

```
int renderSceneTreeUI(
  const Scene& scene, int node, int selectedNode)
{
  const std::string name = getNodeName(scene, node);
  const std::string label = name.empty() ?
    (std::string("Node") + std::to_string(node)) :
    name;
```

2. Depending on whether the scene node is a leaf node – meaning it has no child nodes – we set different colors and ImGui tree node flags. This setup enables ImGui to "unfold" a non-leaf tree node when you click on the bullet next to its name:

```
  const bool isLeaf  = scene.hierarchy[node].firstChild < 0;
  ImGuiTreeNodeFlags flags =
    isLeaf ? ImGuiTreeNodeFlags_Leaf |
             ImGuiTreeNodeFlags_Bullet : 0;
  if (node == selectedNode)
    flags |= ImGuiTreeNodeFlags_Selected;
  ImVec4 color = isLeaf ?
    ImVec4(0, 1, 0, 1) : ImVec4(1, 1, 1, 1);
```

3. Certain nodes, like the root node and a few others, should always be open by default. Let's update the ImGui flags for those nodes accordingly:

```
  if (name.starts_with("Root")) {
    flags |= ImGuiTreeNodeFlags_DefaultOpen |
             ImGuiTreeNodeFlags_Leaf |
             ImGuiTreeNodeFlags_Bullet;
    color = ImVec4(0.9f, 0.6f, 0.6f, 1);
  }
  if (name == "sun" || name == "sun_0" ||
      name.ends_with(".stk") ||
      name.starts_with("p3.earth")) {
    flags |= ImGuiTreeNodeFlags_DefaultOpen;
  }
```

4. Now, we can apply the appropriate color and call `ImGui::TreeNodeEx()` to handle the actual rendering of the tree UI:

```
ImGui::PushStyleColor(ImGuiCol_Text, color);
const bool isOpened = ImGui::TreeNodeEx(
  &scene.hierarchy[node], flags, "%s", label.c_str());
ImGui::PopStyleColor();
ImGui::PushID(node);
```

5. If we click on a leaf node that represents a mesh, we mark it as the selected node. If a non-leaf node is opened, we recursively render the UI for all of its children. Here, we leverage our **First Child, Next Sibling** scene graph representation by starting with the first child and traversing through all its sibling nodes:

```
  if (ImGui::IsItemHovered() && isLeaf)
    selectedNode = node;
  if (isOpened) {
    for (int ch = scene.hierarchy[node].firstChild;
         ch != -1;
         ch = scene.hierarchy[ch].nextSibling)
    {
      if (int subNode =
            renderSceneTreeUI(scene, ch, selectedNode);
          subNode>-1) selectedNode = subNode;
    }
    ImGui::TreePop();
  }
  ImGui::PopID();
  return selectedNode;
}
```

This function serves as a central building block for creating custom scene editing UIs of any kind. You can use it as a foundation to render more complex UIs tailored to your needs. Here is a screenshot showing what the scene tree UI looks like:

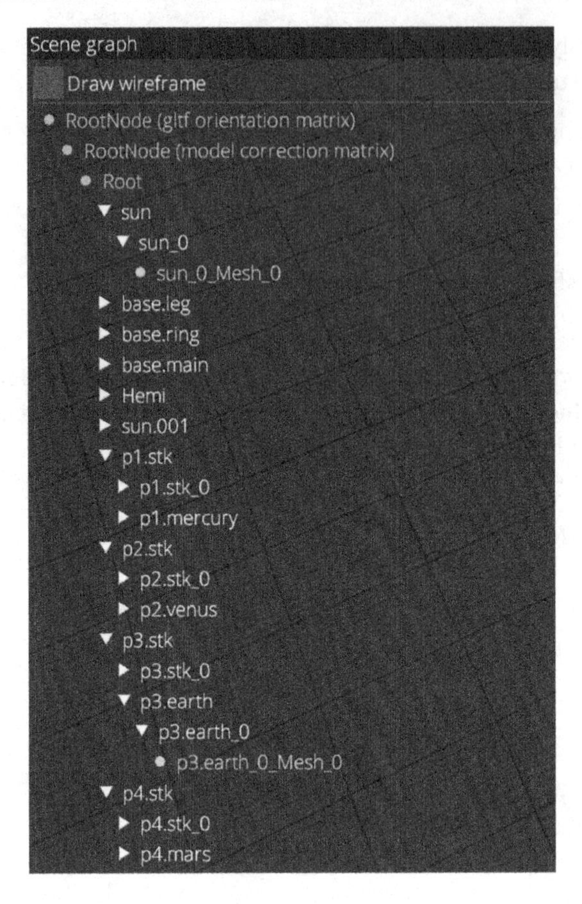

Figure 8.4: The scene tree user interface

Now, let's move on to explore another helper function, `editNodeUI()`, which offers similar foundational tools for creating an editing interface.

1. The `editNodeUI()` function takes additional parameters, as it requires access to `MeshData` and textures. The view and projection matrices are also needed to render transformation gizmos on top of the mesh within the window:

   ```
   void editNodeUI(Scene& scene, MeshData& meshData,
     const mat4& view, const mat4 proj,
     int node,
     int& outUpdateMaterialIndex,
     const TextureCache& textureCache)
   {
   ```

2. We use the **ImGuizmo** library to render transformation gizmos, which provide an intuitive way for users to manipulate the position, rotation, and scale of objects directly in the scene. ImGuizmo allows us to display interactive 3D handles, such as arrows for translation and circles for rotation, making it easier for users to visualize and control transformations in real time. You can check it out at `https://github.com/CedricGuillemet/ImGuizmo`:

```
ImGuizmo::SetOrthographic(false);
ImGuizmo::BeginFrame();
std::string name  = getNodeName(scene, node);
std::string label = name.empty() ?
  (std::string("Node") + std::to_string(node)) :
  name;
label = "Node: " + label;
```

3. We draw the editor window on the right side of the screen:

```
if (const ImGuiViewport* v = ImGui::GetMainViewport()) {
  ImGui::SetNextWindowPos(ImVec2(v->WorkSize.x * 0.83f, 200));
  ImGui::SetNextWindowSize(
    ImVec2(v->WorkSize.x / 6,
           v->WorkSize.y - 210));
}
ImGui::Begin("Editor", nullptr,
  ImGuiWindowFlags_NoFocusOnAppearing |
  ImGuiWindowFlags_NoCollapse |
  ImGuiWindowFlags_NoResize);
if (!name.empty())
  ImGui::Text("%s", label.c_str());
if (node >= 0) {
  ImGui::Separator();
  ImGuizmo::SetID(1);
```

4. In our editing window, we first retrieve the global and local node transformations from the `scene`. Then, we call the `editTransformUI()` helper function to edit the model-to-world transformation of the selected node. We will take a closer look at how this function works shortly:

```
  mat4 globalTransform = scene.globalTransform[node];
  mat4 srcTransform = globalTransform;
  mat4 localTransform = scene.localTransform[node];
  if (editTransformUI(view, proj, globalTransform))
  {
```

5. If the transformation was edited, we need to calculate the delta value, taking into account the node's global transform, and then modify the local transform in the `scene` accordingly. This ensures that the changes are properly applied and that the node's transformation remains consistent within the `scene`'s coordinate system. The `markAsChanged()` function triggers an update to the scene tree, as explained in the recipe *Implementing transformation trees*:

```
    mat4 deltaTransform = glm::inverse(srcTransform)*globalTransform;
    scene.localTransform[node] = localTransform * deltaTransform;
    markAsChanged(scene, node);
  }
```

6. The second part of our scene editor is the material editor. We call the `editMaterialUI()` function to handle the material editing process. We will dive into the details of this function shortly:

```
        ImGui::Separator();
        ImGui::Text("%s", "Material");
        editMaterialUI(scene, meshData, node,
          outUpdateMaterialIndex, textureCache);
      }
    ImGui::End();
  }
```

The UI of the node editor appears as shown in the following screenshot. In this case, we have selected a node named `sun_0_Mesh_0`.

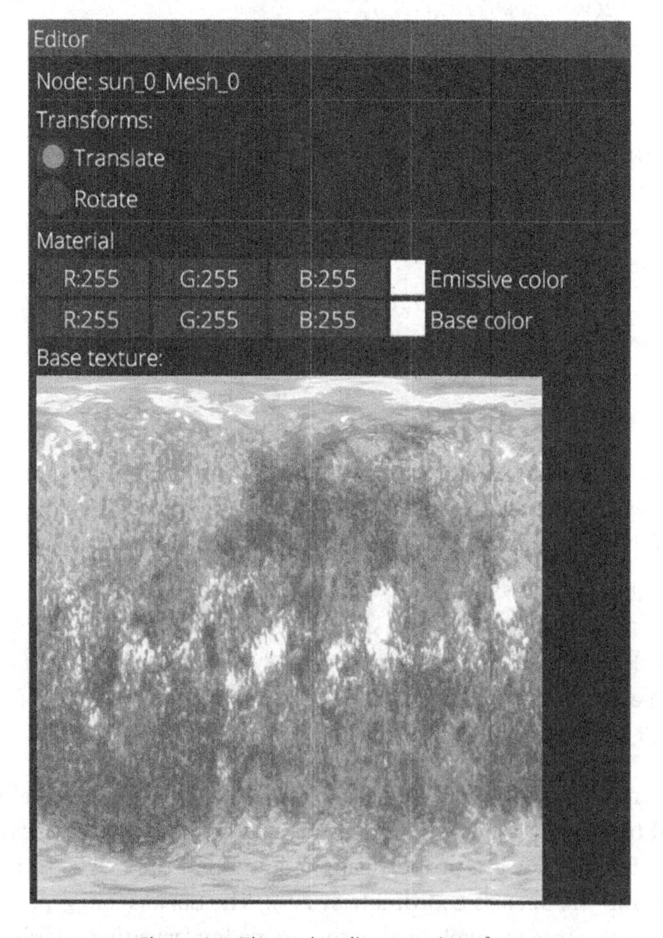

Figure 8.5: The node editor user interface

Let's take a closer look at the `editMaterialUI()` function, which is responsible for rendering the material section of the UI shown in this screenshot.

1. The `editMaterialUI()` function accepts the `Scene`, `MeshData`, a node ID, a reference to output the index of the modified material, and the texture cache. To simplify the texture editing process, we introduce a global variable called `textureToEdit`. This allows us to open a modal window, making it easier to replace textures in the material:

```
int* textureToEdit = nullptr;
bool editMaterialUI(
  Scene& scene, MeshData& meshData,
  int node, int& outUpdateMaterialIndex,
  const TextureCache& textureCache)
{
  if (!scene.materialForNode.contains(node))
    return false;
  const uint32_t matIdx = scene.materialForNode[node];
  Material& material = meshData.materials[matIdx];
```

2. We introduce a Boolean flag, `updated`, to indicate if any of the material properties have been updated. Then, we use ImGui functions to edit individual properties, such as `emissiveFactor` or `baseColorFactor`. You can easily extend this system to include additional properties of your choice by adding more relevant ImGui controls:

```
  bool updated = false;
  updated |= ImGui::ColorEdit3("Emissive color",
    glm::value_ptr(material.emissiveFactor));
  updated |= ImGui::ColorEdit3("Base color",
    glm::value_ptr(material.baseColorFactor));
```

3. Let's add one more cool editing trick: the ability to replace a texture in a material by simply clicking on it. To achieve this, we implement a helper lambda function that opens a modal popup when a texture image is clicked:

```
  const char* ImagesGalleryName = "Images Gallery";
  auto drawTextureUI =
    [&textureCache, ImagesGalleryName](
    const char* name, int& texture)
  {
    if (texture == -1) return;
    ImGui::Text(name);
    ImGui::Image(textureCache[texture].index(),
      ImVec2(512, 512), ImVec2(0, 1), ImVec2(1, 0));
    if (ImGui::IsItemClicked()) {
      textureToEdit = &texture;
      ImGui::OpenPopup(ImagesGalleryName);
    }
  };
```

4. Now, we can render all the material textures using this lambda function:

```
drawTextureUI("Base texture:", material.baseColorTexture);
drawTextureUI("Emissive texture:", material.emissiveTexture);
drawTextureUI("Normal texture:", material.normalTexture);
drawTextureUI("Opacity texture:", material.opacityTexture);
```

5. Once all the material textures are rendered in the UI, we display a modal pop-up window featuring a gallery of all the loaded textures, allowing the user to easily pick and replace a texture. The modal pop-up window is centered on the screen:

```
if (const ImGuiViewport* v = ImGui::GetMainViewport())
{
  ImGui::SetNextWindowPos(
    ImVec2(v->WorkSize.x * 0.5f,
           v->WorkSize.y * 0.5f),
    ImGuiCond_Always, ImVec2(0.5f, 0.5f));
}
```

6. The Images Gallery displays all the loaded textures arranged in a 4x4 grid:

```
if (ImGui::BeginPopupModal(
      ImagesGalleryName, nullptr, ImGuiWindowFlags_AlwaysAutoResize))
{
  for (int i = 0; i != textureCache.size(); i++) {
    if (i && i % 4 != 0) ImGui::SameLine();
    ImGui::Image(textureCache[i].index(),
      ImVec2(256, 256), ImVec2(0, 1), ImVec2(1, 0));
```

7. When a texture is hovered over, we add a transparent color overlay to visually highlight the selected texture. The color value `0x66ffffff` represents a semi-transparent white:

```
    if (ImGui::IsItemHovered()) {
      ImGui::GetWindowDrawList()->AddRectFilled(
        ImGui::GetItemRectMin(),
        ImGui::GetItemRectMax(),
        0x66ffffff);
    }
```

8. When a texture is clicked, we assign its index to our `textureToEdit` global variable, set the `updated` flag, and then close the pop-up window:

```
    if (ImGui::IsItemClicked()) {
      *textureToEdit = i;
      updated        = true;
      ImGui::CloseCurrentPopup();
    }
```

```
        }
        ImGui::EndPopup();
    }
```

9. If any of the material properties were updated, the function outputs the index of the updated material:

```
    if (updated) outUpdateMaterialIndex = matIdx;
    return updated;
}
```

The Images Gallery rendered by this function appears as shown in the following screenshot. It displays all the loaded textures in a grid layout, allowing easy selection and replacement of textures.

Figure 8.6: The Images Gallery user interface

That was the complete C++ code for this application. The GLSL shaders will be covered in the last recipe, *Rendering large scenes*.

The running application should render the following image:

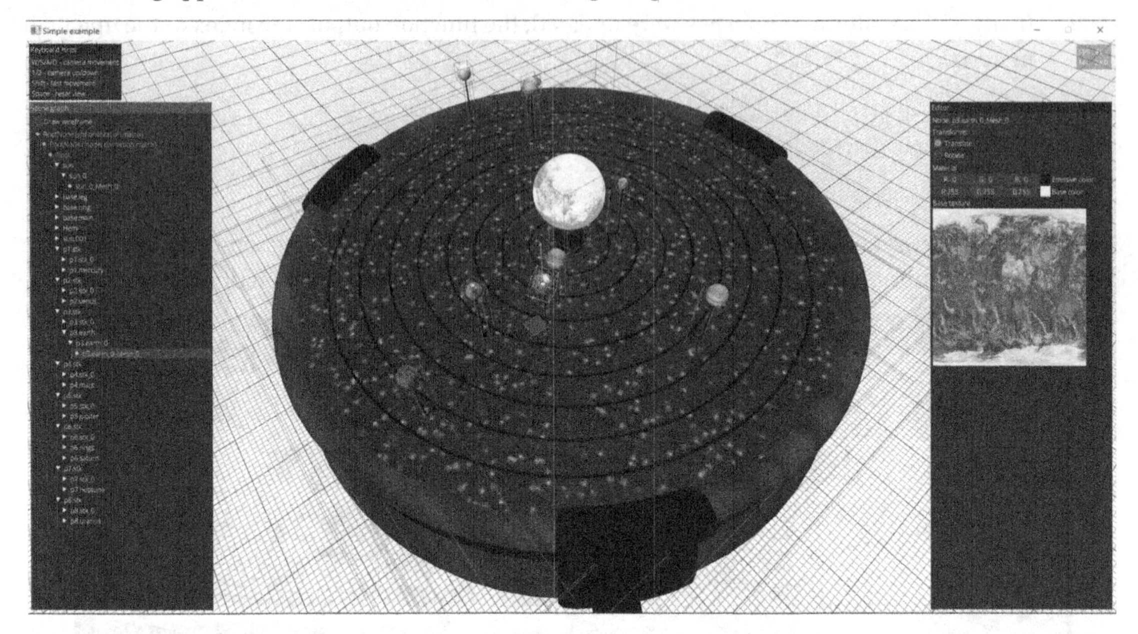

Figure 8.7: The scene graph application

Try clicking on different scene nodes in the scene tree on the right and editing their material properties and transformations using the 3D gizmos.

Now, let's jump to the next recipe to learn some advanced scene graph manipulations.

There's more...

The scene editing methods provided in this recipe serve as basic building blocks for creating a 3D scene editing user interface. To build more advanced UIs, we recommend exploring additional features of the **ImGuizmo** library `https://github.com/CedricGuillemet/ImGuizmo`.

Deleting nodes and merging scene graphs

Our scene graph management routines, as described earlier in this chapter, are incomplete without a few additional operations:

- Deleting scene nodes
- Merging multiple scenes into one (in our case, combining the exterior and interior objects of the Bistro scene)
- Merging material and mesh data lists
- Merging multiple static meshes with the same material into a single, larger mesh

These operations are crucial for the final demo application of this chapter. In the original Lumberyard Bistro scene, a large tree in the backyard contains thousands of small orange and green leaves, which represent nearly two-thirds of the total draw call count in the scene - approximately `18'000` out of `27'000` objects.

This recipe describes the `deleteSceneNodes()` and `mergeNodesWithMaterial()` routines, which are used for manipulating the scene graph. Along with the next recipe *Rendering large scenes*, these functions complete our scene graph data management.

Getting ready

Let's recall the recipe *Using data-oriented design for a scene graph*, where we packed all the scene data into continuous arrays wrapped in `std::vector` for convenience, making them directly usable by the GPU. In this recipe, we use STL's partitioning algorithms to keep everything tightly packed while efficiently deleting and merging scene nodes.

To start, let's implement a utility function to delete a collection of items from an array. While it may seem unnecessary to have a routine that removes a collection of nodes all at once, even in the simplest case of deleting a single node, we must also ensure that all of its child nodes are properly removed.

The idea behind this approach is to use the `std::stable_partition()` algorithm to move all nodes marked for deletion to the end of the array. Once the nodes to be deleted are relocated, we can simply resize the container to remove them from the active scene. This method ensures that we maintain the relative order of the nodes that are not deleted, while efficiently cleaning up those that are. The following diagram illustrates the deletion process:

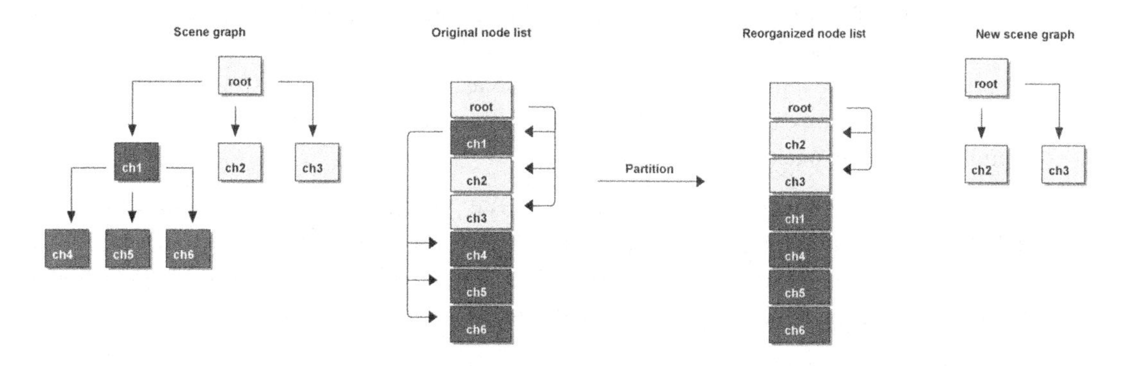

Figure 8.8: Deleting nodes from a scene graph

On the left, we see the original scene graph, where nodes `ch1`, `ch4`, `ch5`, and `ch6` are highlighted for deletion (dark red). In the middle, the nodes are arranged in a linear list, with arrows showing the references between them. On the right, the nodes have been reorganized after partitioning, and the updated scene graph is displayed.

How to do it...

Let's explore how to implement this functionality.

1. The `eraseSelected()` routine from `shared/Utils.h` is designed to be flexible enough to delete any type of item from an array, so we use template arguments for its implementation:

```
template <typename T, typename Index = int>
void eraseSelected(
  std::vector<T>& v, const std::vector<Index>& selection)
{
  v.resize( std::distance(v.begin(),
    std::stable_partition(v.begin(), v.end(),
      [&selection, &v](const T& item) {
        return !std::binary_search(
          selection.begin(), selection.end(),
          static_cast<Index>(
            static_cast<const T*>(&item) - &v[0]));
        })) );
}
```

   The function "chops off" the elements moved to the end of the vector by using `resize()`. The number of items to retain is determined by calculating the distance from the start of the array to the iterator returned by the partition function. The `std::stable_partition()` algorithm takes a lambda function that determines whether an element should be moved to the end of the array. In this case, the lambda checks if the item is in the selection container passed as an argument. While the typical way to find an item's index in an array is by using `std::distance()` and `std::find()`, we can also rely on the more straightforward pointer arithmetic, since the container is tightly packed.

2. Now that we have our workhorse, the `eraseSelected()` routine, we can implement scene node deletion. When deleting a node, all of its children must also be marked for deletion. We achieve this by using a recursive routine that iterates over all the child nodes, adding each node's index to the deletion array. This function is located in `shared/Scene/Scene.cpp`:

```
void collectNodesToDelete(
  const Scene& scene, int node,
  std::vector<uint32_t>& nodes)
{
  for (int n = scene.hierarchy[node].firstChild;
       n != -1;
       n = scene.hierarchy[n].nextSibling) {
    addUniqueIdx(nodes, n);
    collectNodesToDelete(scene, n, nodes);
  }
}
```

3. A helper function `addUniqueIndex()` ensures that items are not added multiple times:

```
void addUniqueIdx(
  std::vector<uint32_t>& v, uint32_t index)
{
  if (!std::binary_search(v.begin(), v.end(), index))
    v.push_back(index);
}
```

   One subtle requirement here is that the array must be sorted. When all the children follow their parents in order, this is not an issue. However, if that's not the case, `std::find()` must be used, which naturally increases the runtime cost of the algorithm.

4. Our deletion routine, `deleteSceneNodes()`, begins by adding all child nodes to the deleted nodes container. To keep track of the moved nodes, we create an array of node indices starting from zero. The `std::iota()` algorithm fills the range with sequentially increasing values, starting with the specified value `0`:

```
void deleteSceneNodes(
  Scene& scene,
  const std::vector<uint32_t>& nodesToDelete)
{
  auto indicesToDelete = nodesToDelete;
  for (uint32_t i : indicesToDelete)
    collectNodesToDelete(scene, i, indicesToDelete);
  std::vector<int> nodes(scene.hierarchy.size());
  std::iota(nodes.begin(), nodes.end(), 0);
```

5. After that, we store the original node count and remove all the indices from our linear index list. To adjust the child node indices, we generate a linear mapping table that maps the old node indices to the new ones:

```
  const size_t oldSize = nodes.size();
  eraseSelected(nodes, indicesToDelete);
  std::vector<int> newIndices(oldSize, -1);
  for (int i = 0; i < nodes.size(); i++)
    newIndices[nodes[i]] = i;
```

6. Before deleting nodes from the `Hierarchy` array, we first remap all node indices. The following lambda function modifies each `Hierarchy` item by locating the corresponding non-null node in the `newIndices` container:

```
  auto nodeMover =
    [&scene, &newIndices](Hierarchy& h)
  {
    return Hierarchy{
```

```
        .parent = (h.parent != -1) ? newIndices[h.parent] : -1,
        .firstChild = findLastNonDeletedItem(scene, newIndices, h.firstChild),
        .nextSibling = findLastNonDeletedItem(scene, newIndices, h.nextSibling),
        .lastSibling = findLastNonDeletedItem(scene, newIndices, h.lastSibling),
      };
    };
```

7. The `std::transform()` algorithm updates all the nodes in the hierarchy. Once the node indices are fixed, we're ready to delete the data. Three calls to `eraseSelected()` remove the unused hierarchy and transform items:

```
    std::transform(scene.hierarchy.begin(),
                   scene.hierarchy.end(),
                   scene.hierarchy.begin(),
                   nodeMover);
    eraseSelected(scene.hierarchy, indicesToDelete);
    eraseSelected(scene.localTransform, indicesToDelete);
    eraseSelected(scene.globalTransform, indicesToDelete);
```

8. Finally, we need to adjust the indices in the mesh, material, and name maps. To do this, we use the `shiftMapIndices()` function, which is described below:

```
    shiftMapIndices(scene.meshForNode, newIndices);
    shiftMapIndices(scene.materialForNode, newIndices);
    shiftMapIndices(scene.nameForNode, newIndices);
  }
```

9. The search for node replacement, used during node index shifting, is implemented recursively. The `findLastNonDeletedItem()` function returns the index of a non-deleted node replacement:

```
int findLastNonDeletedItem(const Scene& scene,
    const std::vector<int>& newIndices, int node)
  {
    if (node == -1) return -1;
    return newIndices[node] == -1 ?
      findLastNonDeletedItem(scene, newIndices,
        scene.hierarchy[node].nextSibling) :
      newIndices[node];
  }
```

If the input is empty, no replacement is necessary. If no replacement is found for the node, we recursively move to the next sibling of the deleted node.

10. The final function, `shiftMapIndices()`, updates the `pair::second` value in each item in the map:

```
void shiftMapIndices(
  std::unordered_map<uint32_t, uint32_t>& items,
  const std::vector<int>& newIndices)
{
  std::unordered_map<uint32_t, uint32_t> newItems;
  for (const auto& m : items) {
    int newIndex = newIndices[m.first];
    if (newIndex != -1)
      newItems[newIndex] = m.second;
  }
  items = newItems;
}
```

The `deleteSceneNodes()` routine helps us compress and optimize the scene graph, while also merging multiple meshes with the same material. Now, we need a method to combine multiple meshes into a single one and remove any scene nodes referring to the merged meshes. The process of merging mesh data mainly involves modifying the index data. Let's go through the steps:

1. The `mergeNodesWithMaterial()` function relies on two helper functions. The first one calculates the total number of merged indices. We keep track of the starting vertex offset for each mesh. Then, in a loop, we shift the indices within the individual mesh blocks of the `meshData.indexData` array. Additionally, for each `Mesh` object, a new `minVtxOffset` value is assigned to the vertex data offset field. The function returns the difference between the original and merged index counts, which also represents the offset where the merged index data begins:

```
uint32_t shiftMeshIndices(
  MeshData& meshData,
  const vector<uint32_t>& meshesToMerge)
{
  uint32_t minVtxOffset = std::numeric_limits<uint32_t>::max();
  for (uint32_t i : meshesToMerge) {
    minVtxOffset = std::min(
     meshData.meshes[i].vertexOffset, minVtxOffset);
  }
  uint32_t mergeCount = 0;
  for (uint32_t i : meshesToMerge) {
    Mesh& m = meshData.meshes[i];
    const uint32_t delta = m.vertexOffset - minVtxOffset;
    const uint32_t idxCount = m.getLODIndicesCount(0);
    for (uint32_t ii = 0; ii < idxCount; ii++) {
      meshData.indexData[m.indexOffset + ii] += delta;
```

```
    }
    m.vertexOffset = minVtxOffset;
    mergeCount += idxCount;
  }
  return meshData.indexData.size() - mergeCount;
}
```

2. The `mergeIndexArray()` function copies the indices from each mesh into a new, consolidated indices array:

```
void mergeIndexArray(MeshData& md,
  const vector<uint32_t>& meshesToMerge,
  unordered_map<uint32_t, uint32_t>& oldToNew)
{
  vector<uint32_t> newIndices(md.indexData.size());
  uint32_t copyOffset = 0;
  uint32_t mergeOffset = shiftMeshIndices(md, meshesToMerge);
```

For each mesh, we determine where to copy its index data. The `copyOffset` value is used for meshes that are not merged, while the `mergeOffset` value starts at the beginning of the merged index data returned by the `shiftMeshIndices()` function.

3. Two variables store the mesh indices of the merged mesh and the copied mesh. We iterate through all the meshes to determine if the current one should be merged:

```
  const size_t mergedMeshIndex = md.meshes.size() - meshesToMerge.size();
  uint32_t newIndex = 0u;
  for (auto midx = 0u; midx < md.meshes.size(); midx++) {
    const bool shouldMerge = std::binary_search(
      meshesToMerge.begin(), meshesToMerge.end(), midx);
```

4. Each index is recorded in the old-to-new correspondence map:

```
    oldToNew[midx] = shouldMerge ?
      mergedMeshIndex : newIndex; newIndex += shouldMerge ? 0 : 1;
```

5. The offset of the index block for this mesh is adjusted by first calculating the source offset for the index data:

```
    Mesh& mesh = md.meshes[midx];
    const uint32_t idxCount = mesh.getLODIndicesCount(0);
    const auto start = md.indexData.begin() + mesh.indexOffset;
    mesh.indexOffset = copyOffset;
```

6. We select between the two offsets and copy the index data from the original array to the new one. The updated index array is then integrated into the `Mesh` data structure:

```
        uint32_t* const offsetPtr =
          shouldMerge ? &mergeOffset : &copyOffset;
        std::copy(start, start + idxCount,
          newIndices.begin() + *offsetPtr);
        *offsetPtr += idxCount;
      }
      md.indexData = newIndices;
```

7. The final step in the merge process is the creation of the merged mesh. We copy the mesh descriptor of the first of the merged meshes and assign the new LOD offsets:

```
      Mesh lastMesh = md.meshes[meshesToMerge[0]];
      lastMesh.indexOffset  = copyOffset;
      lastMesh.lodOffset[0] = copyOffset;
      lastMesh.lodOffset[1] = mergeOffset;
      lastMesh.lodCount     = 1;
      md.meshes.push_back(lastMesh);
    }
```

The `mergeNodesWithMaterial()` routine omits a few key details. First, we only merge the finest LOD. This is sufficient for our purposes, as our scene consists of many simple meshes (1-2 triangles) with only a single LOD. Second, we assume that all merged meshes have the same transformation. This works for our test scene, but if proper transformations are needed, each vertex should be transformed into the global coordinate system and then back into the local coordinates of the node where the merged mesh is placed. Let's explore the implementation:

1. To avoid string comparisons, we convert material names into their corresponding indices in the material name array:

```
void mergeNodesWithMaterial(Scene& scene,
  MeshData& meshData,
  const std::string& materialName)
{
  const int oldMaterial = (int)std::distance(
    std::begin(scene.materialNames),
    std::find(std::begin(scene.materialNames),
              std::end(scene.materialNames),
              materialName));
```

2. Once you have the material index, gather all the scene nodes that need to be deleted:

```
std::vector<uint32_t> toDelete;
for (size_t i = 0u;
     i < scene.hierarchy.size(); i++) {
  if (scene.meshForNode.contains(i) &&
       scene.materialForNode.contains(i) &&
       (scene.materialForNode.at(i)==oldMaterial))
    toDelete.push_back(i);
}
```

3. The number of meshes to merge matches the number of scene nodes marked for deletion (at least in our case), so convert the scene node indices into mesh indices:

```
vector<uint32_t> meshesToMerge(toDelete.size());
std::transform(
  toDelete.begin(), toDelete.end(),
  meshesToMerge.begin(),
  [&scene](uint32_t i)
    { return scene.meshForNode.at(i); });
```

4. A key part of this code merges the index data and assigns the updated mesh indices to the scene nodes:

```
std::unordered_map<uint32_t, uint32_t> oldToNew;
mergeIndexArray(meshData, meshesToMerge, oldToNew);
for (auto& n : scene.meshForNode)
  n.second = oldToNew[n.second];
```

5. Finally, remove the individual merged meshes and attach a new node containing the merged meshes to the scene graph:

```
  int newNode = addNode(scene, 0, 1);
  scene.meshForNode[newNode] = (int)meshData.meshes.size() - 1;
  scene.materialForNode[newNode] = (uint32_t)oldMaterial;
  deleteSceneNodes(scene, toDelete);
}
```

The `mergeNodesWithMaterial()` function is used in the next recipe, *Rendering large scenes*, to merge multiple meshes in the Lumberyard Bistro scene.

Now, let's jump to the next recipe to learn how to render a much more complex 3D model where these techniques will come in very handy.

Rendering large scenes

In the previous recipes, we learned how to manipulate scene graphs and render a scene with some editing capabilities added to it. Now, let's move on to rendering a more complex scene, Lumberyard Bistro, along with all its materials. All the scene storage and optimization techniques we've covered in this chapter will really come into play, especially given the large number of objects we need to handle in this 3D scene. Let's dive deeper and take a closer look.

Getting ready

Make sure to revisit the previous recipe, *Putting it all together into a scene editing application*, to refresh how our scene rendering is done.

The source code for this recipe is located in `Chapter08/03_LargeScene`.

How to do it...

The rendering in this recipe is quite similar to the previous one, with a few key differences required to render the Bistro scene. The entire scene is made up of two separate mesh files: `exterior.obj` and `interior.obj`. We need to combine them into a single `Scene` object and then render them. Additionally, the scene includes some transparent objects, such as windows and bottles, which will require special handling.

Let's begin by taking a look at the C++ code in `Chapter08/03_LargeScene/src/main.cpp` to understand the details.

1. The source code begins with the `renderSceneTreeUI()` function, which is very similar to the one from the previous recipe, *Putting it all together into a scene editing application*. The only difference here is the name of the scene nodes we want to keep open by default. The `NewRoot` node is created when combining the two different scenes from `exterior.obj` and `interior.obj`. We will delve into this further in a moment:

```
int renderSceneTreeUI(
  const Scene& scene, int node, int selectedNode)
{
  // ...
  if (name == "NewRoot") {
    flags |= ImGuiTreeNodeFlags_DefaultOpen |
             ImGuiTreeNodeFlags_Leaf |
             ImGuiTreeNodeFlags_Bullet;
    color = ImVec4(0.9f, 0.6f, 0.6f, 1);
  }
  // ...
}
```

2. Despite originating from two different `.obj` files, the scene is combined and cached into a single file, which makes subsequent loading almost instantaneous:

```
const char* fileNameCachedMeshes = ".cache/ch08_bistro.meshes";
const char* fileNameCachedMaterials = ".cache/ch08_bistro.materials";
const char* fileNameCachedHierarchy = ".cache/ch08_bistro.scene";
int main() {
  if (!isMeshDataValid(fileNameCachedMeshes) ||
      !isMeshHierarchyValid(fileNameCachedHierarchy)||
      !isMeshMaterialsValid(fileNameCachedMaterials))
  {
    printf("No cached mesh data found. Precaching...");
```

3. Here, we load two `.obj` files into two entirely separate sets of `Scene` and `MeshData` objects:

```
MeshData meshData_Exterior;
MeshData meshData_Interior;
Scene ourScene_Exterior;
Scene ourScene_Interior;
loadMeshFile("deps/src/bistro/Exterior/exterior.obj",
  meshData_Exterior, ourScene_Exterior, false);
loadMeshFile("deps/src/bistro/Interior/interior.obj",
  meshData_Interior, ourScene_Interior, false);
```

4. Before merging the two scenes into one, we perform some preprocessing. The mesh `exterior.obj` contains a tree with over 10K individual leaves, each represented as a separate object. Rendering each leaf individually would be inefficient, so we merge them into larger submeshes for improved performance. The `mergeNodesWithMaterial()` function was covered in the previous recipe, *Deleting nodes and merging scene graphs*. It combines all mesh nodes that share a specified material name into a single large mesh:

```
mergeNodesWithMaterial(
  ourScene_Exterior, meshData_Exterior, "Foliage_Linde_Tree_Large_Orange_Leaves");
mergeNodesWithMaterial(
  ourScene_Exterior, meshData_Exterior, "Foliage_Linde_Tree_Large_Green_Leaves");
mergeNodesWithMaterial(
  ourScene_Exterior, meshData_Exterior, "Foliage_Linde_Tree_Large_Trunk");
```

5. Now that we have two optimized scenes, we can merge them into one large scene. To achieve this, we need a few helper functions: `mergeScenes()`, `mergeMeshData()`, and `mergeMaterialLists()`. We will take a closer look at each of these functions shortly:

```
MeshData meshData;
Scene ourScene;
mergeScenes(ourScene,
```

```
    { &ourScene_Exterior, &ourScene_Interio, },
    {},
    { meshData_Exterior.meshes.size(),
      meshData_Interior.meshes.size() });
  mergeMeshData(meshData,
    { &meshData_Exterior, &meshData_Interior });
  mergeMaterialLists(
    { &meshData_Exterior.materials,
      &meshData_Interior.materials },
    { &meshData_Exterior.textureFiles,
      &meshData_Interior.textureFiles },
    meshData.materials, meshData.textureFiles);
```

6. Once the scene is merged, we want to scale the entire Bistro scene by `0.01` to fit it into our 3D world bounds, recalculate its bounding boxes, and then save everything into our cached format, as explained in the recipe *Loading and saving a scene graph*:

```
  ourScene.localTransform[0] = glm::scale(vec3(0.01f));
  markAsChanged(ourScene, 0);
  recalculateBoundingBoxes(meshData);
  saveMeshData(fileNameCachedMeshes, meshData);
  saveMeshDataMaterials(
    fileNameCachedMaterials, meshData);
  saveScene(fileNameCachedHierarchy, ourScene);
}
```

7. The loading code for cached scene data is the same as in the previous recipe:

Note

Because the Bistro mesh is quite large and initially stored in a textual `.obj` format, the first run to parse the `.obj` data can be extremely slow in `Debug` builds. We recommend running this demo app in the `Release` build for the first time. After the precaching is complete, subsequent runs in `Debug` mode will be very fast, enabling quick restarts even with this large mesh.

```
MeshData meshData;
const MeshFileHeader header =
  loadMeshData(fileNameCachedMeshes, meshData);
loadMeshDataMaterials(fileNameCachedMaterials, meshData);
Scene scene;
loadScene(fileNameCachedHierarchy, scene);
VulkanApp app({
  .initialCameraPos    = vec3(-19.26f, 8.47f,-7.32f),
```

```
    .initialCameraTarget = vec3(0, 2.5f, 0),
  });
  LineCanvas3D canvas3d;
  unique_ptr<lvk::IContext> ctx(app.ctx_.get());
```

8. Next, let's load skybox cube maps and GLSL shaders, and create a rendering pipeline:

```
  Holder<TextureHandle> texSkybox = loadTexture(ctx,
      "data/immenstadter_horn_2k_prefilter.ktx",
      TextureType_Cube);
  Holder<TextureHandle> texSkyboxIrradiance = loadTexture(ctx,
      "data/immenstadter_horn_2k_irradiance.ktx",
      TextureType_Cube);
  Holder<lvk::ShaderModuleHandle> vertSkybox = loadShaderModule(ctx,
      "Chapter08/02_SceneGraph/src/skybox.vert");
  Holder<ShaderModuleHandle> fragSkybox = loadShaderModule(ctx,
      "Chapter08/02_SceneGraph/src/skybox.frag");
  Holder<RenderPipelineHandle> pipelineSkybox =
    ctx->createRenderPipeline({
      .smVert      = vertSkybox,
      .smFrag      = fragSkybox,
      .color       = { { .format = ctx->getSwapchainFormat() } },
      .depthFormat = app.getDepthFormat(),
  });
```

9. Node selection works similarly to the previous recipe. The only additional parameter is `drawBoundingBoxes`, which allows toggling the rendering of bounding boxes through the ImGui UI:

```
  bool drawWireframe     = false;
  bool drawBoundingBoxes = false;
  int selectedNode       = -1;
  const VKMesh mesh(ctx, meshData, scene,
    ctx->getSwapchainFormat(), app.getDepthFormat());
  app.run([&](uint32_t width, uint32_t height,
    float aspectRatio, float deltaSeconds) {
    const mat4 view = app.camera_.getViewMatrix();
    const mat4 proj = glm::perspective(
      45.0f, aspectRatio, 0.01f, 1000.0f);
```

10. The entire scene is rendered in a single render pass:

```
    const lvk::RenderPass renderPass = {
      .color = {
        { .loadOp = lvk::LoadOp_Clear,
```

```
        .clearColor = { 1.f, 1.f, 1.f, 1.f } } },
    .depth = { .loadOp = lvk::LoadOp_Clear,
               .clearDepth = 1.f }
  };
  const lvk::Framebuffer framebuffer = {
    .color        =
     { { .texture = ctx->getCurrentSwapchainTexture() } },
    .depthStencil =
      { .texture = app.getDepthTexture() },
  };
  ICommandBuffer& buf = ctx->acquireCommandBuffer();
  buf.cmdBeginRendering(renderPass, framebuffer);
```

11. Let's render the skybox using its own rendering pipeline. Notice how the model-view-projection (`mvp`) matrix is constructed by discarding the translation part of the `view` matrix, ensuring the skybox always remains centered on the camera:

```
  buf.cmdBindRenderPipeline(pipelineSkybox);
  const struct {
    mat4 mvp;
    uint32_t texSkybox;
  } pc = {
    .mvp       = proj * mat4(mat3(view)),
    .texSkybox = texSkybox.index(),
  };
  buf.cmdPushConstants(pc);
  buf.cmdDraw(36);
```

12. Then, let's render the Bistro scene. The entire mesh is rendered with just one indirect draw command. Here, we provide a sky irradiance texture to add some basic image-based lighting to our scene, enhancing the lighting in the environment. The remaining part of the C++ code is similar to the previous recipe. We render a grid and an FPS counter, draw the scene graph tree, and use our 3D canvas to render bounding boxes for the objects in the scene:

```
  mesh.draw(buf, view, proj, texSkyboxIrradiance, drawWireframe);
  app.imgui_->beginFrame(framebuffer);
  // ...
  canvas3d.clear();
  canvas3d.setMatrix(proj * view);
  if (drawBoundingBoxes)
    for (auto& p : scene.meshForNode) {
      const BoundingBox box = meshData.boxes[p.second];
      canvas3d.box(scene.globalTransform[p.first],
```

```
            box, vec4(1, 0, 0, 1));
        }
      // … skipped the scene graph tree rendering here
      canvas3d.render(*ctx.get(), framebuffer, buf);
      app.imgui_->endFrame(buf);
      buf.cmdEndRendering();
      ctx->submit(buf, ctx->getCurrentSwapchainTexture());
    });
    ctx.release();
    return 0;
}
```

Now let's explore the GLSL shaders. They are shared with the previous demo application and located in `Chapter08/02_SceneGraph/src/`, with a common section of the code containing various buffer declarations used between the vertex and fragment shaders. This shared code can be found in `Chapter08/02_SceneGraph/src/common.sp`. Let's take a closer look at it first.

1. The shared GLSL code reuses the `common_material.sp` material declarations, which were explained in *Chapter 6*. The `DrawData` struct corresponds to a similar C++ struct described in the recipe *Implementing indirect rendering with Vulkan*:

```
#include <data/shaders/gltf/common_material.sp>
struct DrawData {
  uint transformId;
  uint materialId;
};
```

2. The buffer declarations and push constants correspond to the C++ buffers created in `VKMesh`, which were described in the recipe *Implementing indirect rendering with Vulkan*:

```
layout(std430, buffer_reference)
  readonly buffer TransformBuffer {
  mat4 model[];
};
layout(std430, buffer_reference)
  readonly buffer DrawDataBuffer {
  DrawData dd[];
};
layout(std430, buffer_reference)
  readonly buffer MaterialBuffer {
  MetallicRoughnessDataGPU material[];
};
layout(push_constant) uniform PerFrameData {
  mat4 viewProj;
```

```
  TransformBuffer transforms;
  DrawDataBuffer drawData;
  MaterialBuffer materials;
  uint texSkyboxIrradiance;
} pc;
```

Now, let's explore the vertex shader located in `Chapter08/02_SceneGraph/src/main.vert`.

1. The vertex shader inputs include `vec3` for vertex position, `vec2` for texture coordinates, and a normal vector. The outputs contain a world position and an integer material ID:

```
#include <Chapter08/02_SceneGraph/src/common.sp>
layout (location=0) in vec3 in_pos;
layout (location=1) in vec2 in_tc;
layout (location=2) in vec3 in_normal;
layout (location=0) out vec2 uv;
layout (location=1) out vec3 normal;
layout (location=2) out vec3 worldPos;
layout (location=3) out flat uint materialId;
```

2. As explained in the recipe *Implementing indirect rendering with Vulkan*, each indirect draw command stores the draw data ID (which is a mesh ID) in the `firstInstance` member field, which is passed into GLSL shaders as `gl_BaseInstance`. We now use this built-in variable to index into the `DrawData` buffer and retrieve the corresponding values for `transformId` and `materialId`:

```
void main() {
  mat4 model = pc.transforms.model[
    pc.drawData.dd[gl_BaseInstance].transformId];
  gl_Position = pc.viewProj * model * vec4(in_pos, 1.0);
  uv = vec2(in_tc.x, 1.0-in_tc.y);
  normal = transpose( inverse(mat3(model)) ) * in_normal;
  vec4 posClip = model * vec4(in_pos, 1.0);
  worldPos = posClip.xyz/posClip.w;
  materialId = pc.drawData.dd[gl_BaseInstance].materialId;
}
```

The fragment shader is more complex. First, let's take a look at the helper function `runAlphaTest()`, which is defined in `data/shaders/AlphaTest.sp`. This function helps us handle transparency by discarding fragments that fall below a certain alpha threshold, ensuring that only visible parts of transparent objects are rendered.

Note

To simplify things and avoid sorting the scene, alpha transparency is simulated using a technique called dithering, combined with punch-through transparency. This approach avoids the complexity of traditional transparency handling. For more insights, you can refer to this article by Alex Charlton: `http://alex-charlton.com/posts/Dithering_on_the_GPU`. Below, we provide the final implementation for your convenience:

```
void runAlphaTest(float alpha, float alphaThreshold) {
  if (alphaThreshold > 0.0) {
    mat4 thresholdMatrix = mat4(
      1.0 /17.0,  9.0/17.0,  3.0/17.0, 11.0/17.0,
      13.0/17.0,  5.0/17.0, 15.0/17.0,  7.0/17.0,
      4.0 /17.0, 12.0/17.0,  2.0/17.0, 10.0/17.0,
      16.0/17.0,  8.0/17.0, 14.0/17.0,  6.0/17.0 );
    alpha = clamp(alpha - 0.5 *
      thresholdMatrix[int(mod(gl_FragCoord.x, 4.0))]
                     [int(mod(gl_FragCoord.y, 4.0))],
      0.0, 1.0);
    if (alpha < alphaThreshold) discard;
  }
}
```

Now let's dive into the actual GLSL fragment shader, which handles material and lighting calculations. It can be found at `Chapter08/02_SceneGraph/src/main.frag`.

1. The shared GLSL header files `data/shaders/common.sp` and `data/shaders/AlphaTest.sp`, mentioned earlier, provide key declarations and functions for material handling and alpha transparency. The `data/shaders/UtilsPBR.sp` file contains the `perturbNormal()`function, which adjusts the normal vectors for bump mapping. This function was described in *Chapter 6*:

```
#include <Chapter08/02_SceneGraph/src/common.sp>
#include <data/shaders/AlphaTest.sp>
#include <data/shaders/UtilsPBR.sp>
layout (location=0) in vec2 uv;
layout (location=1) in vec3 normal;
layout (location=2) in vec3 worldPos;
layout (location=3) in flat uint materialId;
layout (location=0) out vec4 out_FragColor;
```

2. In this chapter, we use a simplified lighting model to keep the focus on the scene structure. However, we still use the same material struct `MetallicRoughnessDataGPU`, which ensures that we can reuse this code in later chapters with more complex lighting models:

```
void main() {
  MetallicRoughnessDataGPU mat = pc.materials.material[materialId];
  vec4 emissiveColor =
    vec4(mat.emissiveFactorAlphaCutoff.rgb, 0) *
    textureBindless2D(mat.emissiveTexture, 0, uv);
  vec4 baseColor = mat.baseColorFactor *
    (mat.baseColorTexture > 0 ?
       textureBindless2D( mat.baseColorTexture, 0, uv) :
       vec4(1.0));
```

3. Here, we use `runAlphaTest()` to simulate transparency by combining alpha-test and punch-through transparency. A naïve check against a constant alpha-cutoff value would cause alpha-tested foliage geometry to disappear at greater distances from the camera. To prevent this, we use a simple trick described in `https://bgolus.medium.com/anti-aliased-alpha-test-the-esoteric-alpha-to-coverage-8b177335ae4f`, where we scale the alpha-cutoff value using `fwidth()` to achieve better anti-aliasing for alpha-tested geometry:

```
  runAlphaTest(baseColor.a,
    mat.emissiveFactorAlphaCutoff.w / max(32.0 * fwidth(uv.x), 1.0));
```

4. This approach is a "poor man's" version of normal mapping, providing a simplified implementation. Here, `n` is the world-space normal vector:

```
  vec3 n = normalize(normal);
  vec3 normalSample =
    textureBindless2D(mat.normalTexture, 0, uv).xyz;
  if (length(normalSample) > 0.5) {
    n = perturbNormal(n, worldPos, normalSample, uv);
  }
```

If you want to learn a more accurate method for transforming normal vectors, be sure to check out this article *Transforming Normals* by Eric Lengyel: `https://terathon.com/blog/transforming-normals.html`.

5. We hardcode two directional lights directly into the shader code for simplicity. We use an ad hoc scaling factor if there's a skybox irradiance map available:

```
  float NdotL1 = clamp(
    dot(n, normalize(vec3(-1, 1, +0.5))), 0.1, 1.0);
  float NdotL2 = clamp(
    dot(n, normalize(vec3(+1, 1, -0.5))), 0.1, 1.0);
  const bool hasSkybox = pc.texSkyboxIrradiance > 0;
  float NdotL =
    (hasSkybox ? 0.2 : 1.0) * (NdotL1 + NdotL2);
```

6. The **IBL** component for diffuse lighting here is kept simple and shiny, without aiming for any PBR accuracy. The primary goal of this chapter is to demonstrate scene rendering techniques, without adding the complexity of realistic material setups. We manually rotate the skybox to adjust the incoming IBL for a more aesthetically pleasing result. This same hack is applied to the skybox rendering fragment shader found in `Chapter08/02_SceneGraph/src/skybox.frag`:

```
  const vec4 f0 = vec4(0.04);
  vec3 sky = vec3(-n.x, n.y, -n.z); // rotate skybox
  vec4 diffuse = hasSkybox ?
    (textureBindlessCube(pc.texSkyboxIrradiance, 0, sky) +
      vec4(NdotL)) * baseColor * (vec4(1.0) - f0) :
    NdotL * baseColor;
  out_FragColor = emissiveColor + diffuse;
}
```

That's it for the GLSL shaders. Before running the sample application, let's revisit the scene merging functions mentioned earlier – `mergeScenes()`, `mergeMeshData()`, and `mergeMaterialLists()` – to explore how they work.

How it works...

The `mergeMeshData()` routine takes a vector of `MeshData` instances and creates a new `MeshFileHeader` instance while simultaneously copying all the indices and vertices into the output object `MeshData m`:

1. First, we merge the `std::vector` containers for the index buffers, vertex buffers, meshes, and bounding boxes. We retrieve the size of the vertex format for this mesh using the function `MeshData::streams::getVertexSize()`:

```
MeshFileHeader mergeMeshData(
  MeshData& m, const std::vector<MeshData*> md)
{
  uint32_t numTotalVertices = 0;
  uint32_t numTotalIndices  = 0;
  if (!md.empty())m.streams = md[0]->streams;
  const uint32_t vertexSize = m.streams.getVertexSize();
  uint32_t offset    = 0;
  uint32_t mtlOffset = 0;
  for (const MeshData* i : md) {
    mergeVectors(m.indexData, i->indexData);
    mergeVectors(m.vertexData, i->vertexData);
    mergeVectors(m.meshes, i->meshes);
    mergeVectors(m.boxes, i->boxes);
```

2. After merging the containers, we need to shift the index offset by the total size of the already merged indices and do the same for materials using the count of the already merged materials. The values of `Mesh::vertexCount`, `Mesh::lodCount`, and `Mesh::vertexOffset` remain unchanged because the vertex offsets are local and have already been baked into the indices:

```
      for (size_t j = 0; j != i->meshes.size(); j++) {
        m.meshes[offset + j].indexOffset += numTotalIndices;
        m.meshes[offset + j].materialID += mtlOffset;
      }
```

3. Next, we shift the individual indices by the total number of vertices that have already been merged:

```
      for (size_t j = 0;
           j != i->indexData.size(); j++) {
        m.indexData[numTotalIndices + j] += numTotalVertices;
      }
```

4. After processing each `MeshData` instance, we update the number of already merged meshes, materials, indices, and vertices. Since the vertex data size is calculated in bytes, we need to divide that number by `vertexSize`:

```
      offset += (uint32_t)i->meshes.size();
      mtlOffset += (uint32_t)i->materials.size();
      numTotalIndices  += (uint32_t)i->indexData.size();
      numTotalVertices += (uint32_t)i->vertexData.size() / vertexSize;
    }
```

5. The resulting `MeshFileHeader` instance contains the total size of the index and vertex data arrays:

```
    return MeshFileHeader {
      .magicValue     = 0x12345678,
      .meshCount      = (uint32_t)offset,
      .indexDataSize  = numTotalIndices * sizeof(uint32_t),
      .vertexDataSize = m.vertexData.size(),
    };
  }
```

6. The `mergeVectors()` helper function is a templated one-liner that appends the second vector `v2` to the end of the first vector `v1`:

```
  template <typename T> void mergeVectors(
    vector<T>& v1, const vector<T>& v2) {
    v1.insert(v1.end(), v2.begin(), v2.end());
  }
```

Along with merging mesh data, we need to aggregate the material descriptions from different meshes into a single collection.

1. The `mergeMaterialLists()` function creates one combined vector of texture filenames and a vector of material descriptions, ensuring that texture indices are correctly updated:

```
void mergeMaterialLists(
  const vector<vector<Material>*>& oldMaterials,
  const vector<vector<std::string>*>& oldTextures,
  vector<Material>& allMaterials,
  vector<std::string>& newTextures)
{
```

2. The merge process begins by creating a unified list of materials. Each material list index is linked with a texture, allowing us to later determine the list the texture belongs to:

```
  unordered_map<size_t, size_t> materialToTextureList;
  for (size_t midx = 0; midx != oldMaterials.size(); midx++) {
    for (const Material& m : *oldMaterials[midx]) {
      allMaterials.push_back(m);
      materialToTextureList[allMaterials.size()-1] = midx;
    }
  }
```

3. The combined texture container `newTextureNames` holds only unique texture filenames. The indices of these texture files are stored in a map, which is then used to update the texture references in the material descriptors:

```
  unordered_map<std::string, int> newTextureNames;
  for (const vector<std::string>* tl : oldTextures)
  {
    for (const std::string& file : *tl) {
      newTextureNames[file] = addUnique(newTextures, file);
    }
  }
```

4. The `replaceTexture()` lambda function takes a texture index from an old texture container and assigns it a texture index from the `newTextureNames` array:

```
  auto replaceTexture = [&materialToTextureList,
    &oldTextures, &newTextureNames](
    int mtlId, int* textureID) {
    if (*textureID == -1)return;
    const size_t listIdx = materialToTextureList[mtlId];
    const std::vector<std::string>& texList = *oldTextures[listIdx];
```

```
    const std::string& texFile = texList[*textureID];
    *textureID = newTextureNames[texFile];
  };
```

5. The final loop iterates over all materials and adjusts the texture indices accordingly:

```
  for (size_t i = 0; i < allMaterials.size(); i++) {
    Material& m = allMaterials[i];
    replaceTexture(i, &m.baseColorTexture);
    replaceTexture(i, &m.emissiveTexture);
    replaceTexture(i, &m.normalTexture);
    replaceTexture(i, &m.opacityTexture);
  }
}
```

To merge multiple object collections, we need one more routine that combines multiple scene hierarchies into a single large scene graph. The scene data is defined by the `Hierarchy` item array, local and global transforms, and associative arrays for meshes, materials, and scene nodes. Similar to merging mesh index and vertex data, this merge routine essentially involves merging individual arrays and then adjusting the indices within each scene node.

1. The `shiftNodes()` routine increments individual fields of the scene hierarchy structure by the given `shiftAmount` value:

```
void shiftNodes(
  Scene& scene, int startOffset, int nodeCount,
  int shiftAmount)
{
  auto shiftNode = [shiftAmount](Hierarchy& node) {
    if (node.parent > -1) node.parent +=shiftAmount;
    if (node.firstChild > -1 ) node.firstChild  += shiftAmount;
    if (node.nextSibling > -1) node.nextSibling += shiftAmount;
    if (node.lastSibling > -1) node.lastSibling += shiftAmount;
  };
  for (int i = 0; i < nodeCount; i++)
    shiftNode(scene.hierarchy[i + startOffset]);
}
```

2. The `mergeMaps()` helper routine adds the contents of unordered map `otherMap` to the output map `m`, while shifting its integer values by the specified `itemOffset` amount:

```
using ItemMap = std::unordered_map<uint32_t, uint32_t>;
void mergeMaps(ItemMap& m, const ItemMap& otherMap,
  int indexOffset, int itemOffset)
{
```

```
    for (const auto& i : otherMap)
      m[i.first + indexOffset] = i.second + itemOffset;
}
```

Now that we have all the utility functions in place, we can merge two `Scene` objects. The `mergeScenes()` routine creates a new root scene node called `"NewRoot"` and attaches all the root scene nodes from the scenes being merged as child nodes of the `"NewRoot"` node. Let's examine the implementation in `shared/Scene/Scene.cpp`:

1. This source code bundle with this routine has two extra parameters, `mergeMeshes` and `mergeMaterials`, which allow the creation of composite scenes with shared mesh and material data. We omit these inessential parameters to shorten the description:

```
void mergeScenes(
  Scene& scene, const std::vector<Scene*>& scenes,
  const std::vector<glm::mat4>& rootTransforms,
  const std::vector<uint32_t>& meshCounts,
  bool mergeMeshes, bool mergeMaterials)
{
  scene.hierarchy = { {
    .parent      = -1,
    .firstChild  =  1,
    .nextSibling = -1,
    .lastSibling = -1,
    .level       =  0 } };
```

2. The array of names, along with the local and global transforms, initially contains a single element, `"NewRoot"`:

```
  scene.nameForNode[0] = 0;
  scene.nodeNames       = { "NewRoot" };
  scene.localTransform.push_back(glm::mat4(1.f));
  scene.globalTransform.push_back(glm::mat4(1.f));
  if (scenes.empty()) return;
```

3. While iterating through the scenes, we merge and shift all the arrays and maps. The following variables keep track of the item counts in the output scene:

```
  int offs          = 1;
  int meshOffs      = 0;
  int nameOffs      = (int)scene.nodeNames.size();
  int materialOfs = 0;
  auto meshCount  = meshCounts.begin();
  if (!mergeMaterials)
    scene.materialNames = scenes[0]->materialNames;
```

4. This implementation isn't the most efficient one, mainly because it combines the merging of all scene graph components into a single routine. However, it's simple enough:

```
for (const Scene* s : scenes) {
  mergeVectors(scene.localTransform, s->localTransform);
  mergeVectors(scene.globalTransform, s->globalTransform);
  mergeVectors(scene.hierarchy, s->hierarchy);
  mergeVectors(scene.nodeNames, s->nodeNames);
  if (mergeMaterials)
    mergeVectors(scene.materialNames, s->materialNames);
  const int nodeCount = (int)s->hierarchy.size();
  shiftNodes(scene, offs, nodeCount, offs);
  mergeMaps(scene.meshForNode,
    s->meshForNode, offs, mergeMeshes ? meshOffs : 0);
  mergeMaps(scene.materialForNode,
    s->materialForNode, offs, mergeMaterials ? materialOfs : 0);
  mergeMaps(scene.nameForNode, s->nameForNode, offs, nameOffs);
```

5. During each iteration, we add the sizes of the current arrays to the global offsets:

```
  offs += nodeCount;
  materialOfs += (int)s->materialNames.size();
  nameOffs += (int)s->nodeNames.size();
  if (mergeMeshes) {
    meshOffs += *meshCount;
    meshCount++;
  }
}
```

6. Logically, the routine is complete, but there is one more step. Each scene node has a cached index of its last sibling, which we need to update for the new root nodes. We also assign a new local transform to each root node in the following loop:

```
offs    = 1;
int idx = 0;
for (const Scene* s : scenes) {
  const int nodeCount = (int)s->hierarchy.size();
  const bool isLast   = (idx == scenes.size()-1);
```

7. Calculate the new "next sibling" for the old scene roots and attach them to the new root:

```
  const int next = isLast ? -1 : offs + nodeCount;
  scene.hierarchy[offs].nextSibling = next;
  scene.hierarchy[offs].parent = 0;
```

8. Transform the old root nodes, if the transforms are provided:

```
    if (!rootTransforms.empty())
      scene.localTransform[offs] =
        rootTransforms[idx] * scene.localTransform[offs];
    offs += nodeCount;
    idx++;
  }
```

9. At the end of the routine, increment all the depth-from-root levels of the scene nodes, but leave the "NewRoot" node unchanged, hence the +1:

```
  for (auto i = scene.hierarchy.begin() + 1;
       i != scene.hierarchy.end(); i++)
    i->level++;
}
```

Now that our arsenal of scene graph management routines is complete, we've reached the end of what is likely to be the most complex chapter in this book, with the highest cognitive load and a focus on data structures. The upcoming chapters will be more graphics-oriented, we promise.

Note

As a keen reader might have noticed while going through this chapter, we focused more on runtime loading and rendering performance rather than preprocessing performance. That's true. The reason is simple: using STL containers and algorithms often leads to shorter but slower code. Making every scene manipulation faster would add complexity and make that already quite complex code even harder to understand.

By the way, before we move on to the next chapter, let's revisit our sample application in Chapter08/03_LargeScene and run it. You should see the Lumberyard Bistro scene, as shown in the following screenshot.

Figure 8.9: The Lumberyard Bistro scene with textures

Fly around the scene and explore the transparent windows and the interior of the bistro. You can use the scene tree UI on the left side to select individual meshes, and they will be highlighted with a green bounding box.

Now, let's move on to the next chapter and explore glTF animations to bring some dynamics to our static scenes.

There's more...

One might wonder why we went through all these pages of data structures. Was the cost of fast scene loading really worth it? We believe it was. As proof, try running samples from other books that use the same Bistro 3D model – you'll likely be disappointed.

The real benefit of this approach is that the scene loading is now two or even three orders of magnitude faster compared to a naïve approach. This makes it possible to run Debug builds on much larger datasets than just the Bistro scene. Imagine loading hundreds of different meshes the size of Bistro and still being able to run your rendering engine in Debug mode. This is one of those things that sets hobby projects apart from professional 3D engines.

9

glTF Animations

In this chapter, we will explore advanced glTF core specification features, including animations, skinning, and morphing. We will cover the basics of each technique and implement skinning animations and morphing using compute shaders. In the final recipe, we will introduce animation blending, showing how to create complex animation sequences.

Most of the C++ code provided here can be applied not only to these recipes but also throughout the rest of the book.

In this chapter, we will cover the following recipes:

- Introduction to node-based animations
- Introduction to skeletal animations
- Importing skeleton and animation data
- Implementing the glTF animation player
- Doing skeletal animations in compute shaders
- Introduction to morph targets
- Loading glTF morph target data
- Adding morph targets support
- Animation blending

Note

We use the official Khronos Sample Viewer as a reference for applying glTF skinning and morphing data, though our implementation doesn't rely on this project. The official Khronos Sample Viewer uses JavaScript and vertex shaders to implement animation features, while in our recipes, we use compute shaders and a streamlined approach that favors simplicity of implementation over feature completeness.

Technical requirements

We assume readers have a basic understanding of linear algebra and calculus, and we recommend becoming familiar with the glTF 2.0 specification: `https://registry.khronos.org/glTF/specs/2.0/glTF-2.0.html#animations`.

In this chapter, we continue refining our unified glTF Viewer sample code that we built in previous chapters. The variations in each recipe's `main.cpp` file include using different model files to demonstrate the specific glTF animation feature covered in each recipe, adjusting initial camera positions for optimal model presentation, and making minor UI changes to highlight the glTF features explored in each recipe.

The source code for our glTF Viewer is located in `shared/UtilsGLTF.cpp`. The GLSL vertex and fragment shaders can be found in the `data/shaders/gltf/` folder. The code for animation-specific utilities can be found in `shared/UtilsAnim.cpp` and `shared/UtilsAnim.h`.

Introduction to node-based animations

In this recipe, we will explore node-based animations, how they're organized, and what the key principles of the glTF specification are. This builds on the code from *Chapters 6* and *7*. Let's start with an overview of computer animation. At its core, computer animation is the process of creating moving images. This can be achieved in different ways, from playing a sequence of still images to using physics-based simulations for each pixel. Our focus will be on real-time computer graphics, which typically use efficient methods such as scene hierarchy transformations, skeletal animation, and morphing. Efficiency is key, as it allows the data to be represented and processed in real time on today's hardware.

In *Chapters 6* and *7*, we looked at different scene graph representations and how a scene is represented in the glTF specification. Each scene is structured like a tree, with transformations applied to each node. By changing these node transformations, we can create motion, which is the basis of node-based animation. In addition to the scene graph hierarchy, glTF includes a way to store animation data. All animations in glTF are stored in the asset's animations array.

An **animation** is defined by channels and samplers. **Samplers** refers to accessors that contain keyframe data and specify interpolation methods. **Channels** link these keyframe outputs to specific nodes in the hierarchy.

Keyframes assign specific transformations to particular time points. glTF supports translations, quaternion-based rotations, scaling, and weights to represent these transformations.

In glTF, animation **interpolation** is the process of calculating the intermediate parameter values between keyframes, which creates specific transitions between distinct animation poses.

glTF supports several interpolation types:

- **LINEAR**: This is the simplest type, where the value at a given time is linearly interpolated between the values of the previous and next keyframes.
- **STEP**: In this type, the value stays constant until the next keyframe, at which point it jumps to the new value.

- **CUBICSPLINE:** This interpolation type uses cubic splines for smoother, more natural-looking animations and requires extra control points to shape the curve.

Note

Unfortunately, Assimp only supports **linear** interpolation at the time of this book's writing; thus, our examples use this mode.

The glTF animation interpolation process works as follows:

1. Identify the keyframes surrounding the current time.
2. Calculate the weights for each keyframe based on the current time and the distance between them.
3. Interpolate the keyframe values using the specified interpolation type and calculated weights.

Here's a simple example with two keyframes:

Keyframe 1: Time = 0, Value = 10

Keyframe 2: Time = 1, Value = 20

If the current time is `0.4`, the interpolated value would be `(1 - 0.4) * 10 + 0.4 * 20 = 14`.

glTF animations can contain multiple channels, each controlling a different property of a node (such as translation, rotation, or scale). Interpolation is applied independently to each channel. In hierarchical animations, transformations of parent nodes influence the transformations of child nodes, with interpolation applied recursively throughout the hierarchy.

Before diving into the code, let's take a look at how glTF handles skeletal animations.

Introduction to skeletal animations

Vertex skinning animation is a technique for animating 3D characters by linking a polygonal mesh to a hierarchical skeleton. Bones in the skeleton control the mesh deformation, allowing for realistic movement. This method is widely used in animation, enabling animators to control complex characters through intuitive tools.

The process begins with creating a skeleton of bones, each with a defined position, scale, and orientation. These bones are arranged hierarchically, resembling a human or animal skeleton. The mesh is then associated with specific bones, often using weights to indicate how much influence each bone has on particular vertices of the mesh. As the bones move, the mesh deforms naturally, resulting in lifelike animations.

To animate a mesh, we need to dynamically transform its vertices. Each bone defines a transformation, and the most flexible method is to store a 4x4 affine transformation matrix for each bone. This matrix encapsulates the combined effects of scaling, rotation, and translation. While it's possible to store separate translation, rotation, and scaling components, using a matrix simplifies shader operations, particularly when dealing with multiples of 4 components.

These transformations should already account for the effects of parent bones. We'll refer to these as **global transforms**, in contrast to **local transforms**, which are relative to their parent and do not take a parent transform into account. The global transform of a bone is calculated recursively as follows:

$$\text{globalTransform}(\text{bone}) = \text{globalTransform}(\text{parent}) * \text{localTransform}(\text{bone})$$

If a bone has no parent, its global transform is the same as its local transform.

Local bone transformations are usually defined in a specific coordinate system rather than in world coordinates. For example, when rotating an arm, it makes sense to use a local coordinate system centered around the shoulder joint, which eliminates the need for explicit translation.

Each bone has its own local coordinate system. To transform vertices from the model's coordinate system into bone-local space, we use an **inverse bind matrix**. This matrix effectively repositions the vertices relative to the origin of the bone.

To render hierarchical mesh data after applying the animation transformation in the bone-local coordinate system, we need to transform the vertices back to the parent bone's local coordinate system. This can be done using inverse bind matrices, as follows:

$$\text{convertToParentCS}(\text{node}) = \text{inverseBindMatrix}(\text{parent}) * \text{inverseBindMatrix}(\text{node})^{-1}$$

This process effectively transforms vertices from one bone-local coordinate system to another. Additionally, bones can be connected to vertices in a transformed state known as the **bind pose**.

Note

It is a pose, in which the mesh is in its original, undeformed state, typically serving as the basis for skinning (attaching the mesh to a skeleton).

This requires an extra transformation to position the vertices correctly in the expected local coordinate system for each bone.

To transform each vertex, we follow these steps:

1. Transform it to the model's bind pose.
2. Use the **inverse bind matrix** to convert the vertex to the bone-local coordinate system.
3. Apply the animation transformations in the bone-local coordinate system.
4. Transform the vertex back from the bone-local space to the model's coordinate system.
5. If the bone has a parent, repeat steps 2-4 for the parent bone.

As discussed in *Chapter 6*, a glTF scene consists of nodes, which can represent various elements, like meshes, cameras, light sources, and skeleton bones, or simply act as parent containers for other nodes. Each node has its own transformation, defining its position, rotation, and scale relative to its parent or the world origin.

If a mesh node uses skeletal animation, it includes a list of joints, which are glTF node IDs representing the skeleton's bones. These joints form a hierarchy. Note that there is no explicit armature or skeleton node; instead, bone nodes are used directly.

For a mesh primitive using skeletal animation, each vertex requires bone ID and weight attributes. Alongside the skinned mesh data, glTF files can include animations, which provide instructions for updating transformations in the node hierarchy.

glTF organizes transformations in the following way:

1. **Model Bind Pose:** This is either applied to the model directly or pre-multiplied into the inverse bind matrices. In glTF, the bind pose can essentially be ignored.
2. **Inverse Bind Matrices:** These are provided as an accessor containing an array of 4x4 matrices.
3. **Animations:** Animations can be defined externally or stored as keyframe splines for rotation, translation, and scale. These animations are combined with the transformation from their local coordinate system to the parent's local coordinate system.
4. **Parent Relationships:** While parent nodes are defined in the node hierarchy, the conversion to the parent's coordinate system has already been applied. Therefore, you only need to repeat steps 3-5 above for parent nodes if they exist.

Thus, the final global transform for a glTF bone is constructed as follows:

$$
\begin{aligned}
\mathrm{globalTransform}(\mathrm{bone}) = \mathrm{globalTransform}(\mathrm{animation}(\mathrm{root})) * \ldots \\
\ldots * \mathrm{globalTransform}(\mathrm{animation}(\mathrm{parent})) * \mathrm{localTransform}(\mathrm{animation}(\mathrm{bone})) \\
* \mathrm{invBind}(\mathrm{bone})
\end{aligned}
$$

We will implement this approach in the following recipes. Now, let's look into the actual code to learn how to store skeleton and animation data and load it from glTF files.

Note

Here's a link to the glTF 2.0 specification, which provides a detailed explanation of skinning:

`https://registry.khronos.org/glTF/specs/2.0/glTF-2.0.html#skins`

Importing skeleton and animation data

This recipe introduces two key improvements to our glTF Viewer: it now supports storing data for skeletons and animations, and it includes functionality to load this data from glTF assets using Assimp.

These enhancements are put to use in the following recipes, where they enable the implementation of mesh skinning and animations.

Getting ready

This recipe does not include a separate code example. Instead, it explains the modifications made to the utility code, located in the files `shared/UtilsGLTF.h` and `shared/UtilsGLTF.cpp`.

How to do it...

Let's begin with the skeleton by defining the essential data structures needed to store it.

1. First, we should add a way to reference transformations. The identifier `modelMtxId` is used to point to a specific matrix within the `GLTFContext::matricesBuffer`. This approach generalizes how glTF transformations are managed and accessed.

```
using GLTFNodeRef = uint32_t;
using GLTFMeshRef = uint32_t;
struct GLTFTransforms {
  uint32_t modelMtxId;
  uint32_t matId;
  GLTFNodeRef nodeRef; // for CPU only
  GLTFMeshRef meshRef; // for CPU only
  uint32_t sortingType;
};
```

2. We need a struct to store vertex skeleton attribution data, specifically the bone indices and bone weights. This data will allow us to define how each vertex is influenced by the skeleton's bones. Here, we use `vec4` for positions and normal vectors to ensure padding compatibility, as this structure will be shared between the CPU and GPU.

```
#define MAX_BONES_PER_VERTEX 8
struct VertexBoneData {
  vec4 position;
  vec4 normal;
  uint32_t boneId[MAX_BONES_PER_VERTEX] = {
    ~0u, ~0u, ~0u, ~0u, ~0u, ~0u, ~0u, ~0u };
  float weight[MAX_BONES_PER_VERTEX] = {};
  uint32_t meshId = ~0u;
};
```

3. Next, we need a way to store bone data. While we could reuse glTF nodes, we create an additional structure, `GLTFBone`, to keep things simpler and more efficient.

```
struct GLTFBone {
  uint32_t boneId = ~0u;
  mat4 transform = mat4(1.0f);
};
```

4. To store this data, we'll need to add a few new member fields to `GLTFContext` in `shared/UtilsGLTF.h`:

```
class GLTFContext
  // ...
```

```
  unordered_map<std::string, GLTFBone> bonesByName;
  std::vector<MorphTarget> morphTargets;
  Holder<BufferHandle> vertexSkinningBuffer;
  Holder<BufferHandle> vertexMorphingBuffer;
  // ...
};
```

Note

Here, we're using a simpler approach with `std::unordered_map` to map bone names to unique `GLTFBone` objects. This choice was made to keep things straightforward and make the material in this chapter easier to follow. When performance becomes critical, it is recommended to replace string identifiers with indices and avoid accessing hash tables at runtime, similar to how scene nodes were managed in *Chapter 8*.

With the key data structures now declared, let's examine the new code responsible for loading the bone data.

1. This code, added to the `loadGLTF()` function and located in `shared/UtilsGLTF.cpp`, will handle importing and organizing the bones.

```
void loadGLTF(GLTFContext& gltf,
  const char* glTFName, const char* glTFDataPath)
{
    // ...
    uint32_t numBones = 0;
    uint32_t vertOffset = 0;
    // ...
    gltf.hasBones = mesh->mNumBones > 0;
    for (uint32_t id = 0; id < mesh->mNumBones;id++) {
      const aiBone& bone  = *mesh->mBones[id];
      const char* boneName = bone.mName.C_Str();
```

2. For convenience, all bones are stored in an `std::unordered_map` container called `bonesByName`.

```
      const bool hasBone = gltf.bonesByName.contains(boneName);
      const uint32_t boneId = hasBone ?
        gltf.bonesByName[boneName].boneId : numBones++;
      if (!hasBone)
        gltf.bonesByName[boneName] = {
          .boneId = boneId,
          .transform = aiMatrix4x4ToMat4(bone.mOffsetMatrix) };
```

3. For each bone, we gather all weights provided by Assimp and store them in the `VertexBoneData::weight` array.

```
        for (uint32_t w = 0; w < bone.mNumWeights; w++) {
          const uint32_t vertexId = bone.mWeights[w].mVertexId;
          VertexBoneData& vtx = skinningData[vertexId + vertOffset];
          vtx.position = vec4(vertices[vertexId +
            vertOffset].position, 1.0f);
          vtx.normal = vec4(vertices[vertexId + vertOffset].normal, 0.0f);
          vtx.meshId = m;
          for (uint32_t i = 0; i < MAX_BONES_PER_VERTEX; i++) {
            if (vtx.boneId[i] == ~0u) {
              vtx.weight[i] = bone.mWeights[w].mWeight;
              vtx.boneId[i] = boneId;
              break;
            }
          }
        }
      }
      vertOffset += mesh->mNumVertices;
```

It's a lengthy function but very straightforward. We store a vector of `VertexBoneData` structures, matching the base vertices from the glTF mesh. For each vertex, we store the `meshId`, copy the original positions, and assign the `boneIds` and `weights`. Assimp provides a list of bones, and we use the bone names to link them to the correct glTF mesh index. The entire `VertexBoneData` vector is then uploaded into a GPU buffer called `vertexSkinningBuffer`, ensuring that all vertex skinning data is accessible on the GPU.

1. An additional modification in the `loadGLTF()` function is the node matrix mapping. For each node, we assign a matrix ID and store its transformation, allowing efficient access to each node's transformation matrix:

```
uint32_t nonBoneMtxId = numBones;
// ...
const char* rootName =
  scene->mRootNode->mName.C_Str() ?
    scene->mRootNode->mName.C_Str() : "root";
gltf.nodesStorage.push_back({
    .name        = rootName,
    .modelMtxId =
      getNextMtxId(gltf, rootName, nonBoneMtxId,
      aiMatrix4x4ToMat4(scene->mRootNode->mTransformation)),
    .transform  = aiMatrix4x4ToMat4(
      scene->mRootNode->mTransformation),
});
```

2. The idea behind the `getNextMtxId()` helper function is straightforward. If a node has an associated bone, we assign the bone ID as the node's matrix position; otherwise, we allocate the next available ID, starting from `numBones` passed as `nextEmptyId`.

```
uint32_t getNextMtxId(GLTFContext& gltf,
  const char* name,
  uint32_t& nextEmptyId,
  const mat4& mtx)
{
  const auto it = gltf.bonesByName.find(name);
  const uint32_t mtxId =
    it == gltf.bonesByName.end() ?
      nextEmptyId++ : it->second.boneId;
  if (gltf.matrices.size() <= mtxId)
    gltf.matrices.resize(mtxId + 1);
  gltf.matrices[mtxId] = mtx;
  return mtxId;
}
```

3. After the loading is complete, we can proceed to upload the corresponding data to a GPU buffer.

```
gltf.vertexSkinningBuffer = ctx->createBuffer({
  .usage     = lvk::BufferUsageBits_Vertex |
               lvk::BufferUsageBits_Storage,
  .storage   = lvk::StorageType_Device,
  .size      = sizeof(VertexBoneData) * skinningData.size(),
  .data      = skinningData.data(),
  .debugName = "Buffer: skinning vertex data",
});
```

With the skeleton data loaded, which is essential for mesh skinning, we can now move on to animations and see how the loading process works for them.

How it works...

Before we can start with the loading of glTF animations, let's define various animation data structures that we will require. These declarations are located in `shared/UtilsAnim.h`.

1. First, we define the structures for animation keyframes. These structures will serve as the building blocks for organizing and managing our animation data. Each `AnimationKey`... structure represents a keyframe of transformation data linked to a specific timestamp.

```
struct AnimationKeyPosition {
  vec3 pos;
  float time;
};
```

```
struct AnimationKeyRotation {
  quat rot;
  float time;
};
struct AnimationKeyScale {
  vec3 scale;
  float time;
};
```

2. Here we declare the `AnimationChannel` structure, which simply holds three collections of keyframes for position, rotation, and scale. It combines these keys into a complete transformation representation.

```
struct AnimationChannel {
  std::vector<AnimationKeyPosition> pos;
  std::vector<AnimationKeyRotation> rot;
  std::vector<AnimationKeyScale> scale;
};
```

3. Lastly, we declare the `Animation` structure. To keep things simple, we utilize a hashmap to map the bones to their corresponding animation channels. The `Animation` structure associates these channels with a particular glTF transformation node and includes additional metadata: the animation's length in seconds as `duration`, the time unit for the animation `ticksPerSecond`, and animation `name` for debugging purposes.

```
struct Animation {
  unordered_map<int, AnimationChannel> channels;
  std::vector<MorphingChannel> morphChannels;
  float duration; // in seconds
  float ticksPerSecond;
  std::string name;
};
```

The vector of `morphChannels` is used for animation blending and morph targets. We will explore this topic later in this chapter in the recipes *Loading morph targets data* and *Adding morph targets support*.

With these data structures established, the code for loading animations from Assimp becomes quite straightforward, as outlined below. Check `shared/UtilsAnim.cpp` for the complete source code:

1. A helper function called `initAnimations()` is responsible for loading and initializing all glTF animations from an `aiScene` object. This code is quite mechanical and straightforward; we simply load the data and attribute it according to the node names.

```
void initAnimations(GLTFContext& glTF, const aiScene* scene) {
  glTF.animations.resize(scene->mNumAnimations);
  for (uint32_t i = 0; i < scene->mNumAnimations; ++i) {
```

```
    Animation& anim     = glTF.animations[i];
    anim.name = scene->mAnimations[i]->mName.C_Str();
    anim.duration = scene->mAnimations[i]->mDuration;
    anim.ticksPerSecond = scene->mAnimations[i]->mTicksPerSecond;
    for (uint32_t c = 0;
         c < scene->mAnimations[i]->mNumChannels; c++) {
      const aiNodeAnim* channel = scene->mAnimations[i]->mChannels[c];
      uint32_t boneId = glTF.bonesByName[
        channel->mNodeName.data].boneId;
```

2. If a bone wasn't already populated during the bones initialization process, we create it here. For now, we skip the morph targets loading code, which we will explore later in this chapter in the recipes *Introduction to morph targets* and *Loading glTF morph targets data*.

```
      if (boneId == ~0u) {
        for (const GLTFNode& node : glTF.nodesStorage) {
          if (node.name != channel->mNodeName.data)
            continue;
          boneId = node.modelMtxId;
          glTF.bonesByName[boneName] = {
            .boneId    = boneId,
            .transform = glTF.hasBones ?
              inverse(node.transform):mat4(1) };
          break;
        }
      }
      anim.channels[boneId] = initChannel(channel);
    }
    // … we skip the morph targets loading code here
  }
}
```

3. Another helper function mentioned above, `initChannel()`, is also simple and mechanical. It retrieves the data from an Assimp object `aiNodeAnim` and transforms it into our `AnimationChannel` representation.

```
AnimationChannel initChannel(const aiNodeAnim* anim)
{
  AnimationChannel channel;
  channel.pos.resize(anim->mNumPositionKeys);
  for (uint32_t i = 0; i < anim->mNumPositionKeys; ++i) {
    channel.pos[i] = {
     .pos  = aiVector3DToVec3(anim->mPositionKeys[i].mValue),
     .time = (float)anim->mPositionKeys[i].mTime };
```

```
  }
  channel.rot.resize(anim->mNumRotationKeys);
  for (uint32_t i = 0; i < anim->mNumRotationKeys; ++i) {
    channel.rot[i] = {
      .rot  = aiQuaternionToQuat(anim->mRotationKeys[i].mValue),
      .time = (float)anim->mRotationKeys[i].mTime };
  }
  channel.scale.resize(anim->mNumScalingKeys);
  for (uint32_t i = 0; i < anim->mNumScalingKeys; ++i) {
    channel.scale[i] = {
      .scale = aiVector3DToVec3(anim->mScalingKeys[i].mValue),
      .time  = (float)anim->mScalingKeys[i].mTime };
  }
  return channel;
}
```

4. The final modification to `loadGLTF()` is to call `initAnimations()` immediately after loading the glTF nodes to load all the animation data.

```
void loadGLTF(GLTFContext& gltf,
  const char* glTFName, const char* glTFDataPath)
{
  const aiScene* scene =
    aiImportFile(glTFName, aiProcess_Triangulate);
  // ...
  traverseTree(scene->mRootNode, gltf.root);
  initAnimations(gltf, scene);
  // ...
}
```

This ensures that the animations are properly initialized and ready for use. Now, let's learn how to implement an actual glTF animation player using the animation data we've just loaded.

Implementing the glTF animation player

Let's start by building a foundation for our glTF animation player. Before we do skinned animations, advanced morphing, or complex blending, we first need to implement the basic animation rendering capabilities using the glTF data we loaded in the previous recipes. Let's walk through how to set up and extend our glTF viewer to support this essential animation rendering.

Getting ready

Make sure to read the previous recipe, *Importing skeleton and animation data*, to recap the data structures we use to store glTF animation data.

The source code for this recipe is located in `Chapter09/01_AnimationPlayer`.

How to do it...

To begin, let's introduce the data structures and helper functions needed to support animation playback. We will use them to implement all further animation rendering functionality in our glTF viewer.

1. First, we need a way to track the state of an active animation. This includes remembering the time of the last update, whether the animation should loop, and its active status. The struct `AnimationState` declared in `shared/UtilsAnim.h` keeps track of this data.

```
struct AnimationState {
  uint32_t animId   = ~0u;
  float currentTime = 0.0f;
  bool playOnce     = false;
  bool active       = false;
};
```

2. We need a helper function, `updateAnimation()`, which is responsible for applying changes to the animation over time. This function performs checks to ensure that the animation is active and valid before updating. The function is defined in `shared/UtilsAnim.cpp`.

```
void updateAnimation(
  GLTFContext& glTF, AnimationState& anim, float dt) {
  if (!anim.active || (anim.animId == ~0u)) {
    glTF.morphing = false;
    glTF.skinning = false;
    return;
  }
```

3. The function calculates the next timestamp for the active animation by incrementing the `currentTime` variable by the time elapsed since the last frame. It then checks if this timestamp is within the animation's total duration. If the animation is set to play only once, `currentTime` is clamped to the duration, and the animation is disabled, updating it to its final state with no further advancement needed. For looping animations, `currentTime` is wrapped back to the start using the floating-point modulo function `fmodf()`, ensuring it consistently stays within the `0...duration` range for smooth looping.

```
  const Animation& activeAnim = glTF.animations[anim.animId];
  anim.currentTime += activeAnim.ticksPerSecond *dt;
  if (anim.playOnce && anim.currentTime > activeAnim.duration) {
    anim.currentTime = activeAnim.duration;
    anim.active = false;
  } else {
    anim.currentTime =
      fmodf(anim.currentTime, activeAnim.duration);
  }
```

4. The next step is to apply the newly calculated animation timestamp `currentTime` to the actual glTF node transformations. To achieve this, we need to traverse the glTF scene, iterating through all registered nodes and channels, and then apply the corresponding transformations. We define a local lambda function, `traverseTree()`, to handle this recursive traversal. It takes the current node, `gltfNode`, and a parent node transformation, `parentTransform`.

```
std::function<void(GLTFNodeRef gltfNode,
                   const mat4& parentTransform)>
  traverseTree = [&](GLTFNodeRef gltfNode,
                     const mat4& parentTransform)
{
  const GLTFBone& bone = glTF.bonesByName[
    glTF.nodesStorage[gltfNode].name];
  const uint32_t boneId = bone.boneId;
```

5. If this node is part of the bone hierarchy and has an active animation channel, we call the `animationTransform()` function to calculate the appropriate transformation. If the node has no associated animation channels, its transformation doesn't need updating. Instead, we simply inherit and apply the parent's transformation. The `skinning` flag will be explained in the next recipe, *Doing skeletal animations in compute shaders*.

```
  if (boneId != ~0u) {
    auto channel = activeAnim.channels.find(boneId);
    const bool hasActiveChannel =
      channel != activeAnim.channels.end();
    glTF.matrices[glTF.nodesStorage[gltfNode].
      modelMtxId] =
        parentTransform * (hasActiveChannel ?
          animationTransform(channel->second, anim.currentTime) :
        glTF.nodesStorage[gltfNode].transform);
    glTF.skinning = true;
  } else {
    glTF.matrices[glTF.nodesStorage[gltfNode].
      modelMtxId] = parentTransform *
        glTF.nodesStorage[gltfNode].transform;
  }
```

6. We then make a recursive call for all child nodes to continue applying transformations throughout the glTF nodes hierarchy.

```
  for (uint32_t i = 0;
       i < glTF.nodesStorage[gltfNode].children.size(); i++) {
    const GLTFNodeRef child =
      glTF.nodesStorage[gltfNode].children[i];
```

```
        traverseTree(child,
          glTF.matrices[glTF.nodesStorage[gltfNode].modelMtxId]);
      }
    };
```

7. The helper function `animationTransform()`, which we just mentioned, is quite simple. It calculates `mat4` transformations for each transformation channel and combines them into one final transformation.

Note

It is important to note that the order of these transformations is defined by the glTF specification, ensuring consistent application across the animations.

```
mat4 animationTransform(
  const AnimationChannel& channel, float time)
{
  mat4 translation = glm::translate(mat4(1.0f),
    interpolatePosition(channel, time));
  mat4 rotation = glm::toMat4(
    glm::normalize(interpolateRotation(channel, time)));
  mat4 scale = glm::scale(mat4(1.0f),
    interpolateScaling(channel, time));
  return translation * rotation * scale;
}
```

8. Before we dive into our interpolation functions, let's introduce two more helper functions. The first one is called `getTimeIndex()`, which performs a search within a vector of sorted keyframe data, `t`. We utilize binary search to locate the nearest element with a timestamp that is not less than the specified `time` value and clamp it to the maximum index of elements. As a result, we obtain our starting interpolation index, which will be used for interpolation between keyframes.

```
template <typename T> uint32_t getTimeIndex(
  const std::vector<T>& t, float time) {
  return std::max(0,
    (int)std::distance( t.begin(), std::lower_bound(
      t.begin(), t.end(), time,
      [&](const T& lhs, float rhs)
      { return lhs.time < rhs; })
    ) - 1);
}
```

9. The second function is `interpolationVal()`, which returns the interpolation coefficient between two keyframe timestamps. This coefficient helps determine how to blend between the values of the two keyframes based on the current time within the animation. In our book, we use only linear interpolation for animations.

```
float interpolationVal(
  float lastTimeStamp, float nextTimeStamp,
  float animationTime)
{
  return (animationTime - lastTimeStamp) /
         (nextTimeStamp - lastTimeStamp);
}
```

Now, let's dive into the implementation details for each transformation channel, beginning with translations.

1. The function `interpolatePosition()` is straightforward. If there is only one interpolation keyframe, we can simply use it as is. If there are many keyframes, we find two adjacent ones and calculate the interpolation factor between them using `interpolationVal()`. This value is then used to perform linear interpolation between the actual position data contained in the keyframes using the `glm::mix()` function.

```
vec3 interpolatePosition(
  const AnimationChannel& channel, float time)
{
  if (channel.pos.size() == 1)
    return channel.pos[0].pos;
  const uint32_t start = getTimeIndex<>(channel.pos, time);
  const uint32_t end = start + 1;
  const float factor = interpolationVal(
    channel.pos[start].time,
    channel.pos[end].time, time);
  return glm::mix(channel.pos[start].pos,
                  channel.pos[end  ].pos, factor);
}
```

2. Let's take a look at a similar function, `interpolateRotation()`. This function operates in a comparable manner to handle rotation keyframes.

```
quat interpolateRotation(
  const AnimationChannel& channel, float time)
{
  if (channel.rot.size() == 1)
    return channel.rot[0].rot;
  const uint32_t start = getTimeIndex<>(channel.rot, time);
```

```
  const uint32_t end = start + 1;
  const float factor = interpolationVal(
    channel.rot[start].time,
    channel.rot[end].time, time);
```

3. Instead of linear interpolation between values, we use `slerp()` to interpolate between the rotation quaternions. This approach ensures that the rotation transitions are smooth and maintain the correct orientation throughout the animation.

```
  return glm::slerp(channel.rot[start].rot,
                    channel.rot[end  ].rot, factor);
}
```

Note

Slerp is a shorthand for spherical linear interpolation. It is an algorithm that interpolates a new quaternion between two given quaternions. `https://en.wikipedia.org/wiki/Slerp`.

4. Here, we have the scale interpolator. It copies the structure of `interpolatePosition()`; but, as expected, it utilizes scale keyframe data to perform the interpolation.

```
vec3 interpolateScaling(
  const AnimationChannel& channel, float time)
{
  if (channel.scale.size() == 1)
    return channel.scale[0].scale;
  const uint32_t start = getTimeIndex<>(channel.scale, time);
  const uint32_t end = start + 1;
  const float factor = interpolationVal(
    channel.scale[start].time,
    channel.scale[end  ].time, time);
  return glm::mix( channel.scale[start].scale,
    channel.scale[end].scale, factor );
}
```

At this point, we have covered how we calculate the updated transformations. Let's return to the function `updateAnimation()` where we still need to address one important step: updating by applying the parent inverse transform. This step is crucial for ensuring that the transformations are correctly applied relative to the parent node's space.

We discussed the process of hierarchy updates earlier, in the recipe *Introduction to skeletal animations*. Here, we will apply the bind transformation.

```
void updateAnimation(
  GLTFContext& glTF, AnimationState& anim, float dt)
{
  // … the skipped code here was explained in the step 5
  traverseTree(glTF.root, mat4(1.0f));
  for (const std::pair<std::string, GLTFBone>& b : glTF.bonesByName)
  {
    if (b.second.boneId != ~0u)
      glTF.matrices[b.second.boneId] =
        glTF.matrices[b.second.boneId] * b.second.transform;
  }
  // …
}
```

After this step, our matrices are updated and ready to be used for rendering.

How it works...

All that's left to do now is modify the `renderGLTF()` function in `shared/UtilsGLTF.cpp` to upload the new matrices to the GPU buffer `GLTFContext::matricesBuffer`. The skinning and morphing code is not necessary here and will be explored in the next recipe, *Doing skeletal animations in compute shaders*.

```
void renderGLTF(GLTFContext& gltf,
   const mat4& model, const mat4& view, const mat4& proj,
   bool rebuildRenderList)
{
  // …
  if (gltf.animated)
    buf.cmdUpdateBuffer(gltf.matricesBuffer, 0,
      gltf.matrices.size() * sizeof(mat4), gltf.matrices.data());
  // …
}
```

We have all the necessary code to execute node-based animation. Now, all that's left is to load a glTF file, `data/meshes/medieval_fantasy_book/scene.gltf`, that contains a 3D model with node-based animations, and run the animation player. Here's the complete `main()` function from the sample app located in `Chapter09/01_AnimationPlayer/src/main.cpp`, which accomplishes this:

```
int main() {
  VulkanApp app(
    { .initialCameraPos    = vec3(7.0f,  6.8f, -13.6f),
      .initialCameraTarget = vec3(1.7f, -1.0f,   0.0f),
      .showGLTFInspector   = true });
  GLTFContext gltf(app);
  loadGLTF(gltf,
    "data/meshes/medieval_fantasy_book/scene.gltf",
    "data/meshes/medieval_fantasy_book/");
  gltf.enableMorphing = false;
  const mat4 t = glm::translate(mat4(1.0f),
    vec3(0.0f, 2.1f, 0.0f)) * glm::scale(mat4(1.0f), vec3(0.2f));
  AnimationState anim = { .animId      = 0,
                          .currentTime = 0.0f,
                          .playOnce    = false,
                          .active      = true };
  gltf.inspector = { .activeAnim = { 0 },
                     .showAnimations = true  };
  app.run([&](uint32_t width, uint32_t height,
    float aspectRatio, float deltaSeconds)
  {
    const mat4 m = t * glm::rotate(
      mat4(1.0f), glm::radians(180.0f), vec3(0, 1, 0));
    const mat4 v = app.camera_.getViewMatrix();
    const mat4 p = glm::perspective(45.f, aspectRatio, 0.01f, 100.f);
    animateGLTF(gltf, anim, deltaSeconds);
    renderGLTF(gltf, m, v, p);
    if (gltf.inspector.activeAnim[0] != anim.animId)
      anim = { .animId = gltf.inspector.activeAnim[0],
               .currentTime = 0.0f,
               .playOnce    = false,
               .active      = true };
  });
  return 0;
}
```

The running application should render an animated scene as in the following screenshot:

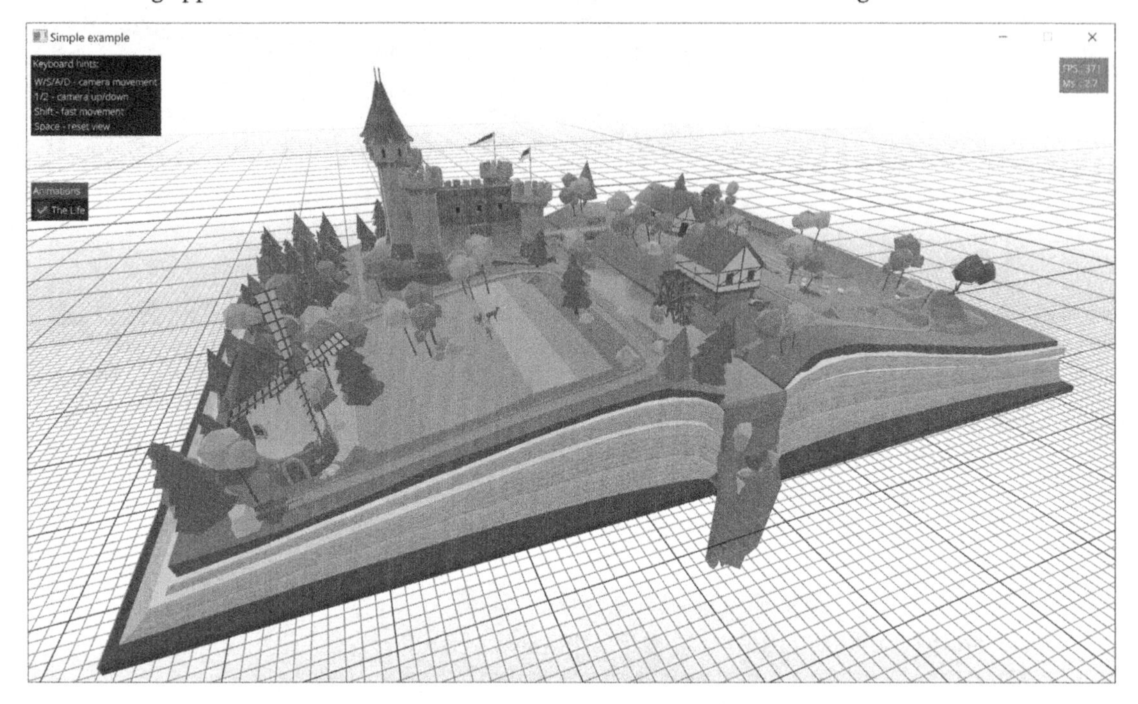

Figure 9.1: Node-based animations

In fact, we've already implemented much of the necessary C++ code to support not only node-based animations but also skeletal animations. However, we still need to implement skinning to calculate the mesh vertices based on an animated skeleton. Let's move on to the next recipe and learn how to achieve this using compute shaders.

There's more...

The 3D model used in this recipe is based on "Medieval Fantasy Book" (`https://sketchfab.com/3d-models/medieval-fantasy-book-06d5a80a04fc4c5ab552759e9a97d91a`) by Pixel (`https://sketchfab.com/stefan.lengyel1`) licensed under CC-BY-4.0 (`http://creativecommons.org/licenses/by/4.0/`).

In addition to node-based animations, this 3D model includes morph targets for animating the flags. However, at the time of writing this chapter, Assimp lacks full support for loading morph targets from glTF files, making it impossible to animate the flags here.

Doing skeletal animations in compute shaders

Skeletal animations imply moving vertices based on the weighted influence of multiple matrices, which represent a skeleton of bones. Each bone is represented as a matrix that influences nearby vertices based on its weight. For instance, the hand bone would heavily influence hand vertices, while it would have no effect on foot vertices. Vertices near the wrist might be affected by both the hand and arm bones. In essence, skinning requires bones, a hierarchy of matrices, and weights.

Weights, like vertex colors, are assigned per vertex and range from `0` to `1`, indicating how much a specific bone influences that vertex's position. These weights are stored in a buffer and passed as vertex attributes. In this recipe, we cover the final part of the animation player with skinning support.

In the recipe *Importing skeleton and animation data*, we learned how to load bone weights and hierarchy. Now, let's look at how to use this data to skin a glTF mesh. We will do this with a compute shader that updates the vertex positions in our vertex buffer.

An older approach to applying skinning data to each vertex of a mesh is to use a vertex shader. This was once common practice, and the official Khronos glTF viewer still uses it. However, vertex shaders can be less efficient when vertices are reused multiple times with different indices, leading to redundant complex calculations. To improve efficiency, we can perform these calculations in a compute shader, avoiding the need to process each vertex instance individually. This reduces redundancy but adds complexity and requires an extra compute pass, which can create extra dependencies in the rendering pipeline. We chose this approach because traditional vertex shader implementations are more commonly seen, while compute shader-based solutions are less typical.

Getting ready

Be sure to review the recipe *Importing skeleton and animation data* before moving forward.

The demo source code for this recipe is located in `Chapter09/02_Skinning`, and the compute shader is located in `data/shaders/gltf/animation.comp`.

How to do it...

Before diving into our GLSL skinning and morphing compute shader code, let's start with the C++ side and see how the data is passed into the shader. These changes are applied to `shared/UtilsGLTF.h` and `shared/UtilsGLTF.cpp`.

1. Let's add a compute pipeline state and an animation shader module to the `GLTFContext` class.

```
struct GLTFContext {
  // ...
  Holder<ComputePipelineHandle> pipelineComputeAnimations;
  // ...
  Holder<ShaderModuleHandle> animation;
  // ...
};
```

2. In the `loadGLTF()` function, we load the compute shader:

```
  // ...
  gltf.animation = loadShaderModule(ctx,
    "data/shaders/gltf/animation.comp");
  // ...
```

3. The compute pipeline that uses this shader is created lazily in the `animateGLTF()` function, only when animation is needed.

```
void animateGLTF(
  GLTFContext& gltf, AnimationState& anim, float dt)
{
  if (gltf.transforms.empty()) return;
  if (gltf.pipelineComputeAnimations.empty())
    gltf.pipelineComputeAnimations =
      gltf.app.ctx_->createComputePipeline({
        .smComp = gltf.animation });
  anim.active   = anim.animId != ~0;
  gltf.animated = anim.active;
  if (anim.active) updateAnimation(gltf, anim, dt);
}
```

Now, let's examine additions to the `renderGLTF()` function, which handles all the rendering. We need to add a new code snippet to update animation-related data and invoke the compute shader.

1. First, we need to update the matrices buffer. We calculated these earlier in the previous recipe, *Implementing the glTF animation player*. The morphing data is also updated here. We will cover the missing morphing parts in the upcoming recipes, including *Introduction to morph targets*.

```
void renderGLTF(GLTFContext& gltf,
  const mat4& model, const mat4& view, const mat4& proj,
  bool rebuildRenderList)
{
  // ...
  if (gltf.animated) {
    buf.cmdUpdateBuffer(gltf.matricesBuffer, 0,
      gltf.matrices.size() * sizeof(mat4), gltf.matrices.data());
    if (gltf.morphing){
      buf.cmdUpdateBuffer(gltf.morphStatesBuffer, 0,
        gltf.morphStates.size() * sizeof(MorphState),
        gltf.morphStates.data());
    }
    updateLights(gltf, buf);
```

2. The compute shader should only be run if skinning is enabled with valid bones, or if mesh morphing is enabled. We use Vulkan push constants to pass all inputs to the compute shader, including the scene node matrices and input vertices. We prepared all these buffers in the recipes *Importing skeleton and animation data* and *Implementing the glTF animation player*. The output buffer is assigned to `gltf.vertexBuffer`, which is then used as vertex input for our main rendering pass.

```
    if ((gltf.skinning && gltf.hasBones) ||
         gltf.morphing)
    {
      struct ComputeSetup {
        uint64_t matrices;
        uint64_t morphStates;
        uint64_t morphVertexBuffer;
        uint64_t inBuffer;
        uint64_t outBuffer;
        uint32_t numMorphStates;
      } pc = {
        .matrices    = ctx->gpuAddress(gltf.matricesBuffer),
        .morphStates = ctx->gpuAddress(gltf.morphStatesBuffer),
        .morphVertexBuffer = ctx->gpuAddress(gltf.vertexMorphingBuffer),
        .inBuffer  = ctx->gpuAddress(gltf.vertexSkinningBuffer),
        .outBuffer = ctx->gpuAddress(gltf.vertexBuffer),
        .numMorphStates = gltf.morphStates.size(),
      };
      buf.cmdBindComputePipeline(gltf.pipelineComputeAnimations);
      buf.cmdPushConstants(pc);
```

3. Note that we specify all mutable buffers as dependencies in the dispatch command. This ensures that **LightweightVK** places all necessary Vulkan buffer memory barriers before and after buffers are used, based on their usage flags and previous use. We need to use barriers in this case to ensure the buffer for the previous frame is fully processed before we start preparing the next one. The size of our vertex buffer is always padded to `16` vertices, so we don't have to worry about alignment.

Note

Padding with dummy vertices is done in `loadGLTF()` right before we allocate the vertex buffer:

```
gltf.maxVertices = (1 + (vertices.size() / 16)) * 16;
vertices.resize(gltf.maxVertices);
gltf.vertexBuffer = ctx->createBuffer({
    .usage     = lvk::BufferUsageBits_Vertex |
                 lvk::BufferUsageBits_Storage,
    .storage   = lvk::StorageType_Device,
    .size      = sizeof(Vertex) * vertices.size(),
    .data      = vertices.data(),
    .debugName = "Buffer: vertex" });
```

```
      buf.cmdDispatchThreadGroups(
        { .width = gltf.maxVertices / 16, },
        { .buffers = {
            BufferHandle(gltf.vertexBuffer),
            BufferHandle(gltf.morphStatesBuffer),
            BufferHandle(gltf.matricesBuffer),
            BufferHandle(gltf.vertexSkinningBuffer)}
      });
    }
  }
```

The output buffer `gltf.vertexBuffer` is never read back to the CPU; it is used solely as vertex input for subsequent rendering passes. This means we have completed the C++ code. Now, let's take a look at the compute shader and see how everything works.

How it works...

The compute shader is located in `data/shaders/gltf/animation.comp`.

1. It begins by declaring the necessary GLSL extensions because we want to avoid injecting any automatic LightweightVK preamble here. The `GL_EXT_buffer_reference` extension is used to handle buffer device addresses passed via push constants. The extension `GL_EXT_scalar_block_layout` is required to simplify access to tightly packed vertex data, which is stored as a combination of different `vec3` and scalar values.

```
#version 460
#extension GL_EXT_buffer_reference : require
#extension GL_EXT_scalar_block_layout : require
layout (local_size_x=16, local_size_y=1, local_size_z=1) in;
```

2. The shader uses several structures that resemble the data in various buffers. Some of these structures correspond to their C++ counterparts described earlier in this chapter in the recipe *Importing skeleton and animation data*, while others should be familiar from previous chapters. We list them all here for your reference.

```
struct TransformsBuffer {
  uint mtxId;
  uint matId;
  uint nodeRef; // for CPU only
  uint meshRef; // for CPU only
  uint opaque;  // for CPU only
};
struct VertexSkinningData {
  vec4 pos;
  vec4 norm;
```

```
  uint bones[8];
  float weights[8];
  uint meshId;
};
struct VertexData {
  vec3 pos;
  vec3 norm;
  vec4 color;
  vec4 uv;
  float padding[2];
};
#define MAX_WEIGHTS 8
struct MorphState {
  uint meshId;
  uint morphTarget[MAX_WEIGHTS];
  float weights[MAX_WEIGHTS];
};
```

3. Now, let's declare buffer references using these structures. This is where the `scalar` GLSL block layout comes into play, ensuring that `vec3` and `float` values are not padded to larger data sizes. This significantly simplifies our data storage scheme and interactions between the C++ and GLSL code. Each buffer here contains an unbounded array of the corresponding items.

```
layout (std430, buffer_reference) readonly buffer Matrices {
  mat4 matrix[];
};
layout (scalar, buffer_reference) readonly buffer MorphStates {
  MorphState morphStates[];
};
layout (scalar, buffer_reference) readonly buffer VertexSkinningBuffer {
  VertexSkinningData vertices[];
};
layout (scalar, buffer_reference) writeonly buffer VertexBuffer {
  VertexData vertices[];
};
layout (scalar, buffer_reference) readonly buffer MorphVertexBuffer {
  VertexData vertices[];
};
```

4. The final piece of our input data is the push constants declaration, which combines all the buffers mentioned above. The `numMorphStates` parameter holds the number of morph targets.

```
layout (push_constant) uniform PerFrameData {
  Matrices matrices;
  MorphStates morphStates;
  MorphVertexBuffer morphTargets;
  VertexSkinningBuffer inBufferId;
  VertexBuffer outBufferId;
  uint numMorphStates;
} pc;
```

5. This is where the `main()` function begins, starting by retrieving the bind pose vertices and normal vectors. The bind pose just holds the initial data loaded from the glTF file.

```
void main() {
  uint index = gl_GlobalInvocationID.x;
  VertexSkinningData inVtx = pc.inBufferId.vertices[index];
  vec4 inPos  = vec4(inVtx.pos.xyz, 1.0);
  vec4 inNorm = vec4(inVtx.norm.xyz, 0.0);
```

6. If vertex morphing is enabled for this `meshId`, we handle it here. We will explore this further in the upcoming recipes, *Introduction to morph targets* and *Adding morph targets support*. We can skip it for now.

```
  if (inVtx.meshId < pc.numMorphStates) {
    MorphState ms = pc.morphStates.morphStates[inVtx.meshId];
    if (ms.meshId != ~0)
      for (int m = 0; m != MAX_WEIGHTS; m++)
        if (ms.weights[m] > 0) {
          VertexData mVtx = pc.morphTargets.vertices[
            ms.morphTarget[m] + index];
          inPos.xyz  += mVtx.pos  * ms.weights[m];
          inNorm.xyz += mVtx.norm * ms.weights[m];
        }
  }
```

7. The skinning process itself is straightforward. We iterate through all the bones associated with the vertex and accumulate the weighted sum of their transformed positions, which are obtained by applying the `boneMat` transformation matrix.

```
vec4 pos = vec4(0);
vec4 norm = vec4(0);
int i = 0;
for (; i != MAX_WEIGHTS; i++) {
  if (inVtx.bones[i] == ~0) break;
  mat4 boneMat = pc.matrices.matrix[inVtx.bones[i]];
  pos += boneMat * inPos * inVtx.weights[i];
```

8. Normal vectors are treated differently. We transform them using the inverse transpose of the bone matrices. This approach ensures that the normals retain their original unit length and direction, even under non-uniform scaling that can occur with bone transformations.

Note

For readers seeking a deeper understanding of normal transformation, you can refer to this resource: `https://terathon.com/blog/transforming-normals.html`.

```
  norm += transpose(inverse(boneMat)) * inNorm * inVtx.weights[i];
}
```

9. If there were no valid weights for this vertex, we fall back to the bind pose vertex position and normal vector. The calculated position and normal are then written to the output buffer.

```
  if (i == 0) {
    pos.xyz = inPos.xyz;
    norm.xyz = inNorm.xyz;
  }
  pc.outBufferId.vertices[index].pos  = pos.xyz;
  pc.outBufferId.vertices[index].norm = normalize(norm.xyz);
}
```

Now we can run the demo application, `Chapter09/02_Skinning`, and observe the results of the vertex skinning process. We use the Khronos model located at `deps/src/glTF-Sample-Assets/Models/Fox/glTF/Fox.gltf`, which should be rendered walking, as shown in the following screenshot.

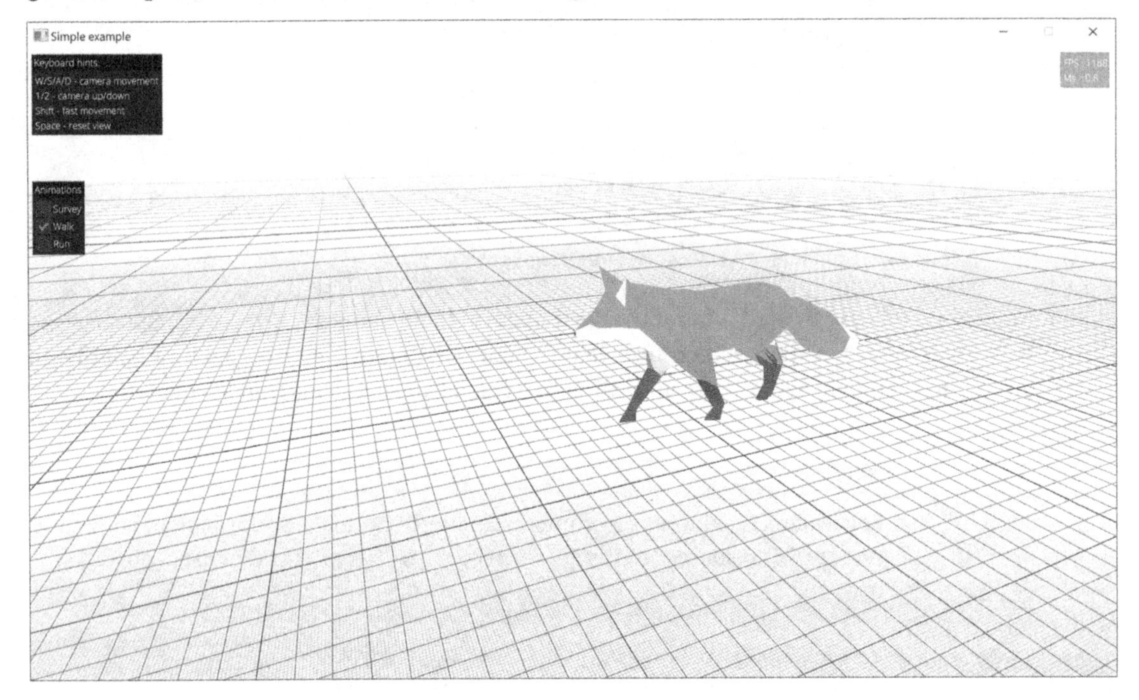

Figure 9.2: Skeletal animation with vertex skinning

Try switching between different animations: `Survey`, `Walk`, and `Run`. When you click to switch animations, you will notice that the transition occurs quite abruptly and in an unnatural way. This can be improved by interpolating between multiple animations using a floating-point transition factor. We will learn how to achieve this in the recipe *Animation blending*.

For now, let's wrap up our animation player and learn how to implement morph targets in the upcoming recipes.

Introduction to morph targets

A **morph target** is a deformed version of a mesh. In glTF, morph targets are used to create mesh deformations by blending different sets of vertex positions, or other attributes, like normal vectors, according to specified weights. They are especially useful for facial animations, character expressions, and other types of mesh deformation in 3D models.

- **Base Mesh:** This is the original, unaltered mesh defined by its vertices, normals, texture coordinates, and other attributes.
- **Morph Targets:** These are modified versions of the base mesh, where the positions of some or all vertices have been adjusted to create a specific deformation. Multiple morph targets can be defined for the same mesh, each representing a different shape variation.

- **Vertex Data:** Morph targets include offset data for vertices. A morph target usually adjusts the vertex positions (and sometimes other attributes like normals) of the base mesh. These offsets specify how much each vertex should move when the morph target is applied.
- **Weights:** Each morph target has an associated weight that controls its influence on the mesh. The weight is a floating-point value between `0` and `1`, where `0` means no influence (the mesh appears identical to the base mesh); `1` means full influence (the mesh fully deforms to match the morph target); and values between `0` and `1` blend the base mesh with the morph target.

When multiple morph targets are used, the mesh's final shape is calculated as a weighted sum of the morph targets. For each vertex, its final position is determined by blending the corresponding vertex positions from the base mesh and the morph targets, according to their weights.

The formula for calculating a vertex's final `position` with applied morph targets is:

$$\text{position} = \text{basePosition} + (w_1 * \text{offset}_1) + (w_2 * \text{offset}_2) + \ldots + (w_N * \text{offset}_N)$$

Where:

- `basePosition` is the original position of the vertex in the base mesh.
- w_1, w_2, ..., w_N are the weights for each morph target.
- $offset_1$, $offset_2$, ..., $offset_N$ are the vertex offsets from the morph targets.

Now, let's look at how to load glTF morph targets data using the Assimp library.

Loading glTF morph targets data

Similar to the recipe *Importing skeleton and animation data*, this one doesn't include a standalone example either. Instead, it focuses on the modifications needed in our glTF loading code to load morph target data from glTF files using the Assimp library.

Getting ready

Refer to the recipe *Importing skeleton and animation data* for a quick refresher on the data structures needed to support animations. You can also review the recipe *Doing skeletal animations in compute shaders* and the shader code in `data/shaders/gltf/animation.comp` to revisit our animation compute shader.

How to do it...

To load morph targets from glTF files, we will need to add some new data structures and update the loading code. Let's go through how to do this.

1. We begin, as usual, with the data structures. `MorphTarget` represents a reference for all morph targets attached to a specific mesh. All morph targets are stored in a single buffer, needing only a `meshId` and an `offset` in this buffer. The actual delta positions use the same base vertex format as the rest of our glTF code, so no new changes are necessary. The structure is declared in `shared/UtilsGLTF.h`.

```
struct MorphTarget {
  uint32_t meshId = ~0u;
  std::vector<uint32_t> offset;
};
```

2. The `MorphState` structure in `shared/UtilsAnim.h` holds the morphing state for a given mesh. The values of `morphTarget[]` and `weights[]` are updated at each animation frame in the `morphTransform()` function, which we will cover in the next recipe, *Adding morph targets support*.

```
struct MorphState {
  uint32_t meshId = ~0u;
  uint32_t morphTarget[MAX_MORPH_WEIGHTS] = {};
  float weights[MAX_MORPH_WEIGHTS] = {};
};
```

3. Let's add a few new member fields to the `GLTFContext` class. The `morphTargets` vector indexed by `meshId` provides the mapping information between meshes and the `vertexMorphingBuffer`. The `vertexMorphingBuffer` holds the vertex data for the morph targets.

```
struct GLTFContext {
  // ...
  std::vector<MorphTarget> morphTargets;
  // ...
  Holder<BufferHandle> morphStatesBuffer;
  // ...
  Holder<BufferHandle> vertexMorphingBuffer;
  // ...
  std::vector<MorphState> morphStates;
  // ...
  bool morphing = false;
  // ...
};
```

With all the data structures and declarations set up, let's examine the loading code in `loadGLTF()` located in `shared/UtilsGLTF.cpp`.

1. First, let's load the morph target data for each mesh from the Assimp `aiScene` object and store it in `GLTFContext::morphTargets`.

```
void loadGLTF(GLTFContext& gltf,
  const char* glTFName, const char* glTFDataPath)
{
  const aiScene* scene = aiImportFile(glTFName, aiProcess_Triangulate);
  // ...
  for (uint32_t meshId = 0;
```

```
    meshId != scene->mNumMeshes; meshId++) {
  const aiMesh* m = scene->mMeshes[meshId];
  if (!m->mNumAnimMeshes) continue;
  MorphTarget& morphTarget = gltf.morphTargets[meshId];
  morphTarget.meshId = meshId;
```

2. Next, we populate the vertex data for each morph target associated with the mesh. An `aiAnimMesh`, which corresponds to a glTF morph target, is an attachment to an `aiMesh` and stores per-vertex animations for a specific frame.

```
  for (uint32_t a = 0; a<m->mNumAnimMeshes; a++) {
    const aiAnimMesh* mesh = m->mAnimMeshes[a];
    for (uint32_t i = 0; i < mesh->mNumVertices; i++) {
      const aiVector3D srcNorm = m->mNormals ?
        m->mNormals[i] : aiVector3D(0, 1, 0);
      const aiVector3D v = mesh->mVertices[i];
      const aiVector3D n = mesh->mNormals ?
        mesh->mNormals[i] : aiVector3D(0, 1, 0);
      const aiColor4D c = mesh->mColors[0] ?
        mesh->mColors[0][i] : aiColor4D(1);
      const aiVector3D uv0 = mesh->mTextureCoords[0] ?
          mesh->mTextureCoords[0][i] : aiVector3D(0);
      const aiVector3D uv1 = mesh->mTextureCoords[1] ?
          mesh->mTextureCoords[1][i] : aiVector3D(0);
```

3. We subtract the original vertex data from the `aiAnimMesh` data for two reasons. First, the glTF specification defines morph targets as deltas between original mesh attributes, whereas the Assimp library converts them into complete mesh data. Second, this simplification makes the shader computations a bit more efficient.

```
      morphData.push_back({
        .position = vec3(v.x - m->mVertices[i].x,
                         v.y - m->mVertices[i].y,
                         v.z - m->mVertices[i].z),
        .normal   = vec3(n.x - srcNorm.x,
                         n.y - srcNorm.y,
                         n.z - srcNorm.z),
        .color    = vec4(c.r, c.g, c.b, c.a),
        .uv0      = vec2(uv0.x, 1.0f - uv0.y),
        .uv1      = vec2(uv1.x, 1.0f - uv1.y),
      });
    }
    morphTarget.offset.push_back(morphTargetsOffset);
```

```
        morphTargetsOffset += mesh->mNumVertices;
      }
    }
    // ...
  }
```

4. After extracting the data from Assimp, we can store the morph target vertices in a GPU buffer. It's important to note that we always include a dummy value in this buffer for one vertex, even if there's no morphing data in the mesh. This simplifies the shader code by eliminating the need for unnecessary range checking.

```
    const bool hasMorphData = !morphData.empty();
    gltf.vertexMorphingBuffer = ctx->createBuffer({
      .usage = lvk::BufferUsageBits_Vertex |
               lvk::BufferUsageBits_Storage,
      .storage = lvk::StorageType_Device,
      .size = hasMorphData ?
        sizeof(Vertex) * morphData.size() : sizeof(Vertex),
      .data = hasMorphData ? morphData.data() : nullptr,
      .debugName = "Buffer: morphing vertex data",
    });
```

While the `morphData` values are intended for the GPU to morph individual mesh vertices, the `morphTargets` values are used by the CPU to update animation states.

This concludes our code for loading morph targets data. Now, let's explore how to use it for animations.

Adding morph targets support

In the previous recipe, we loaded the morph target data from glTF. Now, let's use that data to create a demo application and render morphing meshes.

Getting ready

Be sure to read the previous recipe, *Loading glTF morph targets data*, to review all the necessary data structures for morph targets.

The source code for this recipe is located in `Chapter09/03_Morphing`.

How to do it...

To display the results rendered on the screen, we need to enable support for mesh morph targets in our glTF animation player and incorporate this functionality into our compute shader.

1. Let's begin with the rendering code in the `renderGLTF()` function. Here's the complete setup for our compute pass, which was mentioned earlier in the recipe *Doing skeletal animations in compute shaders*.

```
    if (gltf.animated) {
      buf.cmdUpdateBuffer(gltf.matricesBuffer, 0,
        gltf.matrices.size() * sizeof(mat4),
        gltf.matrices.data());
```

2. The current value of `GLTFContext::morphStates`, which we will learn how to calculate shortly, is uploaded to the GPU buffer `GLTFContext::morphStatesBuffer` for processing by our compute shader.

```
      if (gltf.morphing)
        buf.cmdUpdateBuffer(gltf.morphStatesBuffer, 0,
          gltf.morphStates.size() * sizeof(MorphState),
          gltf.morphStates.data());
      updateLights(gltf, buf);
      if ((gltf.skinning && gltf.hasBones) || gltf.morphing)
      {
        struct ComputeSetup {
          uint64_t matrices;
          uint64_t morphStates;
          uint64_t morphVertexBuffer;
          uint64_t inBuffer;
          uint64_t outBuffer;
          uint32_t numMorphStates;
        } pc = {
          .matrices    = ctx->gpuAddress(gltf.matricesBuffer),
          .morphStates = ctx->gpuAddress(gltf.morphStatesBuffer),
```

3. The `morphVertexBuffer`, which contains all the base mesh morphing data described in the previous recipe, *Loading glTF morph targets data*, is passed to the compute shader.

```
          .morphVertexBuffer = ctx->gpuAddress(gltf.vertexMorphingBuffer),
          .inBuffer  = ctx->gpuAddress(gltf.vertexSkinningBuffer),
          .outBuffer = ctx->gpuAddress(gltf.vertexBuffer),
          .numMorphStates = static_cast<uint32_t>(gltf.morphStates.size()),
        };
        buf.cmdBindComputePipeline(gltf.pipelineComputeAnimations);
        buf.cmdPushConstants(pc);
        buf.cmdDispatchThreadGroups(
          { .width = gltf.maxVertices / 16, },
          { .buffers = {
              BufferHandle(gltf.vertexBuffer),
              BufferHandle(gltf.morphStatesBuffer),
              BufferHandle(gltf.matricesBuffer),
```

```
                BufferHandle(gltf.vertexSkinningBuffer)}
          });
        }
      }
```

As you can see, the entire morphing code path is seamlessly integrated with the skinning code path described in the recipe *Doing skeletal animations in compute shaders.*

Let's examine how the current morphing data, which goes into the `morphStatesBuffer`, is calculated based on the animation states.

1. First, let's declare the necessary data structures to hold interpolation values for our morphing animations in `shared/UtilsAnim.h`. We support up to `8` meshes, `8` weights for vertex attributes, and `100` morph targets.

```
#define MAX_MORPH_WEIGHTS 8
#define MAX_MORPHS 100
struct MorphingChannelKey {
  float time = 0.0f; // in seconds
  uint32_t mesh[MAX_MORPH_WEIGHTS] = {};
  float  weight[MAX_MORPH_WEIGHTS] = {};
};
struct MorphingChannel {
  std::string name;
  std::vector<MorphingChannelKey> key;
};
```

2. We need to add a new member field, `morphChannels`, to our `Animation` structure. Each animation can include morphing channels, consisting of a vector of pairs that represent mesh and weight data, along with a key time.

```
struct Animation {
  unordered_map<int, AnimationChannel> channels;
  std::vector<MorphingChannel> morphChannels;
  float duration;
  float ticksPerSecond;
  std::string name;
};
```

3. The `initAnimations()` function in `shared/UtilsAnim.cpp` needs to initialize morphing channels data. Here's the new code snippet that was added to it:

```
void initAnimations(GLTFContext& glTF, const aiScene* scene) {
  glTF.animations.resize(scene->mNumAnimations);
  for (uint32_t i = 0; i < scene->mNumAnimations; ++i)
  {
```

```
// ...
const uint32_t numMorphTargetChannels =
  scene->mAnimations[i]->mNumMorphMeshChannels;
anim.morphChannels.resize(numMorphTargetChannels);
```

4. Morph target channels are extracted from the `aiScene` object in the same way as we handled node-based animations in the recipe *Importing skeleton and animation data*.

```
for (int c = 0; c<numMorphTargetChannels; c++) {
  const aiMeshMorphAnim* channel =
    scene->mAnimations[i]->mMorphMeshChannels[c];
  MorphingChannel& morphChannel = anim.morphChannels[c];
```

5. Assimp uses string names to bind channels to meshes, which doesn't fully align with the glTF specification.

```
    morphChannel.name = channel->mName.C_Str();
    morphChannel.key.resize(channel->mNumKeys);
    for (uint32_t k = 0; k < channel->mNumKeys; ++k)
    {
      MorphingChannelKey& key = morphChannel.key[k];
      key.time = channel->mKeys[k].mTime;
      for (uint32_t v = 0;
           v < std::min(MAX_MORPH_WEIGHTS,
           channel->mKeys[k].mNumValuesAndWeights); v++)
      {
        key.mesh[v] = channel->mKeys[k].mValues[v];
        key.weight[v] = channel->mKeys[k].mWeights[v];
      }
    }
}}}
```

6. To play morphing animations, we need a structure, `MorphStates`, to maintain the state. This structure is used to run the player and upload data to the GPU. As mentioned above, we support a maximum of `100` states.

```
struct MorphState {
  uint32_t meshId = ~0u;
  uint32_t morphTarget[MAX_MORPH_WEIGHTS] = {};
  float weights[MAX_MORPH_WEIGHTS] = {};
};
```

7. Let's add code to the `updateAnimation()` function in `shared/UtilsAnim.cpp` to process morphing animations. This is much simpler than skinning, as we don't need to update the actual scene hierarchy. All we need to do here is calculate new `MorphState` data:

```
void updateAnimation(
  GLTFContext& glTF, AnimationState& anim, float dt)
{
  // ...
  glTF.morphStates.resize(glTF.meshesStorage.size());
  if (glTF.enableMorphing) {
    if (!activeAnim.morphChannels.empty()) {
      for (size_t i = 0; i < activeAnim.morphChannels.size(); ++i) {
        const MorphingChannel& channel = activeAnim.morphChannels[i];
        const uint32_t meshId = glTF.meshesRemap[channel.name];
        const MorphTarget& morphTarget = glTF.morphTargets[meshId];
        if (morphTarget.meshId != ~0u)
          glTF.morphStates[morphTarget.meshId] =
            morphTransform(morphTarget, channel, anim.currentTime);
      }
      glTF.morphing = true;
    }
}}
```

8. As usual, all the details are hidden in a helper function called `morphTransform()`:

```
MorphState morphTransform(
  const MorphTarget& target,
  const MorphingChannel& channel,
  float time)
{
  MorphState ms;
  ms.meshId = target.meshId;
  float mix = 0.0f;
  int start = 0;
  int end   = 0;
```

9. Here, we use the same helper function, `interpolationVal()`, to find the blending coefficient, just as we did for skeletal animations in the recipe *Implementing the glTF animation player*.

```
  if (channel.key.size() > 0) {
    start = getTimeIndex(channel.key, time);
    end   = start + 1;
    mix   = interpolationVal(
```

```
      channel.key[start].time,
      channel.key[end].time, time);
  }
```

10. This function is quite similar to the other interpolation functions we described earlier, but instead of mixing transformations, we mix blend weights. Additionally, we copy the `offset` to the actual buffer along with the weights, which we use later in our shader to retrieve the actual morph target vertices.

```
    for (uint32_t i = 0;
         i < std::min((uint32_t)target.offset.size(),
                      (uint32_t)MAX_MORPH_WEIGHTS); ++i)
    {
      ms.morphTarget[i] = target.offset[channel.key[start].mesh[i]];
      ms.weights[i] =
        glm::mix(channel.key[start].weight[i],
                 channel.key[end  ].weight[i], mix);
    }
    return ms;
  }
```

That's everything for the C++ portion. Now, let's take a quick look at the modifications needed for our GLSL compute shader located in `data/shaders/gltf/animation.comp`.

1. First, we assign the `inPos` and `inNorm` values to the corresponding base mesh values, regardless of whether morphing is enabled.

```
void main() {
  uint index = gl_GlobalInvocationID.x;
  VertexSkinningData inVtx = pc.inBufferId.vertices[index];
  vec4 inPos  = vec4(inVtx.pos.xyz, 1.0);
  vec4 inNorm = vec4(inVtx.norm.xyz, 0.0);
```

2. Then, we perform vertex morphing using the `MorphState` for this mesh that we calculated earlier in C++. We check if each vertex matches a morph target's `meshId` and is valid. If it does, we calculate the weighted sum of the positions and normals using the active morph target offset and weight.

```
  if (inVtx.meshId < pc.numMorphStates) {
    MorphState ms =
      pc.morphStates.morphStates[inVtx.meshId];
    if (ms.meshId != ~0) {
      for (int m = 0; m != MAX_WEIGHTS; m++) {
        if (ms.weights[m] > 0) {
          VertexData mVtx = pc.morphTargets.
```

```
          vertices[ms.morphTarget[m] + index];
        inPos.xyz  += mVtx.pos  * ms.weights[m];
        inNorm.xyz += mVtx.norm * ms.weights[m];
      }
  }}}
  // ...
}
```

The shader then continues with vertex skinning, as described in the recipe *Doing skeletal animations in compute shaders*, using the morphed values of `inPos` and `inNorm`.

How it works...

Now, let's take a quick look at the demo application `Chapter09/03_Morphing/src/main.cpp`, specifically focusing on its `main()` function.

1. We use one of the sample Khronos 3D models, `AnimatedMorphCube.gltf`, which is a simple example to demonstrate the morphing capabilities we've implemented.

```
int main() {
  VulkanApp app(
    { .initialCameraPos = vec3(2.13f, 2.44f, -3.5f),
      .initialCameraTarget = vec3(0, 0, 2),
      .showGLTFInspector   = true });
  GLTFContext gltf(app);
  loadGLTF(gltf, "deps/src/glTF-Sample-Assets/
      Models/AnimatedMorphCube/glTF/AnimatedMorphCube.gltf",
    "deps/src/glTF-Sample-Assets/Models/AnimatedMorphCube/glTF/");
  const mat4 t = glm::translate(mat4(1.0f), vec3(0.0f, 1.1f, 0.0f));
```

2. Let's play the very first morphing animation from the `.gltf` file.

```
  AnimationState anim = { .animId      = 0,
                          .currentTime = 0.0f,
                          .playOnce    = false,
                          .active      = true };
  gltf.inspector = { .activeAnim     = { 0 },
                     .showAnimations = true };
  app.run([&](uint32_t width, uint32_t height,
    float aspectRatio, float deltaSeconds)
  {
    const mat4 m = t * glm::rotate(mat4(1.0f),
      glm::radians(180.0f), vec3(0, 1, 0));
    const mat4 v = app.camera_.getViewMatrix();
    const mat4 p = glm::perspective(45, aspectRatio, 0.01, 100);
```

```
    animateGLTF(gltf, anim, deltaSeconds);
    renderGLTF(gltf, m, v, p);
```

3. We reassign the current animation to our `GLTFIntrospective` introspector, allowing it to be controlled through the ImGui UI.

```
    if (gltf.inspector.activeAnim[0] != anim.animId) {
      anim = { .animId      = gltf.inspector.activeAnim[0],
               .currentTime = 0.0f,
               .playOnce    = false,
               .active      = true };
    }
  });
  return 0;
}
```

When we run this demo application, you should see an animated deforming cube, similar to the one shown in the following screenshot.

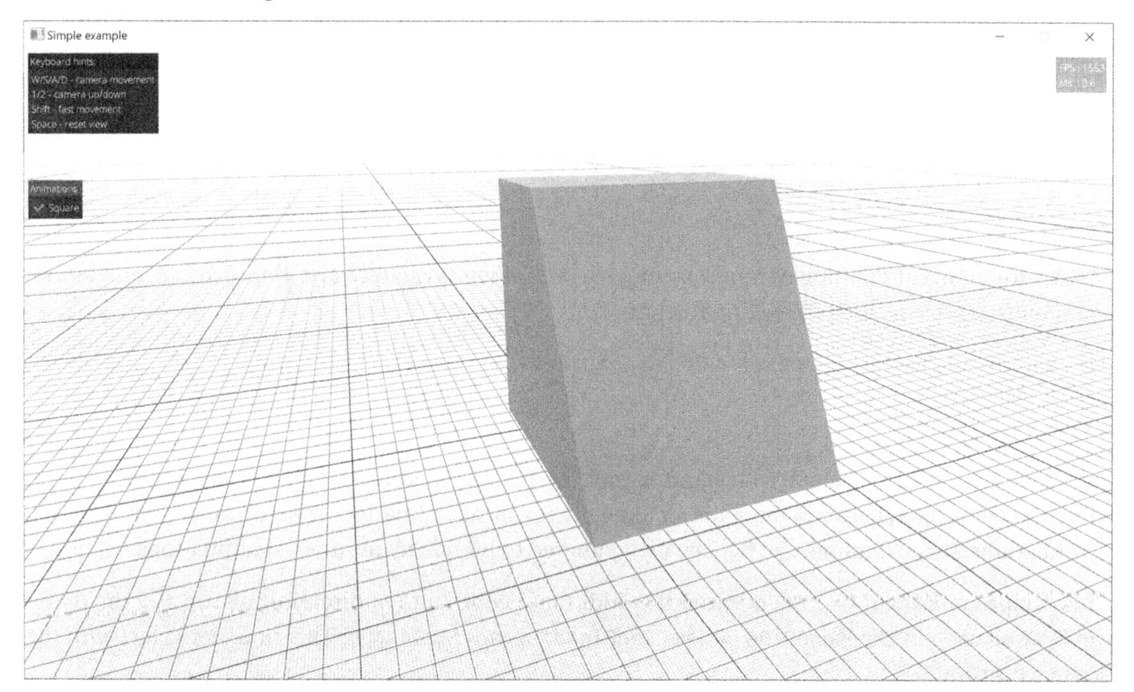

Figure 9.3: Morph targets animation

With this, we conclude our glTF morphing implementation. As you can see from all the code above, there are no differences in the API calls between skinning and morphing animations. In fact, skinning and morphing animations often complement each other, allowing for more dynamic and expressive character movements.

There's more...

It's important to note that at the time this book was written, Assimp used string names to bind animation channels to meshes, which does not fully align with the glTF specification. Additionally, it lacks comprehensive support for loading morph targets from glTF files, making it impossible to animate them here using the Assimp library. This limitation affected how we handled morphing animations in our implementation.

For a deeper understanding of animation concepts, we recommend the book *Computer Animation: Algorithms and Techniques* by Rick Parent. This comprehensive guide covers the principles and methodologies of this dynamic field.

Animation blending

In the recipes *Implementing the glTF animation player* and *Doing skeletal animations in compute shaders*, we implemented an animation player with the capability to switch between skeletal animations. However, when switching between animations, you may have noticed that the transition happens rather abruptly and unnaturally.

We can improve this by interpolating between animations using a floating-point transition factor. In this recipe, we'll demonstrate a more advanced approach that enables smooth blending between two animations. This technique requires not only the interpolation of per-channel transformations but also a gradual blending between different animations, softening transitions and making them look more natural. Let's dive in and learn how to implement it!

Getting ready

Make sure to revisit the recipes *Implementing the glTF animation player* and *Doing skeletal animations in compute shaders* before proceeding further.

The source code for this recipe is located in `Chapter09/04_AnimationBlending`.

How to do it...

The basic idea of this approach is straightforward: we have 2 different animations and a floating-point blending factor, `0...1`, between them to calculate the resulting animation. If we can successfully blend 2 animations, this technique can be trivially extended to blend any number of animations.

This recipe consists of two main parts: the additional code in the animation player and the demo application code in `Chapter09/04_AnimationBlending/src/main.cpp`. This time, let's start with the `main()` function.

1. Here, we include the entire `main()` function, as it is quite small. We enable the `showGLTFInspector` configuration flag for `VulkanApp` to add a simple UI that allows control over the running animations and the blend factor between them. We are using the `Fox.gltf` model from Khronos, which includes multiple skeletal animations that we can blend together.

```
int main() {
  VulkanApp app(
    { .initialCameraPos = vec3(2.13f, 2.44f, -3.5f),
      .initialCameraTarget = vec3(0, 0, 2),
      .showGLTFInspector   = true });
  GLTFContext gltf(app);
  loadGLTF(gltf, "deps/src/glTF-Sample-Assets/Models/Fox/glTF/Fox.gltf",
    "deps/src/glTF-Sample-Assets/Models/Fox/glTF/");
```

2. Translate (`t`) and scale (`s`) the model appropriately for convenience. Then, create two `AnimationState` objects for the two animations we want to play.

```
  const mat4 t = glm::translate(
    mat4(1.0f), vec3(0.0f, 1.1f, 0.0f));
  const mat4 s = glm::scale(
    mat4(1.0f), vec3(0.01f, 0.01f, 0.01f));
  AnimationState anim1 = { .animId      = 1,
                           .currentTime = 0.0f,
                           .playOnce    = false,
                           .active      = true };
  AnimationState anim2 = { .animId      = 2,
                           .currentTime = 0.0f,
                           .playOnce    = false,
                           .active      = true };
```

3. Enable skinning in our glTF player and register both animations in our GLTF inspector, allowing us to switch between them later using the ImGui UI and to manually adjust the blending factor.

```
  gltf.skinning  = true;
  gltf.inspector = { .activeAnim         = {1, 2},
                     .blend              = 0.5f,
                     .showAnimations     = true,
                     .showAnimationBlend = true };
  app.run([&](uint32_t width, uint32_t height,
    float aspectRatio, float deltaSeconds)
  {
    const mat4 m = t * glm::rotate(mat4(1.0f),
      glm::radians(90.0f), vec3(0, 1, 0)) * s;
    const mat4 v = app.camera_.getViewMatrix();
    const mat4 p = glm::perspective(45, aspectRatio, 0.01, 100);
```

4. Call the helper function `animateBlendingGLTF()`, which performs the animation blending and updates both `AnimationState` objects, `anim1` and `anim2`. Use the blending factor set in the inspector UI. Then render our animated glTF mesh normally.

```
animateBlendingGLTF(gltf, anim1, anim2,
  gltf.inspector.blend, deltaSeconds);
renderGLTF(gltf, m, v, p);
```

5. If we switch animations in the inspector, assign the new animations to both `AnimationState` objects.

```
      if (gltf.inspector.activeAnim[0]!=anim1.animId||
          gltf.inspector.activeAnim[1]!=anim2.animId) {
        anim1 = { .animId = gltf.inspector.activeAnim[0],
                  .currentTime = 0.0f,
                  .playOnce = false,
                  .active = gltf.inspector.activeAnim[0] != ~0u };
        anim2 = { .animId = gltf.inspector.activeAnim[1],
                  .currentTime = 0.0f,
                  .playOnce = false,
                  .active = gltf.inspector.activeAnim[1] != ~0u };
      }
    });
    return 0;
}
```

That's all for the demo application C++ code. As you will see, the real magic happens behind the `animateBlendingGLTF()` helper function in `shared/UtilsGLTF.cpp`. Let's take a closer look at how it works.

How it works...

The function `animateBlendingGLTF()` is very similar to `animateGLTF()`, which we introduced earlier, in the recipe *Doing skeletal animations in compute shaders*.

1. We use the same lazy-loaded compute shader and check the states of both provided animations.

```
void animateBlendingGLTF(GLTFContext& gltf,
   AnimationState& anim1,
   AnimationState& anim2,
   float weight, float dt)
{
  if (gltf.transforms.empty()) return;
  if (gltf.pipelineComputeAnimations.empty())
    gltf.pipelineComputeAnimations =
      gltf.app.ctx_->createComputePipeline({
```

```
        .smComp = gltf.animation });
    anim1.active   = anim1.animId != ~0;
    anim2.active   = anim2.animId != ~0;
    gltf.animated = anim1.active || anim2.active;
```

2. We only do animation blending when both animations are assigned and active. If that's not the case, we update only one of the remaining animations or none at all.

```
    if (anim1.active && anim2.active) {
      updateAnimationBlending(gltf, anim1, anim2, weight, dt);
    } else if (anim1.active) {
      updateAnimation(gltf, anim1, dt);
    } else if (anim2.active) {
      updateAnimation(gltf, anim2, dt);
    }
  }
```

3. The function `updateAnimationBlending()` is the blending counterpart to the `updateAnimation()` function, which we introduced earlier, in the recipe *Implementing the glTF animation player*. Both functions are located in `shared/UtilsAnim.cpp`. Based on the combination of active animations, we advance the animation time for both of them or just one, depending on the situation. Remember, we never enter this function if no animations are playing.

```
void updateAnimationBlending(GLTFContext& glTF,
  AnimationState& anim1,
  AnimationState& anim2,
  float weight, float dt)
{
  if (anim1.active && anim2.active) {
    const Animation& activeAnim1 = glTF.animations[anim1.animId];
    anim1.currentTime += activeAnim1.ticksPerSecond * dt;
    if (anim1.playOnce && anim1.currentTime > activeAnim1.duration) {
      anim1.currentTime = activeAnim1.duration;
      anim1.active      = false;
    } else {
      anim1.currentTime = fmodf(
        anim1.currentTime, activeAnim1.duration);
    }
    const Animation& activeAnim2 = glTF.animations[anim2.animId];
    anim2.currentTime += activeAnim2.ticksPerSecond * dt;
    if (anim2.playOnce && anim2.currentTime > activeAnim2.duration) {
      anim2.currentTime = activeAnim2.duration;
      anim2.active      = false;
    } else {
```

```
    anim2.currentTime = fmodf(
      anim2.currentTime, activeAnim2.duration);
  }
```

4. Next, we update the skinning animation states in a way similar to how we did it in the recipe *Implementing the glTF animation player*.

```
std::function<void(GLTFNodeRef gltfNode,
  const mat4& parentTransform)> traverseTree =
    [&](GLTFNodeRef gltfNode,
    const mat4& parentTransform)
{
  const GLTFBone& bone  = glTF.bonesByName[
    glTF.nodesStorage[gltfNode].name];
  const uint32_t boneId = bone.boneId;
  if (boneId != ~0u) {
```

5. The main difference here is that we retrieve animation channels for both animations and blend them using the function `animationTransformBlending()`, based on the provided `weight` and separate animation times for each animation. When one of the animations stops playing, we switch to using the helper function `animationTransform()`, described in the recipe *Implementing the glTF animation player*, which calculates the transformation for just one animation channel.

```
    auto channel1 = activeAnim1.channels.find(boneId);
    auto channel2 = activeAnim2.channels.find(boneId);
    if (channel1 != activeAnim1.channels.end() &&
        channel2 != activeAnim2.channels.end())
    {
      glTF.matrices[glTF.nodesStorage[
        gltfNode].modelMtxId] = parentTransform *
          animationTransformBlending(
            channel1->second, anim1.currentTime,
            channel2->second, anim2.currentTime, weight);
    } else if (channel1 != activeAnim1.channels.end()) {
      glTF.matrices[glTF.nodesStorage[
        gltfNode].modelMtxId] = parentTransform *
          animationTransform(channel1->second, anim1.currentTime);
    } else if (channel2 != activeAnim2.channels.end()) {
      glTF.matrices[glTF.nodesStorage[
        gltfNode].modelMtxId] = parentTransform *
          animationTransform(channel2->second, anim2.currentTime);
    } else {
```

```
          glTF.matrices[glTF.nodesStorage[
            gltfNode].modelMtxId] = parentTransform *
              glTF.nodesStorage[gltfNode].transform;
        }
        glTF.skinning = true;
      }
```

6. We make a recursive call to this same lambda `traverseTree()` to process all child nodes.

```
      for (uint32_t i = 0;
           i < glTF.nodesStorage[gltfNode].children.size(); i++)
      {
        const uint32_t child =
          glTF.nodesStorage[gltfNode].children[i];
        traverseTree(child,
          glTF.matrices[glTF.nodesStorage[
          gltfNode].modelMtxId]);
      }
    }; // the end of the local lambda declaration
```

7. We invoke the lambda function `traverseTree()` from the root node to calculate all the transformations and update the bone matrices.

```
    traverseTree(glTF.root, mat4(1.0f));
    for (const std::pair<std::string, GLTFBone>& b : glTF.bonesByName) {
      if (b.second.boneId != ~0u)
        glTF.matrices[b.second.boneId] =
          glTF.matrices[b.second.boneId] * b.second.transform;
    }
  } else {
    glTF.morphing = false;
    glTF.skinning = false;
  }
}
```

8. Last but not least, we have the `animationTransformBlending()` helper function in `shared/UtilsAnim.cpp`. After all the concepts covered in this chapter, we hope this function is fairly self-explanatory – we retrieve the positions, rotations, and scales for each animation channel, and then use LERP and SLERP mixing operations to combine them based on the blend factor `weight`. For the individual interpolation functions, refer to the recipe *Implementing the glTF animation player*.

```
mat4 animationTransformBlending(
  const AnimationChannel& channel1, float time1,
  const AnimationChannel& channel2, float time2,
```

```
    float weight)
{
  mat4 trans1 = glm::translate(mat4(1.0f),
    interpolatePosition(channel1, time1));
  mat4 trans2 = glm::translate(mat4(1.0f),
    interpolatePosition(channel2, time2));
  mat4 translation = glm::mix(trans1, trans2, weight);
  quat rot1 = interpolateRotation(channel1, time1);
  quat rot2 = interpolateRotation(channel2, time2);
  mat4 rotation = glm::toMat4(
    glm::normalize(glm::slerp(rot1, rot2, weight)));
  vec3 scl1  = interpolateScaling(channel1, time1);
  vec3 scl2  = interpolateScaling(channel2, time2);
  mat4 scale = glm::scale(mat4(1.0f), glm::mix(scl1, scl2, weight));
  return translation * rotation * scale;
}
```

And that's the complete implementation for the animation blending feature in our glTF animation player. You can now run the demo application `Chapter09/04_AnimationBlending` and see an animated fox, similar to the one shown in the following screenshot.

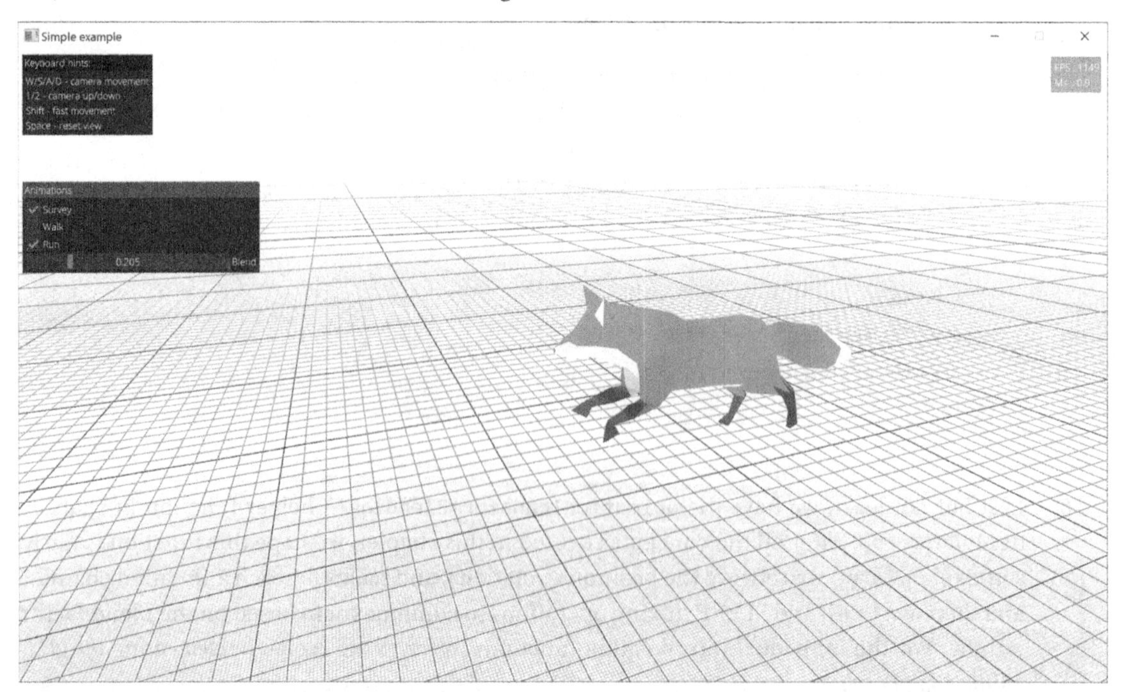

Figure 9.4: Animation blending

Feel free to experiment with the demo. For example, try selecting both the `Survey` and `Run` animations, then adjust the blending factor so that the fox slowly turns its head while running. Moving the `Blend` slider slowly from `0` to `1` will give you a smooth transition between surveying and running.

The main opportunity here, of course, is not just manual adjustment but the ability to automatically create smooth transitions between different animations. In a 3D game, you could have a state machine (commonly called a **blend tree**) to switch animations based on high-level game logic. This state machine would rely on the animation blending mechanism to ensure seamless, automatic transitions between animations.

With this chapter, we conclude our brief exploration of the glTF file format and its capabilities. Let's now shift gears and go back to rendering, so we can learn more about image-based rendering techniques and post-processing effects.

There's more...

To round out our exploration of the glTF file format, we've added two more code examples. The first, `Chapter09/05_ImportLights`, demonstrates how to import lights from a glTF file. The second, `Chapter09/06_ImportCameras`, shows how to import cameras. Feel free to explore them on your own!

The `06_ImportCameras` app is particularly interesting because it showcases how to import different camera definitions from a glTF file and switch between them within the app. Here's a screenshot:

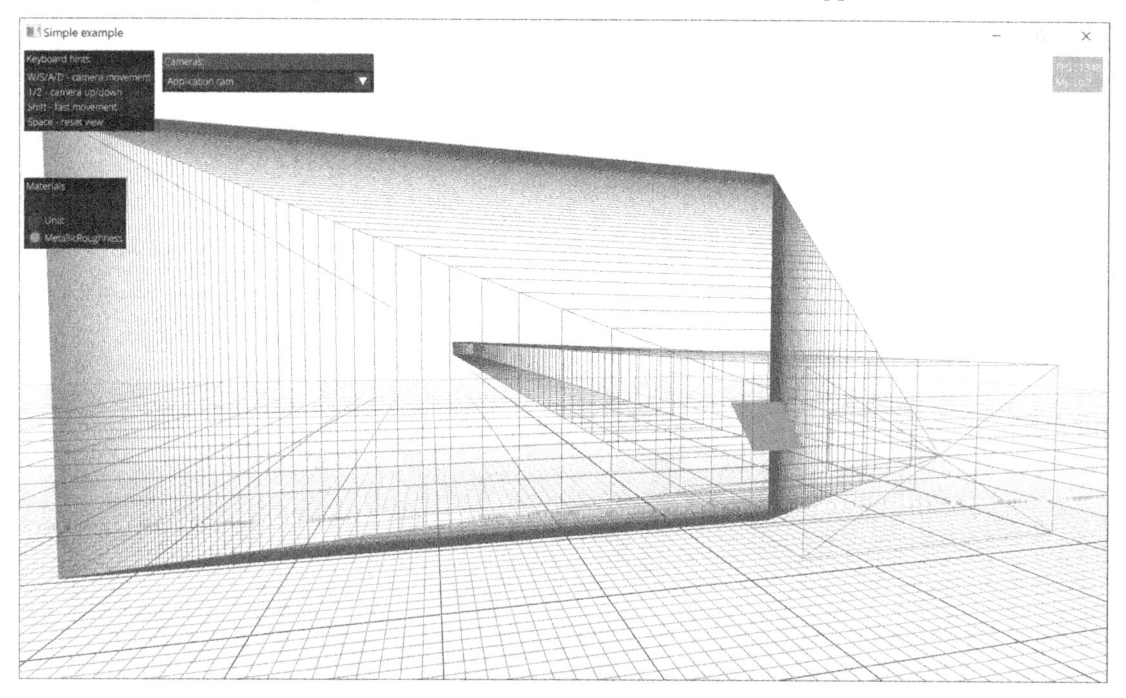

Figure 9.5: Switching between different glTF cameras

10

Image-Based Techniques

In this chapter, we will learn how to build a post-processing pipeline and implement various image-based effects in Vulkan using *LightweightVK*, and how to integrate them with our scene rendering code. We will examine *LightweightVK*'s implementation and review the underlying Vulkan code. Many of these techniques (such as ambient occlusion, **high dynamic range** (**HDR**) rendering and tone mapping, light adaptation, and temporal anti-aliasing) are important pieces of any modern post-processing pipeline. The idea is to render the scene to an offscreen image and then apply these effects, which is why they are referred to as "image-based" techniques.

In addition to these techniques, projective shadow mapping relies on somewhat similar offscreen rendering machinery. We will implement a very basic projective shadow mapping algorithm as one of our first examples, and then revisit the topic of shadow mapping in the next chapter with a more advanced approach.

This chapter covers the post-processing pipeline and has the following recipes:

- Implementing offscreen rendering in Vulkan
- Implementing full-screen triangle rendering
- Implementing shadow maps
- Implementing MSAA in Vulkan
- Implementing screen space ambient occlusion
- Implementing HDR rendering and tone mapping
- Implementing HDR light adaptation

Technical requirements

To run the recipes from this chapter, you need a computer with a video card and drivers supporting Vulkan 1.3. Read *Chapter 1*, if you want to learn how to build and run demo applications from this book.

This chapter relies on the scene and geometry loading code explained in *Chapter 8*, so make sure you read it before proceeding any further.

Implementing offscreen rendering in Vulkan

Before we dive into general post-processing effects, let's learn how to implement some foundational Vulkan code for offscreen rendering. We will use this functionality throughout the rest of the book to implement various rendering and post-processing techniques.

In this recipe, we will learn how to render directly into specific mip levels of an image and access each mip level individually. As a demonstration, we will render a textured icosahedron where each mip level of the texture is painted in a different color, allowing us to clearly see how the GPU handles mip level transitions.

Getting ready

It is recommended to review the recipe *Using texture data in Vulkan* from *Chapter 3*.

The demo app for this recipe is located in `Chapter10/01_OffscreenRendering`.

How to do it...

Let's start by examining the high-level code of our example app in `Chapter10/01_OffscreenRendering/src/main.cpp`.

1. First, we create our standard `VulkanApp` application and vertex and index buffers for our icosahedron. Here, `t` represents the golden ratio, as described at `https://en.wikipedia.org/wiki/Regular_icosahedron`. The following code to generate an icosahedron mesh might come in handy.

```
struct VertexData { float pos[3]; };
const float t = (1.0f + sqrtf(5.0f)) / 2.0f;
const VertexData vertices[] = {
  {-1, t, 0}, {1, t, 0}, {-1, -t, 0}, { 1, -t,  0},
  { 0,-1, t}, {0, 1, t}, { 0, -1,-t}, { 0,  1, -t},
  { t, 0,-1}, {t, 0, 1}, {-t,  0,-1}, {-t,  0,  1} };
const uint16_t indices[] = { 0, 11, 5, 0, 5, 1, 0, 1,
  7, 0, 7, 10, 0, 10, 11, 1, 5, 9, 5, 11, 4,  11, 10,
  2, 10, 7, 6, 7, 1, 8, 3, 9,  4, 3, 4, 2, 3, 2, 6,
  3, 6, 8,  3, 8,  9,  4, 9, 5, 2, 4,  11, 6,  2, 10,
  8, 6, 7, 9, 8, 1 };
int main() {
  VulkanApp app({ … });
  std::unique_ptr<lvk::IContext> ctx(app.ctx_.get());
  Holder<BufferHandle> bufferIndices = ctx->createBuffer({
      .usage     = lvk::BufferUsageBits_Index,
      .storage   = lvk::StorageType_Device,
      .size      = sizeof(indices),
      .data      = indices });
  Holder<BufferHandle> bufferVertices = ctx->createBuffer({
```

```
    .usage      = lvk::BufferUsageBits_Vertex,
    .storage    = lvk::StorageType_Device,
    .size       = sizeof(vertices),
    .data       = vertices });
```

2. Here are the vertex and fragment shaders, along with a graphics pipeline to render the mesh.

```
Holder<ShaderModuleHandle> vert =
  loadShaderModule(ctx, "Chapter10/01_OffscreenRendering/src/main.vert");
Holder<ShaderModuleHandle> frag =
  loadShaderModule(ctx, "Chapter10/01_OffscreenRendering/src/main.frag");
const lvk::VertexInput vdesc = {
  .attributes = {{.location = 0,
                  .format = VertexFormat::Float3 }},
  .inputBindings = {{.stride = sizeof(VertexData)}},
};
Holder<lvk::RenderPipelineHandle> pipeline =
  ctx->createRenderPipeline({
    .vertexInput = vdesc,
    .smVert      = vert,
    .smFrag      = frag,
    .color       = { { .format = ctx->getSwapchainFormat() } },
    .depthFormat = app.getDepthFormat(),
});
```

3. Let's create a 512x512 texture with a full mipmap pyramid.

```
constexpr uint8_t numMipLevels = lvk::calcNumMipLevels(512, 512);
Holder<TextureHandle> texture = ctx->createTexture({
    .type        = lvk::TextureType_2D,
    .format      = lvk::Format_RGBA_UN8,
    .dimensions  = {512, 512},
    .usage       = lvk::TextureUsageBits_Attachment |
                   lvk::TextureUsageBits_Sampled,
    .numMipLevels = numMipLevels });
```

4. Now, let's create a set of texture views, each targeting individual mip levels of this texture. As we learned in the recipe *Using texture data in Vulkan* from *Chapter 3*, a `TextureHandle` in LightweightVK is simply a wrapper around `VkImage` and `VkImageView`, along with additional metadata. The `IContext::createTextureView()` function creates a new `VkImageView` object while keeping the original `VkImage` from the initial texture. We will take a closer look at its implementation in the *How it works...* section.

```
Holder<TextureHandle> mipViews[numMipLevels];
for (uint32_t mip = 0; mip != numMipLevels; mip++)
  mipViews[mip] = ctx->createTextureView(texture, { .mipLevel = mip });
```

This allows the newly created texture views to function seamlessly in any texture mapping process as if they were complete textures.

5. Now, let's fill each of these individual mip levels with data. For simplicity, we will clear each of the 10 mip levels with a distinct color from the following table:

```
const vec3 colors[10] = {
  {1, 0, 0}, {0, 1, 0}, {0, 0, 1},
  {1, 1, 0}, {0, 1, 1}, {1, 0, 1},
  {1, 0, 0}, {0, 1, 0}, {0, 0, 1}, {0, 0, 0} };
```

6. We can use offscreen rendering to achieve this. Since we are not rendering any geometry, all we need to do is specify rendering for a specific mip level with `LoadOp_Clear` and provide a color value from the table.

```
ICommandBuffer& buf = ctx->acquireCommandBuffer();
for (uint8_t i = 0; i != numMipLevels; i++) {
  buf.cmdBeginRendering(RenderPass{
    .color = { {.loadOp = lvk::LoadOp_Clear,
                .level = i,
                .clearColor = {colors[i].r, …, 1}}} },
    lvk::Framebuffer{.color = { { .texture = texture }} });
  buf.cmdEndRendering();
}
ctx->submit(buf);
```

Since the colors are essentially static, we only need to do this once, before entering the main application loop.

7. The main application loop is quite standard. We rotate the icosahedron model and render it to the current swapchain image.

```
app.run([&](uint32_t width, uint32_t height,
  float aspectRatio, float deltaSeconds) {
  const mat4 view = app.camera_.getViewMatrix();
  const mat4 proj = glm::perspective(…);
  const mat4 model = glm::rotate(…);
  const lvk::Framebuffer framebuffer = {
    .color = {{ .texture = ctx->getCurrentSwapchainTexture() }},
    .depthStencil ={.texture=app.getDepthTexture()},
  };
  ICommandBuffer& buf = ctx->acquireCommandBuffer();
  buf.cmdBindVertexBuffer(0, bufferVertices);
  buf.cmdBindIndexBuffer(bufferIndices, lvk::IndexFormat_UI16);
```

8. We don't need any additional Vulkan image layout transitions or memory barriers here because our command buffer submissions are strictly ordered, as explained in *Chapter 2*. Moreover, the render-to-texture code in `cmdEndRendering()` (step 6) has already transitioned the texture to `VK_IMAGE_LAYOUT_SHADER_READ_ONLY_OPTIMAL` after rendering.

```
  buf.cmdBeginRendering( lvk::RenderPass{
      .color = {{.loadOp = lvk::LoadOp_Clear,
                 .clearColor = {1.0f, 1.0f, 1.0f, 1.0f}}},
      .depth = { .loadOp = lvk::LoadOp_Clear,
                 .clearDepth = 1.0f } },
    framebuffer );
  buf.cmdBindRenderPipeline(pipeline);
  buf.cmdBindDepthState(
    { .compareOp = lvk::CompareOp_Less,
      .isDepthWriteEnabled = true });
```

9. We specify the texture that we populated earlier with the custom mip pyramid of colored mip levels and draw the mesh.

```
  const struct PushConstants {
    mat4 mvp;
    uint32_t texture;
  } pc = {
    .mvp     = proj * view * model,
    .texture = texture.index(),
  };
```

```
    buf.cmdPushConstants(pc);
    buf.cmdDrawIndexed(LVK_ARRAY_NUM_ELEMENTS(indices));
```

10. Next, we have our standard grid rendering along with UI elements, such as the FPS counter and the memo. The most interesting part here will be the ImGui UI. Here, we display the entire contents of the mip pyramid within the ImGui UI. We use the texture views we created to access the individual mip levels of the parent texture object.

```
    // … skipped drawing the grid, FPS, memo…
    app.imgui_->beginFrame(framebuffer);
    const ImGuiViewport* v = ImGui::GetMainViewport();
    ImGui::Begin("Control", nullptr, ImGuiWindowFlags_AlwaysAutoResize);
    ImGui::Text("Mip-pyramid 512x512");
    const float windowWidth = v->WorkSize.x / 5;
    for (uint32_t l = 0; l != numMipLevels; l++)
      ImGui::Image(mipViews[l].index(),
        ImVec2((int)windowWidth >> l,
               ((int)windowWidth >> l)));
    ImGui::End();
    app.imgui_->endFrame(buf);
    buf.cmdEndRendering();
    ctx->submit(buf, ctx->getCurrentSwapchainTexture());
  });
  // …
}
```

We conclude the main loop by submitting our command buffer and presenting the swapchain image.

11. Here is the GLSL vertex shader for our rendering. The key part here is that we use ad hoc cylindrical mapping to calculate the uv coordinates from the vertex positions.

```
layout (location = 0) in  vec3 pos;
layout (location = 0) out vec2 uv;
layout(push_constant) uniform PushConstants {
  mat4 mvp;
};
#define PI 3.1415926
float atan2(float y, float x) {
  return x == 0.0 ? sign(y) * PI/2 : atan(y, x);
}
void main() {
  gl_Position = mvp * vec4(pos, 1.0);
  float theta = atan2(pos.y, pos.x) / PI + 0.5;
```

```
    uv = vec2(theta, pos.z);
  }
```

12. And here is the fragment shader, which is trivial. Note that the only color information applied here comes from the texture.

```
layout (location=0) in vec2 uv;
layout (location=0) out vec4 out_FragColor;
layout(push_constant) uniform PushConstants {
  mat4 mvp;
  uint texture;
};
void main() {
  out_FragColor = textureBindless2D(texture, 0, uv);
}
```

The running demo application should display a colorful, rotating icosahedron, as shown in the following screenshot.

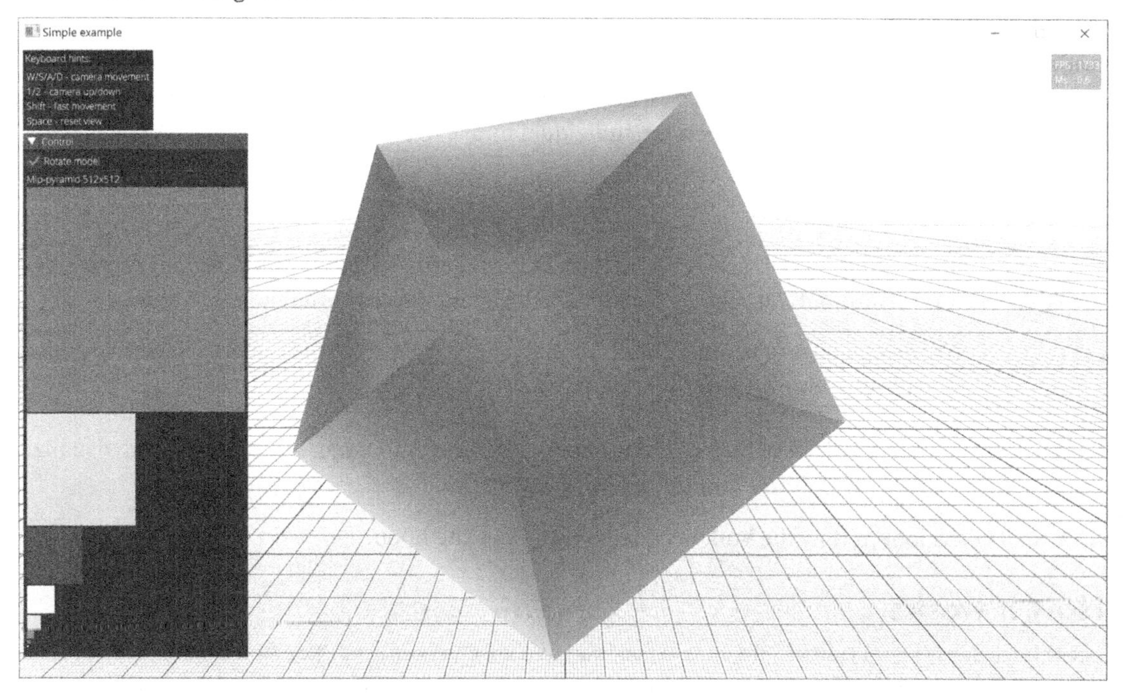

Figure 10.1: A textured, rotating icosahedron

Quick tip: Need to see a high-resolution version of this image? Open this book in the next-gen Packt Reader or view it in the PDF/ePub copy.

The next-gen Packt Reader and a **free PDF/ePub copy** of this book are included with your purchase. Unlock them by scanning the QR code below or visiting `https://www.packtpub.com/unlock/9781803248110`.

While this rendering may seem extremely simple, and it is, let's take a moment to step back and understand what's happening here, as it is well worth it.

The color information comes from a single texture. Based on the distance from the camera and the angle of each triangle to the screen plane, we sample from different mip levels of that texture. Since our texture has a completely custom, and in some sense, unusual, mip pyramid, with each mip level containing independent color information, we can observe a distinctive color pattern on the mesh. Notice the smooth transitions between the different colors, which occur due to trilinear filtering, where the GPU linearly blends between the individual mip levels.

Note

Back in the days of early GPUs, in the late 1990s and early 2000s, this technique was used to visualize and compare the quality of trilinear filtering across different GPUs and drivers.

Try using a custom sampler with `lvk::SamplerMip_Nearest` to disable trilinear filtering and observe a completely different visual pattern.

For us, this approach will serve as a building block to render into offscreen textures and visualize intermediate data. We will utilize it in the subsequent recipes.

Now we'll take a look at how the Vulkan code works under the hood.

How it works...

Let's check `deps/src/lightweightvk/lvk/vulkan/VulkanClasses.cpp`. The function of interest is `VulkanContext::createTextureView()`, which is responsible for key work with texture views.

1. It accepts a texture handle and a `TextureViewDesc` object, which specifies the parameters of a view we want to create. In our example, we only used the `mipLevel` field to extract a single mip level.

```
struct TextureViewDesc {
  TextureType type = TextureType_2D;
  uint32_t layer = 0;
```

```
  uint32_t numLayers = 1;
  uint32_t mipLevel = 0;
  uint32_t numMipLevels = 1;
  ComponentMapping swizzle = {};
};
Holder<TextureHandle> VulkanContext::createTextureView(
    TextureHandle texture,
    const TextureViewDesc& desc,
    const char* debugName,
    Result* outResult) {
```

2. The function creates a copy of the `VulkanImage` wrapper object from the existing texture and marks it as non-owning. As a result, the texture view remains valid only as long as the original texture is still in place. Check the recipe *Using texture data in Vulkan* from *Chapter 3* to recap how Vulkan objects are stored in pools in LightweightVK. In our copy of `VulkanImage`, we erase all previously created `VkImageView` objects because they belong to the parent texture.

```
  VulkanImage image = *texturesPool_.get(texture);
  image.isOwningVkImage_ = false;
  memset(&image.imageViewStorage_, 0, sizeof(image.imageViewStorage_));
  memset(&image.imageViewForFramebuffer_, 0,
    sizeof(image.imageViewForFramebuffer_));
```

3. Let's choose `VkImageAspectFlags` for the new image view based on the image format. This can be done manually by providing the necessary aspect mask as part of our `TextureViewDesc`. However, this is how it is currently handled in *LightweightVK* for simplicity.

```
  VkImageAspectFlags aspect = 0;
  if (image.isDepthFormat_||image.isStencilFormat_) {
    if (image.isDepthFormat_)
      aspect |= VK_IMAGE_ASPECT_DEPTH_BIT;
    else if (image.isStencilFormat_)
      aspect |= VK_IMAGE_ASPECT_STENCIL_BIT;
  } else {
    aspect = VK_IMAGE_ASPECT_COLOR_BIT;
  }
```

4. Let's select the `VkImageViewType` based on the provided `TextureViewDesc::type` value.

```
  VkImageViewType vkImageViewType = VK_IMAGE_VIEW_TYPE_MAX_ENUM;
  switch (desc.type) {
  case TextureType_2D:
    vkImageViewType = desc.numLayers > 1 ?
                          VK_IMAGE_VIEW_TYPE_2D_ARRAY :
                          VK_IMAGE_VIEW_TYPE_2D;
```

```
    break;
  case TextureType_3D:
    vkImageViewType = VK_IMAGE_VIEW_TYPE_3D;
    break;
  case TextureType_Cube:
    vkImageViewType = desc.numLayers > 1 ?
                          VK_IMAGE_VIEW_TYPE_CUBE_ARRAY :
                          VK_IMAGE_VIEW_TYPE_CUBE;
    break;
  }
```

5. One useful feature of Vulkan image views is the ability to swizzle the components of a texture before returning the samples to the shader. For example, it allows loading a 1-channel R texture as a 4-channel RGBA texture without reordering the underlying data. The value of the R channel can be broadcasted into all three RGB channels, rendering it white instead of red, while the A channel can be explicitly set to 1. This will be helpful later in this chapter, in the recipe *Implementing HDR rendering and tone mapping*, to visualize intermediate buffers with ImGui.

```
  const VkComponentMapping mapping = {
    .r = VkComponentSwizzle(desc.swizzle.r),
    .g = VkComponentSwizzle(desc.swizzle.g),
    .b = VkComponentSwizzle(desc.swizzle.b),
    .a = VkComponentSwizzle(desc.swizzle.a),
  };
```

6. Let's create the actual `VkImageView` object. The function `VulkanImage::createImageView()` was explained in the recipe *Using texture data in Vulkan* from *Chapter 3*.

```
  image.imageView_ = image.createImageView(vkDevice_,
    vkImageViewType, image.vkImageFormat_, aspect,
    desc.mipLevel, desc.numMipLevels,
    desc.layer, desc.numLayers,
    mapping, nullptr, debugName);
```

7. One important extra step: If the original `VkImage` object was created with the flag `VK_IMAGE_USAGE_STORAGE_BIT` and we want it to be accessible as a storage image from a compute shader, the swizzle must be identity. Therefore, we need another `VkImageView` object to be packed into the descriptor set used for storage images. Let's create it here. While this might seem like an ad hoc decision, it is a trade-off we make to simplify the high-level LightweightVK interface.

```
  if (image.vkUsageFlags_& VK_IMAGE_USAGE_STORAGE_BIT) {
    if (!desc.swizzle.identity()) {
      image.imageViewStorage_ = image.createImageView(
        vkDevice_, vkImageViewType,
        image.vkImageFormat_, aspect, 0,
```

```
          VK_REMAINING_MIP_LEVELS, 0, desc.numLayers,
          {}, nullptr, debugName);
    }
  }
  TextureHandle handle = texturesPool_.create(std::move(image));
  awaitingCreation_ = true;
  return {this, handle};
}
```

8. As for offscreen rendering, there's another set of `VkImageView` objects maintained inside `VulkanImage`, as was explained in the recipe *Using texture data in Vulkan* from *Chapter 3*. These are lazily created from the member function `CommandBuffer::cmdBeginRendering()` by calling `VulkanImage::getOrCreateVkImageViewForFramebuffer()`. One `VkImageView` object is required for each mip level or layer that we intend to render into. Here's the code:

```
VkImageView VulkanImage::getOrCreateVkImageViewForFramebuffer(
    VulkanContext& ctx, uint8_t level, uint16_t layer)
{
  if (imageViewForFramebuffer_[level][layer])
    return imageViewForFramebuffer_[level][layer];
  imageViewForFramebuffer_[level][layer] = createImageView(
    ctx.getVkDevice(), VK_IMAGE_VIEW_TYPE_2D, vkImageFormat_,
      getImageAspectFlags(), level, 1u, layer, 1u);
  return imageViewForFramebuffer_[level][layer];
}
```

9. Here's a quick look at the relevant parts of the `VulkanImage` struct for your convenience. The full declaration was explained in the recipe *Using texture data in Vulkan* from *Chapter 3*. In this struct, `imageView_` is the view we created, `imageViewStorage_` is the image view with the identity swizzle required for storage images, and `imageViewForFramebuffer_` is a lazily allocated collection of image views for each level and layer we want to render into. We limit the number of framebuffer layers to `6` for rendering into cube maps —another tradeoff made for the sake of simplifying the high-level API:

```
struct VulkanImage final {
  // ...
  VkImageView imageView_ = VK_NULL_HANDLE;
  VkImageView imageViewStorage_ = VK_NULL_HANDLE;
  VkImageView imageViewForFramebuffer_[LVK_MAX_MIP_LEVELS][6]= {};
};
```

This offscreen rendering functionality serves as a building block for all our subsequent offscreen rendering Vulkan demos. Before we apply it, however, let's first learn how to render full-screen quads in the next recipe.

Implementing full-screen triangle rendering

In most of the post-processing techniques covered in this chapter, you'll need to render a textured rectangle that spans the entire screen and apply a specific fragment shader to it. To make this easier, we can use a single vertex shader that works across all of these cases. While it is straightforward to render a quad using the traditional vertex and index buffers approach, managing this can become complex when you need to handle many shader combinations across different stages of the rendering pipeline. In this recipe, we will demonstrate how to render a full-screen quad by generating geometry directly in a vertex shader. Additionally, instead of using quad geometry, we will use a single triangle covering the entire screen. Let's explore how this is done.

Getting ready

Check out the recipe *Implementing programmable vertex pulling* from *Chapter 5*.

How to do it...

Let's go through the code for our full-screen triangle vertex shader. The GLSL shader is located in the `data/shaders/Quad.vert` file and is as follows:

1. This shader has no inputs except the built-in GLSL variable `gl_VertexIndex`, which ranges from `0` to `2` for a triangle. Instead of calculating vertex positions in world space, we define them directly in homogeneous clip-space coordinates, which Vulkan uses. These clip-space coordinates range from `-1.0` to `+1.0`, covering the entire screen. Simple arithmetic with `gl_VertexIndex` allows us to achieve this. Our `uv` texture coordinates range from `0.0` to `1.0`, from top to bottom and left to right.

```
#version 460
layout (location=0) out vec2 uv;
void main() {
  uv = vec2((gl_VertexIndex << 1) & 2, gl_VertexIndex & 2 );
  gl_Position = vec4(uv * 2.0 + -1.0, 0.0, 1.0);
}
```

2. There is an alternate version of this vertex shader in `data/shaders/QuadFlip.vert` that generates vertically flipped `uv` coordinates, as shown below:

```
#version 460
layout (location=0) out vec2 uv;
void main() {
  uv = vec2((gl_VertexIndex << 1) & 2, gl_VertexIndex & 2 );
  gl_Position = vec4(uv * vec2(2, -2) + vec2(-1, 1), 0.0, 1.0);
}
```

3. Although rendering a full-screen quad is straightforward and useful for learning purposes, using a full-screen triangle can often be faster in many real-world scenarios. Here's how a triangle can cover the entire screen: the top-right and bottom-left portions are clipped, leaving only the rectangular area visible.

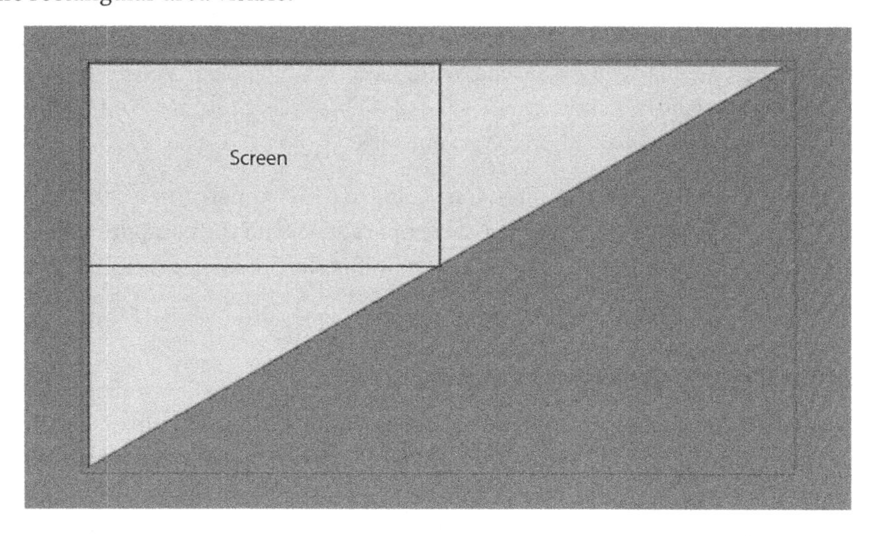

Figure 10.2: A full-screen triangle

4. Here's how to set up a rendering pipeline using one of these vertex shaders. This code snippet is from `Chapter10/04_SSAO/src/main.cpp` in the recipe *Implementing screen space ambient occlusion*.

```
Holder<ShaderModuleHandle> vertCombine =
  loadShaderModule(ctx, "data/shaders/QuadFlip.vert");
Holder<ShaderModuleHandle> fragCombine =
  loadShaderModule(ctx, "Chapter10/04_SSAO/src/combine.frag");
HolderRenderPipelineHandle> pipelineCombine =
  ctx->createRenderPipeline({
    .smVert = vertCombine,
    .smFrag = fragCombine,
    .color  = { { .format = ctx->getSwapchainFormat()}} });
```

5. To render our full-screen "quad" (actually a triangle), we can use the following commands, as shown in `Chapter10/04_SSAO/src/main.cpp`. The content of the push constants depends entirely on the fragment shader in use, as they are not needed for our vertex shaders.

```
buf.cmdBindRenderPipeline(pipelineCombine);
buf.cmdPushConstants(pcCombine);
buf.cmdBindDepthState({});
buf.cmdDraw(3);
```

In the upcoming recipes, we'll learn how to pair this vertex shader with various fragment shaders to create different post-processing effects. Now, let's move on to another "basic" image-based technique: projective shadows.

There's more...

This technique was popularized by Sascha Willems in his blog post, *Vulkan tutorial on rendering a fullscreen quad without buffers*: `https://www.saschawillems.de/blog/2016/08/13/vulkan-tutorial-on-rendering-a-fullscreen-quad-without-buffers`.

An alternative to the full-screen quad rendering approach is to perform a full image compute pass. However, this method requires image load-store operations, and the compute shader must be aware of the output image format. We explored this technique in the recipe *Generating textures in Vulkan using compute shaders* in *Chapter 5*.

Implementing shadow maps

As we've seen in previous chapters, we can render complex scenes with diverse materials, including physically based rendering (PBR) materials. While these techniques can create visually appealing images, the realism of our scenes was severely lacking. Adding shadows can significantly enhance the depth and authenticity of a scene, making it feel more grounded and believable. Shadow mapping, in particular, is a foundational technique for achieving realistic lighting and shading effects. In this recipe, we will go through the basics of projective shadow mapping techniques.

Getting ready

The demo application for this recipe, located in `Chapter10/02_ShadowMapping`, implements all the essential steps for setting up a projective shadow mapping rendering pipeline. It is recommended to briefly review the code before reading this recipe.

Be sure to review the recipe *Implementing offscreen rendering in Vulkan* to refresh your understanding of rendering to offscreen images, as this technique is essential for the shadow mapping process.

How to do it...

The idea behind projective shadow mapping is fairly straightforward: the scene is rendered from the light's point of view. Objects closest to the light are illuminated, while everything else falls into shadow. To identify these closest objects, a depth buffer is used, storing the distance of each visible surface from the light. This requires rendering the scene into an offscreen depth framebuffer.

The rendering process has three main phases: calculating the light's projection and view matrices; rendering the entire scene from the light's perspective into an offscreen depth shadow map; and finally, rendering the scene again from the camera's view, using the depth shadow map to apply shadows.

Let's walk through the code in `Chapter10/02_ShadowMapping/src/main.cpp` to see how each step of this process is implemented:

1. First, let's set up our `VulkanApp` and load some mesh data. We will need at least two objects: one to cast the shadow and a ground or surface to receive the shadow.

```
int main() {
  VulkanApp app({ … });
  LineCanvas3D canvas3d;
  std::unique_ptr<lvk::IContext> ctx(app.ctx_.get());
  struct VertexData {
    vec3 pos;
    vec3 n;
    vec2 tc;
  };
```

2. The first object is a duck model, loaded from a `.gltf` file.

```
  std::vector<VertexData> vertices;
  std::vector<uint32_t> indices;
  const aiScene* scene = aiImportFile(
    "data/rubber_duck/scene.gltf", aiProcess_Triangulate);
  const aiMesh* mesh = scene->mMeshes[0];
  for (uint32_t i = 0; i != mesh->mNumVertices; i++) {
    // … Load vertices into `vertices` …
  }
  for (uint32_t i = 0; i != mesh->mNumFaces; i++) {
    for (uint32_t j = 0; j != 3; j++)
      indices.push_back(mesh->mFaces[i].mIndices[j]);
  }
  aiReleaseImport(scene);
```

3. The second object is a simple plane that will receive the shadow. This plane is created directly from vertices. We'll store both models, the duck and the plane, in the same vertex and index buffers. So, we need to keep track of how many indices the duck mesh has and the offset where the plane mesh begins. Our quad plane is defined by 2 triangles, 4 vertices, and 6 indices. We use the `mergeVectors()` function, introduced in the recipe *Rendering large scenes* from *Chapter 5* to combine `std::vector` containers.

```
  const uint32_t duckNumIndices = indices.size();
  const uint32_t planeVertexOffset = vertices.size();
  mergeVectors( indices, { 0, 1, 2, 2, 3, 0 } );
  mergeVectors( vertices, {
    {vec3(-4, -4, 0), vec3(0, 0, 1), vec2(0, 0)},
    {vec3(-4, +4, 0), vec3(0, 0, 1), vec2(0, 1)},
    {vec3(+4, +4, 0), vec3(0, 0, 1), vec2(1, 1)},
    {vec3(+4, -4, 0), vec3(0, 0, 1), vec2(1, 0)} } );
```

4. This vertex and index data is immutable, meaning we can upload it into the GPU buffers and then forget about it for a while.

```
Holder<BufferHandle> bufferIndices = ctx->createBuffer({
    .usage     = lvk::BufferUsageBits_Index,
    .storage   = lvk::StorageType_Device,
    .size      = sizeof(uint32_t) * indices.size(),
    .data      = indices.data() });
Holder<BufferHandle> bufferVertices = ctx->createBuffer({
    .usage     = lvk::BufferUsageBits_Vertex,
    .storage   = lvk::StorageType_Device,
    .size      = sizeof(VertexData) * vertices.size(),
    .data      = vertices.data() });
```

5. Next, let's load the textures for our 3D models.

```
Holder<TextureHandle> duckTexture = loadTexture(ctx,
  "data/rubber_duck/textures/Duck_baseColor.png");
Holder<TextureHandle> planeTexture = loadTexture(ctx, "data/wood.jpg");
```

6. We need some per-frame data for our shaders. The `view` and `proj` matrices are the standard camera ones we've already used. The `light` matrix will store the product of the light's view and projection matrices, multiplied by the scale-bias matrix. We will explain how this is calculated shortly. For now, one light matrix is enough since we are using only a single light in this recipe. The `lightAngles` field stores the cosines of our spotlight's inner and outer angles, which we'll discuss shortly. `lightPos` is a shortcut for the light's position in world space. The shadow texture and sampler fields will be used in our GLSL fragment shader. Let's allocate a GPU buffer to store this structure. We will populate the buffer later, inside the rendering loop.

```
struct PerFrameData {
  mat4 view;
  mat4 proj;
  mat4 light;
  vec4 lightAngles; // cos(inner), cos(outer)
  vec4 lightPos;
  uint32_t shadowTexture;
  uint32_t shadowSampler;
};
Holder<BufferHandle> bufferPerFrame = ctx->createBuffer({
    .usage     = lvk::BufferUsageBits_Uniform,
    .storage   = lvk::StorageType_Device,
    .size      = sizeof(PerFrameData) });
```

7. Now, let's create our 16-bit depth shadow map texture and a sampler. This sampler is slightly different from the one we used earlier. It enables the Vulkan depth compare operation, where the depth comparison is applied to the fetched texture data before filtering. As a result, the depth texture lookup will return `1.0` if the comparison is `true`, and `0.0` if it is `false`. This is what we will need in our fragment shader to apply shadows. We will revisit this when we discuss the GLSL shaders in a few moments.

```
Holder<TextureHandle> shadowMap = ctx->createTexture({
    .type       = lvk::TextureType_2D,
    .format     = lvk::Format_Z_UN16,
    .dimensions = {1024, 1024},
    .usage      = lvk::TextureUsageBits_Attachment |
                  lvk::TextureUsageBits_Sampled });
Holder<SamplerHandle> samplerShadow = ctx->createSampler({
    .wrapU               = lvk::SamplerWrap_Clamp,
    .wrapV               = lvk::SamplerWrap_Clamp,
    .depthCompareOp      = lvk::CompareOp_LessEqual,
    .depthCompareEnabled = true });
```

8. Now, we can create a graphics pipeline for our main render pass. Let's load the shaders and set up the pipeline.

```
Holder<ShaderModuleHandle> vert =
  loadShaderModule(ctx, "Chapter10/02_ShadowMapping/src/main.vert");
Holder<ShaderModuleHandle> frag =
  loadShaderModule(ctx, "Chapter10/02_ShadowMapping/src/main.frag");
```

9. The `VertexInput` format corresponds to the `VertexData` structure declared earlier in step *1*.

```
const lvk::VertexInput vdesc = {
  .attributes    =
    {{ .location = 0,
       .format = lvk::VertexFormat::Float3,
       .offset = offsetof(VertexData, pos) },
     { .location = 1,
       .format = lvk::VertexFormat::Float3,
       .offset = offsetof(VertexData, n) },
     { .location = 2,
       .format = lvk::VertexFormat::Float2,
       .offset = offsetof(VertexData, tc) }, },
  .inputBindings = {{.stride = sizeof(VertexData)}},
};
Holder<RenderPipelineHandle> pipeline =
  ctx->createRenderPipeline({
```

```
    .vertexInput = vdesc,
    .smVert      = vert,
    .smFrag      = frag,
    .color       = { { .format = ctx->getSwapchainFormat() } },
    .depthFormat = app.getDepthFormat() });
```

10. A separate graphics pipeline with its own set of shaders is required to render the scene into the shadow map. The `vertexInput` field here contains only `vec3` positions, as this is the only data our offscreen rendering vertex shader requires.

```
Holder<ShaderModuleHandle> vertShadow =
  loadShaderModule(ctx, "Chapter10/02_ShadowMapping/src/shadow.vert");
Holder<ShaderModuleHandle> fragShadow =
  loadShaderModule(ctx, "Chapter10/02_ShadowMapping/src/shadow.frag");
Holder<RenderPipelineHandle> pipelineShadow =
  ctx->createRenderPipeline({
    .vertexInput = lvk::VertexInput{
       .attributes = {
         { .location = 0,
           .format = lvk::VertexFormat::Float3,
           .offset = offsetof(VertexData, pos) } },
       .inputBindings = { { .stride = sizeof(VertexData) } } },
    .smVert      = vertShadow,
    .smFrag      = fragShadow,
    .depthFormat = ctx->getFormat(shadowMap)
  });
```

11. We're almost done with the setup. The light's view and projection matrices can be calculated using the following variables. Here, `g_LightFOV` represents the field-of-view angle of the light, `g_LightInnerAngle` defines the light's inner cone, and `g_LightNear` and `g_LightFar` are the near and far clip planes for the light, respectively. All these variables are controlled from the ImGui UI.

```
float g_LightFOV        = 45.0f;
float g_LightInnerAngle = 10.0f;
float g_LightNear       = 0.8f;
float g_LightFar        = 8.0f;
```

12. The value `g_LightDist` represents the distance of the light from the origin, while `g_LightXAngle` and `g_LightYAngle` are the rotation angles around the X and Y axes, respectively. The depth values of shadow fragments are adjusted by depth bias factors to reduce shadow acne artifacts that occur when self-shadowing scene objects, which are caused by Z-fighting. These parameters are somewhat ad hoc and tricky, requiring significant fine-tuning in real-world applications. You can tweak them from the ImGui UI.

```
float g_LightDist   = 4.0f;
float g_LightXAngle = 240.0f;
float g_LightYAngle = 0.0f;
float g_LightDepthBiasConst = 1.5f;
float g_LightDepthBiasSlope = 3.0f;
```

A few other parameters are needed to control the scene from the UI. We skip them here.

Now, we can enter the rendering loop. Let's go through it step by step to understand how the entire rendering process unfolds.

1. Rotate the scene and the light around the vertical axis in opposite directions.

```
app.run([&](uint32_t width, uint32_t height,
  float aspectRatio, float deltaSeconds)
{
  g_ModelAngle  = fmodf(g_ModelAngle-50*deltaSeconds, 360);
  g_LightYAngle = fmodf(g_LightYAngle+50*deltaSeconds, 360);
```

2. Calculate the light's position, as well as its view and projection matrices.

```
  mat4 rotY = glm::rotate(mat4(1.f),
    glm::radians(g_LightYAngle), vec3(0, 1, 0));
  mat4 rotX = glm::rotate(rotY,
    glm::radians(g_LightXAngle), vec3(1, 0, 0));
  vec4 lightPos = rotX * vec4(0, 0, g_LightDist, 1);
  const mat4 lightProj = glm::perspective(
    glm::radians(g_LightFOV), 1.0f, g_LightNear, g_LightFar);
  const mat4 lightView = glm::lookAt(
    vec3(lightPos), vec3(0), vec3(0, 1, 0));
```

3. Calculate the main camera's view and projection matrices. The main camera can be toggled via ImGui to switch between the regular view and the view from the light source.

```
  const bool showLightCamera = cameraType == comboBoxItems[1];
  mat4 view = showLightCamera ? lightView : app.camera_.getViewMatrix();
  mat4 proj = showLightCamera ?
    glm::perspective(glm::radians(g_LightFOV),
      aspectRatio, g_LightNear, g_LightFar) :
    glm::perspective(glm::radians(60.0f), aspectRatio, 0.1f, 1000.0f);
  mat4 m1 = glm::rotate(mat4(1.0f), glm::radians(-90.0f), vec3(1, 0, 0));
  mat4 m2 = glm::rotate(mat4(1.0f),
    glm::radians(g_ModelAngle), vec3(0, 1, 0));
```

4. Now, we can begin filling in the command buffer. The vertex and index buffers will be shared between both graphics pipelines, and the push constants will be the same for convenience.

```
ICommandBuffer& buf = ctx->acquireCommandBuffer();
buf.cmdBindVertexBuffer(0, bufferVertices);
buf.cmdBindIndexBuffer(bufferIndices, lvk::IndexFormat_UI32);
struct PushConstants {
  mat4 model;
  uint64_t perFrameBuffer;
  uint32_t texture;
};
```

5. Let's render the scene into the shadow map. The per-frame data buffer `bufferPerFrame` needs to be updated with the data corresponding to the light matrices. Our shadow rendering shaders consume only `view` and `proj` matrices.

```
buf.cmdUpdateBuffer(bufferPerFrame, PerFrameData{
  .view = lightView, .proj = lightProj });
buf.cmdBeginRendering(
  lvk::RenderPass{
    .depth = {.loadOp = LoadOp_Clear,
              .clearDepth = 1.f} },
  lvk::Framebuffer{
    .depthStencil = { .texture = shadowMap } });
```

6. Note that we only draw the duck model into the shadow map. The `model` matrix is passed via push constants. Depth bias factors are configured using `cmdSetDepthBias()`. The `g_LightDepthBiasConst` factor determines the constant depth offset added to each fragment, while `g_LightDepthBiasSlope` is a scalar factor applied based on the fragment's slope.

```
buf.cmdBindRenderPipeline(pipelineShadow);
buf.cmdPushConstants(PushConstants{
  .model          = m2 * m1,
  .perFrameBuffer = ctx->gpuAddress(bufferPerFrame) });
buf.cmdBindDepthState(
  { .compareOp = lvk::CompareOp_Less,
    .isDepthWriteEnabled = true });
buf.cmdSetDepthBias(g_LightDepthBiasConst,
                    g_LightDepthBiasSlope);
buf.cmdSetDepthBiasEnable(true);
buf.cmdDrawIndexed(duckNumIndices);
buf.cmdEndRendering();
```

At this point, our shadow map texture should contain a view of the scene from the light's perspective, which should resemble the following image.

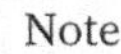

Note

Here, the content of the single-channel shadow map is rendered as the R channel. You can change that to white by using texture views and swizzling, as described earlier in the recipe *Implementing offscreen rendering in Vulkan*.

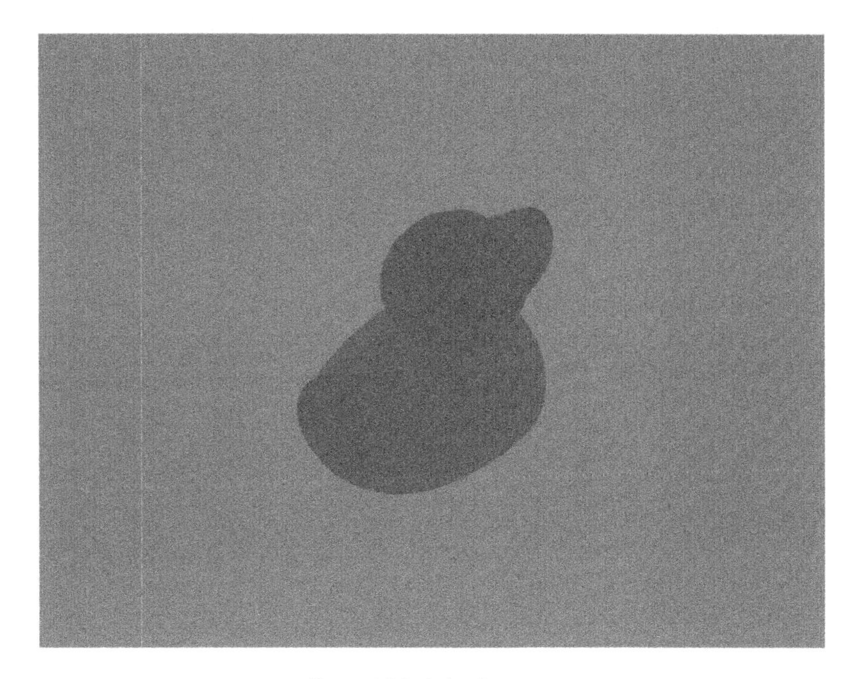

Figure 10.3: A shadow map

Now we are ready to continue rendering and render the main camera view while applying the shadow map to the scene.

1. The scale-bias matrix scales the UV texture coordinates by `0.5` and then translates them by `vec2(0.5, 0.5)`. This step is necessary because the light's clip-space `xy` coordinates are in the range `-1...+1`, while the texture coordinates are sampled using the range `0...1`. Therefore, we need to multiply them by `0.5` and add `0.5`.

```
const mat4 scaleBias = mat4(0.5, 0.0, 0.0, 0.0,
                            0.0, 0.5, 0.0, 0.0,
                            0.0, 0.0, 1.0, 0.0,
                            0.5, 0.5, 0.0, 1.0);
```

2. Our main rendering shaders require the entire set of values in the `bufferPerFrame` buffer to be filled in. Notice how the `scaleBias` matrix is multiplied with the `lightProj` and `lightView` matrices to prebake all texture coordinate transformations into a single `mat4` matrix.

```
buf.cmdUpdateBuffer(bufferPerFrame,
  PerFrameData{
    .view  = view,
    .proj  = proj,
    .light = scaleBias * lightProj * lightView,
    .lightAngles = vec4(
      cosf(glm::radians(0.5f * g_LightFOV)),
      cosf(glm::radians(0.5f * (g_LightFOV - g_LightInnerAngle))),
      1.0f, 1.0f),
    .lightPos      = lightPos,
    .shadowTexture = shadowMap.index(),
    .shadowSampler = samplerShadow.index() });
```

3. When we begin rendering, we must specify the `shadowMap` texture as a dependency to ensure the necessary Vulkan image layout transitions are applied. This framebuffer is also reused later for ImGui rendering.

```
const lvk::Framebuffer framebuffer = {
  .color = { { .texture = ctx->getCurrentSwapchainTexture() } },
  .depthStencil = { .texture = app.getDepthTexture() },
};
buf.cmdBeginRendering(
  lvk::RenderPass{
    .color =
      {{.loadOp = lvk::LoadOp_Clear,
        .clearColor = { 1.f, 1.f, 1.f, 1.f } }},
    .depth = { .loadOp = lvk::LoadOp_Clear,
               .clearDepth = 1.0f } },
  framebuffer,
  { .textures = {{TextureHandle(shadowMap)}} });
```

4. The scene rendering itself is straightforward. We bind the required pipeline and then render the duck mesh, followed by the plane. The push constants store the texture ID for each mesh, while the shadow map texture ID is stored in the `bufferPerFrame` buffer.

```
buf.cmdBindRenderPipeline(pipeline);
buf.cmdBindDepthState({
  .compareOp = lvk::CompareOp_Less,
  .isDepthWriteEnabled = true });
buf.cmdPushConstants(PushConstants{       // Duck
```

```
    .model         = m2 * m1,
    .perFrameBuffer = ctx->gpuAddress(bufferPerFrame),
    .texture       = duckTexture.index() });
  buf.cmdDrawIndexed(duckNumIndices);
  buf.cmdPushConstants(PushConstants{      // Plane
    .model         = m1,
    .perFrameBuffer = ctx->gpuAddress(bufferPerFrame),
    .texture       = planeTexture.index() });
  buf.cmdDrawIndexed(6, 1, 0, planeVertexOffset);
```

At this point, our shadow mapping code is complete. The remaining part of the C++ code handles drawing the traditional grid and FPS counter, followed by the ImGui UI to control various light and shadow mapping parameters. We'll skip that section here.

5. There's one more thing to mention. When rendering the scene from a regular first-person camera, we can render the light frustum using the 3D drawing canvas introduced in the recipe *Implementing immediate mode 3D drawing canvas* from *Chapter 5*. This is incredibly useful for debugging various shadow mapping algorithms. You will see it in action while running the demo application.

```
  if (!showLightCamera && g_DrawFrustum) {
    canvas3d.clear();
    canvas3d.setMatrix(proj * view);
    canvas3d.frustum(lightView, lightProj, vec4(1, 0, 0, 1));
    canvas3d.render(*ctx.get(), framebuffer, buf);
  }
```

Now, let's take a look at both sets of GLSL shaders to understand how everything works.

How it works...

All of our GLSL shaders here share the same set of common declarations, `Chapter10/02_ShadowMapping/src/common.sp`, for the per-frame buffer and push constants, just for convenience.

```
layout(std430, buffer_reference) readonly buffer PerFrameData {
  mat4 view;
  mat4 proj;
  mat4 light;
  vec4 lightAngles;
  vec4 lightPos;
  uint shadowTexture;
  uint shadowSampler;
};
layout(push_constant) uniform PushConstants {
  mat4 model;
```

```
  PerFrameData perFrame;
  uint texture;
} pc;
struct PerVertex {
  vec2 uv;
  vec3 worldNormal;
  vec3 worldPos;
  vec4 shadowCoords;
};
```

The first set of GLSL shaders is necessary to render our scene into the shadow map.

1. The vertex shader `Chapter10/02_ShadowMapping/src/shadow.vert` takes the `view` and `proj` matrices from the buffer and the `model` matrix from the push constants.

```
#include <Chapter10/02_ShadowMapping/src/common.sp>
layout (location = 0) in vec3 pos;
void main() {
  gl_Position = pc.perFrame.proj * pc.perFrame.view *
    pc.model * vec4(pos, 1.0);
}
```

2. The fragment shader for shadow map rendering is just empty because we do not output any color information, only the depth values. The source code is located in `Chapter10/02_ShadowMapping/src/shadow.frag`.

```
void main() {
}
```

The second set of GLSL shaders is where the actual projective texture mapping occurs. Let's take a closer look.

1. The vertex shader `Chapter10/02_ShadowMapping/src/main.vert` does the same transformation for `gl_Position`.

```
#include <Chapter10/02_ShadowMapping/src/common.sp>
layout (location = 0) in vec3 pos;
layout (location = 1) in vec3 normal;
layout (location = 2) in vec2 uv;
layout (location=0) out PerVertex vtx;
void main() {
  gl_Position = pc.perFrame.proj * pc.perFrame.view *
    pc.model * vec4(pos, 1.0);
```

2. Next, we populate all the remaining data necessary for rendering. Notice how the vertex position is first transformed into the light's coordinate system, then into normalized device coordinates, and finally, into texture coordinates. This is done using the model matrix and the combined light matrix we calculated in C++, which is the product of the light's view and projection matrices, multiplied by the scale-bias matrix. The resulting `vec4` value is stored in `shadowCoords` and passed to the fragment shader, where we will perform the perspective division.

```
    mat3 normalMatrix = transpose( inverse(mat3(pc.model)) );
    vtx.uv = uv;
    vtx.worldNormal = normalMatrix * normal;
    vtx.worldPos = (pc.model * vec4(pos, 1.0)).xyz;
    vtx.shadowCoords = pc.perFrame.light * pc.model * vec4(pos, 1.0);
  }
```

3. The fragment shader `Chapter10/02_ShadowMapping/src/main.frag` does several things. It implements a simple soft shadow effect using a technique called **percentage-closer filtering** (**PCF**). The helper function `PCF()`, which is defined in the include file `data/shaders/Shadow.sp`, performs a 3x3 averaging of 9 depth comparison operations. Note that we average not the results of depth map sampling at adjacent locations but the results of multiple comparisons between the depth value of the current fragment (in the light space) and the sampled depth values obtained from the shadow map. This is done automatically using the shadow sampler we created earlier in the C++ code.

Note

Naïve averaging of 9 depth comparison operations can lead to light leaking and soft shadows that are too blurry. Using more advanced PCF kernels or variance shadow mapping can achieve better results.

```
#include <Chapter10/02_ShadowMapping/src/common.sp>
layout (location=0) in PerVertex vtx;
layout (location=0) out vec4 out_FragColor;
float PCF3x3(vec3 uvw, uint textureid, uint samplerid) {
  float size = 1.0 / textureBindlessSize2D(textureid).x;
  float shadow = 0.0;
  for (int v=-1 ; v<=+1 ; v++) for (int u=-1 ; u<=+1 ; u++) {
    shadow += textureBindless2DShadow(textureid,
      samplerid, uvw + size * vec3(u, v, 0));
  }
  return shadow / 9;
}
```

4. The helper function `shadow()` encapsulates all the shadowing logic and returns a single shadow factor for the current fragment. The `shadowCoord` value passed as the argument `s` represents the position of the current fragment in the light's clip space, interpolated from the vertex shader. We perform the perspective division by the `w` component and check if the fragment lies within the `-1.0…+1.0` clip space `Z` range, then call the `PCF3()` function to evaluate the result. Our viewport is flipped to be similar to OpenGL, hence the flipped `Y` coordinate `1.0-s.y`.

```
float shadow(vec4 s, uint textureid, uint samplerid) {
  s = s / s.w;
  if (s.z > -1.0 && s.z < 1.0) {
    float shadowSample = PCF3x3(
      vec3(s.x, 1.0-s.y, s.z), textureid, samplerid);
```

5. Note that the surface in full shadow receives an ad hoc factor of `0.3` instead of `0` to prevent over-darkening of the shadowed areas.

```
    return mix(0.3, 1.0, shadowSample);
  }
  return 1.0;
}
```

6. The `spotLightFactor()` function calculates the spot light shadowing coefficient based on the inner and outer cone angles of the spot light. This allows for a gradual shadow falloff between these two angles.

Note

The inner angle is the angle within which the spotlight emits its maximum intensity. The light within this cone is bright and uniform. The outer angle is the angle at which the light completely fades out to zero intensity. The region between the inner and outer angles forms a gradual falloff zone, creating a smooth transition from bright light to darkness.

Both angles are controlled from our UI, so you can adjust them and observe the results. The values of `outerAngle` and `innerAngle` are actually the cosines of the angles, which we calculated earlier in C++ for performance reasons. The light direction `dirSpot` is calculated as `-lightPos` in this case because our spot light is always aimed at the origin `(0, 0, 0)`.

```
float spotLightFactor(vec3 worldPos) {
  vec3 dirLight = normalize(worldPos - pc.perFrame.lightPos.xyz);
  vec3 dirSpot  = normalize(-pc.perFrame.lightPos.xyz);
  float rho = dot(dirLight, dirSpot);
  float outerAngle = pc.perFrame.lightAngles.x;
  float innerAngle = pc.perFrame.lightAngles.y;
  if (rho > outerAngle)
    return smoothstep(outerAngle, innerAngle, rho);
```

```
    return 0.0;
  }
```

7. With all the helper functions in place, the `main()` function becomes straightforward. We calculate the simple diffuse lighting contribution `NdotL` based on the light position and surface normal, then multiply it by the results of the `shadow()` and `spotLightFactor()` function calls. The resulting value is then multiplied by a diffuse texture sample.

```
void main() {
  vec3 n = normalize(vtx.worldNormal);
  vec3 l = normalize(pc.perFrame.lightPos.xyz);
  float NdotL = clamp(dot(n, l), 0.1, 1.0);
  float Ka = 0.1;
  float Kd = NdotL * shadow(vtx.shadowCoords,
      pc.perFrame.shadowTexture,
      pc.perFrame.shadowSampler) *
    spotLightFactor(vtx.worldPos);
  out_FragColor = textureBindless2D(
    pc.texture, 0, vtx.uv) * clamp(Ka + Kd, 0.3, 1.0);
}
```

This is all the code necessary to implement simple PCF shadow mapping. The demo application should render a shadowed scene, as shown in the following screenshot. Make sure to experiment with all the ImGui controls to see how different light parameters affect the shadow mapping.

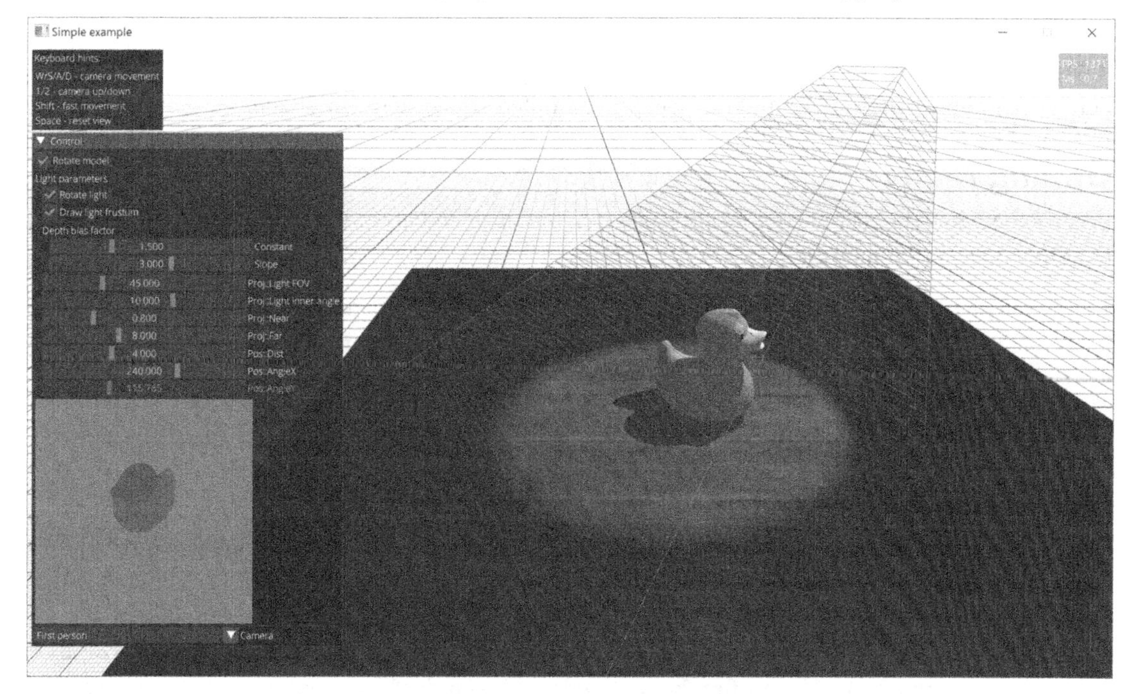

Figure 10.4: Rendering a shadow-mapped duck

This concludes our basic Vulkan shadow mapping example and the first use case of offscreen rendering. Let's now shift focus back to the main topic of this chapter and continue exploring image-based rendering techniques by learning how to implement **multisample anti-aliasing** (MSAA) in Vulkan.

There's more...

The technique described in this recipe for calculating the light's view and projection matrices is suitable only for spot lights and, to some extent, omni lights. For directional lights, which impact the entire visible scene, the view and projection matrices depend on the geometry of the scene and how it intersects the main camera's frustum. We will cover this topic in *Chapter 11*, where we will create an outdoor shadow map for a directional light and the Bistro scene.

Implementing MSAA in Vulkan

Multisample anti-aliasing (MSAA) is an anti-aliasing technique used in computer graphics to reduce jagged edges and improve the visual quality of rendered images. Multisampling is a specialized form of supersampling where the fragment shader runs only once per pixel, with only the depth (and stencil) values being supersampled within a pixel. More precisely, with multisampling, the fragment shader only needs to be evaluated once per pixel for each triangle that covers at least one sample point.

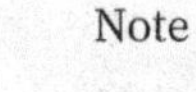

Note

Although MSAA is gradually being replaced by more advanced anti-aliasing techniques like **temporal anti-aliasing** (TAA), it remains relevant on mobile GPUs. When paired with tiled rendering, its performance cost is minimal.

In Vulkan, multiple sample values can later be resolved into non-multisampled color and depth images. Let's go over how to use MSAA and resolve multisampled attachments in Vulkan.

Getting ready

Be sure to review the recipe *Implementing offscreen rendering in Vulkan* to refresh your knowledge on handling offscreen rendering.

To find the Bistro scene rendering code used here, refer to the recipe *Rendering large scenes* in *Chapter 8*. We placed the Bistro loading code into a helper function `loadBistro()`, located in `Chapter10/Bistro.h`, which will be used in all upcoming demo applications in this chapter and the next.

The sample app for this recipe is located in `Chapter10/03_MSAA`.

How to do it...

In most of the previous recipes, we rendered the main scene, with some shaders applied, directly into a swapchain image, which holds only one sample per pixel. Multisample rendering in Vulkan works differently. Here, we need to create an offscreen render target capable of storing multiple samples per pixel and render our scene into that target. Before this multisampled image can be displayed or used in post-processing effects, it must first be resolved into a standard non-multisampled render target. Vulkan provides several ways to perform this resolution.

One option is to use the Vulkan command `vkCmdResolveImage()`, which takes a fully rendered multisampled image and resolves it into a non-multisampled image. While straightforward, this approach requires the multisampled image to be completely rendered and stored in memory before the resolve operation can begin. Storing multiple samples per pixel consumes a large amount of memory, putting significant pressure on memory bandwidth – a limited resource on power-constrained devices like smartphones. As a result, this approach has very limited practical use.

The Vulkan API offers another method that is well-suited for mobile devices with tiled GPUs. As covered in the recipe *Initializing Vulkan pipelines* in *Chapter 2*, the command `vkCmdBeginRendering()` requires a list of framebuffer attachments, each of which can include a resolve attachment, referred to as `VkRenderingAttachmentInfo::resolveImageView` in Vulkan. By specifying this resolve attachment and following some additional setup steps, the Vulkan driver can resolve multisampled images into non-multisampled images without storing intermediate multisampled data in memory. Let's see how to implement this. We will first review the high-level LightweightVK wrapper code, then explore the Vulkan implementation details.

1. We begin by examining the app's `main()` function, located in the `Chapter10/03_MSAA/src/main.cpp` file. Here, we perform the usual setup and load the Bistro mesh using the `loadBistro()` helper function from `Chapter10/Bistro.h`. The full process was explained in the recipe *Rendering large scenes* in *Chapter 8*.

```
int main() {
  MeshData meshData;
  Scene scene;
  loadBistro(meshData, scene);
  VulkanApp app({ … });
  // …
```

2. Next, we choose the number of MSAA samples to use and create an offscreen color render target with the specified number of samples. The dimensions of this render target should match those of our swapchain.

```
  const uint32_t kNumSamples = 8;
  const lvk::Dimensions sizeFb = ctx->getDimensions(
    ctx->getCurrentSwapchainTexture());
  Holder<TextureHandle> msaaColor =
    ctx->createTexture({
      .format     = ctx->getSwapchainFormat(),
      .dimensions = sizeFb,
      .numSamples = kNumSamples,
      .usage      = lvk::TextureUsageBits_Attachment });
```

3. The same applies to the depth buffer: we need a multisampled depth buffer, so we can no longer use the default one from `VulkanApp`.

```
Holder<TextureHandle> msaaDepth =
  ctx->createTexture({
    .format      = app.getDepthFormat(),
    .dimensions = sizeFb,
    .numSamples = kNumSamples,
    .usage       = lvk::TextureUsageBits_Attachment });
```

4. We want to control rendering from the ImGui UI and create two `VKMesh` objects for our Bistro scene. Since Vulkan requires the number of samples to be specified when creating the graphics pipeline, these two separate `VKMesh` instances are necessary. The `VKMesh` class is declared in `Chapter08/VKMesh08.h` and was used in the recipe *Rendering large scenes* in *Chapter 8*.

```
bool enableMSAA = true;
const VKMesh mesh(ctx, meshData, scene,
  ctx->getSwapchainFormat(), app.getDepthFormat());
const VKMesh meshMSAA(ctx, meshData, scene,
  ctx->getSwapchainFormat(), app.getDepthFormat(),
  kNumSamples);
app.run([&](uint32_t width, uint32_t height,
  float aspectRatio, float deltaSeconds) {
  const mat4 view = app.camera_.getViewMatrix();
  const mat4 proj = glm::perspective( … );
  ICommandBuffer& buf = ctx->acquireCommandBuffer();
```

5. Entering the rendering loop, we define a list of color and depth attachments for our framebuffer. We use the `enableMSAA` flag to toggle between MSAA and non-MSAA rendering. When MSAA is enabled, we set `msaaColor` as the offscreen texture and the current swapchain texture as `resolveTexture`. When MSAA is disabled, we use the current swapchain texture as the framebuffer texture, leaving the resolve texture empty. Note that we don't need to resolve the depth texture, so we can skip filling in the `resolveTexture` field for it.

```
  const lvk::Framebuffer framebufferOffscreen = {
    .color = { {
      .texture = enableMSAA ?
        msaaColor : ctx->getCurrentSwapchainTexture(),
      .resolveTexture = enableMSAA ?
        ctx->getCurrentSwapchainTexture() : {} } },
    .depthStencil = {
      .texture = enableMSAA ? msaaDepth : app.getDepthTexture() },
  };
```

6. At the start of rendering, we need to set the render pass parameters and the appropriate load and store operations. For the color attachment, we use `StoreOp_MsaaResolve`, while for the depth attachment, which we don't need to resolve, we keep the implicit store operation `StoreOp_DontCare`.

```
    buf.cmdBeginRendering( lvk::RenderPass{
        .color = { { .loadOp  = lvk::LoadOp_Clear,
                     .storeOp = enableMSAA ?
                       lvk::StoreOp_MsaaResolve : lvk::StoreOp_Store,
                     .clearColor = {1, 1, 1, 1} } },
        .depth = { .loadOp = lvk::LoadOp_Clear,
                   .clearDepth = 1.0f } },
      framebufferOffscreen);
```

7. During rendering, we must specify the number of MSAA samples for all our rendering helper functions. This is required so they can update their internal graphics pipelines, which are created lazily. We will look into the details shortly.

```
    (enableMSAA ? meshMSAA : mesh).draw(
       buf, view, proj, texSkyboxIrradiance, drawWireframe);
    app.drawGrid(buf, proj, vec3(0, -1.0f, 0), enableMSAA ? kNumSamples : 1);
    canvas3d.render(*ctx.get(),
      framebufferOffscreen, buf, enableMSAA ? kNumSamples : 1);
    buf.cmdEndRendering();
```

8. We use a separate framebuffer description for ImGui rendering because it is rendered on top of the resolved, non-multisampled swapchain image. Once the command buffer is submitted, we present the swapchain image.

```
    const lvk::Framebuffer framebufferMain = {
      .color = { { .texture = ctx->getCurrentSwapchainTexture() } } };
    buf.cmdBeginRendering( lvk::RenderPass{
        .color = { { .loadOp = lvk::LoadOp_Load,
                     .clearColor = {1, 1, 1, 1}} } },
      framebufferMain);
    // ...skipped canvas and ImGui rendering code here...
    buf.cmdEndRendering();
    ctx->submit(buf, ctx->getCurrentSwapchainTexture());
  });
  ctx.release();
  return 0;
}
```

This covers the high-level rendering code. Now, let's take a look into the internals to understand how it works and interacts with the Vulkan API.

How it works...

To get the big picture, we will examine some key parts of the function `lvk::CommandBuffer::cmdBeginRendering()`, located in the `deps/src/lightweightvk/lvk/vulkan/VulkanClasses.cpp` file. Its main code flow was described in the recipe *Initializing Vulkan pipelines* in *Chapter 2*. Now, let's focus on how to construct an array of `VkRenderingAttachmentInfo` structs.

1. First, we should recap on the loop that converts all LightweightVK color attachments into Vulkan `VkRenderingAttachmentInfo` structs.

```
void CommandBuffer::cmdBeginRendering(
  const lvk::RenderPass& renderPass,
  const lvk::Framebuffer& fb,
  const Dependencies& deps)
{
  const uint32_t numPassColorAttachments = renderPass.getNumColorAttachments();
  // ...
  VkRenderingAttachmentInfo colorAttachments[LVK_MAX_COLOR_ATTACHMENTS];
  for (uint32_t i=0; I != numFbColorAttachments;i++) {
    const AttachmentDesc& attachment = fb.color[i];
    lvk::VulkanImage& colorTexture =
      *ctx_->texturesPool_.get(attachment.texture);
    const auto& descColor = renderPass.color[i];
    // ...
    samples = colorTexture.vkSamples_;
    colorAttachments[i] = {
      .sType = VK_STRUCTURE_TYPE_RENDERING_ATTACHMENT_INFO,
      .imageView = colorTexture.getOrCreateVkImageViewForFramebuffer(
        *ctx_, descColor.level, descColor.layer),
      .imageLayout = VK_IMAGE_LAYOUT_COLOR_ATTACHMENT_OPTIMAL,
```

2. Note how the resolve mode is selected for MSAA attachments. The resolve mode defines how multisampled data is resolved. `VK_RESOLVE_MODE_NONE` indicates no resolve operation is performed, while `VK_RESOLVE_MODE_AVERAGE_BIT` means the result of the resolve operation is the average of the sample values. There are other resolve modes, such as `VK_RESOLVE_MODE_MIN_BIT` for the minimum of the sample values and `VK_RESOLVE_MODE_MAX_BIT` for the maximum. These have specific use cases with a depth buffer, but we do not use them in our book.

```
      .resolveMode = (samples > 1) ?
        VK_RESOLVE_MODE_AVERAGE_BIT :
        VK_RESOLVE_MODE_NONE,
      .resolveImageView = VK_NULL_HANDLE,
      .resolveImageLayout = VK_IMAGE_LAYOUT_UNDEFINED,
      .loadOp = loadOpToVkAttachmentLoadOp(descColor.loadOp),
```

```
        .storeOp = storeOpToVkAttachmentStoreOp(descColor.storeOp),
        .clearValue =
          {.color={.float32={descColor.clearColor[…]}}},
      };
```

3. The next interesting part comes when we have a render pass description for an attachment with a store operation set to `StoreOp_MsaaResolve`. This is where we take the `VkImageView` of our resolve texture and assign it to the field `VkRenderingAttachmentInfo::resolveImageView`.

```
      if (descColor.storeOp == StoreOp_MsaaResolve) {
        lvk::VulkanImage& colorResolveTexture = *ctx_->
          texturesPool_.get(attachment.resolveTexture);
        colorAttachments[i].resolveImageView =
          colorResolveTexture.getOrCreateVkImageViewForFramebuffer(
            *ctx_, descColor.level, descColor.layer);
        colorAttachments[i].resolveImageLayout =
          VK_IMAGE_LAYOUT_COLOR_ATTACHMENT_OPTIMAL;
      }
    }
```

4. The depth attachment is very similar. For brevity, we focus only on the MSAA part in the book text. The only difference here is that the image layout was set to `VK_IMAGE_LAYOUT_DEPTH_STENCIL_ATTACHMENT_OPTIMAL`.

```
  VkRenderingAttachmentInfo depthAttachment = {};
  if (fb.depthStencil.texture) {
    lvk::VulkanImage& depthTexture = *ctx_->texturesPool_.get(
      fb.depthStencil.texture);
    const RenderPass::AttachmentDesc& descDepth = renderPass.depth;
    depthAttachment = { … }; // same as for color
    if (descDepth.storeOp == StoreOp_MsaaResolve) {
      const AttachmentDesc& attachment = fb.depthStencil;
      lvk::VulkanImage& depthResolveTexture =
        *ctx_->texturesPool_.get(attachment.resolveTexture);
      depthAttachment.resolveImageView =
        depthResolveTexture.getOrCreateVkImageViewForFramebuffer(
          *ctx_, descDepth.level, descDepth.layer);
      depthAttachment.resolveImageLayout =
        VK_IMAGE_LAYOUT_DEPTH_STENCIL_ATTACHMENT_OPTIMAL;
      depthAttachment.resolveMode = VK_RESOLVE_MODE_AVERAGE_BIT;
    }
    // …
  }
```

5. The only remaining part is the graphics pipeline creation in the helper class `VKMesh`, located in `Chapter08/VKMesh.h`. This is where the number of MSAA samples is passed to the pipeline. The `minSampleShading` value is passed into Vulkan as `VkPipelineMultisampleStateCreateInfo::minSampleShading` and is explained below.

```
VKMesh(const std::unique_ptr<lvk::IContext>& ctx,
       const MeshData& meshData, const Scene& scene,
       lvk::Format colorFormat,
       lvk::Format depthFormat,
       uint32_t numSamples = 1,
       Holder<ShaderModuleHandle>&& vert = {},
       Holder<ShaderModuleHandle>&& frag = {}) {
  // ...
  pipeline_  = ctx->createRenderPipeline({
    .vertexInput  = meshData.streams,
    .smVert       = vert_,
    .smFrag       = frag_,
    .color        = { { .format  = colorFormat } },
    .depthFormat  = depthFormat,
    .cullMode     = lvk::CullMode_None,
    .samplesCount  = numSamples,
    .minSampleShading  = numSamples > 1 ? 0.25f : 0.0f,
  });
  // ...
}
```

6. And here is the corresponding Vulkan code responsible for creating the graphics pipeline in `VulkanPipelineBuilder::VulkanPipelineBuilder()`, which populates the `VkPipelineMultisampleStateCreateInfo` struct and is located in the `deps/src/lightweightvk/lvk/vulkan/VulkanClasses.cpp` file.

```
multisampleState_({
  .sType = VK_STRUCTURE_TYPE_PIPELINE_MULTISAMPLE_STATE_CREATE_INFO,
  .rasterizationSamples = getVulkanSampleCountFlags(desc.samplesCount),
  .sampleShadingEnable = desc.minSampleShading > 0 ? VK_TRUE : VK_FALSE,
  .minSampleShading = desc.minSampleShading,
  .pSampleMask = nullptr,
  .alphaToCoverageEnable = VK_FALSE,
  .alphaToOneEnable = VK_FALSE,
}),
```

Color sampling occurs once per pixel, which can lead to "shader aliasing" or Moiré patterns on high-frequency textures. However, this can be mitigated using the `minSampleShading` parameter. This parameter controls the minimum number of samples that should be shaded for each pixel. A value of `1.0` ensures that every sample is shaded independently. The implementation guarantees at least as many color values per pixel as the product of `minSampleShading` and `rasterizationSamples`. In our case, with `minSampleShading` set to `0.25` and `rasterizationSamples` at `8`, this results in at least 2 samples per pixel (`0.25 * 8 = 2`).

Note

As per Vulkan 1.3 specification *28.8. Sample Shading*, if a fragment shader uses the built-in `gl_SampleID` or `gl_SamplePosition` variables, sample shading is automatically enabled, and the `minSampleShading` value is overridden with `1.0`. Please note that it can have a significant performance impact.

This code is enough to enable MSAA in our demo application. If you run it, you should see the textured Bistro mesh rendered as shown in the following screenshot.

Figure 10.5: MSAA anti-aliased scene

Try toggling the `Enable MSAA` checkbox in the ImGui UI and observe the difference between anti-aliased and non-anti-aliased rendering. The difference is most noticeable on the geometry edges and when the camera is slowly moving.

From this point on, all subsequent recipes in the book will have MSAA enabled. Now, let's continue exploring post-processing effects and learn how to add more realism to our scenes with screen space ambient occlusion.

There's more...

While the material in this recipe is sufficient to set up MSAA rendering, it misses one important aspect. When we allocate our offscreen textures, `msaaColor` and `msaaDepth`, they occupy valuable GPU memory, even though they are never used for actual rendering on tiled GPUs. Vulkan provides a solution to this problem.

When we create Vulkan images, we can set `VK_MEMORY_PROPERTY_LAZILY_ALLOCATED_BIT` in `VkMemoryPropertyFlags` and the image usage flag `VK_IMAGE_USAGE_TRANSIENT_ATTACHMENT_BIT`, which tells the Vulkan driver not to allocate memory for the image unless it is actually needed. This can be done in LightweightVK using the `StorageType_Memoryless` storage type. You can check `deps/src/lightweightvk/lvk/vulkan/VulkanClasses.cpp` to see how these Vulkan flags are set up.

Implementing screen space ambient occlusion

Screen space ambient occlusion (SSAO) is an image-based technique used to approximate global illumination in real time. Ambient occlusion, at its core, provides a simplified representation of global illumination. It can be thought of as the amount of open "sky" visible from a point on a surface, unobstructed by nearby geometry. In its most basic implementation, this is estimated by sampling several points around the surface point of interest and determining how visible they are from that central point. Let's take a look at a simple implementation of this technique.

Getting ready

The demo application for this recipe, `Chapter10/04_SSAO`, implements basic steps for SSAO. Before going into this recipe, make sure to read the earlier ones in this chapter: *Implementing offscreen rendering in Vulkan* and *Implementing full-screen triangle rendering*.

How to do it...

Instead of tracing the depth buffer's height field, we use a simpler method where each selected neighboring point is projected onto the depth buffer. This projected point acts as a potential occluder. The occlusion factor $O(dZ)$ for each point is calculated based on the difference dZ between the projected depth value and the current fragment's depth using the following formula:

$$O(dZ) = (dZ > 0)\ ?\ 1/(1+dZ*dZ) : 0$$

These occlusion factors are averaged to determine the SSAO value for the current fragment. Before applying the final SSAO to the scene, it is blurred to minimize aliasing artifacts.

The SSAO shader works exclusively with the depth buffer, requiring no additional scene data. This makes it an easy, standalone snippet to kickstart your exploration of SSAO. Let's review the C++ portion of the code in `Chapter10/04_SSAO/src/main.cpp` to understand how it's implemented. We will skip the setup and the Bistro loading code here since it is identical to what was covered in the previous recipe, *Implementing MSAA in Vulkan*.

1. First, we load the GLSL shaders for SSAO calculation and blurring.

```
int main() {
  MeshData meshData;
  Scene scene;
  loadBistro(meshData, scene);
  // ...
  Holder<ShaderModuleHandle> compSSAO =
    loadShaderModule(ctx, "Chapter10/04_SSAO/src/SSAO.comp");
  Holder<ComputePipelineHandle> pipelineSSAO =
    ctx->createComputePipeline({.smComp = compSSAO});
  Holder<ShaderModuleHandle> compBlur =
    loadShaderModule(ctx, "data/shaders/Blur.comp");
```

2. The blurring shader employs a separable Gaussian blur, performing horizontal and vertical passes separately. However, the same compute shader is used for both passes, with specialization constants creating two distinct compute pipelines: one for horizontal blur and one for vertical blur.

```
  const uint32_t kHorizontal = 1;
  const uint32_t kVertical   = 0;
  Holder<ComputePipelineHandle> pipelineBlurX =
    ctx->createComputePipeline({
      .smComp    = compBlur,
      .specInfo = {
        .entries = { { .constantId = 0,
                       .size = sizeof(uint32_t) } },
        .data = &kHorizontal,
        .dataSize = sizeof(uint32_t)} });
  Holder<ComputePipelineHandle> pipelineBlurY =
    ctx->createComputePipeline({
      .smComp    = compBlur,
      .specInfo = {
        .entries = { { .constantId = 0,
                       .size = sizeof(uint32_t) } },
        .data = &kVertical,
        .dataSize = sizeof(uint32_t)} });
```

3. Next, we create a graphics pipeline to combine the generated SSAO and blur with the rendered color image. This process utilizes the full-screen quad (or triangle) shader we discussed earlier in the recipe *Implementing full-screen triangle rendering*. We use a graphics pipeline here instead of a compute pipeline to avoid dealing with a fixed swapchain pixel format required when using the `imageStore()` command in compute shaders.

```
Holder<ShaderModuleHandle> vertCombine =
 loadShaderModule(ctx,"data/shaders/QuadFlip.vert");
Holder<ShaderModuleHandle> fragCombine =
  loadShaderModule(ctx, "Chapter10/04_SSAO/src/combine.frag");
Holder<RenderPipelineHandle> pipelineCombine =
  ctx->createRenderPipeline({
    .smVert = vertCombine,
    .smFrag = fragCombine,
    .color  = {{ .format = ctx->getSwapchainFormat() }} });
```

4. Now, let's create a texture for SSAO and two auxiliary textures for the blur. These blur textures will be used in a ping-pong fashion to perform multipass Gaussian blur. We will get to the details of how this works in a moment.

```
Holder<TextureHandle> texSSAO =
  ctx->createTexture({
    .format     = ctx->getSwapchainFormat(),
    .dimensions = ctx->getDimensions(
      ctx->getCurrentSwapchainTexture()),
    .usage      = lvk::TextureUsageBits_Sampled |
                  lvk::TextureUsageBits_Storage });
Holder<TextureHandle> texBlur[] = {
  ctx->createTexture({
    .format     = ctx->getSwapchainFormat(),
    .dimensions = ctx->getDimensions(
      ctx->getCurrentSwapchainTexture()),
    .usage      = lvk::TextureUsageBits_Sampled |
                  lvk::TextureUsageBits_Storage }),
  ctx->createTexture({
    .format     = ctx->getSwapchainFormat(),
    .dimensions = ctx->getDimensions(
      ctx->getCurrentSwapchainTexture()),
    .usage      = lvk::TextureUsageBits_Sampled |
                  lvk::TextureUsageBits_Storage }),
};
```

Note

We use the same pixel format for the SSAO and blur textures as the swapchain format. This makes debug rendering into the main framebuffer simpler by allowing the use of `vkCmdCopyImage()` without the need for component swizzling or `vkCmdBlitImage()`.

5. Now, we can create the multisampled textures and offscreen resolve color and depth textures, as explained in the previous recipe, *Implementing MSAA in Vulkan*.

```
const lvk::Dimensions sizeFb =
  ctx->getDimensions(ctx->getCurrentSwapchainTexture());
const lvk::Dimensions sizeOffscreen = { sizeFb.width, sizeFb.height };
const uint32_t kNumSamples = 8;
Holder<TextureHandle> msaaColor = ctx->createTexture({
    .format      = ctx->getSwapchainFormat(),
    .dimensions = sizeFb,
    .numSamples = kNumSamples,
    .usage       = lvk::TextureUsageBits_Attachment });
Holder<TextureHandle> msaaDepth = ctx->createTexture({
    .format      = app.getDepthFormat(),
    .dimensions = sizeFb,
    .numSamples = kNumSamples,
    .usage       = lvk::TextureUsageBits_Attachment });
Holder<TextureHandle> offscreenColor = ctx->createTexture({
    .format      = ctx->getSwapchainFormat(),
    .dimensions = sizeOffscreen,
    .usage       = lvk::TextureUsageBits_Attachment |
                   lvk::TextureUsageBits_Sampled |
                   lvk::TextureUsageBits_Storage });
Holder<TextureHandle> offscreenDepth =
  ctx->createTexture({
    .format      = app.getDepthFormat(),
    .dimensions = sizeOffscreen,
    .usage       = lvk::TextureUsageBits_Attachment |
                   lvk::TextureUsageBits_Sampled |
                   lvk::TextureUsageBits_Storage });
```

6. We also need a special rotation texture, loaded from the `data/rot_texture.bmp` file, and a `Clamp` sampler to prevent edge texels from wrapping.

```
Holder<TextureHandle> texRotations = loadTexture(ctx, "data/rot_texture.bmp");
Holder<SamplerHandle> samplerClamp = ctx->createSampler({
    .wrapU = lvk::SamplerWrap_Clamp,
    .wrapV = lvk::SamplerWrap_Clamp,
    .wrapW = lvk::SamplerWrap_Clamp });
```

The rotation texture `rot_texture.bmp` is 4x4 pixels and contains 16 random `vec3` vectors. It looks like this. We will explore how it is used a bit later.

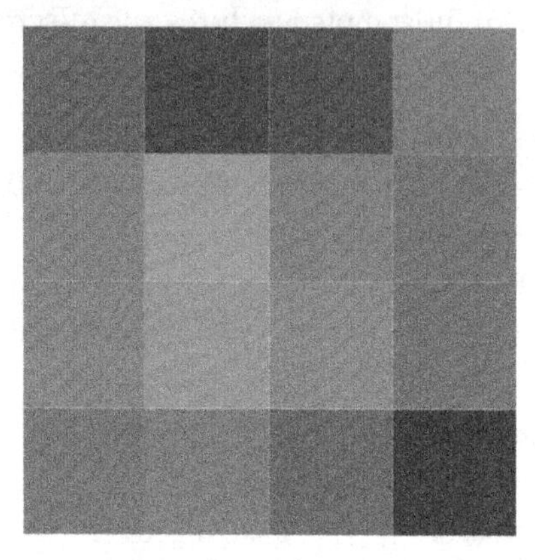

Figure 10.6: Random vectors texture (4x4 pixels)

7. We also need a few parameters for the ImGui UI that will allow us to tweak the SSAO effect and conveniently inspect our intermediate textures.

```
enum DrawMode {
  DrawMode_ColorSSAO = 0,
  DrawMode_Color     = 1,
  DrawMode_SSAO      = 2,
};
int drawMode             = DrawMode_ColorSSAO;
float depthThreshold = 30.0f;
```

8. Before we can start rendering, we need to define push constants for our SSAO and Combine shaders. The SSAO parameters are chosen arbitrarily and can be adjusted through the ImGui interface.

```
struct {
  uint32_t texDepth;
  uint32_t texRotation;
  uint32_t texOut;
  uint32_t sampler;
  float zNear;
  float zFar;
  float radius;
```

```
    float attScale;
    float distScale;
  } pcSSAO = {
    .texDepth    = offscreenDepth.index(),
    .texRotation = texRotations.index(),
    .texOut      = texSSAO.index(),
    .sampler     = samplerClamp.index(),
    .zNear       = 0.01f,
    .zFar        = 1000.0f,
    .radius      = 0.03f,
    .attScale    = 0.95f,
    .distScale   = 1.7f,
  };
```

9. The push constants for the Combine shader define the `scale` and `bias` parameters, which can be tweaked through ImGui. The Bistro and Skybox meshes are created here as well. The `Skybox` helper class encapsulates our previous skybox rendering code from *Chapter 8*, and is defined in `Chapter10/Skybox.h`.

```
  struct {
    uint32_t texColor;
    uint32_t texSSAO;
    uint32_t sampler;
    float scale;
    float bias;
  } pcCombine = {
    .texColor = offscreenColor.index(),
    .texSSAO  = texSSAO.index(),
    .sampler  = samplerClamp.index(),
    .scale    = 1.5f,
    .bias     = 0.16f,
  };
  const Skybox skyBox(ctx, …, kNumSamples);
  const VKMesh mesh(ctx, …, kNumSamples);
```

The preparation work is now complete. The rendering loop is fairly straightforward. Let's go through it step by step to highlight the different parts of our SSAO implementation.

1. The first step is to render the multisampled scene and resolve it into our offscreen render targets. We covered this process in the previous recipe, *Implementing MSAA in Vulkan*. This time, we need to resolve both color and depth attachments, as the SSAO algorithm requires the scene's depth information.

```
app.run([&](uint32_t width, uint32_t height,
  float aspectRatio, float deltaSeconds) {
  const mat4 view = app.camera_.getViewMatrix();
  const mat4 proj = glm::perspective(
    45.0f, aspectRatio, pcSSAO.zNear, pcSSAO.zFar);
  ICommandBuffer& buf = ctx->acquireCommandBuffer();
  buf.cmdBeginRendering(
    lvk::RenderPass{
      .color = { { .loadOp = LoadOp_Clear,
                   .storeOp = StoreOp_MsaaResolve,
                   .clearColor = {1, 1, 1, 1} } },
      .depth = { .loadOp = lvk::LoadOp_Clear,
                 .storeOp = StoreOp_MsaaResolve,
                 .clearDepth = 1.0f } },
    lvk::Framebuffer{
      .color = {{.texture = msaaColor,
                 .resolveTexture = offscreenColor}},
      .depthStencil = {
        .texture = msaaDepth,
        .resolveTexture = offscreenDepth }
  });
  skyBox.draw(buf, view, proj);
  mesh.draw(buf, view, proj, skyBox.texSkyboxIrradiance, drawWireframe);
  app.drawGrid(buf, proj, vec3(0, -1.0f, 0), kNumSamples);
  buf.cmdEndRendering();
```

2. Once the scene is rendered and resolved, we run our compute shader to generate SSAO using the scene's depth buffer information.

```
  buf.cmdBindComputePipeline(pipelineSSAO);
  buf.cmdPushConstants(pcSSAO);
  buf.cmdDispatchThreadGroups(
    { .width  = 1 + (uint32_t)sizeFb.width  / 16,
      .height = 1 + (uint32_t)sizeFb.height / 16 },
    { .textures = {
        TextureHandle(offscreenDepth),
        TextureHandle(texSSAO) } });
```

3. While the resulting raw SSAO texture in *Figure 10.7* is already useful, it can be significantly improved with a simple blur effect.

Figure 10.7: Raw SSAO texture

We apply our multipass separable Gaussian blur to it.

```
if (enableBlur) {
  const lvk::Dimensions blurDim = {
    .width  = 1 + sizeFb.width  / 16,
    .height = 1 + sizeFb.height / 16,
  };
  struct BlurPC {
    uint32_t texDepth;
    uint32_t texIn;
    uint32_t texOut;
    float depthThreshold;
  };
  struct BlurPass {
    TextureHandle texIn;
    TextureHandle texOut;
  };
```

4. The interesting part here is the ping-pong passes for the blur. We run the shader twice per pass – once for horizontal blur and once for vertical blur. We use push constants to specify the inputs and outputs for each shader run.

```
std::vector<BlurPass> passes;
passes.reserve(2 * numBlurPasses);
passes.push_back({ texSSAO, texBlur[0] });
for (int i = 0; i != numBlurPasses - 1; i++) {
  passes.push_back({ texBlur[0], texBlur[1] });
  passes.push_back({ texBlur[1], texBlur[0] });
}
passes.push_back({ texBlur[0], texSSAO });
```

5. Next, we go through these pre-made push constants and invoke the shaders, alternating between the horizontal and vertical blur shaders for even and odd invocations. Note how the `texIn` and `texOut` textures are passed as dependencies to `cmdDispatchThreadGroups()` to allow LightweightVK to insert the appropriate image memory barriers.

```
for (uint32_t i = 0; i != passes.size(); i++) {
  const BlurPass p = passes[i];
  buf.cmdBindComputePipeline(i & 1 ? pipelineBlurX : pipelineBlurY);
  buf.cmdPushConstants(BlurPC{
    .texDepth       = offscreenDepth.index(),
    .texIn          = p.texIn.index(),
    .texOut         = p.texOut.index(),
    .depthThreshold = pcSSAO.zFar * depthThreshold });
  buf.cmdDispatchThreadGroups(
    blurDim, { .textures = { p.texIn, p.texOut,
      TextureHandle(offscreenDepth)} });
}
}
```

6. At this point, our SSAO effect is fully rendered into the `texSSAO` texture and looks as in the following screenshot. Notice how the blur effect is not applied to areas with depth discontinuities. This is called a bilateral blur, and we will explore it shortly when we examine the GLSL shaders.

Figure 10.8: Blurred SSAO buffer

7. Now, let's draw a full-screen quad, as described in *Implementing full-screen triangle rendering*, to render the combined scene with the SSAO effect into a swapchain image. In addition to normal rendering, we added draw modes to display either the SSAO effect alone or the rendered scene without SSAO.

```
if (drawMode == DrawMode_SSAO) {
  buf.cmdCopyImage(texSSAO,
    ctx->getCurrentSwapchainTexture(),
    ctx->getDimensions(offscreenColor));
} else if (drawMode == DrawMode_Color) {
  buf.cmdCopyImage(offscreenColor,
    ctx->getCurrentSwapchainTexture(),
    ctx->getDimensions(offscreenColor));
}
const lvk::RenderPass renderPassMain = {
  .color = { { .loadOp = lvk::LoadOp_Load,
               .clearColor = { 1, 1, 1, 1 } } } };
const lvk::Framebuffer framebufferMain = {
  .color = { { .texture = ctx->getCurrentSwapchainTexture() } } };
buf.cmdBeginRendering(
  renderPassMain,
  framebufferMain,
  { .textures = {TextureHandle(texSSAO)} });
```

8. Here is our actual SSAO Combine shader. We skipped the ImGui rendering code for the sake of brevity.

```
      if (drawMode == DrawMode_ColorSSAO) {
        buf.cmdBindRenderPipeline(pipelineCombine);
        buf.cmdPushConstants(pcCombine);
        buf.cmdBindDepthState({});
        buf.cmdDraw(3);
      }
      // ... ImGui rendering skipped here ...
      buf.cmdEndRendering();
      ctx->submit(buf, ctx->getCurrentSwapchainTexture());
    });
    // ...
  }
```

That's all for the C++ code. Now, let's take a look at the GLSL shaders to understand how the SSAO effect works.

How it works...

Now that we have an overall sense of the rendering process, we can start examining the GLSL shader code.

The compute shader `Chapter10/04_SSAO/src/SSAO.comp` takes the scene depth buffer and the rotation vectors texture as input. This texture contains 16 random `vec3` vectors, a technique originally proposed by Crytek in the early days of real-time SSAO algorithms.

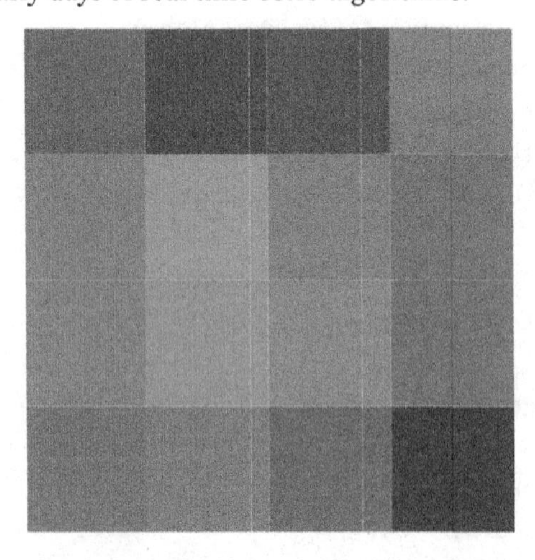

Figure 10.9: Random vectors texture (4x4 pixels)

1. We start by declaring shader inputs and texture access helper functions.

```
layout (local_size_x = 16, local_size_y = 16) in;
layout (set = 0, binding = 0) uniform texture2D kTextures2D[];
layout (set = 0, binding = 1) uniform sampler   kSamplers[];
layout (set = 0, binding = 2, rgba8) uniform writeonly image2D kTextures2DOut[];
ivec2 textureBindlessSize2D(uint textureid) {
  return textureSize(nonuniformEXT(kTextures2D[textureid]), 0);
}
vec4 textureBindless2D(uint textureid, vec2 uv) {
  return textureLod(
    nonuniformEXT(sampler2D(kTextures2D[textureid],
    kSamplers[pc.smpl])), uv, 0);
}
```

2. The helper function `scaleZ()` normalizes the depth buffer value into the `0…1` range.

```
layout(push_constant) uniform PushConstants {
  uint texDepth;
  uint texRotation;
  uint texOut;
  uint smpl;
  float zNear;
  float zFar;
  float radius;
  float attScale;
  float distScale;
} pc;
float scaleZ(float smpl) {
  return (pc.zFar * pc.zNear) /
         (smpl * (pc.zFar - pc.zNear) - pc.zFar);
}
```

3. Let's define a table with 3D offsets for 8 points to sample around the current fragment.

```
const vec3 offsets[8] = vec3[8](
  vec3(-0.5, -0.5, -0.5), vec3( 0.5, -0.5, -0.5),
  vec3(-0.5,  0.5, -0.5), vec3( 0.5,  0.5, -0.5),
  vec3(-0.5, -0.5,  0.5), vec3( 0.5, -0.5,  0.5),
  vec3(-0.5,  0.5,  0.5), vec3( 0.5,  0.5,  0.5) );
```

4. The `main()` function calculates the xy texel position and the corresponding uv coordinates, shifted to texel centers by `0.5` pixels.

```
void main() {
  const vec2 size = textureBindlessSize2D(pc.texDepth).xy;
```

```
const vec2 xy = gl_GlobalInvocationID.xy;
const vec2 uv = (gl_GlobalInvocationID.xy + vec2(0.5)) / size;
if (xy.x > size.x || xy.y > size.y) return;
```

5. Next, we take the aforementioned random rotations 4x4 texture, tile it across the entire framebuffer, and sample a `vec3` value from it corresponding to the current fragment. This value serves as a normal vector to a random plane. In the attenuation accumulation loop, we reflect each of our `vec3` offsets from this plane, producing a new sampling point, `rSample`, in the neighborhood of our area of interest, defined by the `pc.radius` value. The depth value `zSample` for this point is sampled from the depth texture and immediately converted to eye space. Afterward, the value is zero-clipped and scaled using an ad hoc `pc.distScale` parameter, which can be controlled through ImGui.

```
const float Z = scaleZ( textureBindless2D(pc.texDepth, uv).x );
const vec3 plane = textureBindless2D(
  pc.texRotation, xy / 4.0).xyz - vec3(1.0);
float att = 0.0;
for ( int i = 0; i < 8; i++ ) {
  vec3  rSample = reflect( offsets[i], plane );
  float zSample = scaleZ( textureBindless2D(
    pc.texDepth, uv + pc.radius*rSample.xy/Z ).x );
  float dist = max(zSample - Z, 0.0) / pc.distScale;
```

6. The distance difference `occl` is scaled by an arbitrarily selected weight. Further averaging is performed using quadratic attenuation, according to the formula `O(dZ)= (dZ > 0) ? 1/(1+dZ^2) : 0`. The final scale factor `pc.attScale` is controlled through ImGui:

```
    float occl = 15.0 * max( dist * (2.0-dist), 0.0 );
    att += 1.0 / (1.0 + occl*occl);
  }
  att = clamp(att*att/64 + 0.45, 0, 1) * pc.attScale;
  imageStore( kTextures2DOut[pc.texOut], ivec2(xy), vec4( vec3(att), 1.0 ) );
}
```

While this method does not even come close to the best screen-space ambient occlusion implementations, it is very simple in terms of its input parameters and can operate using just a naked depth buffer.

Let's quickly take a look at how to blur the SSAO values. The blurring compute shader is located in `data/shaders/Blur.comp`.

1. There is a table containing `17` weights for the Gaussian blur, which add up to `1.0`. You can compute your own set of weights for any number of values using this online tool: `https://drdesten.github.io/web/tools/gaussian_kernel`.

```
const int kFilterSize = 17;
const float gaussWeights[kFilterSize] = float[](
  0.00001525878906, 0.0002441406250, … );
```

2. We use a specialization constant to distinguish between horizontal and vertical blur. All shader parameters are passed as push constants. This bilateral blur shader requires the depth buffer to calculate depth discontinuities.

```
layout(constant_id = 0) const bool kIsHorizontal=true;
layout(push_constant) uniform PushConstants {
  uint texDepth;
  uint texIn;
  uint texOut;
  float depthThreshold;
} pc;
```

3. The main() function calculates the input texture size, the xy texel position, and the normalized texture coordinates, texCoord.

```
void main() {
  const vec2 size = textureBindlessSize2D(pc.texIn).xy;
  const vec2 xy = gl_GlobalInvocationID.xy;
  if (xy.x > size.x || xy.y > size.y) return;
  const vec2 texCoord = (gl_GlobalInvocationID.xy + vec2(0.5)) / size;
  const float texScaler = 1.0 / (kIsHorizontal ? size.x : size.y);
  vec3 c = vec3(0.0);
  vec3  fragColor = textureBindless2D(pc.texIn,    texCoord).rgb;
  float fragDepth = textureBindless2D(pc.texDepth, texCoord).r;
```

4. Gaussian averaging is performed inside this for loop.

```
  for ( int i = 0; i != kFilterSize; i++ ) {
    float offset = float(i - kFilterSize/2);
    vec2 uv = texCoord + texScaler * (kIsHorizontal ?
      vec2(offset, 0) : vec2(0, offset));
    vec3  color = textureBindless2D(pc.texIn, uv).rgb;
    float depth = textureBindless2D(pc.texDepth, uv).r;
```

5. The most interesting part here is the bilateral blur calculation. We compute the depth difference weight by comparing the depth value difference between the current offset pixel and the central pixel `fragDepth`. This difference is then scaled by an ad hoc `pc.depthThreshold` factor, which is controlled through ImGui. The resulting `weight` value is used to interpolate between the center pixel color `fragColor` and the offset pixel color, effectively reducing the amount of blur between pixels with significantly different depth values.

```
    float w = clamp(abs(depth - fragDepth) * pc.depthThreshold, 0.0, 1.0);
    c += mix(color, fragColor, w) * gaussWeights[i];
  }
  imageStore(kTextures2DOut[pc.texOut], ivec2(xy), vec4(c, 1.0) );
}
```

This simple bilateral blur technique greatly improves the visual quality of our SSAO effect.

To combine the resulting SSAO effect with the rendered scene, the following full-screen GLSL fragment shader should be used, which is located in the `Chapter10/04_SSAO/src/combine.frag` file. The `scale` and `bias` values are controlled through ImGui.

```
layout (location=0) in vec2 uv;
layout (location=0) out vec4 out_FragColor;
layout(push_constant) uniform PushConstants {
  uint texColor;
  uint texSSAO;
  uint smpl;
  float scale;
  float bias;
} pc;
void main() {
  vec4 color = textureBindless2D(pc.texColor, pc.smpl, uv);
  float ssao = clamp(textureBindless2D(
    pc.texSSAO,  pc.smpl, uv).x + pc.bias, 0.0, 1.0 );
  out_FragColor = vec4(mix(color, color * ssao, pc.scale).rgb, 1.0 );
}
```

Now, our SSAO implementation is complete. The demo application should render the following image:

Figure 10.10: SSAO demo

With all this knowledge, you should be able to add a similar SSAO effect to your rendering engine. In the next recipe, we will learn how to implement a more complex post-processing scheme for HDR rendering and tone mapping.

There's more...

The SSAO algorithm used here is very naïve and basic. For more advanced techniques, look into modern implementations like **horizon-based ambient occlusion** (**HBAO**) or **ground truth ambient occlusion** (**GTAO**).

While running the demo, you may have noticed that the SSAO effect behaves strangely on transparent surfaces. This is understandable, as our transparency rendering is done using punch-through transparency, where part of the transparent surface pixels are discarded based on their transparency value. These holes expose the depth values beneath the transparent surface, causing the SSAO effect to work inconsistently. In a real-world rendering engine, you would likely want to calculate the SSAO effect after the opaque objects have been fully rendered and before any transparent objects affect the depth buffer.

Implementing HDR rendering and tone mapping

In all our previous examples, the color values in the framebuffer were always clamped between `0.0` and `1.0`. Additionally, we used 1 byte for each color component in the framebuffer, allowing for only 256 levels of brightness. This limits the contrast ratio between the darkest and brightest areas of the image to `255:1`.

While this may be sufficient for many applications, it poses a problem in scenarios with very bright regions, such as areas illuminated by the sun or multiple light sources. In such cases, values exceeding `1.0` are clipped, and any brightness information above this limit is lost. HDR brightness values preserve these details, improving image realism, and can be mapped back into the `0.0` to `1.0` **low dynamic range** (**LDR**) using various tone mapping techniques.

In this recipe, we will walk through setting up HDR rendering and cover all the steps needed to implement a basic tone mapping pipeline.

Getting ready

You can find the source code for this recipe in the `Chapter10/05_HDR` folder.

Before diving in, be sure to review the previous recipes: *Implementing screen space ambient occlusion*, *Implementing full-screen triangle rendering*, and *Implementing offscreen rendering in Vulkan*.

How to do it...

To implement HDR rendering, we need to store HDR values in the framebuffers. This can be achieved using Vulkan floating-point color formats like `VK_FORMAT_R16G16B16A16_SFLOAT`, a 16-bit floating-point RGBA format suitable for rendering.

Other floating-point pixel formats, like `A2B10G10R10_SNORM_PACK32` or `B10G11R11_UFLOAT_PACK32`, are more memory-efficient and can be used in many scenarios.

After rendering the scene into a floating-point HDR framebuffer, we can calculate the average luminance of the HDR image for our tone mapping calculations. Additionally, we can identify high-luminance areas and use them to simulate the bloom effect seen in real-world cameras: https://en.wikipedia.org/wiki/Bloom_(shader_effect).

Let's walk through the C++ code in `Chapter10/05_HDR/src/main.cpp` to understand the complete HDR rendering and tone mapping pipeline:

1. As usual, the `main()` function loads the Bistro mesh and sets up all the necessary objects. The Bistro scene is rendered using the shaders located in `Chapter08/02_SceneGraph/src/main.*`, which apply image space lighting to the Bistro model. This process is similar to what was done in *Chapter 6*, except that this time, only the diffuse component of image-based lighting (IBL) is used. This ensures that the Bistro mesh has some HDR lighting applied to it. Additionally, the skybox uses an HDR image for the cube map, contributing HDR values as well.

```
int main() {
  MeshData meshData;
  Scene scene;
  loadBistro(meshData, scene);
  VulkanApp app({…});
  std::unique_ptr<lvk::IContext> ctx(app.ctx_.get());
  const uint32_t kNumSamples = 8;
  const Format kOffscreenFormat = Format_RGBA_F16;
  const Skybox skyBox(ctx, …);
  const VKMesh mesh(ctx, …);
```

2. Next, we create a compute pipeline using the `BrightPass.comp` compute shader, which identifies the bright areas in the rendered scene and converts the entire HDR scene to 16-bit luminance. We'll explore this further in the *How it works...* section.

```
  Holder<ShaderModuleHandle> compBrightPass =
    loadShaderModule(ctx, "Chapter10/05_HDR/src/BrightPass.comp");
  Holder<ComputePipelineHandle>
    pipelineBrightPass = ctx->createComputePipeline({
      .smComp = compBrightPass });
```

3. The next step is to create two compute pipelines for the Bloom effect using the `Chapter10/05_HDR/src/Bloom.comp` shader. The Bloom effect is somewhat similar to the blur effect described in the previous recipe, *Implementing screen space ambient occlusion*, and is implemented in a similar manner. Two passes are used to generate the horizontal and vertical components of the bloom, utilizing ping-pong buffers. This shader is simpler than the blur shader because it does not perform bilateral filtering.

```
const uint32_t kHorizontal = 1;
const uint32_t kVertical   = 0;
Holder<ShaderModuleHandle> compBloomPass =
  loadShaderModule(ctx, "Chapter10/05_HDR/src/Bloom.comp");
Holder<ComputePipelineHandle> pipelineBloomX =
  ctx->createComputePipeline({
    .smComp   = compBloomPass,
    .specInfo = {.entries = {{
      .constantId = 0, .size = sizeof(uint32_t) }},
    .data = &kHorizontal,
    .dataSize = sizeof(uint32_t)} });
Holder<ComputePipelineHandle> pipelineBloomY =
  ctx->createComputePipeline({
    .smComp   = compBloomPass,
    .specInfo = {.entries = {{
      .constantId = 0, .size = sizeof(uint32_t) }},
    .data = &kVertical,
    .dataSize = sizeof(uint32_t)} });
```

4. A graphics pipeline is needed for tone mapping, which renders a full-screen rectangle onto a swapchain image. The tone mapping shader is particularly interesting, and we will explore it in the *How it works...* section.

```
Holder<ShaderModuleHandle> vertToneMap =
  loadShaderModule(ctx, "data/shaders/QuadFlip.vert");
Holder<ShaderModuleHandle> fragToneMap =
  loadShaderModule(ctx, "Chapter10/05_HDR/src/ToneMap.frag");
Holder<RenderPipelineHandle> pipelineToneMap =
  ctx->createRenderPipeline({
    .smVert = vertToneMap,
    .smFrag = fragToneMap,
    .color  =
      { { .format = ctx->getSwapchainFormat() } }});
```

5. A `Clamp` sampler is needed to prevent edge texels from wrapping around the texture edges.

```
Holder<SamplerHandle> samplerClamp =
  ctx->createSampler({
    .wrapU = lvk::SamplerWrap_Clamp,
    .wrapV = lvk::SamplerWrap_Clamp,
    .wrapW = lvk::SamplerWrap_Clamp });
```

6. Next, let's create textures for our offscreen render targets. The first set of textures is for multisampled rendering, as explained in the recipe *Implementing MSAA in Vulkan*. It is worth noting that `kOffscreenFormat` here is an HDR format, `Format_RGBA_F16`, which corresponds to the Vulkan format `VK_FORMAT_R16G16B16A16_SFLOAT`. It can store color values outside the `0..1` range.

```
const lvk::Dimensions sizeFb =
  ctx->getDimensions(ctx->getCurrentSwapchainTexture());
Holder<TextureHandle> msaaColor = ctx->createTexture({
    .format     = kOffscreenFormat,
    .dimensions = sizeFb,
    .numSamples = kNumSamples,
    .usage      = lvk::TextureUsageBits_Attachment });
Holder<TextureHandle> msaaDepth = ctx->createTexture({
    .format     = app.getDepthFormat(),
    .dimensions = sizeFb,
    .numSamples = kNumSamples,
    .usage      = lvk::TextureUsageBits_Attachment });
```

7. The `offscreenColor` texture is used to resolve the MSAA color information. The depth texture is not used in our HDR post-processing so we do not have to resolve it.

```
Holder<TextureHandle> offscreenColor =
  ctx->createTexture({
    .format     = kOffscreenFormat,
    .dimensions = sizeFb,
    .usage      = lvk::TextureUsageBits_Attachment |
                  lvk::TextureUsageBits_Sampled |
                  lvk::TextureUsageBits_Storage });
```

8. The bright pass and bloom textures are smaller, with a resolution of 512x512. They are supposed to capture only low-frequency image details and don't need to be high-resolution.

```
const lvk::Dimensions sizeBloom = { 512, 512 };
Holder<TextureHandle> texBrightPass = ctx->createTexture({
    .format     = kOffscreenFormat,
    .dimensions = sizeBloom,
```

```
    .usage      = lvk::TextureUsageBits_Sampled |
                  lvk::TextureUsageBits_Storage });
Holder<TextureHandle> texBloomPass = ctx->createTexture({
     .format     = kOffscreenFormat,
     .dimensions = sizeBloom,
     .usage      = lvk::TextureUsageBits_Sampled |
                   lvk::TextureUsageBits_Storage });
```

9. Let's create a texture to store the average scene luminance. The goal is to convert the rendered scene into a single-channel 16-bit image format representing luminance and then downscale it to a `1x1` texture using the mipmapping pyramid. The single pixel in the `1x1` mip level represents the average luminance. For debugging and demonstration purposes, we will display the mip pyramid's contents via ImGui, using the texture views approach described in the recipe *Implementing offscreen rendering in Vulkan*. Component swizzling is applied to render the single-channel R image as a grayscale image. A mip pyramid for a `512x512` texture consists of `10` mip levels.

```
const lvk::ComponentMapping swizzle = {
  .r = lvk::Swizzle_R, .g = lvk::Swizzle_R,
  .b = lvk::Swizzle_R, .a = lvk::Swizzle_1 };
Holder<TextureHandle> texLumViews[10] = {
  ctx->createTexture({
    .format     = lvk::Format_R_F16,
    .dimensions = sizeBloom,
    .usage      = lvk::TextureUsageBits_Sampled |
                  lvk::TextureUsageBits_Storage,
    .numMipLevels =
      lvk::calcNumMipLevels(sizeBloom.width, sizeBloom.height),
    .swizzle = swizzle }) };
```

10. In the `texLumViews[]` array, only the `0`-th element is the actual texture; the rest are texture views.

```
for (uint32_t v = 1; v != LVK_ARRAY_NUM_ELEMENTS(texLumViews); v++)
  texLumViews[v] =
    ctx->createTextureView(texLumViews[0],
      { .mipLevel = v, .swizzle = swizzle });
```

11. Here are two ping-pong textures used to compute the bloom effect over multiple passes.

```
Holder<TextureHandle> texBloom[] = {
  ctx->createTexture({
    .format     = kOffscreenFormat,
    .dimensions = sizeBloom,
    .usage      = lvk::TextureUsageBits_Sampled |
                  lvk::TextureUsageBits_Storage }),
```

```
    ctx->createTexture({
      .format     = kOffscreenFormat,
      .dimensions = sizeBloom,
      .usage      = lvk::TextureUsageBits_Sampled |
                    lvk::TextureUsageBits_Storage }),
  };
```

12. We are almost ready to start rendering. We just need a few variables to tweak the rendering using the ImGui UI. The `ToneMappingMode` enum lets us switch between different tone mapping operators, such as Reinhard, Uchimura, and the Khronos PBR Neutral Tone Mapper, for easy comparison. We also set up an `ImPlotContext` to visualize our tone mapping curves within the ImGui UI using the ImPlot library.

```
bool drawCurves     = true;
bool enableBloom    = true;
float bloomStrength = 0.1f;
int numBloomPasses  = 2;
enum ToneMappingMode {
  ToneMapping_None        = 0,
  ToneMapping_Reinhard    = 1,
  ToneMapping_Uchimura    = 2,
  ToneMapping_KhronosPBR = 3,
};
ImPlotContext* implotCtx = ImPlot::CreateContext();
```

13. The `pcHDR` variable is used to pass push constants to our HDR tone mapping shader. It includes the default values for all the different tone mapping algorithms, and its parameters can be adjusted dynamically through the UI.

```
struct {
  uint32_t texColor;
  uint32_t texLuminance;
  uint32_t texBloom;
  uint32_t sampler;
  int drawMode = ToneMapping_Uchimura;
  float exposure      = 1.0f;
  float bloomStrength = 0.1f;
  // Reinhard
  float maxWhite = 1.0f;
  // Uchimura
  float P = 1.0f;  // max display brightness
  float a = 1.05f; // contrast
  float m = 0.1f;  // linear section start
```

```
    float l = 0.8f;  // linear section length
    float c = 3.0f;  // black tightness
    float b = 0.0f;  // pedestal
    // Khronos PBR
    float startCompression = 0.8f; // highlight compression start
    float desaturation     = 0.15f; // desaturation speed
  } pcHDR = {
    .texColor     = offscreenColor.index(),
    .texLuminance = texLumViews[LVK_ARRAY_NUM_ELEMENTS(texLumViews)-1].index(),
    .texBloom     = texBloomPass.index(),
    .sampler      = samplerClamp.index(),
  };
```

With all these preparations completed, we can proceed to the rendering loop. Let's go through it step by step.

1. We begin by updating the view and projection matrices, rendering the scene into the multi-sampled render targets, and resolving the color data into the `offscreenColor` texture. Since the depth texture is not used in this HDR demo, it is not resolved.

```
app.run([&](uint32_t width, uint32_t height,
  float aspectRatio, float deltaSeconds) {
  const mat4 view = app.camera_.getViewMatrix();
  const mat4 proj = glm::perspective(…);
  ICommandBuffer& buf = ctx->acquireCommandBuffer();
  buf.cmdBeginRendering(
    lvk::RenderPass{
      .color = { { .loadOp  = LoadOp_Clear,
                   .storeOp = StoreOp_MsaaResolve,
                   .clearColor = { 1, 1, 1, 1 } } },
      .depth = { .loadOp = lvk::LoadOp_Clear,
                 .clearDepth = 1.0f } },
    lvk::Framebuffer{
      .color = {{.texture = msaaColor,
                 .resolveTexture = offscreenColor}},
      .depthStencil = { .texture = msaaDepth } });
```

2. We render the skybox, the Bistro mesh, and the infinite grid. To simplify handling transparency, we render the skybox first.

Note

The skybox is rendered first here for simplicity. In a real application, it's usually better to avoid processing sky fragments that will ultimately be occluded.

```
skyBox.draw(buf, view, proj);
mesh.draw(buf, view, proj, skyBox.texSkyboxIrradiance, drawWireframe);
app.drawGrid(buf, proj, vec3(0, -1.0f, 0), kNumSamples, kOffscreenFormat);
buf.cmdEndRendering();
```

3. Once we finish rendering the main scene into a floating-point render target, we start our post-processing pipeline. Bright regions are extracted from the rendered image and stored in the `texBrightPass` texture, and the scene converted to luminance is stored in `texLumViews[0]`. The compute shader used is `Chapter10/05_HDR/src/BrightPass.comp`.

```
const struct {
  uint32_t texColor;
  uint32_t texOut;
  uint32_t texLuminance;
  uint32_t sampler;
  float exposure;
} pcBrightPass = {
  .texColor     = offscreenColor.index(),
  .texOut       = texBrightPass.index(),
  .texLuminance = texLumViews[0].index(),
  .sampler      = samplerClamp.index(),
  .exposure     = pcHDR.exposure,
};
buf.cmdBindComputePipeline(pipelineBrightPass);
buf.cmdPushConstants(pcBrightPass);
```

4. Our compute shader operates on `16x16` texel blocks. Once the luminance texture is ready, we generate the entire mip pyramid down to `1x1`, which represents the average scene luminance.

```
buf.cmdDispatchThreadGroups(
  sizeBloom.divide2D(16), { .textures =
    { TextureHandle(offscreenColor), TextureHandle(texLumViews[0])} });
buf.cmdGenerateMipmap(texLumViews[0]);
```

The resulting luminance mip pyramid is later rendered via ImGui and should appear as shown in the following screenshot.

Figure 10.11: Mip pyramid to compute average luminance

5. Next, we return to the bright areas of the scene, which were extracted into the `texBrightPass` texture, and apply a multipass separable Gaussian filter to simulate the Bloom effect. The bright areas will be heavily blurred, creating halos that leak into the adjacent pixels. The technique used here for the multipass bloom effect is exactly the same as the one used for the blur in the previous recipe, *Implementing screen space ambient occlusion*. We leave the code here without comments to help better understand the flow of the subsequent code.

```
struct BlurPC {
  uint32_t texIn;
  uint32_t texOut;
  uint32_t sampler;
};
struct StreaksPC {
  uint32_t texIn;
  uint32_t texOut;
  uint32_t texRotationPattern;
  uint32_t sampler;
};
struct BlurPass {
  TextureHandle texIn;
  TextureHandle texOut;
};
std::vector<BlurPass> passes;
passes.reserve(2 * numBloomPasses);
passes.push_back({ texBrightPass, texBloom[0] });
for (int i = 0; i != numBloomPasses - 1; i++) {
  passes.push_back({ texBloom[0], texBloom[1] });
  passes.push_back({ texBloom[1], texBloom[0] });
}
passes.push_back({ texBloom[0], texBloomPass });
for (uint32_t i = 0; i != passes.size(); i++) {
  const BlurPass p = passes[i];
  buf.cmdBindComputePipeline(i & 1 ? pipelineBloomX : pipelineBloomY);
  buf.cmdPushConstants(BlurPC{
    .texIn   = p.texIn.index(),
    .texOut  = p.texOut.index(),
    .sampler = samplerClamp.index() });
  if (enableBloom)
    buf.cmdDispatchThreadGroups(
      sizeBloom.divide2D(16), {
        .textures = { p.texIn, p.texOut, TextureHandle(texBrightPass)} });
}
```

6. Now that everything is set up, we can combine all the intermediate textures and perform the most complex part – tone mapping. We render a full-screen triangle into a swapchain image using our tone mapping shader `Chapter10/05_HDR/src/ToneMap.frag`. The input textures for the shader are the resolved HDR scene image, the average luminance value, and the blurred bloom texture, all of which are stored in the `pcHDR` variable and passed into the shader as push constants.

```
const lvk::RenderPass renderPassMain = {
  .color = { { .loadOp = lvk::LoadOp_Load,
               .clearColor = { 1, 1, 1, 1 } } } };
const lvk::Framebuffer framebufferMain = {
  .color = { { .texture = ctx->getCurrentSwapchainTexture() } } };
```

7. The tone mapping shader uses only the last `1x1` level of the mip pyramid. However, LightweightVK implements image memory barriers at the Vulkan resource level `VkImage`, not at the subresource level of individual texture mip levels. This requires us to transition the entire mip pyramid to `VK_IMAGE_LAYOUT_SHADER_READ_ONLY_OPTIMAL`.

```
buf.cmdBeginRendering(
  renderPassMain, framebufferMain,
  { .textures = { TextureHandle(texLumViews[0]) } });
buf.cmdBindRenderPipeline(pipelineToneMap);
buf.cmdPushConstants(pcHDR);
buf.cmdBindDepthState({});
buf.cmdDraw(3);
```

Note

This "greedy" image layout transition is fine for a demo like this, but it can become a significant inefficiency in a real rendering engine, especially when multiple effects are rendered in parallel and reuse different Vulkan subresources. Instead of tracking the Vulkan resource state at the subresource level for individual mip levels, a more generic solution exists: the frame graph. A frame graph helps plan and schedule all required operations on GPU and CPU resources for a frame in the most efficient way. It is a complex topic, and to learn more, we recommend reading the GDC presentation *FrameGraph: Extensible Rendering Architecture in Frostbite* by Yuriy O'Donnell and the book *Mastering Graphics Programming with Vulkan* by Marco Castorina and Gabriel Sassone, published by Packt.

Technically, at this point, the final scene image has been tone-mapped and rendered into a swapchain image, so we can conclude our C++ work here and call it a day. However, in this recipe, we will also walk through some parts of the ImGui UI rendering. Besides controlling the tone mapping parameters, we use it to display our tone mapping curves. This can be quite useful for graphics debugging and will complement the material covered in *Chapter 5*.

1. First, we create an ImGui window with a fixed screen position and a fixed size relative to the size of the screen.

```
const ImGuiViewport* v = ImGui::GetMainViewport();
const float windowWidth = v->WorkSize.x / 5;
ImGui::SetNextWindowPos(ImVec2(10, 200));
ImGui::SetNextWindowSize(ImVec2(0, v->WorkSize.y - 210));
ImGui::Begin("HDR", nullptr, ImGuiWindowFlags_NoFocusOnAppearing);
```

2. Next, we add a series of checkboxes and sliders to control the generic rendering and tone mapping parameters that are shared across all our tone mapping modes.

```
ImGui::Checkbox("Draw tone curves", &drawCurves);
ImGui::SliderFloat("Exposure", &pcHDR.exposure, 0.1f, 2.0f);
ImGui::Checkbox("Enable bloom", &enableBloom);
pcHDR.bloomStrength = enableBloom?bloomStrength:0;
ImGui::SliderFloat("Bloom strength", &bloomStrength, 0.0f, 1.0f);
ImGui::SliderInt("Bloom num passes", &numBloomPasses, 1, 5);
```

3. Here are the radio buttons to select between different tone mapping modes. Each mode has its own set of parameters, which are displayed only when the corresponding mode is selected.

```
ImGui::Text("Tone mapping mode:");
ImGui::RadioButton("None", &pcHDR.drawMode, ToneMapping_None);
ImGui::RadioButton("Reinhard", &pcHDR.drawMode, ToneMapping_Reinhard);
if (pcHDR.drawMode == ToneMapping_Reinhard) {
  ImGui::SliderFloat("Max white", &pcHDR.maxWhite, 0.5f, 2.0f);
}
// … parameters for other tone mapping modes…
ImGui::Separator();
```

4. Then, we display all the intermediate textures: the `1x1` average luminance texture, the bright pass texture, and the bloom pass texture.

```
ImGui::Text("Average luminance 1x1:");
ImGui::Image(pcHDR.texLuminance, ImVec2(128, 128));
ImGui::Text("Bright pass:");
ImGui::Image(texBrightPass.index(),
  ImVec2(windowWidth, windowWidth / aspectRatio));
ImGui::Text("Bloom pass:");
ImGui::Image(texBloomPass.index(),
  ImVec2(windowWidth, windowWidth / aspectRatio));
```

5. The entire luminance mip pyramid can be displayed this way using texture views. Since we set up texture component swizzling, our single-channel R luminance texture is rendered as a grayscale texture. Note how we pass the unsigned integer texture `index()` value directly into ImGui. This is possible starting from ImGui 1.91.4, as the `ImTextureID` type was changed from `void*` to `uint64_t`, making it much more convenient to use with LightweightVK.

```
ImGui::Text("Luminance pyramid 512x512");
for (uint32_t v = 0; v != LVK_ARRAY_NUM_ELEMENTS(texLumViews); v++) {
  ImGui::Image( texLumViews[v].index(),
                ImVec2((int)windowWidth >> v,
                       (int)windowWidth >> v));
}
ImGui::End();
```

6. Now comes the most interesting part: drawing the tone mapping curves. These are the graphs of tone mapping functions that convert HDR image values into final LDR image values.

```
if (drawCurves) {
  const ImGuiWindowFlags flags = ImGuiWindowFlags_NoDecoration | …;
  ImGui::SetNextWindowBgAlpha(0.8f);
  ImGui::SetNextWindowPos({ width * 0.6f, height * 0.7f },
    ImGuiCond_Appearing);
  ImGui::SetNextWindowSize({ width * 0.4f, height * 0.3f });
  ImGui::Begin("Tone mapping curve", nullptr, flags);
```

7. We will sample `1001` points from each function, in the range from `0` to `1`, and store the resulting function values in the arrays `ysUchimura`, `ysReinhard2`, and `ysKhronosPBR`. The tone mapping functions, declared in the `shared/Tonemap.h` file, are C++ mirror images of their GLSL implementations, which are used in the tone mapping shader. We will examine them shortly. The functions take in the same push constant values from `pcHDR`, which are passed into the shader. This means that when we tweak the parameters via ImGui, the tone mapping graphs change their shape accordingly.

```
  const int kNumGraphPoints = 1001;
  float xs[kNumGraphPoints];
  float ysUchimura[kNumGraphPoints];
  float ysReinhard2[kNumGraphPoints];
  float ysKhronosPBR[kNumGraphPoints];
  for (int i = 0; i != kNumGraphPoints; i++) {
    xs[i] = float(i) / kNumGraphPoints;
    ysUchimura[i] = uchimura(xs[i], pcHDR.P,
      pcHDR.a, pcHDR.m, pcHDR.l, pcHDR.c, pcHDR.b);
    ysReinhard2[i] = reinhard2(xs[i], pcHDR.maxWhite);
    ysKhronosPBR[i] = PBRNeutralToneMapping(xs[i],
```

```
        pcHDR.startCompression, pcHDR.desaturation);
    }
```

8. We use the sampled values to draw the graphs using the ImPlot library. You can interact with the graphs in the running demo.

```
    if (ImPlot::BeginPlot("Tone mapping curves",
          { width * 0.4f, height * 0.3f }, ImPlotFlags_NoInputs))
    {
      ImPlot::SetupAxes("Input", "Output");
      ImPlot::PlotLine("Uchimura", xs, ysUchimura, kNumGraphPoints);
      ImPlot::PlotLine("Reinhard", xs, ysReinhard2, kNumGraphPoints);
      ImPlot::PlotLine("Khronos PBR", xs, ysKhronosPBR, kNumGraphPoints);
      ImPlot::EndPlot();
    }
    ImGui::End();
  }
```

9. After the ImGui rendering is complete, we end the frame, submit the entire command buffer for execution, and present the rendered swapchain image.

```
      app.imgui_->endFrame(buf);
      buf.cmdEndRendering();
      ctx->submit(buf, ctx->getCurrentSwapchainTexture());
    });
    ImPlot::DestroyContext(implotCtx);
    // ...
  }
```

The C++ part was relatively long due to the ImGui and ImPlot code. However, the overall flow of the HDR pipeline is quite similar to the previous SSAO one. Now, let's dive into the GLSL shader code to understand how the actual tone mapping works.

How it works...

The real actual work is done in the shaders. The first shader is the compute shader `Chapter10/05_HDR/src/BrightPass.comp`, which is responsible for extracting the bright areas and converting the scene image to luminance. Let's take a look.

1. The shader takes in the HDR scene 16-bit RGBA color texture `texColor` and computes two textures: one colored 16-bit texture `texOut` for the bright RGBA areas, and one 16-bit single-channel luminance image `texLuminance`.

```
layout (local_size_x = 16, local_size_y = 16) in;
layout (set = 0, binding = 2, rgba16)
  uniform writeonly image2D kTextures2DOutRGBA[];
```

```
layout (set = 0, binding = 2, r16)
  uniform writeonly image2D kTextures2DOutR[];
layout(push_constant) uniform PushConstants {
  uint texColor;     // rgba16
  uint texOut;       // rgba16
  uint texLuminance; // r16
  uint smpl;
  float exposure;
} pc;
void main() {
  const vec2 sizeIn  = textureBindlessSize2D(pc.texColor).xy;
  const vec2 sizeOut = textureBindlessSize2D(pc.texOut).xy;
  const vec2 xy = gl_GlobalInvocationID.xy;
  const vec2 uv0 = (gl_GlobalInvocationID.xy + vec2(0)) / sizeOut;
  const vec2 uv1 = (gl_GlobalInvocationID.xy + vec2(1)) / sizeOut;
  if (xy.x > sizeIn.x || xy.y > sizeIn.y) return;
```

2. The bright areas texture is a downscaled `512x512` texture. We use a `3x3` box filter to soften the results.

Note

This 3x3 box filter technique will still result in some visible fireflies – very bright, pulsating pixels caused by high-intensity values in the HDR render target. To reduce this effect, check out the Karis average algorithm from `https://www.iryoku.com/next-generation-post-processing-in-call-of-duty-advanced-warfare`, which adds temporal stability.

```
  vec2 dxdy = (uv1-uv0) / 3;
  vec4 color = vec4(0);
  for (int v = 0; v != 3; v++)
    for (int u = 0; u != 3; u++)
      color += textureBindless2D(pc.texColor, uv0 + vec2(u, v) * dxdy );
```

3. We use the dot product to convert the RGB color information to luminance. The `exposure` scaling factor, which is controlled through ImGui, is applied at this stage. The pixel value ends up in the bright areas image only when its luminance value is greater than `1.0`.

```
  float luminance = pc.exposure * dot(
    color.rgb / 9, vec3(0.2126, 0.7152, 0.0722));
  vec3 rgb = luminance > 1.0 ? color.rgb : vec3(0);
  imageStore(kTextures2DOutRGBA[pc.texOut],     ivec2(xy), vec4( rgb, 1.0 ) );
  imageStore(kTextures2DOutR[pc.texLuminance], ivec2(xy), vec4(luminance ) );
}
```

The second shader is the Bloom shader, `Chapter10/05_HDR/src/Bloom.comp`. It generates bloom using the separable blur technique described in the previous recipe, *Implementing screen space ambient occlusion*. It uses the same Gaussian coefficients but does not require the bilateral blur technique. Here's its `main()` function for reference:

```
void main() {
  const vec2 size = textureBindlessSize2D(pc.texIn).xy;
  const vec2 xy   = gl_GlobalInvocationID.xy;
  if (xy.x > size.x || xy.y > size.y) return;
  const vec2 texCoord = (gl_GlobalInvocationID.xy + vec2(0.5)) / size;
  const float texScaler =  1.0 / (kIsHorizontal ? size.x : size.y);
  vec3 c = vec3(0.0);
  for ( int i = 0; i != kFilterSize; i++ ) {
    float offset = float(i - kFilterSize/2);
    vec2 uv = texCoord + texScaler * (kIsHorizontal ?
      vec2(offset, 0) : vec2(0, offset));
    c += textureBindless2D(pc.texIn, uv).rgb * gaussWeights[i];
  }
  imageStore(kTextures2DOut[pc.texOut], ivec2(xy), vec4(c, 1.0) );
}
```

The last shader is `Chapter10/05_HDR/src/ToneMap.frag`, which performs the tone mapping and renders the resulting tone-mapped scene into a swapchain image. Since the main heavy lifting happens here, let's take a close look.

1. The first step is to declare the shader inputs. The tone mapping modes correspond to the C++ `ToneMappingMode` enum, and the push constant structure corresponds to the `pcHDR` variable from C++.

```
layout (location=0) in vec2 uv;
layout (location=0) out vec4 out_FragColor;
const int ToneMappingMode_None = 0;
const int ToneMappingMode_Reinhard = 1;
const int ToneMappingMode_Uchimura = 2;
const int ToneMappingMode_KhronosPBR = 3;
layout(push_constant) uniform PushConstants {
  uint texColor;
  uint texLuminance;
  uint texBloom;
  uint smpl;
  int drawMode;
  float exposure;
  float bloomStrength;
  // Reinhard
```

```
    float maxWhite;
    // Uchimura
    float P;  // max display brightness
    float a;  // contrast
    float m;  // linear section start
    float l;  // linear section length
    float c;  // black tightness
    float b;  // pedestal
    // Khronos PBR
    float startCompression;  // highlight compression start
    float desaturation;      // desaturation speed
  } pc;
```

2. The core of the tone mapping algorithm consists of the tone mapping functions. We use three different ones. The first is `reinhard2()`, which is described in a blog post titled *Tone Mapping* by Matt Taylor: `https://64.github.io/tonemapping`. This is often referred to as the extended Reinhard tone mapping operator. Here, the `maxWhite` value is adjusted to represent the maximum brightness in the scene. Any values brighter than this will be mapped to `1.0`. The mapping is calculated based on the luminance values and not RGB values.

```
float luminance(vec3 v) {
  return dot(v, vec3(0.2126, 0.7152, 0.0722));
}
vec3 reinhard2(vec3 v, float maxWhite) {
  float l_old = luminance(v);
  float l_new =
    l_old * (1.0 + (l_old / (maxWhite * maxWhite))) / (1.0 + l_old);
  return v * (l_new / l_old);
}
```

3. The second tone mapping function we use was presented in *HDR theory and practice* by Hajime Uchimura in 2017: `https://www.slideshare.net/nikuque/hdr-theory-and-practicce-jp`. You can find more references in this presentation: `http://cdn2.gran-turismo.com/data/www/pdi_publications/PracticalHDRandWCGinGTS_20181222.pdf`. Its input parameters can be tweaked from ImGui.

```
vec3 uchimura(vec3 x, float P,
  float a, float m, float l, float c, float b) {
  float l0 = ((P - m) * l) / a;
  float L0 = m - m / a;
  float L1 = m + (1.0 - m) / a;
  float S0 = m + l0;
  float S1 = m + a * l0;
  float C2 = (a * P) / (P - S1);
```

```
    float CP = -C2 / P;
    vec3 w0 = vec3(1.0 - smoothstep(0.0, m, x));
    vec3 w2 = vec3(step(m + l0, x));
    vec3 w1 = vec3(1.0 - w0 - w2);
    vec3 T = vec3(m * pow(x / m, vec3(c)) + b);
    vec3 S = vec3(P - (P - S1) * exp(CP * (x - S0)));
    vec3 L = vec3(m + a * (x - m));
    return T * w0 + L * w1 + S * w2;
  }
```

4. The third tone mapping function is *Khronos PBR Neutral Tone Mapper* from `https://github.com/KhronosGroup/ToneMapping/blob/main/PBR_Neutral/README.md#pbr-neutral-specification`.

```
vec3 PBRNeutralToneMapping(vec3 color,
  float startCompression, float desaturation) {
  startCompression -= 0.04;
  float x = min(color.r, min(color.g, color.b));
  float offset = x < 0.08 ? x - 6.25 * x * x : 0.04;
  color -= offset;
  float peak = max(color.r, max(color.g, color.b));
  if (peak < startCompression) return color;
  const float d = 1. - startCompression;
  float newPeak = 1. - d * d / (peak + d - startCompression);
  color *= newPeak / peak;
  float g = 1. - 1. / (desaturation * (peak - newPeak) + 1.);
  return mix(color, newPeak * vec3(1, 1, 1), g);
}
```

5. With all the aforementioned tone mapping functions in place, the `main()` function of our fragment shader is quite straightforward. We sample the color and bloom values for the current fragment, along with the average luminance value from our `1x1` luminance texture.

```
void main() {
  vec3 color = textureBindless2D(pc.texColor, pc.smpl, uv).rgb;
  vec3 bloom = textureBindless2D(pc.texBloom, pc.smpl, uv).rgb;
  float avgLuminance = textureBindless2D(
    pc.texLuminance, pc.smpl, vec2(0.5)).r;
```

6. Then, we apply the corresponding tone mapping function to it using the parameters from push constants.

```
  if (pc.drawMode != ToneMappingMode_None)
    color *= pc.exposure * 0.5/(avgLuminance + 0.001);
  if (pc.drawMode == ToneMappingMode_Reinhard)
```

```
    color = reinhard2(pc.exposure * color, pc.maxWhite);
  if (pc.drawMode == ToneMappingMode_Uchimura)
    color = uchimura(pc.exposure * color,
      pc.P, pc.a, pc.m, pc.l, pc.c, pc.b);
  if (pc.drawMode == ToneMappingMode_KhronosPBR)
    color = PBRNeutralToneMapping(pc.exposure * color,
      pc.startCompression, pc.desaturation);
```

7. On top of the tone-mapped image, we add the bloom texture, scaled by the `bloomStrength` factor.

```
  out_FragColor = vec4(color + pc.bloomStrength * bloom, 1.0);
}
```

That's all! The running demo should render the Bistro scene with a skybox, as shown in the following screenshot. Note how the bloom effect causes the bright sky color to bleed over the edges of the buildings:

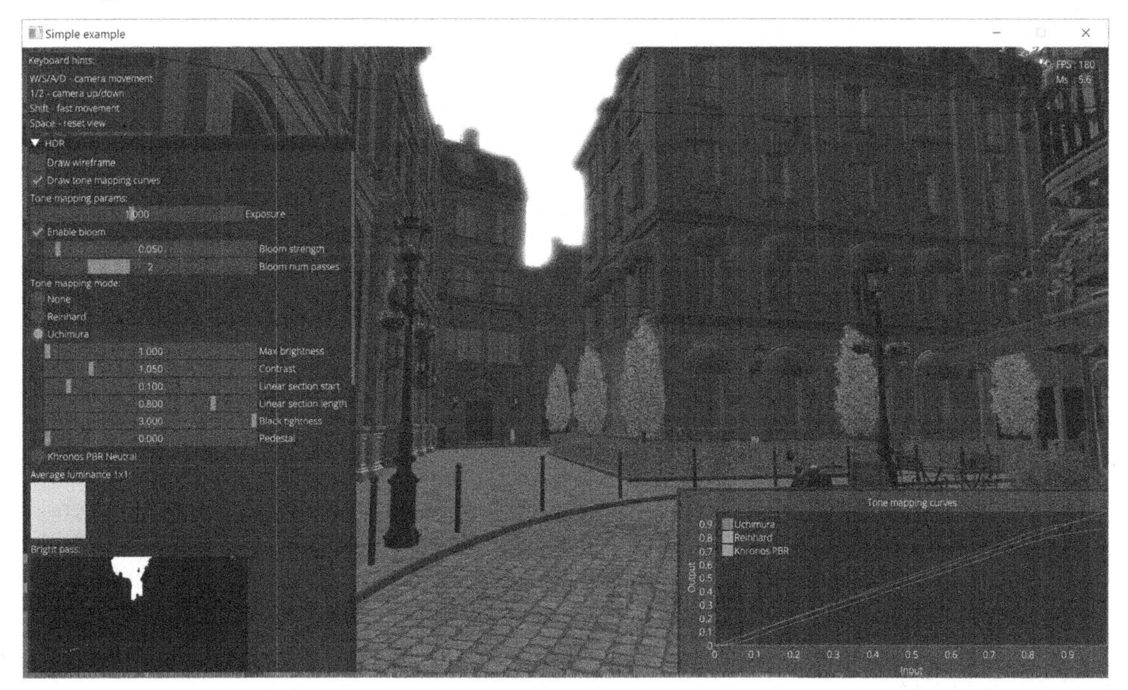

Figure 10.12: A tone-mapped HDR scene

When you move the camera around, you can observe how the scene brightness adjusts based on the current average scene luminance. If you look at the sky, you will see more details in the bright areas, while the rest of the scene becomes darker. If you focus on the dark corners of the buildings, the sky will shift to white. The overall exposure can be manually adjusted in either direction using the ImGui slider.

One downside of this approach is that changes in exposure happen instantly. You look at a different area of the scene, and in the next frame, the exposure changes immediately. This is not how human vision works in reality. It takes time for our eyes to adapt from bright to dark areas. In the next recipe, we will learn how to extend this HDR post-processing pipeline to simulate light adaptation.

There's more...

Strictly speaking, applying a tone mapping operator directly to the RGB channel values is a crude approach. A more accurate model would involve tone mapping the luminance value and then applying it back to the RGB values. However, for many practical purposes, this simple approximation is sufficient. Our goal here is to demonstrate how to build an HDR rendering pipeline, not to strive for perfect physical accuracy.

Certain monitors natively support HDR and can handle swapchain images in Vulkan HDR formats and color spaces, such as `VK_FORMAT_A2B10G10R10_UNORM_PACK32` and `VK_COLOR_SPACE_HDR10_ST2084_EXT`. For more details, check the Vulkan extensions `VK_AMD_display_native_hdr` and `VK_EXT_hdr_metadata`.

Another important remark is regarding how we calculated bloom. The bloom technique used here should be viewed more as a demonstration of how to organize the HDR pipeline in general rather than an attempt to compute a physically accurate bloom effect. The whole problem was described in detail more than 15 years ago in the blog post *Bloom Disasters* by Matthew Gallant: `http://gangles.ca/2008/07/18/bloom-disasters`. A realistic and physically correct post-processing technique with proper bloom can be found in the presentation *Next Generation Post Processing in Call of Duty: Advanced Warfare* by Jorge Jimenez: `https://www.iryoku.com/next-generation-post-processing-in-call-of-duty-advanced-warfare/`. You can also refer to the tutorial *Physically Based Bloom* by Alexander Christensen for an alternative implementation: `https://learnopengl.com/Guest-Articles/2022/Phys.-Based-Bloom`.

Implementing HDR light adaptation

In the previous recipe, *Implementing HDR rendering and tone mapping*, we covered the foundational steps for building a basic HDR tone mapping pipeline. While that was a solid starting point, there is room to make it more realistic by simulating how human vision adapts to changes in brightness. In this recipe, we will extend our HDR post-processing pipeline to incorporate a dynamic light adaptation process, mimicking the way our eyes adjust to both intense light and darker environments over time.

Getting ready

This recipe builds directly on the foundation laid in the previous recipe, *Implementing HDR rendering and tone mapping*. It is strongly recommended to review that recipe first to ensure you have a solid understanding of our HDR rendering pipeline.

The source code for this recipe is located in `Chapter10/06_HDR_Adaptation`.

How to do it...

The light adaptation process operates by gradually transitioning between luminance levels. Rather than using the current average scene luminance value to tone map the scene immediately, we blend it with a previously computed luminance value. This previous value can initially be set to a very high level to simulate an overexposed state. The blending is done gradually in every frame through interpolation in a compute shader, creating a smooth transition between the old and new luminance values.

To incorporate this light adaptation approach into our previous HDR tone mapping example, we need to make a few additions to the C++ code in `Chapter10/06_HDR_Adaptation/src/main.cpp`. Since we are building directly on the previous example, we will focus only on the new code here:

1. First, we need to load a compute shader responsible for performing luminance interpolation and create a pipeline for it.

```
Holder<ShaderModuleHandle> compAdaptationPass =
  loadShaderModule(ctx, "Chapter10/06_HDR_Adaptation/src/Adaptation.comp");
Holder<ComputePipelineHandle> pipelineAdaptationPass =
  ctx->createComputePipeline(
    { .smComp = compAdaptationPass });
```

2. The compute shader will process textures and texture views. Specifically, it will take as input the `1x1` texture view representing the average scene luminance, calculated in the previous recipe, *Implementing HDR rendering and tone mapping*, along with a `1x1` texture holding the previously adapted luminance value. The shader will output a new `1x1` texture containing the updated adapted luminance value. By alternating the roles of the previous and new adapted luminance textures, we can iterate this process in each frame using a ping-pong technique. Let's create two 16-bit single-channel textures and initialize them with a very bright white color value of `50`. This initialization simulates the effect of our vision adjusting from a bright area to a darker one, creating a visually pleasing adaptation effect when the application starts.

```
const uint16_t brightPixel = glm::packHalf1x16(50);
const lvk::TextureDesc luminanceTextureDesc{
  .format     = lvk::Format_R_F16,
  .dimensions = {1, 1},
  .usage      = lvk::TextureUsageBits_Sampled |
                lvk::TextureUsageBits_Storage,
  .swizzle    = swizzle,
  .data       = &brightPixel };
Holder<TextureHandle> texAdaptedLum[2] = {
  ctx->createTexture(luminanceTextureDesc),
  ctx->createTexture(luminanceTextureDesc) };
```

3. Before entering the rendering loop, we need to make two additional initialization changes. First, we will add a variable to control the adaptation speed, allowing us to tweak it interactively via the ImGui UI. Second, we will replace the average scene luminance value in the `pcHDR` push constants, previously used for tone mapping, with the adapted luminance value calculated by the adaptation compute shader.

```
float adaptationSpeed = 3.0f;
struct {
  // ...
  uint32_t texLuminance;
  // ...
} pcHDR = {
  // ...
  .texLuminance = texAdaptedLum[0].index(),
  // ...
};
```

4. Within the main rendering loop, just before performing the tone mapping, we execute the adaptation compute shader to update the adapted luminance value. This involves passing the current luminance of the rendered scene, the previously adapted luminance, and the output texture that will store the new adapted luminance value.

```
const struct {
  uint32_t texCurrSceneLuminance;
  uint32_t texPrevAdaptedLuminance;
  uint32_t texNewAdaptedLuminance;
  float adaptationSpeed;
} pcAdaptationPass = {
  .texCurrSceneLuminance = texLumViews[
    LVK_ARRAY_NUM_ELEMENTS(texLumViews)-1].index(),
  .texPrevAdaptedLuminance = texAdaptedLum[0].index(),
  .texNewAdaptedLuminance  = texAdaptedLum[1].index(),
  .adaptationSpeed = deltaSeconds * adaptationSpeed,
};
buf.cmdBindComputePipeline(pipelineAdaptationPass);
buf.cmdPushConstants(pcAdaptationPass);
```

5. Before invoking the compute shader, we need to transition the Vulkan image layout of the entire mip pyramid to `VK_IMAGE_LAYOUT_GENERAL`, in a similar way as we did in the previous recipe, *Implementing HDR rendering and tone mapping*. The compute shader runs on a single texel.

```
buf.cmdDispatchThreadGroups({1, 1}, { .textures = {
  TextureHandle(texLumViews[0]),
  TextureHandle(texAdaptedLum[0]),
  TextureHandle(texAdaptedLum[1]) } });
```

6. After the compute shader finishes, it is necessary to transition the new adapted luminance texture `texAdaptedLum[1]` back to `VK_IMAGE_LAYOUT_SHADER_READ_ONLY_OPTIMAL`. Once we specify the texture dependency, LightweightVK will automatically insert the necessary Vulkan image layout transition, ensuring correct synchronization.

```
buf.cmdBeginRendering(renderPassMain, framebufferMain,
  { .textures = { TextureHandle(texAdaptedLum[1]) }});
// … render tone-mapped scene into a swapchain image
```

There's one subtle but crucial change to the C++ code. After submitting the command buffer for this frame, we need to swap the ping-pong adapted luminance textures. This ensures that the adapted luminance texture from the current frame, `texAdaptedLum[1]`, becomes the "previous" luminance for the next frame.

```
ctx->submit(buf, ctx->getCurrentSwapchainTexture());
std::swap(texAdaptedLum[0], texAdaptedLum[1]);
```

The rest of the C++ workflow remains unchanged, with the addition of one extra ImGui slider to control the adaptation speed, and displaying the adapted luminance texture. We will leave it to you to explore that and instead focus on the addition to the GLSL part of our HDR pipeline: the light adaptation compute shader. It is located in the `Chapter10/06_HDR_Adaptation/src/Adaptation.comp` file and looks as follows:

1. The local workgroup size is set to `(1, 1, 1)`, as we are processing just a single texel. The input images include the current rendered scene luminance (`texCurrSceneLuminance`), the previous adapted luminance (`texPrevAdaptedLuminance`) from the first ping-pong luminance texture, and the output adapted luminance (`texAdaptedOut`), which is stored in the second ping-pong texture.

```
layout (local_size_x = 1, local_size_y = 1) in;
layout (set = 0, binding = 2, r16)
  uniform readonly image2D kTextures2DIn[];
layout (set = 0, binding = 2, r16)
  uniform writeonly image2D kTextures2DOut[];
layout(push_constant) uniform PushConstants {
  uint texCurrSceneLuminance;
  uint texPrevAdaptedLuminance;
  uint texAdaptedOut;
  float adaptationSpeed;
} pc;
```

2. The `main()` function reads a single texel from each input image and computes the adapted luminance using the *8.1.4.3 Adaptation* equation:

$$L_{avg} = L_{avg} + (L - L_{avg}) \times (1 - e^{-dt \cdot c})$$

This is from https://google.github.io/filament/Filament.md.html#mjx-eqn-adaptation. The `dt` value represents the delta time since the previous frame, and `c` is the parameter controlling the rate of light adaptation. Both values have already been pre-multiplied into `pc.adaptationSpeed` in the C++ code during earlier steps. Finally, the adapted luminance texel is written to the output texture using `imageStore()`.

```
void main() {
  float lumCurr = imageLoad(
    kTextures2DIn[pc.texCurrSceneLuminance], ivec2(0, 0)).x;
  float lumPrev = imageLoad(
    kTextures2DIn[pc.texPrevAdaptedLuminance],  ivec2(0, 0)).x;
  float factor = 1.0 - exp(-pc.adaptationSpeed);
  float newAdaptation = lumPrev + (lumCurr - lumPrev) * factor;
  imageStore(kTextures2DOut[pc.texAdaptedOut],
    ivec2(0, 0), vec4(newAdaptation));
}
```

This technique provides a smooth and visually pleasing light adaptation effect when the scene luminance changes abruptly. Try running the demo application in `Chapter10/06_HDR_Adaptation` and experiment by pointing the camera at both bright areas, such as the sky, and darker areas, like the building. The adaptation speed can be tweaked through an ImGui slider, as shown in the screenshot below:

Figure 10.13: An HDR tone-mapped scene with light adaptation

Both the average scene luminance and the adapted luminance 1x1 textures are displayed in the ImGui UI. Try lowering the adaptation speed and watch how the adapted luminance gradually adjusts to match the scene's luminance.

This recipe concludes our exploration of various image-based effects and post-processing techniques. In the next chapter, we will dive into more advanced rendering techniques and optimizations.

There's more...

HDR rendering is a vast topic, and in this chapter, we barely scratched the surface. If you are interested in learning more about advanced HDR lighting and post-processing techniques, we highly recommend watching the GDC 2010 session *Uncharted 2: HDR Lighting* by John Hable at `https://www.gdcvault.com/play/1012351/Uncharted-2-HDR` and the SIGGRAPH 2014 presentation *Next Generation Post Processing in Call of Duty: Advanced Warfare* by Jorge Jimenez at `https://advances.realtimerendering.com/s2014/index.html#_NEXT_GENERATION_POST`.

11

Advanced Rendering Techniques and Optimizations

In this chapter, we'll scratch the surface of more advanced rendering topics and optimizations. We'll also show how to combine multiple effects and techniques into a single graphical application.

This chapter covers the following recipes:

- Refactoring indirect rendering
- Doing frustum culling on the CPU
- Doing frustum culling on the GPU with compute shaders
- Implementing shadows for directional lights
- Implementing order-independent transparency
- Loading texture assets asynchronously
- Putting it all together into a Vulkan demo

Technical requirements

To run the examples in this chapter, you will need a computer with a graphics card supporting Vulkan 1.3. Read *Chapter 1*, for guidance on how to build demo applications from this book.

This chapter depends on the geometry loading code discussed in *Chapter 8*, and the post-processing effects covered in the previous chapter, *Chapter 10*. Be sure to review both chapters before continuing.

The source code for the examples in this chapter is available on GitHub at `https://github.com/PacktPublishing/3D-Graphics-Rendering-Cookbook-Second-Edition/tree/main/Chapter11`.

Refactoring indirect rendering

In *Chapter 8*, we introduced the `VKMesh` helper class, which provided a straightforward way to encapsulate scene geometry, GPU buffers, and rendering pipelines. This class has been useful in simplifying rendering tasks throughout *Chapters 8* to *10*. However, its simplicity comes with a trade-off: it tightly couples scene data, rendering pipelines, and buffers.

This strong coupling poses challenges when attempting to render the same scene data with different pipelines or when selectively rendering parts of a scene. To address these limitations, let's introduce a new class, `VKMesh11`, designed to resolve these issues and enable more flexible rendering setups. With `VKMesh11`, we will have the new foundation needed to implement more advanced and complex algorithms in this chapter.

Getting ready

As we build on top of the old `VKMesh` class, ensure you've reviewed all the recipes in *Chapter 8* to understand how the old `VKMesh` class works.

The source code for this recipe can be found in `Chapter11/VKMesh11.h`.

How to do it...

In this recipe, we aim to address two main issues with the existing `VKMesh` implementation. The first is to allow the use of custom shaders, depth states, and push constants within the `VKMesh11` rendering pipeline. The second issue is to enable the rendering of specific parts of the mesh by using a custom modified indirect commands buffer. This approach will decouple rendering operations from the default buffer, allowing us to implement frustum culling. Let's go through the steps required to do it:

1. First, we introduce a new helper class, `VKIndirectBuffer11`, designed to manage an indirect commands buffer along with an array of `DrawIndexedIndirectCommand` structures:

```
class VKIndirectBuffer11 final {
public:
  const std::unique_ptr<lvk::IContext>& ctx_;
  Holder<BufferHandle> bufferIndirect_;
  vector<DrawIndexedIndirectCommand> drawCommands_;
```

2. We can define the maximum number of draw commands, `maxDrawCommands`, this buffer can store, though the actual number of commands may be smaller. Additionally, we can specify the buffer's storage type, which will be important later when making the indirect buffer host-visible in the next recipe, *Performing frustum culling on the CPU*:

```
  VKIndirectBuffer11(
    const std::unique_ptr<lvk::IContext>& ctx,
    size_t maxDrawCommands,
    lvk::StorageType indirectBufferStorage = lvk::StorageType_Device)
  : ctx_(ctx)
  , drawCommands_(maxDrawCommands) {
```

3. We allocate the buffer and include a `uint32_t` integer to store the number of commands. It can be used with `vkCmdDrawIndexedIndirectCount()` to enable GPU-side modification of the command count:

Note

Our indirect buffer layout is as follows:

```
uint32_t numCommands;
DrawIndexedIndirectCommand cmd0;
DrawIndexedIndirectCommand cmd1;
…
DrawIndexedIndirectCommand cmdMaxDrawCommands;
```

```
  bufferIndirect_ = ctx->createBuffer(
    { .usage      = lvk::BufferUsageBits_Indirect |
                    lvk::BufferUsageBits_Storage,
      .storage    = indirectBufferStorage,
      .size       = sizeof(uint32_t) +
         sizeof(DrawIndexedIndirectCommand) * maxDrawCommands,
      .debugName = "Buffer: indirect" });
}
```

4. Here's a helper function to upload the updated contents of the indirect buffer using the modified `drawCommands_` array. We store the number of draw commands at the very beginning of the buffer:

```
void uploadIndirectBuffer() {
  const uint32_t numCommands = drawCommands_.size();
  ctx_->upload(bufferIndirect_, &numCommands, sizeof(uint32_t));
  ctx_->upload(bufferIndirect_,
    drawCommands_.data(),
    sizeof(VkDrawIndexedIndirectCommand) * numCommands, sizeof(uint32_t));
};
```

5. Let's introduce another helper function, `selectTo()`, which populates a separate `VKIndirectBuffer11` object with draw commands filtered by a predicate functor, `pred`. This is useful for separating subsets of a scene based on specific criteria. In the *Implementing order-independent transparency* recipe, we'll use it to create separate indirect buffers for opaque and transparent objects:

```
void selectTo(VKIndirectBuffer11& buf,
  const std::function<bool(
    const DrawIndexedIndirectCommand&)>& pred)
{
  buf.drawCommands_.clear();
  for (const auto& c : drawCommands_)
    if (pred(c)) buf.drawCommands_.push_back(c);
  buf.uploadIndirectBuffer();
}
```

6. This helper function creates a pointer to a memory-mapped indirect buffer. This will be used in the next recipe, *Performing frustum culling on the CPU*, to modify the indirect buffer before rendering:

```
  DrawIndexedIndirectCommand*
    getDrawIndexedIndirectCommandPtr() const {
    return std::launder(
      reinterpret_cast<DrawIndexedIndirectCommand*>(
        ctx_->getMappedPtr(bufferIndirect_) + sizeof(uint32_t)));
  }
};
```

Now, let's introduce another class, `VKPipeline11`, to encapsulate the functionality for creating the graphics pipeline.

1. Our pipeline will include vertex and fragment shaders, along with two rendering pipelines: one for standard rendering and one for wireframe rendering, similar to how the old `VKMesh` class handled it:

```
class VKPipeline11 final {
public:
  Holder<ShaderModuleHandle> vert_;
  Holder<ShaderModuleHandle> frag_;
  Holder<RenderPipelineHandle> pipeline_;
  Holder<RenderPipelineHandle> pipelineWireframe_;
```

2. The constructor takes some mental load off the old `VKMesh` constructor. Everything related to the graphics pipeline creation is here:

```
  VKPipeline11(
    const std::unique_ptr<lvk::IContext>& ctx,
    const lvk::VertexInput& streams,
    lvk::Format colorFormat,
    lvk::Format depthFormat,
    uint32_t numSamples = 1,
    Holder<ShaderModuleHandle>&& vert = {},
    Holder<ShaderModuleHandle>&& frag = {}) {
```

3. Custom shaders are optional. By default, we use the shaders from *Chapter 8*:

```
    vert_ = vert.valid() ?
      std::move(vert) : loadShaderModule(
        ctx, "Chapter08/02_SceneGraph/src/main.vert");
    frag_ = frag.valid() ?
      std::move(frag) : loadShaderModule(
        ctx, "Chapter08/02_SceneGraph/src/main.frag");
```

4. Last but not least, two rendering pipelines are created: one for standard rendering and one for wireframe rendering:

```
    pipeline_ = ctx->createRenderPipeline({
      .vertexInput       = streams,
      .smVert            = vert_,
      .smFrag            = frag_,
      .color             = {{ .format = colorFormat }},
      .depthFormat       = depthFormat,
      .cullMode          = lvk::CullMode_None,
      .samplesCount      = numSamples,
      .minSampleShading = numSamples>1 ? 0.25f : 0.f});
    pipelineWireframe_ = ctx->createRenderPipeline({
      .vertexInput  = streams,
      .smVert       = vert_,
      .smFrag       = frag_,
      .color        = { { .format = colorFormat } },
      .depthFormat  = depthFormat,
      .cullMode     = lvk::CullMode_None,
      .polygonMode  = lvk::PolygonMode_Line,
      .samplesCount = numSamples });
  }
};
```

Now, we can use two helper classes, `VKIndirectBuffer11` and `VKPipeline11`, to streamline our new mesh class, `VKMesh11`. Let's see how it works.

How it works...

The `VKMesh11` class is almost a direct drop-in replacement for the old `VKMesh`, with the main difference being that the graphics pipelines are now entirely separate and should be stored elsewhere:

1. All the buffers, except for the indirect command buffer, are still managed directly within this class:

```
class VKMesh11 {
public:
  const std::unique_ptr<lvk::IContext>& ctx;
  uint32_t numIndices_ = 0;
  uint32_t numMeshes_  = 0;
  Holder<BufferHandle> bufferIndices_;
  Holder<BufferHandle> bufferVertices_;
  Holder<BufferHandle> bufferTransforms_;
  Holder<BufferHandle> bufferDrawData_;
```

```
Holder<BufferHandle> bufferMaterials_;
std::vector<DrawData> drawData_;
VKIndirectBuffer11 indirectBuffer_;
```

2. The texture cache is identical to the one in `VKMesh`. We won't discuss it here, so please refer to *Chapter 8* for details. The member fields, `materialsCPU_` and `materialsGPU_`, are used to store the corresponding materials. We will utilize them later, in the *Loading texture assets asynchronously* recipe:

```
TextureFiles textureFiles_;
mutable TextureCache textureCache_;
std::vector<Material> materialsCPU_;
std::vector<GLTFMaterialDataGPU> materialsGPU_;
VKMesh11(
  const std::unique_ptr<lvk::IContext>& ctx,
  const MeshData& meshData,
  const Scene& scene,
  lvk::StorageType indirectBufferStorage = lvk::StorageType_Device,
  bool preloadMaterials = true)
: ctx(ctx)
, numIndices_((uint32_t)meshData.indexData.size())
, numMeshes_((uint32_t)meshData.meshes.size())
, indirectBuffer_(ctx,
    meshData.getMeshFileHeader().meshCount,
    indirectBufferStorage)
, textureFiles_(meshData.textureFiles)
{
  const MeshFileHeader header = meshData.getMeshFileHeader();
  const uint32_t* indices = meshData.indexData.data();
  const uint8_t* vertexData = meshData.vertexData.data();
```

3. Here, we skip the code for creating GPU buffers for materials, vertices, indices, and model-to-world transformation matrices. For details, refer to *Chapter 8*. The materials initialization is now slightly different and can be entirely skipped using the `preloadMaterials` flag. We will use this in the *Loading texture assets asynchronously* recipe:

```
  materialsCPU_ = meshData.materials;
  materialsGPU_.reserve(meshData.materials.size());
  for (const auto& mat : meshData.materials) {
    materialsGPU_.push_back(preloadMaterials ?
      convertToGPUMaterial(ctx, mat, textureFiles_, textureCache_) :
      GLTFMaterialDataGPU{});
  }
```

4. Draw commands are now stored in the default `VKIndirectBuffer11` object. The content is populated in the same way as in the old `VKMesh`, using the same `DrawData` structure. We skip most of that code here for the sake of brevity:

```
    const uint32_t numCommands = header.meshCount;
    indirectBuffer_.drawCommands_.resize(numCommands);
    drawData_.resize(numCommands);
    DrawIndexedIndirectCommand* cmd =
      indirectBuffer_.drawCommands_.data();
    DrawData* dd = drawData_.data();
    uint32_t ddIndex = 0;
    // … prepare indirect commands buffer - skipped
    for (auto& i : scene.meshForNode) { … }
    indirectBuffer_.uploadIndirectBuffer();
    bufferDrawData_ = ctx->createBuffer(
      { .usage     = lvk::BufferUsageBits_Storage,
        .storage   = lvk::StorageType_Device,
        .size      = sizeof(DrawData) * numCommands,
        .data      = drawData_.data(),
        .debugName = "Buffer: drawData" });
  }
```

5. The `draw()` functions have changed, and there are now two of them. The first one takes a mandatory `VKPipeline11` object and uses default push constants to pass data into GLSL shaders, including the familiar `view`, `proj`, and `texSkyboxIrradiance` parameters from `VKMesh`. Note the additional, optional `VKIndirectBuffer11` parameter:

```
  void draw(
    lvk::ICommandBuffer& buf,
    const VKPipeline11& pipeline,
    const mat4& view, const mat4& proj,
    TextureHandle texSkyboxIrradiance = {},
    bool wireframe = false,
    const VKIndirectBuffer11* indirectBuffer = nullptr)
  {
    buf.cmdBindIndexBuffer(bufferIndices_, lvk::IndexFormat_UI32);
    buf.cmdBindVertexBuffer(0, bufferVertices_);
```

6. We use the standard or wireframe pipeline from the provided `VKPipeline11` object:

```
    buf.cmdBindRenderPipeline(wireframe ?
      pipeline.pipelineWireframe_:pipeline.pipeline_);
    buf.cmdBindDepthState(
      { .compareOp = lvk::CompareOp_Less,
        .isDepthWriteEnabled = true });
```

7. These push constants are exactly the same as in the old `VKMesh` class:

```
const struct {
  mat4 viewProj;
  uint64_t bufferTransforms;
  uint64_t bufferDrawData;
  uint64_t bufferMaterials;
  uint32_t texSkyboxIrradiance;
} pc = {
  .viewProj = proj * view,
  .bufferTransforms = ctx->gpuAddress(bufferTransforms_),
  .bufferDrawData   = ctx->gpuAddress(bufferDrawData_),
  .bufferMaterials  = ctx->gpuAddress(bufferMaterials_),
  .texSkyboxIrradiance = texSkyboxIrradiance.index(),
};
buf.cmdPushConstants(pc);
```

8. The last step in the first `draw()` method is to use the optional indirect commands buffer, if provided, or replace it with the default buffer if not. Then, we call `vkCmdDrawIndexedIndirectCount()`, specifying the location of our draw commands counter, which is at the very beginning of the indirect buffer:

```
  if (!indirectBuffer)
    indirectBuffer = &indirectBuffer_;
  buf.cmdDrawIndexedIndirectCount(
    indirectBuffer->bufferIndirect_,
    sizeof(uint32_t),
    indirectBuffer->bufferIndirect_, 0, numMeshes_,
    sizeof(DrawIndexedIndirectCommand));
}
```

9. The second `draw()` function is designed to take custom push constants and depth state, which are populated externally. Everything else remains the same. We will use this function to render our scene with custom shaders in the *Implementing shadows for directional lights* recipe:

```
void draw(
  lvk::ICommandBuffer& buf,
  const VKPipeline11& pipeline,
  const void* pushConstants, size_t pcSize,
  const lvk::DepthState depthState = {
    .compareOp = lvk::CompareOp_Less,
    .isDepthWriteEnabled = true }
  bool wireframe = false,
  const VKIndirectBuffer11* indirectBuffer =nullptr)
```

```
  {
    buf.cmdBindIndexBuffer(bufferIndices_, lvk::IndexFormat_UI32);
    buf.cmdBindVertexBuffer(0, bufferVertices_);
    buf.cmdBindRenderPipeline(wireframe ?
      pipeline.pipelineWireframe_:pipeline.pipeline_);
    buf.cmdBindDepthState(depthState);
    buf.cmdPushConstants(pushConstants, pcSize);
    if (!indirectBuffer)
      indirectBuffer = &indirectBuffer_;
    buf.cmdDrawIndexedIndirectCount(
      indirectBuffer->bufferIndirect_,
      sizeof(uint32_t),
      indirectBuffer->bufferIndirect_, 0, numMeshes_,
      sizeof(DrawIndexedIndirectCommand));
  }
```

10. One last helper function is `getDrawIndexedIndirectCommandPtr()`. It simply returns the corresponding pointer value from our default indirect commands buffer:

```
  DrawIndexedIndirectCommand*
    getDrawIndexedIndirectCommandPtr() const {
      return indirectBuffer_.
        getDrawIndexedIndirectCommandPtr();
  };
};
```

That's it! Now, let's put the `VKMesh11` class to use in the next recipe and learn how to perform frustum culling on the CPU.

Doing frustum culling on the CPU

Frustum culling is a technique used to determine whether a part of the scene is visible within the viewing frustum. While many tutorials online explain how to implement it, most have a significant drawback.

As Inigo Quilez highlighted in his blog `http://www.iquilezles.org/www/articles/frustumcorrect/frustumcorrect.htm`, many frustum culling examples determine whether a mesh is outside the viewing frustum by comparing the mesh's **axis-aligned bounding box** (**AABB**) against the 6 planes of the frustum. They reject any AABB that lies entirely outside any of these planes.

This approach will result in false positives when a large AABB, which is not actually visible, intersects some of the frustum planes. The naïve method will incorrectly classify these AABBs as visible, reducing culling efficiency. For individual meshes, addressing these cases may not be worth the performance cost. However, when culling large bounding boxes that serve as containers, such as sections of an octree, false positives can become a significant issue. The solution is to incorporate a reverse culling test, where the 8 corner points of the viewing frustum are checked against the 6 planes of the AABB.

In this recipe, we will provide self-contained frustum culling code in C++ and show you how to use it to render the Bistro scene.

Getting ready

Make sure you've reviewed all the recipes in *Chapter 8* to understand how the `VKMesh` class works. The example source code for this demo is located in `Chapter11/01_CullingCPU`.

How to do it...

To implement both box-in-frustum and frustum-in-box tests, we need to extract 6 frustum planes and 8 frustum corner points from the view matrix. Here's the code to do that, located in `shared/UtilsMath.h`:

1. By default, GLM uses a right-handed coordinate system. In this case, the 6 frustum planes are extracted from the transposed, premultiplied view-projection matrix, `vp`:

```
void getFrustumPlanes(mat4 vp, vec4* planes) {
  vp = glm::transpose(vp);
  planes[0] = vec4(vp[3] + vp[0]); // left
  planes[1] = vec4(vp[3] - vp[0]); // right
  planes[2] = vec4(vp[3] + vp[1]); // bottom
  planes[3] = vec4(vp[3] - vp[1]); // top
  planes[4] = vec4(vp[3] + vp[2]); // near
  planes[5] = vec4(vp[3] - vp[2]); // far
}
```

2. The 8 frustum corner points can be generated by taking a unit cube, transforming it with the inverse of the premultiplied view-projection matrix, and then performing perspective division. In this way, the unit cube effectively warps into the frustum:

```
void getFrustumCorners(
  glm::mat4 vp, glm::vec4* points) {
  const vec4 corners[] = {
    vec4(-1, -1, -1, 1), vec4( 1, -1, -1, 1),
    vec4( 1,  1, -1, 1), vec4(-1,  1, -1, 1),
    vec4(-1, -1,  1, 1), vec4( 1, -1,  1, 1),
    vec4( 1,  1,  1, 1), vec4(-1,  1,  1, 1)
  };
  const glm::mat4 invVP = glm::inverse(vp);
```

```
  for (int i = 0; i != 8; i++) {
    const vec4 q = invVP * corners[i];
    points[i] = q / q.w;
  }
}
```

3. The AABB culling code is now straightforward. Simply check whether the bounding box is entirely outside any of the 6 frustum planes:

```
bool isBoxInFrustum(
  glm::vec4* frPlanes,
  glm::vec4* frCorners, const BoundingBox& b) {
  using glm::dot;
  for ( int i = 0; i < 6; i++ ) {
    int r = 0;
    r +=(dot(frPlanes[i],
      vec4(b.min_.x,b.min_.y,b.min_.z,1.f)) < 0) ? 1 : 0;
    r +=(dot(frPlanes[i],
      vec4(b.max_.x,b.min_.y,b.min_.z,1.f)) < 0) ? 1 : 0;
    r +=(dot(frPlanes[i],
      vec4(b.min_.x,b.max_.y,b.min_.z,1.f)) < 0) ? 1 : 0;
    r +=(dot(frPlanes[i],
      vec4(b.max_.x,b.max_.y,b.min_.z,1.f)) < 0) ? 1 : 0;
    r +=(dot(frPlanes[i],
      vec4(b.min_.x,b.min_.y,b.max_.z,1.f)) < 0) ? 1 : 0;
    r +=(dot(frPlanes[i],
      vec4(b.max_.x,b.min_.y,b.max_.z,1.f)) < 0) ? 1 : 0;
    r +=(dot(frPlanes[i],
      vec4(b.min_.x,b.max_.y,b.max_.z,1.f)) < 0) ? 1 : 0;
    r +=(dot(frPlanes[i],
      vec4(b.max_.x,b.max_.y,b.max_.z,1.f)) < 0) ? 1 : 0;
    if (r == 8) return false;
  }
```

Note

Here, the terms "inside" and "outside" of a plane are defined by the direction of its normal vector: "inside" lies in the direction of the normal vector, while "outside" is in the opposite direction.

4. Next, invert the test and check whether the frustum is entirely inside the bounding box:

```
  int r = 0;
  r = 0; for (int i = 0; i<8; i++)
```

```
    r+=((frCorners[i].x>b.max_.x) ? 1 : 0);
  if (r == 8) return false;
  r = 0; for (int i = 0; i<8; i++)
    r+=((frCorners[i].x<b.min_.x) ? 1 : 0);
  if (r == 8) return false;
  r = 0; for (int i = 0; i<8; i++)
    r+=((frCorners[i].y>b.max_.y) ? 1 : 0);
  if (r == 8) return false;
  r = 0; for (int i = 0; i<8; i++)
    r+=((frCorners[i].y<b.min_.y) ? 1 : 0);
  if (r == 8) return false;
  r = 0; for (int i = 0; i<8; i++)
    r+=((frCorners[i].z>b.max_.z) ? 1 : 0);
  if (r == 8) return false;
  r = 0; for (int i = 0; i<8; i++)
    r+=((frCorners[i].z<b.min_.z) ? 1 : 0);
  if (r == 8) return false;
  return true;
}
```

Let's take a look at how this culling code can be used in our demos.

How it works...

The main driving code for this demo is located in the `Chapter11/01_CullingCPU/src/main.cpp` file. It is based on the Bistro MSAA rendering example, `Chapter10/03_MSAA`, from *Chapter 10*. Let's now go over the new sections of the code that handle frustum culling:

1. First, we need to add a few parameters to control culling. The `cullingView` matrix represents the view matrix of the culling frustum, which can be detached from the camera view to allow observation of the culling process from an external perspective. The `freezeCullingView` flag enables this functionality. The Boolean flags can be controlled from the ImGui UI:

```
mat4 cullingView        = mat4(1.0f);
bool freezeCullingView = false;
bool drawMeshes         = true;
bool drawBoxes          = true;
bool drawWireframe      = false;
```

2. The remaining setup before the rendering loop is almost identical to the demo in `Chapter10/03_MSAA`, so we'll skip it here. There are two additions. One addition is a keyboard callback that toggles the culling frustum freezing state using the *P* key:

```
// ...
app.addKeyCallback([](GLFWwindow* window,
```

```
    int key, int scancode, int action, int mods)
  {
    const bool pressed = action != GLFW_RELEASE;
    if (key == GLFW_KEY_P && pressed &&
        !ImGui::GetIO().WantCaptureKeyboard)
      freezeCullingView = !freezeCullingView;
  });
```

3. Another change is passing the `StorageType_HostVisible` parameter to `VKMesh11`, ensuring it allocates the indirect commands buffer as host-visible. Check the previous *Refactoring indirect rendering* recipe to understand how the `VKMesh11` class works:

> Note
>
> This is necessary for our CPU culling approach. Later, we will evaluate each mesh's visibility using AABB-to-frustum culling and then directly modify the indirect commands buffer to set the number of instances for that mesh to either `1` or `0`, based on visibility.

```
const VKMesh11 mesh(ctx, meshData, scene,
  lvk::StorageType_HostVisible);
const VKPipeline11 pipeline(ctx, meshData.streams,
  ctx->getSwapchainFormat(), app.getDepthFormat(),
  kNumSamples);
```

4. Let's examine the rendering loop to understand how the culling process is implemented:

```
app.run([&](uint32_t width, uint32_t height,
  float aspectRatio, float deltaSeconds)
{
  const mat4 view = app.camera_.getViewMatrix();
  const mat4 proj = glm::perspective(45.f, aspectRatio, 0.1f, 200.f);
  lvk::ICommandBuffer& buf =
    ctx->acquireCommandBuffer();
```

5. We update the `cullingView` matrix using the camera's view matrix and create a combined view-projection matrix to extract the 6 frustum planes and 8 frustum corners from it using the helper functions defined earlier:

```
if (!freezeCullingView)
  cullingView = app.camera_.getViewMatrix();
vec4 frustumPlanes[6];
getFrustumPlanes(proj * cullingView, frustumPlanes);
vec4 frustumCorners[8];
getFrustumCorners(proj * cullingView, frustumCorners);
```

6. With all the data ready, the actual culling takes place in this `for` loop. We obtain a pointer to the indirect commands buffer from `VKMesh` and iterate through all the meshes. For each mesh, we call `isBoxInFrustum()` for its world-space AABB and set the number of instances to render to either `1` or `0`, depending on its visibility. The `numVisibleMeshes` variable is used to display the number of visible meshes in the UI:

```
int numVisibleMeshes = 0;
DrawIndexedIndirectCommand* cmd =
  mesh.getDrawIndexedIndirectCommandPtr();
for (auto& p : scene.meshForNode) {
  const BoundingBox box = meshData.boxes[p.second].
    getTransformed(scene.globalTransform[p.first]);
  const uint32_t count  = isBoxInFrustum(
    frustumPlanes, frustumCorners, box) ? 1 : 0;
  (cmd++)->instanceCount = count;
  numVisibleMeshes += count;
}
```

Note

As you can see, we transform AABBs from model space to world space in every frame. However, since our Bistro scene is completely static, this extra work is unnecessary. In the next recipe, *Doing frustum culling on the GPU with compute shaders*, we will optimize this process by pre-transforming the bounding boxes, eliminating the need for per-frame transformations.

7. Since our indirect buffer is host-visible and memory-mapped, we need to call `flushMappedMemory()` after updating it to ensure the changes are visible to the GPU. Internally, this function will call `vmaFlushAllocation()` if the Vulkan Memory Allocator is used, or `vkFlushMappedMemoryRanges()` otherwise:

```
ctx->flushMappedMemory(
  mesh.indirectBuffer_.bufferIndirect_, 0,
  mesh.numMeshes_ * sizeof(DrawIndexedIndirectCommand));
```

8. The culling is complete. The remaining code handles drawing the debugging UI. We render the culling frustum (in yellow) so that we can fly around with the camera and observe it from the outside:

```
canvas3d.clear();
canvas3d.setMatrix(proj * view);
if (freezeCullingView)
  canvas3d.frustum(cullingView, proj, vec4(1, 1, 0, 1));
```

9. Next, we render all the bounding boxes. If a mesh is visible, we draw its bounding box in green; otherwise, we draw it in red:

```
if (drawBoxes) {
  const DrawIndexedIndirectCommand* cmd =
    mesh.getDrawIndexedIndirectCommandPtr();
  for (auto& p : scene.meshForNode) {
    const BoundingBox box = meshData.boxes[p.second];
    canvas3d.box(scene.globalTransform[p.first],
      box, (cmd++)->instanceCount ?
        vec4(0, 1, 0, 1) : vec4(1, 0, 0, 1));
  }
}
// ...
if (drawMeshes) {
  mesh.draw(buf, pipeline, view, proj,
    skyBox.texSkyboxIrradiance, drawWireframe);
}
```

The remaining scene rendering code is exactly the same as in `Chapter10/03_MSAA`. Indeed, we only modified the content of the indirect buffer, which controls GPU rendering inside the `VKMesh11` class. There's a small ImGui UI rendering snippet in `Chapter11/01_CullingCPU/src/main.cpp`, which we leave for you to explore.

The running demo application should render something similar to the following screenshot:

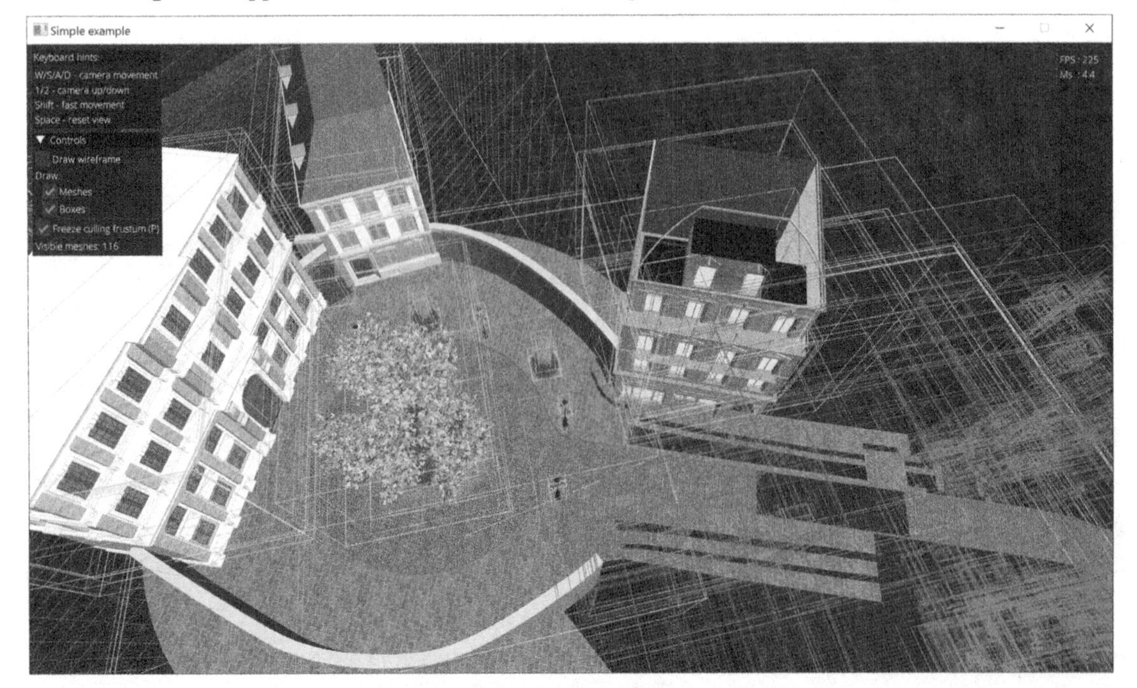

Figure 11.1: Bistro with CPU frustum culling

The culling frustum, rendered with yellow lines, is frozen to allow us to observe its orientation. The green boxes represent visible meshes, while the red boxes indicate invisible ones. Next time the *P* key is pressed, the culling frustum is reattached to the camera view.

There's more...

One might argue that this type of culling is inefficient because modern GPUs can render small meshes much faster than we can cull them on the CPU, and this is mostly true. It doesn't make sense to cull very small objects such as bottles or individual leaves this way. However, the CPU culling pipeline is still extremely useful when culling large objects or even clusters of objects. For instance, if we create a 3D grid that covers the entire Bistro scene with reasonably large cells, we can assign each mesh to the cells that intersect it. This way, CPU culling can be applied to each cell rather than individual meshes, allowing us to skip over entire sections of the scene. On platforms where bandwidth consumption increases power usage and heat—such as mobile phones and handheld consoles—performing culling on the CPU may prevent the costly transfer of uncompressed vertex data from RAM to tile memory, only for it to be discarded.

Another improvement could be to perform this type of CPU culling using multithreading, often referred to as the parallel-brute-force approach. We leave these enhancements as moderately complex exercises for you.

For now, let's port this simple culling technique to the GPU using a compute pipeline and perform some performance measurements.

Doing frustum culling on the GPU with compute shaders

In the previous recipe, we explored how to cull invisible meshes on the CPU using the classic AABB-to-frustum method. However, this CPU-based approach has its limitations. Now, let's learn how to use Vulkan compute shaders to implement a basic frustum culling pipeline on the GPU.

The goal of this recipe is to demonstrate what can be achieved on modern GPUs using compute pipelines and indirect rendering, rather than to create an efficient culling system. Once you are comfortable with the basics, we can explore the limitations and discuss potential directions for improvement.

Getting ready

Be sure to read the previous recipe, *Doing frustum culling on the CPU*, to familiarize yourself with the basics of frustum culling.

The source code for this recipe is available in `Chapter11/02_CullingGPU`.

How to do it...

The concept behind this recipe is simple. We'll take the C++ `isAABBinFrustum()` helper function from the previous recipe, *Doing frustum culling on the CPU*, and port it to GLSL. Next, we'll transform the culling `for` loop from C++ into the body of the `main()` function in our compute shader. In the updated C++ code, we'll retain both the old CPU culling code and the new GPU culling implementation, allowing us to switch between them and compare their performance.

Let's explore how to port the C++ culling code to a GLSL compute shader that is located in `Chapter11/02_CullingGPU/src/FrustumCulling.comp`:

1. Before porting the culling function to GLSL, we need to define some data structures and declare shader inputs. We'll create an `AABB` structure to represent our bounding boxes, corresponding to the `BoundingBox` type in C++. In this structure, 6 floats are used to store the `vec3` min and max values, ensuring we avoid scalar alignment issues in the shader:

```
struct AABB {
  float pt[6];
};
```

2. The GLSL `DrawIndexedIndirectCommand` structure corresponds to the C++ Vulkan `VkDrawIndexedIndirectCommand` struct. This structure will be used to modify the content of the indirect commands buffer and adjust the number of instances based on AABB visibility:

```
struct DrawIndexedIndirectCommand {
  uint count;
  uint instanceCount;
  uint firstIndex;
  int  baseVertex;
  uint baseInstance;
};
```

3. The `DrawData` struct is used to access the `transformId` for each draw call. To recap how it works, refer to the `VKMesh` class and the *Implementing indirect rendering with Vulkan* recipe in *Chapter 8*. The `VKMesh11` class uses the same `DrawData` structure:

```
struct DrawData {
  uint transformId;
  uint materialId;
};
```

4. The `BoundingBoxes` buffer, managed in `Chapter11/02_CullingGPU/src/main.cpp`, is used to pass the bounding boxes to the shader, while the other 2 buffers are managed within the `VkMesh` class:

```
layout(std430, buffer_reference)
  readonly buffer BoundingBoxes {
  AABB boxes[];
};
layout(std430, buffer_reference)
  readonly buffer DrawDataBuffer {
  DrawData dd[];
};
layout(std430, buffer_reference) buffer DrawCommands {
```

```
  uint numDraws; // unused in this compute shader
  DrawIndexedIndirectCommand dc[];
};
```

5. A small `CullingData` buffer is used to pass per-frame data to the compute shader. This includes 6 frustum planes, 8 frustum corners, and the number of meshes to cull. The `numVisibleMeshes` field is initially set to `0` and is atomically incremented for each visible mesh. It will be read back in C++ and displayed in the ImGui UI:

```
layout(std430, buffer_reference) buffer CullingData {
  vec4 planes[6];
  vec4 corners[8];
  uint numShapesToCull;
  uint numVisibleMeshes;
};
```

6. Push constants are used to pass the GPU addresses of all 4 aforementioned buffers to the compute shader:

```
layout(std430, push_constant) uniform PushConstants {
  DrawCommands commands;
  DrawDataBuffer drawData;
  BoundingBoxes AABBs;
  CullingData frustum;
};
```

7. Here's the `isAABBinFrustum()` frustum culling function, which is largely copied from its C++ implementation. It follows the same two-phase approach used in the previous recipe, *Doing frustum culling on the CPU*. In this case, we use macros to provide mnemonic names for the floating-point values in the AABB structure:

```
#define min_x box.pt[0]
#define min_y box.pt[1]
#define min_z box.pt[2]
#define max_x box.pt[3]
#define max_y box.pt[4]
#define max_z box.pt[5]
bool isAABBinFrustum(AABB box) {
  for (int i = 0; i < 6; i++) {
    int r = 0;
    r += (dot( frustum.planes[i],
      vec4(min_x, min_y, min_z, 1.0f)) < 0) ? 1 : 0;
    r += (dot( frustum.planes[i],
      vec4(max_x, min_y, min_z, 1.0f)) < 0) ? 1 : 0;
    r += (dot( frustum.planes[i],
```

```
      vec4(min_x, max_y, min_z, 1.0f)) < 0) ? 1 : 0;
    r += (dot( frustum.planes[i],
      vec4(max_x, max_y, min_z, 1.0f)) < 0) ? 1 : 0;
    r += (dot( frustum.planes[i],
      vec4(min_x, min_y, max_z, 1.0f)) < 0) ? 1 : 0;
    r += (dot( frustum.planes[i],
      vec4(max_x, min_y, max_z, 1.0f)) < 0) ? 1 : 0;
    r += (dot( frustum.planes[i],
      vec4(min_x, max_y, max_z, 1.0f)) < 0) ? 1 : 0;
    r += (dot( frustum.planes[i],
      vec4(max_x, max_y, max_z, 1.0f)) < 0) ? 1 : 0;
    if (r == 8) return false;
  }
  int r = 0;
  r = 0; for ( int i = 0; i < 8; i++ )
    r += ( (frustum.corners[i].x > max_x) ? 1 : 0 );
  if ( r == 8 ) return false;
  r = 0; for ( int i = 0; i < 8; i++ )
    r += ( (frustum.corners[i].x < min_x) ? 1 : 0 );
  if ( r == 8 ) return false;
  r = 0; for ( int i = 0; i < 8; i++ )
    r += ( (frustum.corners[i].y > max_y) ? 1 : 0 );
  if ( r == 8 ) return false;
  r = 0; for ( int i = 0; i < 8; i++ )
    r += ( (frustum.corners[i].y < min_y) ? 1 : 0 );
  if ( r == 8 ) return false;
  r = 0; for ( int i = 0; i < 8; i++ )
    r += ( (frustum.corners[i].z > max_z) ? 1 : 0 );
  if ( r == 8 ) return false;
  r = 0; for ( int i = 0; i < 8; i++ )
    r += ( (frustum.corners[i].z < min_z) ? 1 : 0 );
  if ( r == 8 ) return false;
  return true;
}
```

8. With the culling helper function in place, we can implement the shader's `main()` function as follows. The shader operates on workgroups of 64 elements, so we need to skip any meshes beyond the value of `numMeshesToCull`. The `baseInstance` parameter is extracted from the `DrawCommands` buffer and then used to get an AABB object. Since everything runs in parallel, we use the atomic operation `atomicAdd()` to increment the count of visible meshes:

```
void main() {
  const uint idx = gl_GlobalInvocationID.x;
```

```
  if (idx < frustum.numMeshesToCull) {
    uint baseInstance = commands.dc[idx].baseInstance;
    AABB box = AABBs.boxes[drawData.dd[baseInstance].transformId];
    uint numInstances = isAABBinFrustum(box) ? 1 : 0;
    commands.dc[idx].instanceCount = numInstances;
    atomicAdd(frustum.numVisibleMeshes, numInstances);
  }
}
```

Note

Instead of checking against `numMeshesToCull`, we can pad the underlying buffers to 64 meshes. This will require modifying the `VkMesh11` class, and we leave it as an exercise for you.

More information

In a real application, instead of calling `atomicAdd()` for each value, we should use subgroup memory along with the GLSL commands `subgroupBallot()` and `subgroupBallotBitCount()` to perform a single atomic operation per shader workgroup. Refer to the Khronos Vulkan subgroup tutorial `https://www.khronos.org/blog/vulkan-subgroup-tutorial` to learn more about this approach.

Now, let's take a look at the C++ code to understand how everything works.

How it works...

The C++ code is located in `Chapter11/02_CullingGPU/src/main.cpp`. It is built on the example from the previous recipe, *Doing frustum culling on the CPU*, so here we will focus only on the newly added sections:

1. We need a new `CullingMode` enum to select between different culling modes in ImGui:

```
enum CullingMode {
  CullingMode_None = 0,
  CullingMode_CPU  = 1,
  CullingMode_GPU  = 2,
};
mat4 cullingView = mat4(1.0f);
int cullingMode  = CullingMode_GPU;
// ...
```

2. Let's load our culling compute shader and create a pipeline for it:

```
Holder<ShaderModuleHandle> compCulling = loadShaderModule(ctx,
  "Chapter11/02_CullingGPU/src/FrustumCulling.comp");
```

```
Holder<ComputePipelineHandle> pipelineCulling =
  ctx->createComputePipeline(
    { .smComp = compCulling } );
```

3. In the previous recipe, *Doing frustum culling on the CPU*, we did a lot of extra work by transforming mesh bounding boxes from model space to world space in each frame. Since the scene is static, let's pretransform and reorder all the bounding boxes. The `reorderedBoxes` container can now be indexed by the `DrawData::transformId`, as we saw earlier in the compute shader code. Let's upload it into a GPU buffer:

```
std::vector<BoundingBox> reorderedBoxes;
reorderedBoxes.resize(scene.globalTransform.size());
for (auto& p : scene.meshForNode)
  reorderedBoxes[p.first] = meshData.boxes[p.second]
    .getTransformed(scene.globalTransform[p.first]);
Holder<BufferHandle> bufferAABBs = ctx->createBuffer({
    .usage   = lvk::BufferUsageBits_Storage,
    .storage = lvk::StorageType_Device,
    .size = reorderedBoxes.size() * sizeof(BoundingBox),
    .data = reorderedBoxes.data() });
```

4. Here's a C++ struct that corresponds to the `CullingData` buffer we declared in GLSL. The `CullingData::numVisibleMeshes` member field is used to transfer data to and from the GPU, while the `numVisibleMeshes` variable is used to display this value through ImGui. We will see how to update both of these in a moment:

```
struct CullingData {
  vec4 frustumPlanes[6];
  vec4 frustumCorners[8];
  uint32_t numMeshesToCull  = 0;
  uint32_t numVisibleMeshes = 0; // GPU
};
int numVisibleMeshes = 0; // CPU
```

5. The update of `CullingData::numVisibleMeshes` happens on the GPU, and we need to read it back. If we wait for the GPU to finish before reading the value, it will cause the CPU to stall. Instead, let's read back the value that was updated a frame earlier. To achieve this, we can create multiple buffers to hold `CullingData` for 2 consecutive frames and use them in a round-robin fashion:

```
const lvk::BufferDesc cullingDataDesc = {
  .usage     = lvk::BufferUsageBits_Storage,
  .storage   = lvk::StorageType_HostVisible,
  .size      = sizeof(CullingData) };
```

```
Holder<BufferHandle> bufferCullingData[] = {
  ctx->createBuffer(cullingDataDesc),
  ctx->createBuffer(cullingDataDesc),
};
```

6. We also need 2 corresponding Vulkan fences to wait for each of these previously submitted frames. Each `SubmitHandle` wraps a `VkFence` object, as described in the *Using Vulkan command buffers* recipe from *Chapter 2*. The `currentBufferId` variable ranges from `0` to `1` and is used to track the current buffer ID:

```
vk::SubmitHandle submitHandle[
  LVK_ARRAY_NUM_ELEMENTS(bufferCullingData)] = {};
uint32_t currentBufferId = 0;
```

7. Here's a structure representing push constants. The first 3 buffers are persistent, while the last buffer, `frustum`, holds `CullingData` and is updated each frame in a round-robin fashion:

```
struct {
  uint64_t commands;
  uint64_t drawData;
  uint64_t AABBs;
  uint64_t frustum;
} pcCulling = {
  .commands = ctx->gpuAddress(mesh.bufferIndirect_),
  .drawData = ctx->gpuAddress(mesh.bufferDrawData_),
  .AABBs    = ctx->gpuAddress(bufferAABBs),
};
```

8. Let's now take a look at the rendering loop. In addition to the previously implemented CPU culling, we now have two new code paths: one for GPU culling and one where culling is completely disabled:

```
app.run([&](uint32_t width, uint32_t height,
  float aspectRatio, float deltaSeconds) {
  const mat4 view = app.camera_.getViewMatrix();
  const mat4 proj = glm::perspective(45.f, aspectRatio, 0.1f, 200.f);
  ICommandBuffer& buf = ctx->acquireCommandBuffer();
  if (!freezeCullingView)
    cullingView = app.camera_.getViewMatrix();
  CullingData cullingData = {
    .numMeshesToCull = static_cast<uint32_t>(scene.meshForNode.size()),
  };
  getFrustumPlanes(proj * cullingView, cullingData.frustumPlanes);
  getFrustumCorners(proj * cullingView, cullingData.frustumCorners);
```

9. The no-culling code path simply sets the number of instances to 1 for all meshes and updates `numVisibleMeshes` accordingly. Don't forget to flush the mapped indirect commands GPU buffer:

```
if (cullingMode == CullingMode_None) {
  numVisibleMeshes = scene.meshForNode.size();
  DrawIndexedIndirectCommand* cmd =
    mesh.getDrawIndexedIndirectCommandPtr();
  for (auto& p : scene.meshForNode)
    (cmd++)->instanceCount = 1;
  ctx->flushMappedMemory(
    mesh.indirectBuffer_.bufferIndirect_, 0,
    mesh.numMeshes_ * sizeof(DrawIndexedIndirectCommand));
}
```

10. The CPU culling code path is very similar to the one in the previous recipe, *Doing frustum culling on the CPU*. The main difference here is that we use the pretransformed bounding boxes we calculated earlier:

```
else if (cullingMode == CullingMode_CPU) {
  numVisibleMeshes = 0;
  DrawIndexedIndirectCommand* cmd =
    mesh.getDrawIndexedIndirectCommandPtr();
  for (auto& p : scene.meshForNode) {
    const BoundingBox box = reorderedBoxes[p.first];
    const uint32_t count  = isBoxInFrustum(
      cullingData.frustumPlanes,
      cullingData.frustumCorners, box) ? 1 : 0;
    (cmd++)->instanceCount = count;
    numVisibleMeshes += count;;
  }
  ctx->flushMappedMemory(
    mesh.indirectBuffer_.bufferIndirect_, 0,
    mesh.numMeshes_ * sizeof(DrawIndexedIndirectCommand));
}
```

11. The GPU culling code path is straightforward. We assign the GPU pointer of the current `CullingData` buffer to push constants, bind them, update the contents of the `CullingData` buffer, and dispatch our compute shader. We must specify the dependency to the indirect buffer to ensure a proper Vulkan buffer memory barrier is issued. Our compute shader processes work groups of size `64`, which is why we perform the division and round the result up:

```
else if (cullingMode == CullingMode_GPU) {
  buf.cmdBindComputePipeline(pipelineCulling);
```

```
    pcCulling.meshes = ctx->gpuAddress(bufferCullingData[bufferId]);
    buf.cmdPushConstants(pcCulling);
    buf.cmdUpdateBuffer(
      bufferCullingData[bufferId], cullingData);
    buf.cmdDispatchThreadGroups(
      { 1 + cullingData.numMeshesToCull / 64 },
      { .buffers = { BufferHandle(mesh.bufferIndirect_) } });
  }
```

12. After the compute shader, we can begin rendering our culled scene. We need to specify the indirect buffer as a dependency to ensure proper compute-to-graphics buffer memory barriers:

```
// ...
buf.cmdBeginRendering(
  lvk::RenderPass{
    .color = {{ .loadOp = lvk::LoadOp_Clear,
                .storeOp = lvk::StoreOp_MsaaResolve,
                .clearColor = {1, 1, 1, 1}}},
    .depth = { .loadOp = lvk::LoadOp_Clear,
               .clearDepth = 1.0f } },
  framebufferMSAA,
  { .buffers =
      { BufferHandle(mesh.bufferIndirect_) } });
// ...
```

> Note
>
> The `cmdBeginRendering()` and `cmdDispatchThreadGroups()` functions are located in `deps/src/lightweightvk/lvk/vulkan/VulkanClasses.cpp`. Refer to this file to review which specific Vulkan memory barriers are inserted based on the buffer usage flags.

While the C++ code mentioned above is sufficient to drive our GPU-based culling, there's one more interesting detail worth mentioning.

In the previous recipe, *Doing frustum culling on the CPU*, all culling was performed on the CPU, making it straightforward to compute the number of visible meshes. With GPU-based culling, this becomes a bit more complex, as we need to retrieve the visible mesh count from the GPU buffer back to the CPU. This value is already calculated and stored in the `CullingData` GPU buffer, in the `CullingData::numVisibleMeshes` field.

To read this value back efficiently, we access one of the previously used buffers where the compute shader has already finished execution. To achieve this, we created arrays of `bufferCullingData[]` round-robin buffers and `submitHandle[]` fences, which are wrapped into `lvk::SubmitHandle`. Let's take a look at the code that uses them.

1. After recording all the compute and rendering commands, we submit the command buffer, store the `SubmitHandle` associated with it, and advance to the next `currentBufferId`:

```
// …
buf.cmdEndRendering();
submitHandle[currentBufferId] = ctx->submit(
  buf, ctx->getCurrentSwapchainTexture());
currentBufferId = (currentBufferId + 1) %
  LVK_ARRAY_NUM_ELEMENTS(bufferCullingData);
```

2. Next, we wait on the `SubmitHandle` that follows the submitted `currentBufferId` command buffer. In other words, we check the Vulkan fence corresponding to a command buffer submitted one frame ago. By this point, the GPU has most likely finished processing that buffer, so the fence resolves quickly, allowing us to read the data back. The `VulkanContext::download()` function copies the data from the mapped GPU buffer into a `numVisibleMeshes` variable in RAM, while also properly invalidating non-coherent mapped memory if necessary. You can find its implementation in the `deps/src/lightweightvk/lvk/vulkan/VulkanClasses.cpp` file:

```
if (cullingMode == CullingMode_GPU &&
    app.fpsCounter_.numFrames_ > 1)
{
  ctx->wait(submitHandle[currentBufferId]);
  ctx->download(bufferCullingData[currentBufferId],
    &numVisibleMeshes, sizeof(uint32_t),
    offsetof(CullingData, numVisibleMeshes));
}
```

Once the `numVisibleMeshes` value is updated, we can use it with ImGui to display the number of visible meshes computed on the GPU.

Note

It is important to highlight a significant inefficiency here. While the compute shader completes its task before rendering begins, we wait for the entire submitted command buffer, which includes both the compute and rendering stages. This could be optimized using Vulkan **timeline semaphores**, a feature that became core in Vulkan 1.2. However, at the time of writing this book, the **LightweightVK** framework does not yet support timeline semaphores.

Let's now run the demo application. It should produce an image similar to the one shown in the following screenshot:

Figure 11.2: Bistro with GPU frustum culling

🔍**Quick tip:** Need to see a high-resolution version of this image? Open this book in the next-gen Packt Reader or view it in the PDF/ePub copy.

🔒 **The next-gen Packt Reader** and a **free PDF/ePub copy** of this book are included with your purchase. Unlock them by scanning the QR code below or visiting `https://www.packtpub.com/unlock/9781803248110`.

The culling frustum is frozen, and only the visible geometry is rendered. The visible AABBs are displayed in red, while the invisible ones are shown in green. You can fly around the scene and toggle the frustum freezing state using the *P* key.

You can switch between different culling modes using ImGui. If you run the sample app in a `Release` build and disable Vulkan validation layers using the Vulkan Configurator tool, you can perform basic performance measurements and compare the different culling modes. In most cases, the CPU culling outperforms the non-culled rendering. However, this naïve GPU culling implementation is typically slower than CPU culling. Our `isAABBinFrustum()` function has a lot of branching and is not optimized for SIMD, making it inefficient on the GPU. It is mainly for demonstration purposes, showing how to port C++ code to GLSL. However, performance can be significantly improved by either rewriting the code in a SIMD-friendly way or replacing AABB culling with simpler bounding sphere culling, which can be much more easily calculated using dot products in a branchless fashion. We leave it as an advanced exercise for you.

There's more...

Another optimization for frustum culling, apart from making the code SIMD-friendly, is to use multi-threading on the CPU. This approach, often referred to as parallel-brute-force, can lead to significant performance improvements. For more details on this technique, check out the GDC 2011 *Culling the Battlefield* presentation by Daniel Collin.

It is important to note that the assumption in our GPU culling pipeline—that setting the number of instances in `VkDrawIndexedIndirectCommand` to `0` completely eliminates the penalty of rendering a mesh—is not accurate. In reality, simply setting the number of instances to `0` does not completely remove the associated GPU performance cost. To optimize this, one should consider compacting the indirect buffer by removing culled items or creating a new indirect buffer directly in the compute shader, using the atomic value of `numVisibleMeshes` as an index. This approach is particularly beneficial when culling is not performed every frame, and the buffer is reused across multiple frames.

There is an excellent SIGGRAPH 2015 presentation titled *GPU-Driven Rendering Pipelines* by Ulrich Haar and Sebastian Aaltonen, which provides an in-depth exploration of the use of compute functionality in rendering pipelines. It dives into various strategies and optimizations for efficient GPU-based rendering techniques: `http://advances.realtimerendering.com/s2015/aaltonenhaar_siggraph2015_combined_final_footer_220dpi.pdf`.

Implementing shadows for directional lights

In the previous chapter, we explored setting up a shadow mapping pipeline, focusing primarily on the API and the overall implementation process rather than achieving good shadow accuracy or versatility. In the *Implementing shadow maps* recipe from *Chapter 10*, we learned how to render shadows from spotlights using perspective projection for the shadow-casting light source. For directional light sources, which are often used to simulate sunlight, a single light source can illuminate the entire scene. Implementing shadows for such a light source requires constructing a projection matrix that accounts for the scene's bounds. This recipe will explore how to implement this basic approach and incorporate shadows into our Bistro scene.

Getting ready

Ensure you revisit the *Implementing shadow maps* recipe from *Chapter 10* as we move forward into the shadow mapping topic.

The demo for this recipe can be found in `Chapter11/03_DirectionalShadows`.

How to do it...

To construct a projection matrix for a directional light, we calculate the axis-aligned bounding box of the entire scene, transform it into light-space using the light's view matrix, and then use the bounds of the transformed box to construct an orthographic frustum that fully encloses this box.

We calculate the scene bounding box in two simple steps:

1. First, we pretransform the bounding boxes of individual meshes into world space. This step uses the same approach as outlined earlier for frustum culling in the *Doing frustum culling on the GPU with compute shaders* recipe:

```
std::vector<BoundingBox> reorderedBoxes;
reorderedBoxes.resize(scene.globalTransform.size());
for (auto& p : scene.meshForNode)
  reorderedBoxes[p.first] = meshData.boxes[p.second].
    getTransformed(scene.globalTransform[p.first]);
```

2. Next, we combine all the pretransformed bounding boxes into one single large world-space bounding box:

```
BoundingBox bigBoxWS = reorderedBoxes.front();
for (const auto& b : reorderedBoxes) {
  bigBoxWS.combinePoint(b.min_);
  bigBoxWS.combinePoint(b.max_);
}
```

Since our scene is static, we do it just once, outside of the main rendering loop.

Now we can construct a projection matrix for our directional light source.

1. The light's direction can be adjusted in real time via ImGui controls, using two angles: the azimuthal angle, `theta`, and the elevation angle, `phi`:

```
struct LightParams {
  float theta          = +90.0f;
  float phi            = -26.0f;
  float depthBiasConst = 1.1f;
  float depthBiasSlope = 2.0f;
  bool operator==(const LightParams&) const = default;
} light;
```

2. To construct the light's view matrix, we first create two rotation matrices based on the `theta` and `phi` angles. These matrices are used to rotate a top-down light direction vector, `(0, -1, 0)`, into the desired orientation. Using the resulting light direction, we build the light's view matrix using the `glm::lookAt()` helper function. Since the light source encompasses the entire scene, its origin can be conveniently set to `(0, 0, 0)`:

```
const mat4 rot1 = glm::rotate(mat4(1.f),
  glm::radians(light.theta), vec3(0, 1, 0));
const mat4 rot2 = glm::rotate(rot1,
  glm::radians(light.phi), vec3(1, 0, 0));
const vec3 lightDir = glm::normalize(
  vec3(rot2 * vec4(0.0f, -1.0f, 0.0f, 1.0f)));
const mat4 lightView =
  glm::lookAt(vec3(0, 0, 0), lightDir, vec3(0, 0, 1));
```

3. With the light's view matrix, `lightView`, ready, we transform the scene's world-space bounding box into light-space. Since Vulkan uses a clip coordinate space where the *z* axis ranges from `0` to `1`, we use the `glm::orthoLH_ZO()` helper function to create the projection matrix. However, this function follows the DirectX clip coordinate convention, so we must flip the *z* axis to match Vulkan's expectations:

```
const BoundingBox boxLS = bigBoxWS.getTransformed(lightView);
const mat4 lightProj = glm::orthoLH_ZO(
  boxLS.min_.x, boxLS.max_.x,
  boxLS.min_.y, boxLS.max_.y,
  boxLS.max_.z, boxLS.min_.z);
```

4. The light's view and projection matrices can be premultiplied and stored in a GPU buffer using the `LightData` structure. The associated shadow map texture and sampler are set up similarly to the approach outlined in the *Implementing shadow maps* recipe from *Chapter 10*:

```
struct LightData {
  mat4 viewProjBias;
  vec4 lightDir;
  uint32_t shadowTexture;
  uint32_t shadowSampler;
};
Holder<BufferHandle> bufferLight = ctx->createBuffer({
  .usage     = lvk::BufferUsageBits_Storage,
  .storage   = lvk::StorageType_Device,
  .size      = sizeof(LightData),
  .debugName = "Buffer: light" });
```

5. We create our `VKMesh11` object for the scene and two pipelines: one for rendering the scene into the shadow map and another for rendering the scene normally into a swapchain image. Note that the shadow map pipeline is not multisampled:

```
const VKMesh11 mesh(ctx, meshData, scene);
const VKPipeline11 pipelineMesh(
  ctx, meshData.streams, ctx->getSwapchainFormat(),
  app.getDepthFormat(), kNumSamples,
  loadShaderModule(ctx,
   "Chapter11/03_DirectionalShadows/src/main.vert"),
  loadShaderModule(ctx,
    "Chapter11/03_DirectionalShadows/src/main.frag")
);
const VKPipeline11 pipelineShadow(
  ctx, meshData.streams, lvk::Format_Invalid,
  ctx->getFormat(shadowMap), 1,
  loadShaderModule(ctx,
    "Chapter11/03_DirectionalShadows/src/shadow.vert"),
  loadShaderModule(ctx,
    "Chapter11/03_DirectionalShadows/src/shadow.frag")
);
```

6. The remaining shadow map rendering logic follows the same structure as in the previous *Implementing shadow maps* spotlight recipe from *Chapter 10*. The main difference is that the shadow map is updated only when the light parameters, direction angles, or depth bias, are changed via the ImGui UI, instead of being updated every frame. This is possible because our scene is static:

```
ICommandBuffer& buf = ctx->acquireCommandBuffer();
if (prevLight != light) {
  prevLight = light;
  // … render Bistro into the shadow map here…
  mesh.draw(buf, pipelineShadow, lightView, lightProj);
  // …
  buf.cmdUpdateBuffer(bufferLight, LightData{
    .viewProjBias = scaleBias * lightProj * lightView,
    .lightDir      = vec4(lightDir, 0.0f),
    .shadowTexture = shadowMap.index(),
    .shadowSampler = samplerShadow.index() });
}
```

Here's what a shadow map for the entire scene should look like when the `theta` and `phi` angles are set to `0`:

Figure 11.3: A shadow map for a top-down directional light

Note how the light's viewing frustum is fit to the scene bounds. If you start changing the light direction, it will still be fit to the transformed scene bounding box, but it won't be tightly fit to the scene itself. This is because a transformed axis-aligned bounding box can contain more empty space, as the axes of the original box may no longer be aligned with the axes of the new coordinate system. One simple improvement for this can be calculating the combined scene bounding box directly in light space, from the individual AABBs of meshes transformed to light space, instead of first calculating the scene AABB in world space and then transforming it to light space. We leave this as a simple exercise for you.

The demo application for this recipe should render the following image.

Figure 11.4: The Bistro scene rendered with shadows from a directional light

Try changing the light direction and adjusting the depth bias parameters to observe how these changes affect the shadow quality and help mitigate artifacts. The directional light frustum is rendered in yellow to visualize the bounds of the light's influence on the scene.

There's more...

The preceding screenshot uses an orthographic projection for the shadow map, which leads to aliasing issues despite a high shadow map resolution of 8192x8192. To improve shadow map quality, you can apply techniques such as **Perspective Shadow Maps** (**PSMs**), introduced by Stamminger and Drettakis at SIGGRAPH 2002. PSMs use post-projective space, causing nearby objects to occupy more space in the shadow map, reducing aliasing for objects close to the camera. Another advanced technique is Light Space Perspective Shadow Maps (`https://www.cg.tuwien.ac.at/research/vr/lispsm`), which further refines the concept to achieve better shadow quality across different depths.

To further reduce aliasing in shadow maps, a more generic technique is to divide the light frustum into multiple frustums or cascades, each covering a different range of distances from the camera. This method, known as **Cascaded Shadow Maps** (**CSMs**), involves rendering multiple shadow maps for different depth ranges and then applying the appropriate shadow map based on the viewer's distance. **Parallel Split Shadow Maps** (**PSSMs**) are a variation of this approach, where the split of the frustum is more evenly distributed.

Moreover, perspective shadow maps and cascades can be combined into a single shadow mapping pipeline producing the best results. We recommend the book *Real-Time Shadows* `https://www.realtimeshadows.com`, which, despite its age, remains one of the best "all-in-one-place" references for shadow techniques.

Now, let's move on to the next recipe, where we will explore an interesting technique for rendering transparent objects to improve the visual quality of our Bistro scene.

Implementing order-independent transparency

Until now, we've rendered transparent objects using a basic punch-through transparency method, as described in the *Rendering large scenes* recipe from *Chapter 8*. While this approach was limited in quality, it allowed transparent and opaque objects to be rendered together without any additional care or sorting, greatly simplifying the rendering pipeline. Now, let's take it a step further and implement a method that enables correctly blended transparent objects.

Alpha blending of multiple surfaces requires sorting all transparent surfaces back-to-front, which can be achieved through various methods. These include sorting scene objects back-to-front, using multipass depth peeling techniques, as described in `https://matthewwellings.com/blog/depth-peeling-order-independent-transparency-in-vulkan`, and employing order-independent approximations such as `http://casual-effects.blogspot.com/2014/03/weighted-blended-order-independent.html` and a more recent work, *Phenomenological Transparency*, by Morgan McGuire and Michael Mara (`https://research.nvidia.com/publication/phenomenological-transparency`).

In Vulkan, it is possible to implement **order-independent transparency** (**OIT**) using per-pixel linked lists via atomic counters and load-store-atomic read-modify-write operations on textures. This method is order-independent, meaning it eliminates the need to sort transparent geometry before rendering. Instead, all necessary sorting is performed in a fragment shader at the pixel level after the scene has been rendered. The algorithm works by constructing a linked list of fragments for each screen pixel, with each node storing color and depth values. Once these per-pixel lists are constructed, they can be sorted and blended using a fullscreen fragment shader. Essentially, this is a two-pass algorithm. Our implementation is inspired by `https://fr.slideshare.net/hgruen/oit-and-indirect-illumination-using-dx11-linked-lists`. Let's take a closer look at how it works.

Getting ready

We recommend revisiting the *Implementing a material system* recipe from *Chapter 8* to refresh your understanding of the material system's data structures. This example utilizes multiple indirect buffers with the `VKMesh11` class, as described earlier in this chapter, in the *Refactoring indirect rendering* recipe.

The source code for this recipe is available in `Chapter11/04_OIT`.

How to do it...

The rendering pipeline for this example proceeds as follows: first, we render opaque objects with standard shading, as done in the previous recipe. Next, we render transparent objects, adding shaded fragments to linked lists instead of rendering them directly to the framebuffer. Finally, we sort the linked lists and overlay the blended image onto the opaque framebuffer. To keep things simple, we omit shadows in this example and focus solely on transparency rendering and MSAA. We will combine OIT with shadows and other effects in the *Putting it all together into a Vulkan demo* recipe.

Let's walk through the C++ code to implement this technique:

1. The `main()` function is similar to the one from the `Chapter10/03_MSAA` example and the *Implementing MSAA in Vulkan* recipe from *Chapter 10*. We use a `VKMesh11` object and two pipelines with two sets of shaders: the first set, `main.vert` and `opaque.frag`, is for opaque objects, while the second set, `main.vert` and `transparent.frag`, is for transparent objects:

```
const VKMesh11 mesh(ctx, meshData, scene);
const VKPipeline11 pipelineOpaque(
  ctx, meshData.streams, ctx->getSwapchainFormat(),
  app.getDepthFormat(), kNumSamples,
  loadShaderModule(ctx, "Chapter11/04_OIT/src/main.vert"),
  loadShaderModule(ctx, "Chapter11/04_OIT/src/opaque.frag"));
const VKPipeline11 pipelineTransparent(
  ctx, meshData.streams, ctx->getSwapchainFormat(),
  app.getDepthFormat(), kNumSamples,
  loadShaderModule(ctx, "Chapter11/04_OIT/src/main.vert"),
  loadShaderModule(ctx, "Chapter11/04_OIT/src/transparent.frag"));
```

2. A fullscreen post-processing pass is handled in `GL03_OIT.frag`. We will examine the shaders after reviewing the C++ code:

```
Holder<ShaderModuleHandle> vertOIT =
  loadShaderModule(ctx, "data/shaders/QuadFlip.vert");
Holder<ShaderModuleHandle> fragOIT =
  loadShaderModule(ctx, "Chapter11/04_OIT/src/oit.frag");
Holder<RenderPipelineHandle> pipelineOIT = ctx->createRenderPipeline({
    .smVert = vertOIT,
    .smFrag = fragOIT,
    .color  =
      { { .format = ctx->getSwapchainFormat() } } });
```

3. Let's create two `VKIndirectBuffer11` objects, as described in the *Refactoring indirect rendering* recipe: one for opaque meshes and another for transparent ones. A simple C++ `isTransparent()` lambda can be used to filter draw commands based on the `sMaterialFlags_Transparent` flag in the material flags. This flag was set by the `convertAIMaterial()` helper function from *Chapter 8* during the import of the Bistro mesh:

```
VKIndirectBuffer11 meshesOpaque(ctx, mesh.numMeshes_);
VKIndirectBuffer11 meshesTransparent(ctx, mesh.numMeshes_);
auto isTransparent = [&meshData, &mesh](
  const DrawIndexedIndirectCommand& c) -> bool {
  const uint32_t mtlIndex = mesh.drawData_[c.baseInstance].materialId;
  const Material& mtl = meshData.materials[mtlIndex];
  return (mtl.flags & sMaterialFlags_Transparent) > 0;
};
mesh.indirectBuffer_.selectTo(meshesOpaque,
  [&isTransparent](const DrawIndexedIndirectCommand& c) -> bool {
      return !isTransparent(c);
  });
mesh.indirectBuffer_.selectTo(meshesTransparent,
  [&isTransparent](const DrawIndexedIndirectCommand& c) -> bool {
      return isTransparent(c);
  });
```

4. In addition to multiple indirect buffers, we need a few other buffers to store linked list data. The C++ `TransparentFragment` structure will mirror the same structure in our GLSL shaders, representing a single node of a per-pixel linked list. We need a floating-point color value with an alpha channel to support HDR reflections on transparent objects, along with a depth value and an index for the next node. The RGBA color is stored as a `uint64_t`, which corresponds to the four-component `f16vec4` half-float vector in GLSL. In the C++ code, we don't actually need to worry about these member fields; we only need to know the structure's size in order to allocate a buffer for it:

```
struct TransparentFragment {
  uint64_t rgba;
  float depth;
  uint32_t next;
};
```

Note

In real-world applications, 16-bit floats for color channels are often replaced with 10-bit packed formats, such as `VK_FORMAT_A2R10G10B10_UNORM_PACK32`, to save memory bandwidth. In C++, this can be mimicked using `uint32_t`. However, we avoid using this approach here to simplify our HDR rendering code.

5. The buffer allocation proceeds as follows: we allocate storage to accommodate `kMaxOITFragments` transparent fragments, which corresponds to `4` overdraw layers. This totals roughly `225Mb` of memory in 4K resolution. Any fragments beyond this limit will be discarded by our fragment shader, which performs bounds checking:

```
const uint32_t kMaxOITFragments =
  sizeFb.width * sizeFb.height * 4;
Holder<BufferHandle> bufferTransparencyLists = ctx->createBuffer({
    .usage   = lvk::BufferUsageBits_Storage,
    .storage = lvk::StorageType_Device,
    .size    = sizeof(TransparentFragment) * kMaxOITFragments,
    .debugName = "Buffer: transparency lists",
  });
```

The `textureHeadsOIT` texture is a fullscreen `uint32_t` texture that contains integer indices (which index into the `bufferTransparencyLists` buffer) for the head elements of the per-pixel linked lists corresponding to each screen pixel. Note that this image does not have either the `Sampled` or `Attachment` usage flags, as it is only accessed with image load-store operations:

```
Holder<TextureHandle> textureHeadsOIT = ctx->createTexture({
    .format     = lvk::Format_R_UI32,
    .dimensions = sizeFb,
    .usage      = lvk::TextureUsageBits_Storage,
    .debugName  = "oitHeads" });
```

The `bufferAtomicCounter` contains a single 32-bit integer that tracks the total number of allocated transparent fragments so far and acts as an atomic counter in a linear memory allocator implemented in the GLSL shader. The `bufferTransparencyLists` is the actual memory pool used to store linked lists created by our allocator:

```
Holder<BufferHandle> bufferAtomicCounter = ctx->createBuffer({
    .usage     = lvk::BufferUsageBits_Storage,
    .storage   = lvk::StorageType_Device,
    .size      = sizeof(uint32_t),
    .debugName = "Buffer: atomic counter" });
```

6. Each frame, we should reset the atomic counter to `0` and clear the heads buffer with the value `0xFFFFFFFF`, which acts as an end-of-list guard value to signal that there are no more transparent fragments for that pixel. Here is a C++ lambda to accomplish this:

```
auto clearTransparencyBuffers = [&bufferAtomicCounter,
  &textureHeadsOIT, sizeFb](lvk::ICommandBuffer& buf)
{
  buf.cmdClearColorImage(textureHeadsOIT, { .uint32 = { 0xffffffff } });
  buf.cmdFillBuffer(bufferAtomicCounter, 0, sizeof(uint32_t), 0);
};
```

7. With all the preparations complete, we can now enter the rendering loop and clear the transparency buffers:

```
app.run([&](uint32_t width, uint32_t height,
  float aspectRatio, float deltaSeconds)
{
  const mat4 view = app.camera_.getViewMatrix();
  const mat4 proj = glm::perspective(45.f, aspectRatio, 0.1f, 200.f);
  ICommandBuffer& buf = ctx->acquireCommandBuffer();
  clearTransparencyBuffers(buf);
```

8. Scene rendering can proceed as usual by first drawing the skybox, followed by the opaque meshes of the Bistro scene using the corresponding indirect buffer:

```
  const lvk::Framebuffer framebufferMSAA = {
    .color        = { { .texture = msaaColor,
                        .resolveTexture = offscreenColor } },
    .depthStencil = { .texture = msaaDepth },
  };
  buf.cmdBeginRendering(
    lvk::RenderPass{
      .color = {{.loadOp  = lvk::LoadOp_Clear,
                 .storeOp = lvk::StoreOp_MsaaResolve,
                 .clearColor = {1.f, 1.f, 1.f, 1.f}}},
      .depth = { .loadOp = lvk::LoadOp_Clear,
                 .clearDepth = 1.0f } },
    framebufferMSAA);
  skyBox.draw(buf, view, proj);
```

9. Push constants are shared between the opaque and transparent pipelines:

```
  const struct {
    mat4 viewProj;
    vec4 cameraPos;
    uint64_t bufferTransforms;
    uint64_t bufferDrawData;
    uint64_t bufferMaterials;
    uint64_t bufferAtomicCounter;
    uint64_t bufferTransparencyLists;
    uint32_t texSkybox;
    uint32_t texSkyboxIrradiance;
    uint32_t texHeadsOIT;
    uint32_t maxOITFragments;
  } pc = {
```

```
    .viewProj  = proj * view,
    .cameraPos = vec4(app.camera_.getPosition(), 1.f),
    .bufferTransforms = ctx->gpuAddress(mesh.bufferTransforms_),
    .bufferDrawData = ctx->gpuAddress(mesh.bufferDrawData_),
    .bufferMaterials = ctx->gpuAddress(mesh.bufferMaterials_),
    .bufferAtomicCounter = ctx->gpuAddress(bufferAtomicCounter),
    .bufferTransparencyLists = ctx->gpuAddress(bufferTransparencyLists),
    .texSkybox = skyBox.texSkybox.index(),
    .texSkyboxIrradiance = skyBox.texSkyboxIrradiance.index(),
    .texHeadsOIT = textureHeadsOIT.index(),
    .maxOITFragments = kMaxOITFragments,
  };
```

10. Opaque and transparent meshes use different depth states. The opaque meshes update the depth buffer, while the transparent meshes output everything with overdraw. Later, we will see how to sort them in the fragment shader. The key point is that transparent objects need to be properly depth-culled against opaque objects:

```
if (drawMeshesOpaque)
  mesh.draw(buf, pipelineOpaque, &pc, sizeof(pc),
    { .compareOp = lvk::CompareOp_Less,
      .isDepthWriteEnabled = true },
    drawWireframe, &meshesOpaque);
if (drawMeshesTransparent)
  mesh.draw(buf, pipelineTransparent, &pc, sizeof(pc),
    { .compareOp = lvk::CompareOp_Less,
      .isDepthWriteEnabled = false },
    drawWireframe, &meshesTransparent);
```

11. The grid can be rendered normally after all objects, as the transparent objects don't modify the color and depth buffers; they simply store their output in per-pixel lists:

```
app.drawGrid(buf, proj, vec3(0, -1.0f, 0), kNumSamples);
buf.cmdEndRendering();
```

Now, after the first render pass, we have the MSAA color values resolved in the `offscreenColor` texture and our OIT buffers populated with transparent fragments. Next, we run the second pass, a fullscreen post-processing step, which sorts transparent fragments and blends them together. Here, we combine and render everything directly into a swapchain image:

```
const lvk::Framebuffer framebufferMain = {
  .color = { { .texture =
     ctx->getCurrentSwapchainTexture() } },
};
buf.cmdBeginRendering(
```

```
lvk::RenderPass{
  .color = {{ .loadOp = lvk::LoadOp_Load,
              .storeOp = lvk::StoreOp_Store }} },
framebufferMain,
```

12. We need barriers for textures and buffers. The `offscreenColor` texture will be transitioned from `VK_IMAGE_LAYOUT_ATTACHMENT_OPTIMAL` to `VK_IMAGE_LAYOUT_SHADER_READ_ONLY_OPTIMAL`. The `textureHeadsOIT` texture has the `VK_IMAGE_LAYOUT_GENERAL` image layout and will only require an image memory barrier without any layout transition to ensure execution dependency between the two passes. The `bufferTransparencyLists` storage buffer requires a memory barrier to ensure that subsequent reads from the linked lists stored in it can correctly access the data written by the first render pass. The `bufferAtomicCounter` buffer is not used in the fullscreen pass, so we can skip it:

```
{ .textures = { TextureHandle(textureHeadsOIT),
                TextureHandle(offscreenColor) },
  .buffers  = { BufferHandle(bufferTransparencyLists) }
});
```

13. The fullscreen pass shader uses a different set of push constants. We've added a useful debugging feature—a blinking opacity heatmap overlay—to show how many transparent pixels are drawn at each screen pixel. To implement this, we pass the current time to the shader:

```
const struct {
  uint64_t bufferTransparencyLists;
  uint32_t texColor;
  uint32_t texHeadsOIT;
  float time;
  float opacityBoost;
  uint32_t showHeatmap;
} pcOIT = {
  .bufferTransparencyLists = ctx->gpuAddress(bufferTransparencyLists),
  .texColor      = offscreenColor.index(),
  .texHeadsOIT   = textureHeadsOIT.index(),
  .time          = static_cast<float>(glfwGetTime()),
  .opacityBoost = opacityBoost,
  .showHeatmap   = showHeatmap ? 1u : 0u,
};
```

14. Next, we can draw our fullscreen triangle and render the ImGui UI on top. We'll skip the UI code here for brevity:

```
buf.cmdBindRenderPipeline(pipelineOIT);
buf.cmdPushConstants(pcOIT);
buf.cmdBindDepthState({});
```

```
  buf.cmdDraw(3);
  app.imgui_->beginFrame(framebufferMain);
  app.drawFPS();
  app.drawMemo();
  // … we skip all the ImGui UI rendering here
  app.imgui_->endFrame(buf);
  buf.cmdEndRendering();
}
ctx->submit(buf, ctx->getCurrentSwapchainTexture());
```

That's it for the C++ part. Now, let's switch to the GLSL code and examine how it works to handle transparent objects.

How it works...

There are some shared GLSL declarations in the `common_oit.sp` and `common.sp` files, which are used across our shaders. Let's take a look at them.

1. The `Chapter11/04_OIT/src/common_oit.sp` file is shared between both transparent shaders: the one for rendering transparent meshes and the fullscreen one. It declares the `TransparentFragment` structure, similar to the C++ version, along with a buffer to hold it:

```
struct TransparentFragment {
  f16vec4 color;
  float depth;
  uint next;
};
layout(std430, buffer_reference)
  buffer TransparencyListsBuffer {
  TransparentFragment frags[];
};
```

2. The `Chapter11/04_OIT/src/common.sp` file is shared between the shaders for rendering both opaque and transparent meshes. It uses the `DrawData` structure, which is maintained by the `VKMesh11` class. If you want to recap how these buffers work, refer to the *Refactoring indirect rendering* recipe and *Chapter 8*:

```
#include <data/shaders/gltf/common_material.sp>
#include <Chapter11/04_OIT/src/common_oit.sp>
struct DrawData {
  uint transformId;
  uint materialId;
};
layout(std430, buffer_reference) readonly buffer TransformBuffer {
  mat4 model[];
```

```
};
layout(std430, buffer_reference) readonly buffer DrawDataBuffer {
  DrawData dd[];
};
layout(std430, buffer_reference) readonly buffer MaterialBuffer {
  MetallicRoughnessDataGPU material[];
};
layout(std430, buffer_reference) buffer AtomicCounter {
  uint numFragments;
};
layout(push_constant) uniform PerFrameData {
  mat4 viewProj;
  vec4 cameraPos;
  TransformBuffer transforms;
  DrawDataBuffer drawData;
  MaterialBuffer materials;
  AtomicCounter atomicCounter;
  TransparencyListsBuffer oitLists;
  uint texSkybox;
  uint texSkyboxIrradiance;
  uint texHeadsOIT;
  uint maxOITFragments;
} pc;
```

Now, let's take a look at the actual GLSL shaders used for rendering. The first fragment shader, `Chapter11/04_OIT/src/transparent.frag`, is responsible for rendering transparent meshes and populating the per-pixel linked lists. Let's go through the source code to clarify how it works:

1. The `early_fragment_tests` layout specifier ensures that the fragment shader is not executed unnecessarily if the fragment is discarded based on the depth test. This is crucial because any redundant invocation of this shader could result in corrupted transparency lists. The `kTextures2DInOut[]` array is needed to access our `textureHeadsOIT` storage image in a bindless fashion:

```
#include <Chapter11/04_OIT/src/common.sp>
#include <data/shaders/UtilsPBR.sp>
layout (early_fragment_tests) in;
layout (set = 0, binding = 2, r32ui) uniform uimage2D kTextures2DInOut[];
layout (location=0) in vec2 uv;
layout (location=1) in vec3 normal;
layout (location=2) in vec3 worldPos;
layout (location=3) in flat uint materialId;
```

2. Here's a helper function to slightly improve the ad hoc reflections in the glass. Its sole purpose is to make things a bit shinier. If you want to make reflections more realistic, refer to *Chapter 6*:

```
vec3 fresnelSchlickRoughness(float cosTheta, vec3 F0, float roughness) {
  return F0 +
    (max(vec3(1.0 - roughness), F0) - F0) *
    pow(clamp(1.0 - cosTheta, 0.0, 1.0), 5.0);
}
void main() {
  MetallicRoughnessDataGPU mat = pc.materials.material[materialId];
  vec4 emissiveColor =
    vec4(mat.emissiveFactorAlphaCutoff.rgb, 0) *
    textureBindless2D(mat.emissiveTexture, 0, uv);
  vec4 baseColor = mat.baseColorFactor *
    (mat.baseColorTexture > 0 ?
     textureBindless2D(mat.baseColorTexture, 0, uv) :
     vec4(1.0));
```

3. This includes some simple normal mapping, as described in the *Implementing the glTF 2.0 metallic-roughness shading model* recipe from *Chapter 6* along with two hardcoded directional lights. We will replace these with a proper light source and shadows later, in the *Putting it all together into a Vulkan demo* recipe:

```
  vec3 n = normalize(normal);
  vec3 normalSample = textureBindless2D(mat.normalTexture, 0, uv).xyz;
  if (length(normalSample) > 0.5)
    n = perturbNormal(n, worldPos, normalSample, uv);
  float NdotL1 = clamp(
    dot(n, normalize(vec3(-1, 1,+0.5))), 0.1, 1.0);
  float NdotL2 = clamp(
    dot(n, normalize(vec3(+1, 1,-0.5))), 0.1, 1.0);
  float NdotL = 0.2 * (NdotL1+NdotL2);
```

4. Similarly, we use hacky IBL diffuse lighting here and environment reflections for transparent objects. We're not aiming for any kind of PBR accuracy, but rather keeping it simple and shiny while focusing on transparency:

```
  const vec4 f0 = vec4(0.04);
  vec3 sky = vec3(-n.x, n.y, -n.z); // rotate skybox
  vec4 diffuse = baseColor * (vec4(1.0) - f0) * (
    textureBindlessCube(pc.texSkyboxIrradiance,0,sky) + vec4(NdotL) );
  vec3 v = normalize(pc.cameraPos.xyz - worldPos);
  vec3 reflection = reflect(v, n);
  reflection = vec3(reflection.x, -reflection.y,
```

```
    reflection.z); // rotate reflection
  vec3 colorRefl = textureBindlessCube(
    pc.texSkybox, 0, reflection).rgb;
  vec3 kS = fresnelSchlickRoughness(
    clamp(dot(n, v), 0.0, 1.0), vec3(f0), 0.1);
  vec3 color = emissiveColor.rgb + diffuse.rgb + colorRefl * kS;
```

5. The `color` value we calculated represents the color of our transparent fragment. Now, let's insert it into the corresponding per-pixel linked list based on its transparency. We clamp the alpha value between `0.01` and `0.99` to reduce the number of transparent fragments stored in the transparency lists:

```
  float alpha = clamp(
    baseColor.a * mat.clearcoatTransmissionThickness.z,
    0.0, 1.0);
  bool isTransparent = (alpha > 0.01) && (alpha < 0.99);
```

6. Here comes the interesting part. Our transparent mesh rendering shader runs on an MSAA render target. Instead of storing all multisampled fragments in the transparency lists, which would increase memory consumption and bandwidth for our `8x` MSAA, we store only those samples fully covered by this fragment shader invocation. To do this, we construct the current sample mask using the built-in GLSL `gl_SampleID` variable and compare it with the `gl_SampleMaskIn[0]` array element, which contains the computed sample coverage mask for the current fragment.

Note

A sample bit is set in this mask if and only if the sample is considered covered for this fragment shader invocation. Bit `B` of mask `gl_SampleMask[M]` corresponds to sample `32 * M + B`. The array has `ceil(s/32)` elements, where `s` is the maximum number of color samples supported by the implementation. We use 8x MSAA, so we can take only the `gl_SampleMaskIn[0]` element.

The `gl_HelperInvocation` value indicates whether the fragment shader invocation is considered a helper invocation. We don't want to modify our transparency lists for these helper fragment shader invocations:

Note

A helper invocation is a fragment shader invocation created solely for evaluating derivatives. These derivatives are computed implicitly in the built-in `texture()` GLSL function and explicitly in the `dFdx()` and `dFdy()` derivative functions.

```
  if ( isTransparent &&
      !gl_HelperInvocation &&
       gl_SampleMaskIn[0] == (1 << gl_SampleID) ) {
```

7. Once we decide to add a new element to a per-pixel list corresponding to the current fragment, we need to allocate memory for it. To do this, we use the `bufferAtomicCounter`, which holds a single `uint32_t` value. This value is atomically incremented to obtain the index of the next available slot in the transparent fragments buffer:

```
    uint index =
      atomicAdd(pc.atomicCounter.numFragments, 1);
```

8. If the index we obtained is still within the max bounds of the buffer, we proceed with the write. To do this, we extract the old head of the `prevIndex` linked list and atomically replace it with the new head, which is now stored in the transparency buffer at the location of the index. The `prevIndex` value will become the next pointer of our list:

```
    if (index < pc.maxOITFragments) {
      uint prevIndex = imageAtomicExchange(
        kTextures2DInOut[pc.texHeadsOIT], ivec2(gl_FragCoord.xy), index);
      TransparentFragment frag;
      frag.color = f16vec4(color, alpha);
      frag.depth = gl_FragCoord.z;
      frag.next  = prevIndex;
      pc.oitLists.frags[index] = frag;
    }
  }
}
```

Now the per-pixel list is constructed. Note that this shader performs only image load-store operations without modifying any of the color and depth buffers directly.

Once we have populated the linked lists, let's learn how to sort and blend them to combine everything into the resulting image. This is done in a fullscreen pass in the `Chapter11/04_OIT/src/oit.frag` shader.

1. The fragment shader inputs contain the `texColor` color texture, which holds all opaque objects rendered. The `texHeadsOIT` texture contains the heads of per-pixel lists, and `oitLists` is our transparency buffer:

```
#include <Chapter11/04_OIT/src/common_oit.sp>
layout (set = 0, binding = 2, r32ui)
  uniform uimage2D kTextures2DIn[];
layout (location=0) in vec2 uv;
layout (location=0) out vec4 out_FragColor;
layout (push_constant) uniform PushConstants {
  TransparencyListsBuffer oitLists;
  uint texColor;
  uint texHeadsOIT;
  float time;
```

```
  float opacityBoost;
  uint showHeatmap;
} pc;
```

2. The `main()` function starts by extracting the linked list for the current pixel into a local array. This is done in a loop until we reach the `0xFFFFFFFF` guard marker or the maximal number of overlapping transparent fragments, `MAX_FRAGMENTS`. The number of extracted fragments is stored in `numFragments`:

```
#define MAX_FRAGMENTS 64
void main() {
  TransparentFragment frags[MAX_FRAGMENTS];
  uint numFragments = 0;
  uint idx = imageLoad(kTextures2DIn[pc.texHeadsOIT], ivec2(gl_FragCoord.xy)).r;
  while (idx != 0xFFFFFFFF && numFragments < MAX_FRAGMENTS) {
    frags[numFragments] = pc.oitLists.frags[idx];
    numFragments++;
    idx = pc.oitLists.frags[idx].next;
  }
```

3. Let's sort the array by depth using insertion sort from largest to smallest. This is efficient, given that we have a relatively small number of overlapping fragments:

```
  for (int i = 1; i < numFragments; i++) {
    TransparentFragment toInsert = frags[i];
    uint j = i;
    while (j > 0 && toInsert.depth > frags[j-1].depth) {
      frags[j] = frags[j-1];
      j--;
    }
    frags[j] = toInsert;
  }
```

4. Now we can blend the sorted fragments together. First, retrieve the color of the closest non-transparent object from the framebuffer. Then, traverse the array, blending the colors based on their alpha values. Clamping to the range `0`...1 is necessary to prevent any HDR values from leaking into the alpha channel:

```
  vec4 color = textureBindless2D(pc.texColor, 0, uv);
  for (uint i = 0; i < numFragments; i++) {
    color = mix( color, vec4(frags[i].color),
      clamp(float(frags[i].color.a+pc.opacityBoost),
            0.0, 1.0) );
  }
```

5. Lastly, we can apply our debugging transparency heatmap to visualize the overdraw level for each pixel. This heatmap helps demonstrate how many transparent fragments are being processed at each pixel location:

```
  if (pc.showHeatmap > 0 && numFragments > 0)
    color = (1.0+sin(5.0*pc.time)) * vec4(
      vec3(numFragments, numFragments, 0), 0) / 16.0;
  out_FragColor = color;
}
```

That's all the shader code necessary to implement order-independent transparency using per-pixel linked lists. The running demo application should render an image similar to the one shown in the following screenshot.

Figure 11.5: Looking inside the Bistro with order-independent transparency

Make sure you move the camera inside the bar to check the bottles and glasses on the tables and the windows and observe how all of these objects can overlay each other.

There's more...

While this transparency method gives pixel-perfect blending results, it requires a staggering amount of memory to handle large resolutions and the depth complexity of overlapping transparent objects. This makes it very unfriendly for rendering on mobile and power-limited devices, where other methods involving transparency approximations are more commonly used. For alternative methods to optimize OIT on mobile, search for Per-Fragment Layered Sorting (**PLS**) and Layered Depth Images (**LDI**).

Loading texture assets asynchronously

All our demos so far preload all texture assets at startup, which works fine for applications with small data sizes that can be loaded instantly. Indeed, we downscaled all the Bistro textures to 512x512. However, as content size enters the gigabyte range, it becomes desirable to implement lazy-loading or streaming mechanisms to load assets as needed. Let's enhance our demos with basic lazy-loading functionality that allows textures to load while the application continues rendering the scene. Multithreading will leverage the Taskflow library along with standard C++20 capabilities. To make the example more relevant and the graphics more visually appealing, we'll also increase the dimensions of our compressed textures to 2048x2048.

Getting ready

We recommend revisiting the *Multithreading with Taskflow* recipe from *Chapter 1*.

The source code for this recipe is located in `Chapter11/05_LazyLoading`.

How to do it...

To simplify the implementation, our approach is to replace the `VKMesh11` class, which handled all texture loading synchronously, with a new `VKMesh11Lazy` class, capable of loading textures asynchronously. This new class will subclass `VKMesh11`, replacing the texture loading logic in the constructor with an asynchronous mechanism. It will also include helper functions to create Vulkan textures from the loaded texture data. The new `VKMesh11Lazy` class is designed as a drop-in replacement for `VKMesh11`. Let's examine the declaration of this new scene-loading class in the `Chapter11/VKMesh11Lazy.h` file:

1. First, we need to divide our monolith texture creation process into two distinct parts: loading the texture data and creating a Vulkan texture from it. This separation is essential because textures can only be created on the main thread. To facilitate this, we introduce a data container to store the loaded `.ktx` texture data:

   ```
   struct LoadedTextureData {
     uint32_t index      = 0;
     ktxTexture1* ktxTex = nullptr;
     lvk::TextureDesc desc;
   };
   ```

2. We introduce a `loadTextureData()` helper function, which loads a `.ktx` file using the KTX-Software library and returns the loaded data. Since we precompressed all Bistro textures into BC7 and stored them into the `KTX` format, this KTX-only approach is sufficient for our needs. This function closely resembles the `loadTexture()` function from `shared/Utils.h`, with the key difference being that it does not create a Vulkan texture:

   ```
   LoadedTextureData loadTextureData(const char* fileName)
   {
     ktxTexture1* ktxTex = nullptr;
     ktxTexture1_CreateFromNamedFile(fileName,
   ```

```
    KTX_TEXTURE_CREATE_LOAD_IMAGE_DATA_BIT, &ktxTex);
  const lvk::Format format = [](uint32_t glInternalFormat) {
    switch (glInternalFormat) {
      case GL_COMPRESSED_RGBA_BPTC_UNORM:
        return lvk::Format_BC7_RGBA;
      case GL_RGBA8:
        return lvk::Format_RGBA_UN8;
      // … handle other formats here
    }
    return lvk::Format_Invalid;
  }(ktxTex->glInternalformat);
  return LoadedTextureData{
    .ktxTex = ktxTex,
    .desc = { .type     = lvk::TextureType_2D,
              .format   = format,
              .dimensions = { ktxTex->baseWidth,
                              ktxTex->baseHeight, 1 },
              .usage = lvk::TextureUsageBits_Sampled,
              .numMipLevels     = ktxTex->numLevels,
              .data             = ktxTex->pData,
              .dataNumMipLevels = ktxTex->numLevels,
              .debugName        = fileName} };
}
```

The `LoadedTextureData` structure encapsulates all the information needed to create a texture, as it includes an `lvk::TextureDesc` object—an argument for the `VulkanContext::createTexture()` function, which handles the creation of all necessary Vulkan objects for a texture. This operation will be performed on the main thread.

3. Let's introduce another helper function, `convertToGPUMaterialLazy()`, which will run on the loader thread. This function closely mirrors the existing `convertToGPUMaterial()` function from *Chapter 8*. However, it uses `loadTextureData()` instead of `loadTexture()`. All parameters remain unchanged, except for `loadedTextureData` and `loadingMutex`, which serves as a mutex to safeguard multithreaded access to the `TextureCache` and `LoadedTextureData` objects:

```
GLTFMaterialDataGPU convertToGPUMaterialLazy(
  const std::unique_ptr<lvk::IContext>& ctx,
  const Material& mat,
  const TextureFiles& files,
  TextureCache& cache,
  std::vector<LoadedTextureData>& loadedTextureData,
  std::mutex& loadingMutex)
{
```

4. Populate the `GLTFMaterialDataGPU` structure similarly to how it was handled in `convertToGPUMaterial()`. For brevity, we omit the detailed parameter conversions as there's no value in duplicating them here. By default, the fields of `GLTFMaterialDataGPU` are initialized to `0`. This setup ensures that any textures not yet loaded will be automatically rendered as white textures, thanks to the `0` index, which corresponds to a white dummy texture. This behavior simplifies the rendering process, as no additional checks or logic are needed to handle textures that haven't been loaded yet:

```
GLTFMaterialDataGPU result = {
  .baseColorFactor = mat.baseColorFactor,
  // ...
};
```

5. The local `startLoadingTexture()` lambda function ensures that texture data is loaded only if it isn't already present in the texture cache and isn't currently waiting in the `loadedTextureData` queue:

```
auto startLoadingTexture = [&cache, &ctx, &files,
  &loadedTextureData, &loadingMutex](int textureId)
{
  if (textureId == -1) return;
  if (cache.size() <= textureId)
    cache.resize(textureId+1);
  const bool notInCache = cache[textureId].empty();
  const bool notInQueue = notInCache ?
    std::find_if(loadedTextureData.cbegin(),
      loadedTextureData.cend(),
      [textureId](const LoadedTextureData& d) {
        return d.index == textureId;
    }) == loadedTextureData.end() : false;
  if (notInCache && notInQueue) {
    loadedTextureData.push_back(
      loadTextureData(files[textureId].c_str()));
    loadedTextureData.back().index = textureId;
  }
};
```

6. All manipulations on the texture cache and the list of loaded textures should be performed while holding a locked mutex:

```
std::lock_guard lock(loadingMutex);
startLoadingTexture(mat.baseColorTexture);
startLoadingTexture(mat.emissiveTexture);
startLoadingTexture(mat.normalTexture);
```

```
    startLoadingTexture(mat.opacityTexture);
    return result;
  }
```

Now that we have all the necessary functions, let's take a look at how the new `VKMesh11Lazy` class works to implement asynchronous loading.

How it works...

The new `VKMesh11Lazy` class is derived from `VKMesh11` and adds new data members, along with changes to the constructor and a new member function. Let's take a closer look:

1. The new data members handle multithreading. The mutex protects the entire asynchronous loading process. The `loadedTextureData_` vector holds texture data that has been loaded and is waiting to be processed on the main thread, where the actual textures will be created. The `tf::Taskflow` object drives the **Taskflow** library, and the `tf::Executor` object is a thread pool with 2 loader threads:

```
class VKMesh11Lazy final : public VKMesh11 {
public:
  std::mutex loadingMutex_;
  std::vector<LoadedTextureData> loadedTextureData_;
  tf::Taskflow taskflow_;
  tf::Executor executor_{ size_t(2) };
```

Note

For simplicity, we store an instance of `tf::Executor` here. However, when there are many instances of `VKMesh11Lazy`, it is better to use an external executor shared by all mesh instances.

2. The constructor has the same parameters as `VKMesh11`, except for the last one – the `preloadMaterials` Boolean flag. This flag is not used here, and the one in the base class constructor is set to `false` to disable any texture loading in the base class:

```
  VKMesh11Lazy(
    const std::unique_ptr<lvk::IContext>& ctx,
    const MeshData& meshData, const Scene& scene,
    lvk::StorageType indirectBufferStorage = lvk::StorageType_Device)
  : VKMesh11(ctx, meshData, scene,
             indirectBufferStorage, false)
  {
    materialsGPU_.resize(materialsCPU_.size());
```

3. Instead of loading the materials along with all the textures, let's populate the `Taskflow` object using the `for_each_index()` algorithm for all available materials. For each material, we will call the `convertToGPUMaterialLazy()` function we defined earlier:

```
    taskflow_.for_each_index(
      0u,
      static_cast<uint32_t>(materialsCPU_.size()), 1u,
      [&](int i) {
        materialsGPU_[i] = convertToGPUMaterialLazy(
          ctx, materialsCPU_[i], textureFiles_,
          textureCache_, loadedTextureData_, loadingMutex_);
      });
```

4. At the end of the constructor, we can begin the multithreaded execution:

```
    executor_.run(taskflow_);
  }
```

5. To process the loaded texture data, we need to call the `processLoadedTextures()` method from the main thread. This method processes one loaded texture at a time to avoid stuttering when handling too many textures per frame:

```
  bool processLoadedTextures() {
    LoadedTextureData tex;
```

6. Note the extra C++ scope here, which controls the duration of the mutex lock. We want to extract the loaded texture data while the mutex is locked, but we avoid calling any heavy methods from the main thread while the mutex is locked. Since the order of texture loading doesn't matter, we simply use `pop_back()` instead of managing a proper queue:

```
    {
      std::lock_guard lock(loadingMutex_);
      if (loadedTextureData_.empty()) return false;
      tex = loadedTextureData_.back();
      loadedTextureData_.pop_back();
    }
```

7. Let's create the actual texture and then destroy the loaded texture data, as it is no longer needed:

```
    Holder<TextureHandle> texture = ctx->createTexture(tex.desc);
    ktxTexture_Destroy(ktxTexture(tex.ktxTex));
```

8. To update our texture cache and texture IDs in GPU materials, we need to lock the mutex again. This helper `getTextureFromCache()` local lambda is used to make the code cleaner:

```
    std::lock_guard lock(loadingMutex_);
    textureCache_[tex.index] = std::move(texture);
```

```
auto getTextureFromCache = [this](int textureId) -> uint32_t {
  return textureCache_.size() > textureId ?
    textureCache_[textureId].index() : 0;
};
```

9. Next, we go through all the materials and update them using the texture IDs from the cache. This is necessary because a single loaded texture can be referenced by multiple materials. Since we don't track their relationships, the simplest approach is to update them all—there's not much work involved:

```
for (size_t i = 0; i != materialsCPU_.size(); i++) {
  const Material& mtl = materialsCPU_[i];
  GLTFMaterialDataGPU& m = materialsGPU_[i];
  m.baseColorTexture = getTextureFromCache(mtl.baseColorTexture);
  m.emissiveTexture  = getTextureFromCache(mtl.emissiveTexture);
  m.normalTexture    = getTextureFromCache(mtl.normalTexture);
  m.transmissionTexture = getTextureFromCache(mtl.opacityTexture);
}
```

10. Once the `std::vector materialsGPU_` container with GPU materials is updated, it can be uploaded into the corresponding GPU buffer, which is managed in the `VKMesh11` base class:

```
    ctx->upload(bufferMaterials_,
      materialsGPU_.data(),
      materialsGPU_.size() *
        sizeof(decltype(materialsGPU_)::value_type));
    return true;
  }
};
```

As mentioned at the beginning, the new `VKMesh11Lazy` class is a drop-in replacement for `VKMesh11`. In fact, we only need to change two lines of C++ code:

1. The first line is to replace the mesh declaration as follows:

```
VKMesh11Lazy mesh(ctx, meshData, scene);
```

2. The second line, added at the very beginning of the rendering loop, is needed to poll `processLoadedTextures()` every frame to check for loaded textures:

```
app.run([&](uint32_t width, uint32_t height,
  float aspectRatio, float deltaSeconds)
{
  mesh.processLoadedTextures();
  // ...
```

And that's it. By changing just a few lines of code, our demo now supports asynchronous texture loading. All the code is complete and ready to be tested. Let's run the `Chapter11/05_LazyLoading` demo application to see how the Bistro scene can be rendered with lazy-loaded textures. When you run the demo, the geometry data will be loaded and rendered almost instantly, producing an image like the one in the following screenshot:

Figure 11.6: The Bistro scene rendered with asynchronous texture loading

You can navigate the demo with the keyboard and mouse while all the textures are streamed in within a few moments. This functionality, while essential for any serious rendering engine, is also very useful for smaller-scale applications, as the ability to quickly rerun the app significantly improves debugging capabilities.

There's more...

The asynchronous texture loading approach presented here offers much better responsiveness compared to preloading the entire texture pool at startup. However, a better approach might be to precreate empty textures and load images directly into these memory regions, especially since our data is already compressed into the BC7 format. This could be an interesting exercise for you.

Another improvement could be made in the `main` loop. Instead of polling the `processLoadedTextures()` method every frame, texture updates could be integrated into the events system of your rendering engine. Additionally, rather than loading just one texture per iteration, a "load balancer" could be implemented to upload multiple textures to the GPU, with the number of textures determined by their sizes using heuristics. We recommend increasing the max texture size of the demo app to 4096x4096 so you can experiment with this approach on a much heavier dataset.

One important thing to note is that our asynchronous loading mechanism only loads preprocessed, compressed textures in the `.ktx` format. This means that during the first run of the demo, you'll need to wait until all the textures are compressed. This can be avoided by integrating texture compression with asynchronous loading. You can try this as an exercise, or refer to the **LightweightVK** samples to see how it is done: `https://github.com/corporateshark/lightweightvk/tree/master/samples`.

Putting it all together into a Vulkan demo

To wrap up our minimalist Vulkan rendering engine, we'll bring together the techniques covered in *Chapters 8*, *10*, and *11* into one demo application.

Our final Vulkan demo application showcases the Lumberyard Bistro scene with the following techniques and effects:

- **Multisample anti-aliasing** (**MSAA**)
- **Screen-space ambient occlusion** (**SSAO**)
- **HDR** rendering with light adaptation
- Directional shadow mapping with **percentage-closer filtering** (**PCF**)
- GPU **frustum culling** using compute shaders
- Order-independent transparency (**OIT**)
- Asynchronous loading of textures

In this recipe, we'll focus on the details of how to integrate all these effects rather than on the individual techniques.

Getting ready

Ensure you have a solid understanding of all the recipes from this and previous chapters.

The demo application for this recipe is located in the `Chapter11/06_FinalDemo` folder.

How to do it...

Since we are already familiar with all the rendering techniques used in this recipe, we can quickly review the C++ source code and highlight the key sections that enable these techniques to work together. Different techniques and effects are implemented via different compute and graphics pipeline objects, accessing and modifying various buffers and textures. The following image illustrates an almost complete directed graph of our entire frame, known as a frame graph. Buffers are represented as ovals, textures as white rectangles, graphics pipelines as dark rectangles, and compute pipelines as dark rectangles with a folder icon. The textual labels on the diagram correspond to the names of the variables in the C++ code.

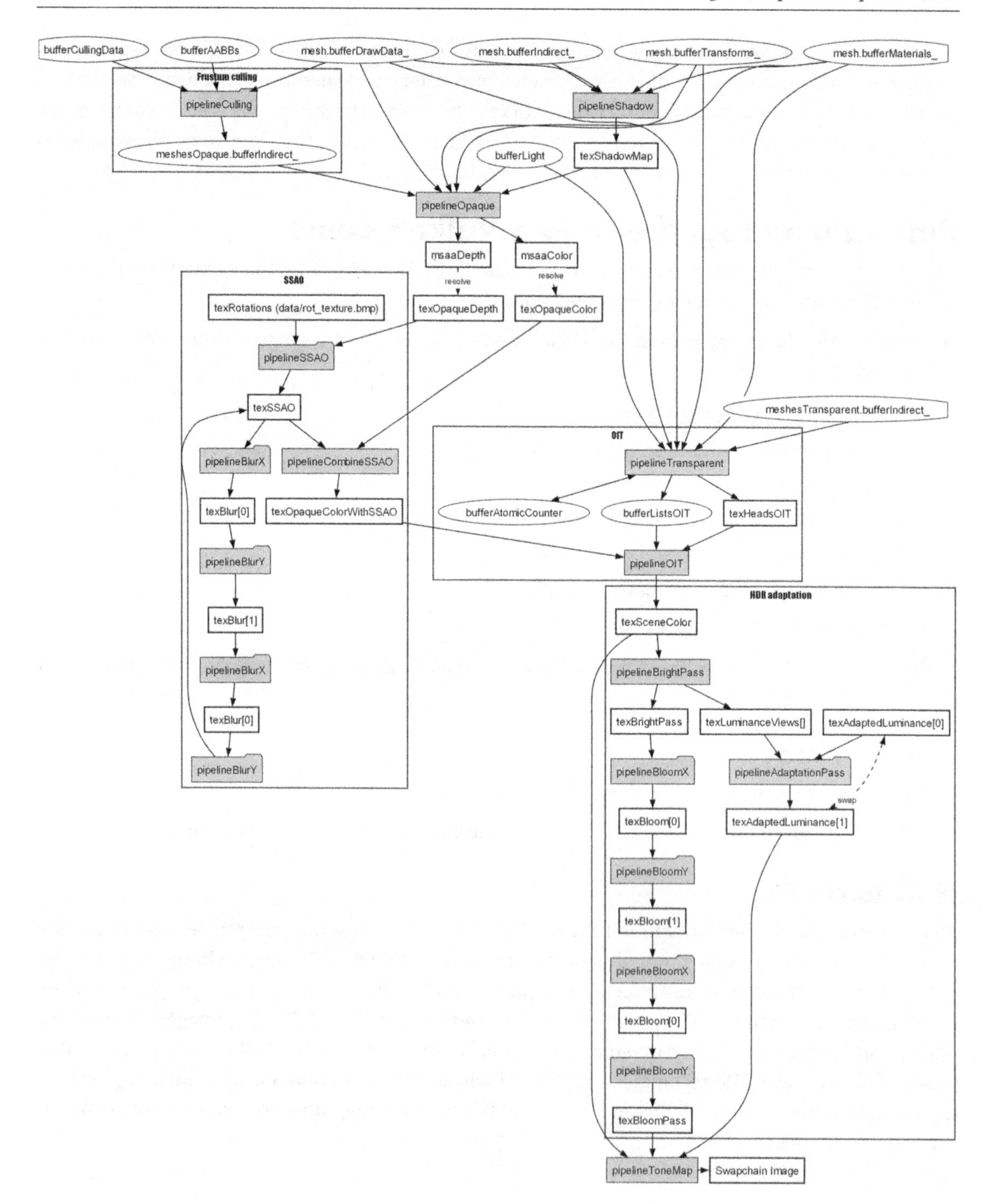

Figure 11.7: Our final demo frame graph

1. We define all the parameters that can be adjusted through the ImGui UI. These should already be familiar to you, but we have made slight tweaks to improve the final appearance of the rendered image:

```
bool drawMeshesOpaque      = true;
bool drawMeshesTransparent = true;
bool drawWireframe         = false;
bool drawBoxes             = false;
bool drawLightFrustum      = false;
// SSAO
bool ssaoEnable          = true;
bool ssaoEnableBlur      = true;
int ssaoNumBlurPasses    = 1;
float ssaoDepthThreshold = 30.0f;
// OIT
bool oitShowHeatmap    = false;
float oitOpacityBoost = 0.0f;
// HDR
bool hdrDrawCurves          = false;
bool hdrEnableBloom         = true;
float hdrBloomStrength      = 0.01f;
int hdrNumBloomPasses       = 2;
float hdrAdaptationSpeed = 1.0f;
// Culling
enum CullingMode {
  CullingMode_None = 0,
  CullingMode_CPU  = 1,
  CullingMode_GPU  = 2,
};
mat4 cullingView        = mat4(1.0f);
int cullingMode         = CullingMode_CPU;
bool freezeCullingView = false;
struct LightParams {
  float theta          = +90.0f;
  float phi            = -26.0f;
  float depthBiasConst = 1.1f;
  float depthBiasSlope = 2.0f;
} light;
```

2. We need to resolve both the MSAA color and depth textures, as the depth buffer will be used for computing SSAO. Since the entire demo uses an HDR rendering pipeline, the offscreen pixel format is set to `Format_RGBA_F16`:

```
std::unique_ptr<lvk::IContext> ctx(app.ctx_.get());
const lvk::Dimensions sizeFb =
  ctx->getDimensions(ctx->getCurrentSwapchainTexture());
const uint32_t kNumSamples           = 8;
const lvk::Format kOffscreenFormat = lvk::Format_RGBA_F16;
Holder<TextureHandle> msaaColor = ctx->createTexture({
    .format      = kOffscreenFormat,
    .dimensions = sizeFb,
    .numSamples = kNumSamples,
    .usage       = lvk::TextureUsageBits_Attachment,
    .storage     = lvk::StorageType_Memoryless });
Holder<TextureHandle> msaaDepth = ctx->createTexture({
    .format      = app.getDepthFormat(),
    .dimensions = sizeFb,
    .numSamples = kNumSamples,
    .usage       = lvk::TextureUsageBits_Attachment,
    .storage     = lvk::StorageType_Memoryless });
```

3. We apply MSAA only when rendering opaque objects, as described in the *Implementing order-independent transparency* recipe, which is why our resolved textures are named `texOpaqueDepth` and `texOpaqueColor`:

```
Holder<TextureHandle> texOpaqueDepth = ctx->createTexture({
    .format      = app.getDepthFormat(),
    .dimensions = sizeFb,
    .usage       = lvk::TextureUsageBits_Attachment |
                   lvk::TextureUsageBits_Sampled |
                   lvk::TextureUsageBits_Storage });
Holder<TextureHandle> texOpaqueColor = ctx->createTexture({
    .format      = kOffscreenFormat,
    .dimensions = sizeFb,
    .usage       = lvk::TextureUsageBits_Attachment |
                   lvk::TextureUsageBits_Sampled |
                   lvk::TextureUsageBits_Storage });
```

4. Two additional HDR color attachments are needed later, closer to the end-of-frame rendering. The first, `texOpaqueColorWithSSAO`, stores the opaque objects with the SSAO effect applied. The second, `texSceneColor`, holds the final HDR-rendered scene just before tone mapping is applied:

```
Holder<TextureHandle> texOpaqueColorWithSSAO = ctx->createTexture({
    .format     = kOffscreenFormat,
    .dimensions = sizeFb,
    .usage      = lvk::TextureUsageBits_Attachment |
                  lvk::TextureUsageBits_Sampled |
                  lvk::TextureUsageBits_Storage });
Holder<TextureHandle> texSceneColor = ctx->createTexture({
    .format     = kOffscreenFormat,
    .dimensions = sizeFb,
    .usage      = lvk::TextureUsageBits_Attachment |
                  lvk::TextureUsageBits_Sampled |
                  lvk::TextureUsageBits_Storage });
```

5. Let's create textures for HDR and light adaptation, as described in the *Implementing HDR rendering and tone mapping* and *Implementing HDR light adaptation* recipes of *Chapter 10*:

```
const lvk::Dimensions sizeBloom = { 512, 512 };
Holder<TextureHandle> texBrightPass = ctx->createTexture({
    .format     = kOffscreenFormat,
    .dimensions = sizeBloom,
    .usage      = lvk::TextureUsageBits_Sampled |
                  lvk::TextureUsageBits_Storage });
Holder<TextureHandle> texBloomPass = ctx->createTexture({
    .format     = kOffscreenFormat,
    .dimensions = sizeBloom,
    .usage      = lvk::TextureUsageBits_Sampled |
                  lvk::TextureUsageBits_Storage });
Holder<TextureHandle> texBloom[] = {
  ctx->createTexture({
      .format     = kOffscreenFormat,
      .dimensions = sizeBloom,
      .usage      = lvk::TextureUsageBits_Sampled |
                    lvk::TextureUsageBits_Storage}),
  ctx->createTexture({
      .format     = kOffscreenFormat,
      .dimensions = sizeBloom,
      .usage      = lvk::TextureUsageBits_Sampled |
                    lvk::TextureUsageBits_Storage})
};
const lvk::ComponentMapping swizzle = {
  .r = lvk::Swizzle_R, .g = lvk::Swizzle_R,
  .b = lvk::Swizzle_R, .a = lvk::Swizzle_1 };
```

```
Holder<TextureHandle> texLumViews[10] = {
  ctx->createTexture({
    .format       = lvk::Format_R_F16,
    .dimensions   = sizeBloom,
    .usage        = lvk::TextureUsageBits_Sampled |
                    lvk::TextureUsageBits_Storage,
    .numMipLevels = lvk::calcNumMipLevels(sizeBloom.width,
                                          sizeBloom.height),
    .swizzle = swizzle }) };
for (uint32_t v = 1; v != LVK_ARRAY_NUM_ELEMENTS(texLumViews); v++) {
  texLumViews[v] = ctx->createTextureView(
    texLumViews[0], { .mipLevel = v, .swizzle = swizzle });
}
```

6. The ping-pong textures for HDR light adaptation are preinitialized with a very bright color value, as explained in the *Implementing HDR light adaptation* recipe from *Chapter 10*:

```
const uint16_t brightPixel = glm::packHalf1x16(50.0f);
const lvk::TextureDesc luminanceTextureDesc{
  .format     = lvk::Format_R_F16,
  .dimensions = {1, 1},
  .usage      = lvk::TextureUsageBits_Sampled |
                lvk::TextureUsageBits_Storage,
  .swizzle    = swizzle };
Holder<TextureHandle> texAdaptedLum[2] = {
  ctx->createTexture(luminanceTextureDesc),
  ctx->createTexture(luminanceTextureDesc) };
```

7. A large 16-bit shadow map is created with a resolution of `4096x4096`, along with a depth-comparison sampler that clamps values to the border texels. To render the shadow map in shades of gray, we use texture swizzling functionality. A single value from the R-channel is duplicated across all three RGB channels:

```
Holder<TextureHandle> texShadowMap = ctx->createTexture({
    .type       = lvk::TextureType_2D,
    .format     = lvk::Format_Z_UN16,
    .dimensions = { 4096, 4096 },
    .usage      = lvk::TextureUsageBits_Attachment |
                  lvk::TextureUsageBits_Sampled,
    .swizzle    = { .r = lvk::Swizzle_R,
                    .g = lvk::Swizzle_R,
                    .b = lvk::Swizzle_R, .a = lvk::Swizzle_1 } });
Holder<SamplerHandle> samplerShadow = ctx->createSampler({
```

```
    .wrapU               = lvk::SamplerWrap_Clamp,
    .wrapV               = lvk::SamplerWrap_Clamp,
    .depthCompareOp      = lvk::CompareOp_LessEqual,
    .depthCompareEnabled = true });
```

8. A separate buffer is used to pass light data to the shaders, as described in the *Implementing shadows for directional lights* recipe:

```
struct LightData {
  mat4 viewProjBias;
  vec4 lightDir;
  uint32_t shadowTexture;
  uint32_t shadowSampler;
};
Holder<BufferHandle> bufferLight = ctx->createBuffer({
    .usage     = lvk::BufferUsageBits_Storage,
    .storage   = lvk::StorageType_Device,
    .size      = sizeof(LightData) });
```

9. Another set of textures and a sampler are required for the SSAO effect, as outlined in the *Implementing screen space ambient occlusion* recipe from *Chapter 10*:

```
Holder<TextureHandle> texSSAO = ctx->createTexture({
    .format     = ctx->getSwapchainFormat(),
    .dimensions = ctx->getDimensions(
      ctx->getCurrentSwapchainTexture()),
    .usage      = lvk::TextureUsageBits_Sampled |
                  lvk::TextureUsageBits_Storage });
Holder<TextureHandle> texBlur[] = {
  ctx->createTexture({
    .format     = ctx->getSwapchainFormat(),
    .dimensions = ctx->getDimensions(
      ctx->getCurrentSwapchainTexture()),
    .usage      = lvk::TextureUsageBits_Sampled |
                  lvk::TextureUsageBits_Storage }),
  ctx->createTexture({
    .format     = ctx->getSwapchainFormat(),
    .dimensions = ctx->getDimensions(
      ctx->getCurrentSwapchainTexture()),
    .usage      = lvk::TextureUsageBits_Sampled |
                  lvk::TextureUsageBits_Storage })
};
Holder<SamplerHandle> samplerClamp = ctx->createSampler({
```

```
    .wrapU = lvk::SamplerWrap_Clamp,
    .wrapV = lvk::SamplerWrap_Clamp,
    .wrapW = lvk::SamplerWrap_Clamp });
```

10. The OIT setup code is identical to the code from the *Implementing order-independent transparency* recipe. The only difference is the floating-point pixel format used for HDR rendering. For brevity, we skip the C++ lambda code here; refer to the recipe for details:

```
const Skybox skyBox(ctx, "data/immenstadter_horn_2k_prefilter.ktx",
  "data/immenstadter_horn_2k_irradiance.ktx",
  kOffscreenFormat, app.getDepthFormat(), kNumSamples);
VKMesh11Lazy mesh(ctx, meshData, scene);
const VKPipeline11 pipelineOpaque(
  ctx, meshData.streams, kOffscreenFormat,
  app.getDepthFormat(), kNumSamples,
  loadShaderModule(ctx, "Chapter11/06_FinalDemo/src/main.vert"),
  loadShaderModule(ctx, "Chapter11/06_FinalDemo/src/opaque.frag"));
const VKPipeline11 pipelineTransparent(
  ctx, meshData.streams, kOffscreenFormat,
  app.getDepthFormat(), kNumSamples,
  loadShaderModule(ctx, "Chapter11/06_FinalDemo/src/main.vert"),
  loadShaderModule(ctx, "Chapter11/06_FinalDemo/src/transparent.frag"));
const VKPipeline11 pipelineShadow(
  ctx, meshData.streams, lvk::Format_Invalid,
  ctx->getFormat(texShadowMap), 1,
  loadShaderModule(ctx, "Chapter11/03_DirectionalShadows/src/shadow.vert"),
  loadShaderModule(ctx, "Chapter11/03_DirectionalShadows/src/shadow.frag"));
VKIndirectBuffer11 meshesOpaque(ctx,
  mesh.numMeshes_, lvk::StorageType_HostVisible);
VKIndirectBuffer11 meshesTransparent(ctx,
  mesh.numMeshes_, lvk::StorageType_HostVisible);
// ... skipped populating indirect buffers here
```

We skip the rest of the initialization code that loads GLSL shaders and creates the corresponding graphics and compute pipelines, and some buffers. Refer to the relevant recipes for instructions on how to do this. Instead, let's focus on some new details in the main rendering loop:

1. We poll `processLoadedTextures()` to handle lazy-loaded textures, as described in the *Loading texture assets asynchronously* recipe, and extract the viewing frustum planes for culling, as explained in the *Doing frustum culling on the GPU with compute shaders* recipe:

```
app.run([&](uint32_t width, uint32_t height,
  float aspectRatio, float deltaSeconds) {
  mesh.processLoadedTextures();
  const mat4 view = app.camera_.getViewMatrix();
  const mat4 proj = glm::perspective(
    45.0f, aspectRatio, pcSSAO.zNear, pcSSAO.zFar);
  if (!freezeCullingView)
    cullingView = app.camera_.getViewMatrix();
  CullingData cullingData = {
    .numMeshesToCull = meshesOpaque.drawCommands_.size() };
  getFrustumPlanes(proj * cullingView, cullingData.frustumPlanes);
  getFrustumCorners(proj * cullingView, cullingData.frustumCorners);
```

2. We update the light view matrix and calculate the scene's AABB in light space to calculate the light projection matrix, as described in the *Implementing shadows for directional lights* recipe:

```
const glm::mat4 rot1 = glm::rotate(mat4(1.f),
  glm::radians(light.theta), glm::vec3(0, 1, 0));
const glm::mat4 rot2 = glm::rotate(rot1,
  glm::radians(light.phi), glm::vec3(1, 0, 0));
const vec3 lightDir  = glm::normalize(
  vec3(rot2 * vec4(0.0f, -1.0f, 0.0f, 1.0f)));
const mat4 lightView = glm::lookAt(
  glm::vec3(0.0f), lightDir, vec3(0, 0, 1));
const BoundingBox boxLS = bigBoxWS.getTransformed(lightView);
const mat4 lightProj = glm::orthoLH_ZO(
  boxLS.min_.x, boxLS.max_.x,
  boxLS.min_.y, boxLS.max_.y,
  boxLS.max_.z, boxLS.min_.z);
```

3. Now we're ready to start filling the command buffer. The first tasks are to clear the OIT buffers, as described in the *Implementing order-independent transparency* recipe, and cull the scene:

```
ICommandBuffer& buf = ctx->acquireCommandBuffer();
clearTransparencyBuffers(buf);
```

4. The culling code here is slightly different. First, we take a shortcut and perform frustum culling only for opaque meshes, as transparent meshes are always rendered due to their low number and very low polygon count. Since we're culling only part of the scene, we can't use the `VKMesh11` indirect buffer directly as we did in the *Doing frustum culling on the GPU with compute shaders* recipe. Instead, we use the indirect buffer from the `VKIndirectBuffer11` object `meshOpaque`, which contains only opaque meshes. Since all transparent meshes are visible, we initialize our `numVisibleMeshes` counter with the number of transparent meshes:

```
if (cullingMode == CullingMode_CPU) {
  numVisibleMeshes =
    meshesTransparent.drawCommands_.size());
  DrawIndexedIndirectCommand* cmd = meshesOpaque.
    getDrawIndexedIndirectCommandPtr();
  for (size_t i = 0; i != meshesOpaque.drawCommands_.size(); i++) {
    const BoundingBox box = reorderedBoxes[
      mesh.drawData_[cmd->baseInstance].transformId];
      const uint32_t count =
        isBoxInFrustum(cullingData.frustumPlanes,
                       cullingData.frustumCorners, box) ? 1 : 0;
      (cmd++)->instanceCount = count;
      numVisibleMeshes += count;
  }
  ctx->flushMappedMemory(
    meshesOpaque.bufferIndirect_, 0,
    meshesOpaque.drawCommands_.size() *
    sizeof(DrawIndexedIndirectCommand));
} else
```

5. The GPU culling code path is similar. We run the compute shader to patch the contents of the `meshesOpaque.bufferIndirect_` buffer. Note the buffer dependency in `cmdDispatchThreadGroups()`, which is necessary to issue proper Vulkan buffer memory barriers:

```
if (cullingMode == CullingMode_GPU) {
  buf.cmdBindComputePipeline(pipelineCulling);
  pcCulling.meshes = ctx->gpuAddress(
    bufferCullingData[currentBufferId]);
  pcCulling.commands = ctx->gpuAddress(
    meshesOpaque.bufferIndirect_);
  cullingData.numVisibleMeshes =
    meshesTransparent.drawCommands_.size());
  buf.cmdPushConstants(pcCulling);
  buf.cmdUpdateBuffer(
```

```
      bufferCullingData[currentBufferId], cullingData);
    buf.cmdDispatchThreadGroups(
      { 1 + cullingData.numMeshesToCull / 64 },
      { .buffers = { BufferHandle(meshesOpaque.bufferIndirect_) } });
  }
```

6. After culling is complete, we update the shadow map if necessary, as described in the *Implementing shadows for directional lights* recipe. The shadow mapping code renders the entire scene using `mesh.draw()` and does not rely on the culling code, so there is no dependency between the two, allowing them to run in parallel, at least in theory:

```
  if (prevLight != light) {
    prevLight = light;
    buf.cmdBeginRendering(
      lvk::RenderPass{
        .depth = {.loadOp = lvk::LoadOp_Clear,
                  .clearDepth = 1.0f} },
      lvk::Framebuffer{
       .depthStencil = {.texture = texShadowMap} });
    buf.cmdSetDepthBias(light.depthBiasConst, light.depthBiasSlope);
    buf.cmdSetDepthBiasEnable(true);
    mesh.draw(buf, pipelineShadow, lightView, lightProj);
    buf.cmdSetDepthBiasEnable(false);
    buf.cmdEndRendering();
    buf.cmdUpdateBuffer(bufferLight, LightData{
      .viewProjBias  = scaleBias * lightProj * lightView,
      .lightDir      = vec4(lightDir, 0.0f),
      .shadowTexture = texShadowMap.index(),
      .shadowSampler = samplerShadow.index(),
    });
  }
```

7. Now we can render the scene using our MSAA render target and resolve it:

```
  const lvk::Framebuffer framebufferMSAA = {
    .color = {{.texture = msaaColor,
               .resolveTexture = texOpaqueColor }},
    .depthStencil = {
      .texture = msaaDepth,
      .resolveTexture = texOpaqueDepth } };
  buf.cmdBeginRendering(
    lvk::RenderPass{
```

```
      .color = { {
        .loadOp = lvk::LoadOp_Clear,
        .storeOp = lvk::StoreOp_MsaaResolve,
        .clearColor = { 1.f, 1.f, 1.f, 1.f } } },
      .depth = {
        .loadOp = lvk::LoadOp_Clear,
        .storeOp = lvk::StoreOp_MsaaResolve,
        .clearDepth = 1.0f } },
    framebufferMSAA, { .buffers = {
      BufferHandle(meshesOpaque.bufferIndirect_)}});
```

8. We draw the skybox and all the meshes, as described in the *Implementing order-independent transparency* recipe. For brevity, we've skipped the code that populates the push constants struct `pc`:

```
skyBox.draw(buf, view, proj);
// ... skipped push constants
if (drawMeshesOpaque) mesh.draw(
  buf, pipelineOpaque, &pc, sizeof(pc),
  { .compareOp = lvk::CompareOp_Less,
    .isDepthWriteEnabled = true },
  drawWireframe, &meshesOpaque);
if (drawMeshesTransparent)
  mesh.draw(buf, pipelineTransparent, &pc, sizeof(pc),
    { .compareOp = lvk::CompareOp_Less,
      .isDepthWriteEnabled = false },
    drawWireframe, &meshesTransparent);
app.drawGrid(buf, proj, vec3(0, -1.0f, 0),
  kNumSamples, kOffscreenFormat);
buf.cmdEndRendering();
```

9. At this point, we have our color and depth buffers resolved into the textures `texOpaqueColor` and `texOpaqueDepth`. Next, let's calculate the SSAO effect using the depth buffer, as described in the *Implementing screen space ambient occlusion* recipe from *Chapter 10*. Note the texture dependencies here:

```
if (ssaoEnable) {
  buf.cmdBindComputePipeline(pipelineSSAO);
  buf.cmdPushConstants(pcSSAO);
  buf.cmdDispatchThreadGroups(
    { .width  = 1 + sizeFb.width / 16,
      .height = 1 + sizeFb.height / 16 },
    { .textures = {
      TextureHandle(texOpaqueDepth),
```

```
            TextureHandle(texSSAO) } });
        if (ssaoEnableBlur) {
          // ... the blur code is skipped
        }
```

10. Now, we can combine the SSAO texture `texSSAO` with `texOpaqueColor` and render the result into `texOpaqueColorWithSSAO`. Note the additional dependency on `texOpaqueColor` for proper Vulkan image layout transition from `VK_IMAGE_LAYOUT_COLOR_ATTACHMENT_OPTIMAL` to `VK_IMAGE_LAYOUT_SHADER_READ_ONLY_OPTIMAL`:

```
        buf.cmdBeginRendering(
          { .color={{.loadOp = lvk::LoadOp_Load,
                     .clearColor={1.f, 1.f, 1.f, 1.f}}},
          },
          { .color = { {
            .texture = texOpaqueColorWithSSAO } },
          },
          { .textures = {
             TextureHandle(texSSAO),
             TextureHandle(texOpaqueColor) } });
        buf.cmdBindRenderPipeline(pipelineCombineSSAO);
        buf.cmdPushConstants(pcCombineSSAO);
        buf.cmdBindDepthState({});
        buf.cmdDraw(3);
        buf.cmdEndRendering();
      }
```

11. Now, we can take `texOpaqueColorWithSSAO` and combine it with transparent objects, as described in the *Implementing order-independent transparency* recipe. The result is rendered into `texSceneColor`, which is our fully rendered HDR scene, ready to be tone-mapped:

```
      const lvk::Framebuffer framebufferOffscreen = {
        .color = { { .texture = texSceneColor } },
      };
      buf.cmdBeginRendering(
        lvk::RenderPass{
          .color = {{ .loadOp = lvk::LoadOp_Load,
                      .storeOp = lvk::StoreOp_Store }},
        },
        framebufferOffscreen,
        { .textures = {
            TextureHandle(texHeadsOIT),
            TextureHandle(texOpaqueColor),
```

```
        TextureHandle(texOpaqueColorWithSSAO) },
      .buffers = { BufferHandle(bufferListsOIT) }});
    // … skipped push constants here
    const struct { … } pcOIT = { … };
    buf.cmdBindRenderPipeline(pipelineOIT);
    buf.cmdPushConstants(pcOIT);
    buf.cmdBindDepthState({});
    buf.cmdDraw(3);
    buf.cmdEndRendering();
    // … skipped tone mapping code
  }
```

12. The skipped tone mapping code depends on `texSceneColor`, but otherwise, it is identical to the code used in the *Implementing HDR light adaptation* recipe from *Chapter 10*. We skip it here, along with the extensive ImGui rendering code, and then submit the following command:

```
  submitHandle[currentBufferId] = ctx->submit(
    buf, ctx->getCurrentSwapchainTexture());
```

13. The following code fragment, which waits for the previous frame, is necessary to retrieve the results of GPU culling for display in ImGui. You can read about it in the *Doing frustum culling on the GPU with compute shaders* recipe:

```
  currentBufferId = (currentBufferId + 1) %
    LVK_ARRAY_NUM_ELEMENTS(bufferCullingData);
  if (cullingMode == CullingMode_GPU &&
      app.fpsCounter_.numFrames_ > 1) {
    ctx->wait(submitHandle[currentBufferId]);
    ctx->download(bufferCullingData[currentBufferId],
      &numVisibleMeshes, sizeof(uint32_t),
      offsetof(CullingData, numVisibleMeshes));
  }
  // swap ping-pong textures
  std::swap(texAdaptedLum[0], texAdaptedLum[1]);
});
```

Now, we can run the demo application, and it should render the Lumberyard Bistro scene with all the aforementioned effects, as shown in the following screenshot:

Figure 11.8: The final Vulkan demo

Try running the demo, moving the camera around, and experimenting with the various UI controls. Every single tweak from the previous recipes should be here.

Note

If you'd like to render high-quality shadows as in the screenshot, set the shadow map resolution to `16384x16384`.

With this example, we conclude *Vulkan 3D Graphics Rendering Cookbook*, and we hope you enjoyed it!

There's more...

The possibilities here are endless. You can use this framework to experiment with more advanced rendering techniques. Adding more screen-space effects, such as temporal antialiasing or screen-space reflections, should be relatively simple. Another easy next step could be replacing the basic SSAO effect implemented here with ray-traced ambient occlusion using Vulkan's raytracing capabilities. The LightweightVK framework is fully capable of handling that.

Adding multiple light sources can be done by storing their parameters in a buffer and iterating over it in the fragment shader. Various optimizations are possible, such as tile deferred shading or clustered shading: `http://www.cse.chalmers.se/~uffe/clustered_shading_preprint.pdf`. The easiest way to incorporate shadows from multiple light sources into this demo would be to use separate per-light shadow maps and access them all via bindless textures. It could be especially cool to implement omnidirectional shadows for light sources inside the Bistro interior.

Going deeper into complex materials rendering could be another exciting direction to explore. A good starting point might be converting the Bistro scene materials to PBR and using our glTF2 rendering code from *Chapter 6* to render them correctly. If you decide to pursue this and dive into the practical side of PBR rendering, the *Filament* engine documentation is an excellent resource to explore: `https://google.github.io/filament/Filament.html`.

One last thing to mention is the complexity of manual resource allocation here. As shown in the frame graph diagram (*Figure 11.7*), the number of buffers and textures used in the demo increases rapidly with each new effect added. This makes it tedious and error-prone to maintain and extend the frame graph manually. If you're interested in writing a generic 3D rendering engine, there are many generalized approaches to automating frame graph construction. A frame graph helps plan and schedule all required operations on the GPU and CPU for a frame, along with all the necessary resources, in the most efficient way. It's a complex topic, and to learn more, we recommend reading the GDC presentation *FrameGraph: Extensible Rendering Architecture in Frostbite* by Yuriy O'Donnell and the book *Mastering Graphics Programming with Vulkan* by Marco Castorina and Gabriel Sassone, published by Packt.

12

Unlock Your Book's Exclusive Benefits

Your copy of *Vulkan 3D Graphics Rendering Cookbook, Second Edition* comes with the following exclusive benefits:

- Next-gen Packt Reader
- AI assistant (beta)
- DRM-free PDF/ePub downloads

Use the following guide to unlock them if you haven't already. The process takes just a few minutes and needs to be done only once.

How to unlock these benefits in three easy steps

Step 1

Have your purchase invoice for this book ready, as you'll need it in *Step 3*. If you received a physical invoice, scan it on your phone and have it ready as either a PDF, JPG, or PNG.

For more help on finding your invoice, visit `https://www.packtpub.com/unlock-benefits/help`.

Note: Bought this book directly from Packt? You don't need an invoice. After completing *Step 2*, you can jump straight to your exclusive content.

Step 2

Scan the following QR code or visit `https://www.packtpub.com/unlock/9781803248110`:

Step 3

Sign in to your Packt account or create a new one for free. Once you're logged in, upload your invoice. It can be in PDF, PNG, or JPG format and must be no larger than 10 MB. Follow the rest of the instructions on the screen to complete the process.

Need help?

If you get stuck and need help, visit `https://www.packtpub.com/unlock-benefits/help` for a detailed FAQ on how to find your invoices and more. The following QR code will take you to the help page directly:

Note: If you are still facing issues, reach out to `customercare@packt.com`.

packt.com

Subscribe to our online digital library for full access to over 7,000 books and videos, as well as industry leading tools to help you plan your personal development and advance your career. For more information, please visit our website.

Why subscribe?

- Spend less time learning and more time coding with practical eBooks and Videos from over 4,000 industry professionals
- Improve your learning with Skill Plans built especially for you
- Get a free eBook or video every month
- Fully searchable for easy access to vital information
- Copy and paste, print, and bookmark content

At www.packt.com, you can also read a collection of free technical articles, sign up for a range of free newsletters, and receive exclusive discounts and offers on Packt books and eBooks.

Other Books You May Enjoy

If you enjoyed this book, you may be interested in these other books by Packt:

The Modern Vulkan Cookbook

None Kakkar, Mauricio Maurer

ISBN: 9781803239989

- Set up your environment for Vulkan development
- Understand how to draw graphics primitives using Vulkan
- Use state-of-the-art Vulkan to implement a wide variety of modern rendering techniques such as DLSS, TAA, OIT, and foveated rendering
- Implement hybrid techniques using rasterization and ray tracing to create photorealistic real-time engines
- Create extended reality (AR/VR/MR) applications using OpenXR and Vulkan
- Explore debugging techniques for graphics applications that use Vulkan

Mastering Graphics Programming with Vulkan

Marco Castorina, Gabriel Sassone

ISBN: 9781803244792

- Understand resources management and modern bindless techniques
- Get comfortable with how a frame graph works and know its advantages
- Explore how to render efficiently with many light sources
- Discover how to integrate variable rate shading
- Understand the benefits and limitations of temporal anti-aliasing
- Get to grips with how GPU-driven rendering works
- Explore and leverage ray tracing to improve render quality

Packt is searching for authors like you

If you're interested in becoming an author for Packt, please visit authors.packtpub.com and apply today. We have worked with thousands of developers and tech professionals, just like you, to help them share their insight with the global tech community. You can make a general application, apply for a specific hot topic that we are recruiting an author for, or submit your own idea.

Share your thoughts

Now you've finished *Vulkan 3D Graphics Rendering Cookbook, Second edition*, we'd love to hear your thoughts! Scan the QR code below to go straight to the Amazon review page for this book and share your feedback or leave a review on the site that you purchased it from.

https://packt.link/r/1803248114

Your review is important to us and the tech community and will help us make sure we're delivering excellent quality content.

Index

Symbols

A

B

C

H

I

K

L

M

N

O

P

R

S

T

U

V

W

Made in the USA
Monee, IL
02 August 2025

22464002R00398